AF618624

# Ultrashort Processes in Condensed Matter

# NATO ASI Series

## Advanced Science Institutes Series

*A series presenting the results of activities sponsored by the NATO Science Committee, which aims at the dissemination of advanced scientific and technological knowledge, with a view to strengthening links between scientific communities.*

The series is published by an international board of publishers in conjunction with the NATO Scientific Affairs Division

| | | |
|---|---|---|
| **A** | **Life Sciences** | Plenum Publishing Corporation |
| **B** | **Physics** | New York and London |
| **C** | **Mathematical and Physical Sciences** | Kluwer Academic Publishers |
| **D** | **Behavioral and Social Sciences** | Dordrecht, Boston, and London |
| **E** | **Applied Sciences** | |
| **F** | **Computer and Systems Sciences** | Springer-Verlag |
| **G** | **Ecological Sciences** | Berlin, Heidelberg, New York, London, |
| **H** | **Cell Biology** | Paris, Tokyo, Hong Kong, and Barcelona |
| **I** | **Global Environmental Change** | |

***Recent Volumes in this Series***

*Volume 310* —Integrable Quantum Field Theories
edited by L. Bonora, G. Mussardo, A. Schwimmer, L. Girardello, and M. Martellini

*Volume 311* —Quantitative Particle Physics: *Cargèse 1992*
edited by Maurice Lévy, Jean-Louis Basdevant, Maurice Jacob, Jean Iliopoulos, Raymond Gastmans, and Jean-Marc Gérard

*Volume 312* —Future Directions of Nonlinear Dynamics in Physical and Biological Systems
edited by P. L. Christiansen, J. C. Eilbeck, and R. D. Parmentier

*Volume 313* —Dissociative Recombination: Theory, Experiment, and Applications
edited by Bertrand R. Rowe, J. Brian A. Mitchell, and André Canosa

*Volume 314* —Ultrashort Processes in Condensed Matter
edited by Walter E. Bron

*Volume 315* —Low-Dimensional Topology and Quantum Field Theory
edited by Hugh Osborn

*Volume 316* —Super-Intense Laser–Atom Physics
edited by Bernard Piraux, Anne L'Huillier, and Kazimierz Rzążewski

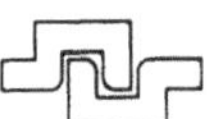

*Series B: Physics*

# Ultrashort Processes in Condensed Matter

Edited by

Walter E. Bron
University of California, Irvine
Irvine, California

Springer Science+Business Media, LLC

Proceedings of a NATO Advanced Study Institute
on Ultrashort Processes in Condensed Matter,
held August 30–September 12, 1992,
in Il Ciocco, Castelvecchio di Pascoli, Lucca, Italy

NATO-PCO-DATA BASE

The electronic index to the NATO ASI Series provides full bibliographical references (with keywords and/or abstracts) to more than 30,000 contributions from international scientists published in all sections of the NATO ASI Series. Access to the NATO-PCO-DATA BASE is possible in two ways:

—via online FILE 128 (NATO-PCO-DATA BASE) hosted by ESRIN, Via Galileo Galilei, I-00044 Frascati, Italy

—via CD-ROM "NATO Science and Technology Disk" with user-friendly retrieval software in English, French, and German (©WTV GmbH and DATAWARE Technologies, Inc. 1989). The CD-ROM also contains the AGARD Aerospace Database.

The CD-ROM can be ordered through any member of the Board of Publishers or through NATO-PCO, Overijse, Belgium.

Library of Congress Cataloging-in-Publication Data

Ultrashort processes in condensed matter / edited by Walter E. Bron.
p. cm. -- (NATO ASI series. Series B, Physics, v. 314)
"Published in cooperation with NATO Scientific Affairs Division."
"Proceedings of a NATO Advanced Study Institute on Ultrashort Processes in Condensed Matter, held August 30-September 12, 1992, in Il Ciocco, Castelvecchio di Pascoli, Lucca, Italy"--T.p. verso.
Includes bibliographical references and index.
ISBN 978-0-306-44574-3 ISBN 978-1-4615-2954-5 (eBook)
DOI 10.1007/978-1-4615-2954-5
1. Condensed matter--Electric properties--Congresses. 2. Laser pulses, Ultrashort--Congresses. 3. Electronic excitation--Congresses. 4. Electrooptics--Congresses. I. Bron, Walter E. II. North Atlantic Treaty Organization. Scientific Affairs Division. III. NATO Advanced Study Institute on Ultrashort Processes in Condensed Matter (1992 : Il Ciocco, Italy) IV. Series.
QC173.4.C65U53 1993
530.4'1--dc20 93-28278
CIP

Additional material to this book can be downloaded from http://extra.springer.com.

ISBN 978-0-306-44574-3


Originally published by Plenum Press, New York in 1993

# PREFACE

The Advanced Study Institute (ASI) considered a number of facets of the very rapidly advancing field of theoretical and experimental aspects of ultrashort processes in condensed matter. Common threads exist between a series of example cases. One major subgroup of topics involves the ultrashort dynamics of excitations of various "particles" produced through the interactions of condensed matter with ultrashort duration laser light. Examples of the excitations include electronic and hole carriers, electron-hole plasma, phonons, vibrons and rotons, two phonon states, and excitons. Experimentation on the dynamics of such excitations, are carried out in the bulk, at surfaces, in thin films, and in quantum wells. The dynamical steps which the excitations usually undergo include photo-excitation, local thermalization, particle-particle interaction, particle phonon interactions and eventual return to true thermal equilibrium.

This ASI was organized to benefit particularly advanced graduate students, specifically, those near the end of their Ph.D. thesis projects, and also for postdoctoral scholars already active in the field. The overall organizational goal was centered around a set of tutorially based lectures intermingled with full scale discussion periods of equal time and importance as the lectures. The general discussion periods were designed to offer to the participants ample time to ask detailed questions and to make comments and contributions of their own. In order to complete the involvement of the participants a full length poster session was also held. A representative set of abstracts of these posters appear as an Appendix to the lectures.

The editor notes with regret that the manuscript of the contribution by Professor J. Ryan was not available at the time of submission for publication.

Some comments about the manuscripts. Modern word processors have become an obvious tool for speeding up the typing of camera-ready manuscripts. However, they lead to a small spread of typing fonts and styles. I have, however, decided to accept this spread in manuscript appearance so as not to undergo both the expense and the consequent delay in large scale retyping a few manuscripts. Content is more important than form.

The ASI was held in the Hotel Il Ciocco in Castelvecchio di Pascoli, near Lucca, Italy, from August 30th to September 12, 1992. Situated in the Tuscany hills and overlooking the picturesque village of Barga amid the countryside of the Garfagnana region, the hotel offered a haven for the heavy schedule of the ASI and I thank Mr. Bruno Giannasi of the Hotel staff for all his help.

We also gratefully acknowledge the support of NATO's Scientific Affairs Division for their organization advice and their financial support. Other support came from various scientific entities which partially supported the travel expenses of many of the participants; the U. S. National Science Foundation is but one example. We would also like to acknowledge the kind funds from the Coherent Laser Corporation and the Spectra Physics Corporation who supported an Institute banquet for which the use of NATO funds were not permitted.

I personally would like to thank the many persons who helped to make this ASI to be the success it turned out to be. Specifically, I acknowledge the organizing committee for its guidance (Prof. Chr. Flytzanis and Dr. J. Kuhl); Professors Schoemaker and Jacoboni who struggled with the vagaries of the European banking system. Many thanks go to my secretary, Cathy Kick, who at times single handedly managed the major share of the correspondence (and still does to this day). By far the greatest thanks goes to my wife, Ann, who took over the job of ASI secretary during our stay in Italy. She not only kept me pointed in the right direction each day, but most of the lecturers and participants as well. Her help and organizational skills were appreciated by all.

Walter E. Bron
Institute Director and
Professor of Physics
University of California, Irvine
Irvine, California 92717

CONTENTS

# FEMTOSECOND PROBING OF PHOTOINDUCED REFRACTIVE INDEX CHANGES IN SEMICONDUCTORS

E. C. Fox and H. M. van Driel
Department of Physics
University of Toronto
Toronto Canada, M5S-1A7

## INTRODUCTION

For decades, information technology has been dominated by electronics. Increasingly, however, the physical limitations of electronics are being or have been reached and scientists are exploring new technologies for transmitting, storing and processing information. Many believe that light or photons will form the new "current" for information in the next century and that photonics could possibly supplant electronics in several devices. Certainly photonics is now making significant inroads in areas such as transmission and storage. However, the same can't be said of routing and switching, since such functions are still carried out using all electronic or hybrid, opto-electronic technologies. Increasing demands for integration call for all-optical switching devices and it has become the "holy grail" of the emerging optical communication technologies to find suitable materials which display *a large enough* and *fast enough* optical response to be considered for such devices. The underlying physical mechanism which is being researched in many of these quests is photo-induced refractive index changes [Shen, 1984; Gibbs,1985]. It is envisioned that a gate optical pulse can be used to alter the local refractive index in a device and thus modify the direction of propagation, phase, or transmission of an optical pulse passing through the device in what is commonly referred to as light-by-light switching. Many different types of materials have been and continue to be investigated for these applications including semiconductors, glasses, semiconductor-doped glasses, and polymers [Miller, 1981; Stegeman, 1985; Haug, 1988; Gibbs, 1990]. Also, several different geometries have been researched for switching applications based on Fabry-Perot interferometers, etalons, waveguides, diffraction, and scattering [Stegeman, 1985]. The field of light-induced optical switching has exploded in recent years and it is difficult to cover the field in a single article. Here, we

*Ultrashort Processes in Condensed Matter*, Edited by
W.E. Bron, Plenum Press, New York, 1993

will concentrate on some fundamental concepts of photo-induced refractive index changes in bulk semiconductors on an ultrashort time scale, and, mainly those which are related to bound and real carrier effects. To illustrate some of the underlying ideas, we review some of our recent experimental work on changes in the refractive index in II-VI semiconductors induced by femtosecond laser pulses.

The photo-induced refractive index changes (PIRIC's) we consider can occur through an induced polarization of the bound electrons by the electric field of an optical pulse (Kerr effect) or through the generation of free carriers or phonons which alter the linear optical susceptibility. The first mechanism is associated with virtual excitation of the material while the second is related to real excitations which are accompanied by the removal of photons from the beam. Contributions to the PIRIC from virtual and real excitations can also be discussed according to their temporal behaviour. In the case of virtual excitation, the contribution to the PIRIC occurs only in the presence of an electric field and will vanish when the optical pulse does. In what follows we will refer to the virtual effect as an instantaneous effect although strictly speaking the response time is dictated by the uncertainty principle. In contrast, contributions to PIRIC from real excitations have a temporal characteristic which follows that of the carriers or phonons induced by the pulse. Carrier induced effects will typically disappear with recombination or diffusion on a time scale between picoseconds and microseconds [Kressel,1977] depending on the excitation conditions while phonon or lattice heating effects disappear with lattice heat diffusion on a microsecond time scale [Wherrett,1990]. For light-by-light switching, the non-instantaneous response limits maximum achievable bit rates, and, since they are associated with real excitation of the medium always lead to optical energy loss. These long lived contributions to PIRIC are generally undesirable and materials and excitation conditions are being sought in which the ratio of the magnitude of the instantaneous to the non-instantaneous contribution is maximized [Friberg, 1987]. There are several figures of merit which one can develop to characterize the suitability of various materials for optical switching applications. In most cases one is interested in maximizing the induced phase shift on a beam during its attenuation length. As a result, it is usually necessary to avoid linear absorption, which for semiconductors, entails operating with photon energies below the band-gap of the material.

Sheik-Bahae et al. [1991] have provided significant insight into the theoretical foundation for PIRIC's in semiconductors from virtual effects. In particular they have shown how an extension of the Kramers-Kronig analysis, normally associated with linear optical properties, can be used to cover nonlinear effects as well. On the other hand, the alteration of refractive indices by changing carrier concentrations and lattice temperature have been investigated over the last several decades and are reasonably well-understood theoretically [Haug, 1988]. Later, we will summarize the general formalisms and expressions that have been developed to estimate the magnitudes of the induced refractive index changes from both real and virtual effects and will use the theory to calculate the expected behavior in II-VI semiconductors.

A number of techniques have been used to study PIRIC's in semiconductors. These include measurements of changes in reflectivity [Downer, 1986], in phase (via interference measurements) [Cotter, 1989], in beam propagation direction [Ding, 1990; Said, 1992], in scattering efficiency (e.g., by induced-grating experiments) [Puls, 1988; Rudolph, 1990 van Lap, 1991], and in beam profile [Sheik-Bahae, 1989]. The z-scan technique pioneered by the group of Van Stryland et al. [Sheik-Bahae, 1989] seems to be particularly simple to use and gives the magnitude and sign of the PIRIC during an optical pulse. The technique is based on measuring the far-field profile of a focused laser beam when a plate of the sample is scanned through the focus region. Typically 25 ps pulses have been used in these investigations but other groups have used even longer pulses. However, despite the obvious technological importance of being able to separately measure the contributions to PIRIC's from real and virtual excitations and despite interest in investigating the physics associated with optical nonlinearities, in few experimental investigations have the instantaneous and long lived contributions to PIRIC been isolated and identified. In the single beam z-scan technique, the fundamental mechanisms which contribute to the net PIRIC can only be separately identified on the basis of their different scaling behaviors with irradiance of the exciting light pulse [Said, 1992]. Although this approach provides some insight into real and virtual effects, it doesn't provide time resolution and requires some assumptions about the underlying physics (e.g. of carrier processes) to yield quantitative information about the various contributions to PIRIC's.

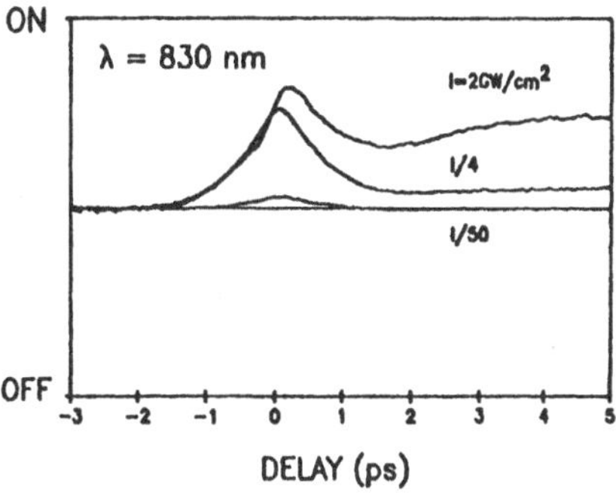


**Figure 1.** Time-resolved refractive index changes in an AlGaAs waveguide [LaGasse, 1989].

LaGasse and co-workers [LaGasse, 1989, Anderson, 1990], were the first to show that the use of subpicosecond pulse techniques could be used to temporally resolve virtual and real carrier effects. In their experiments PIRIC's were measured for AlGaAs waveguides using a pump-probe interferometry experiment. The refractive index change, $\Delta n$, induced by a 480 fs pump beam was measured with a similar probe pulse. A typical time-resolved trace of $\Delta n$ is shown in Fig.1. It can be seen that a peak is observed at zero time delay and has a width comparable to that of the pump pulse. This behaviour is followed by an increase in $\Delta n$ on a time-scale of a few picoseconds. The peak is attributed to an instantaneous PIRIC while the long lived contribution is attributed to carrier bandfilling, although a positive identification based on the scaling behaviour of the two contributions was not made. Despite the limited scope of these experiments, and the limited applicability of the measurement technique which was used (the interferometric technique used is not easily extended to the study of bulk material), the results do illustrate the usefulness of femtosecond techniques for the investigation of refractive index changes and the physics associated with them.

Here, we illustrate how femtosecond beam deflection and induced grating techniques can be used to time resolve refractive index changes in bulk II-VI semiconductors and allow the separation of real and virtual effects in a quantitative manner. Both techniques are sensitive to local gradients in the refractive index, and both can be performed in pump-probe geometries. Moreover, the use of two complementary techniques offers the possibility of distinguishing recombination processes and diffusion processes in high density plasmas.

The beam deflection technique is a two-beam, time-resolved implementation of self-beam deflection techniques. Orthogonally polarized Gaussian pump and probe beams are focused onto a sample with near normal angles of incidence. The spots are overlapped such that the centre of the probe spot is located away from that of the pump spot so that the probe beam interacts with a non uniformly excited region. Refractive index changes induced by the pump beam lead to a lensing phenomenon and change in the direction of propagation of the probe beam. The position of the deflected beam can be monitored and related to the strength of the refractive index change. The beam deflection technique allows the sign of the refractive index change to be determined.

Laser induced grating experiments have been used extensively to study refractive nonlinearities and ultrafast carrier dynamics [Eicher, 1986]. A pair of identical pump pulses is focused onto a sample such that the pulses overlap spatially and temporally. The pulses interfere and produce an excitation grating which in turn leads to a refractive index grating. A time delayed probe pulse is diffracted from this grating and detected. The strength of the diffracted beam is proportional to the modulation of the induced grating. The technique is only sensitive to the absolute value of the change in the (complex) refractive index.

With both these techniques we have used intense sub-band gap radiation to study time resolved refractive index changes in the II-VI semiconductors CdS, $CdS_{0.75}Se_{0.25}$, and ZnSe [Fox, 1991,1992a,1992b]. At low intensities one observes that only virtual effects are observed, while at high pump intensities, two photon absorption for which the absorption coefficient varies linearly with laser intensity, gives rise to significant real carrier creation which then dominates $\Delta n$ and leads to long-lived changes. These experiments have allowed the nature of the refractive nonlinearities displayed by these materials to be studied on a femtosecond time scale for the first time. It will also be seen that above a certain pump irradiance stimulated emission dominates the kinetic behaviour of the carriers leading to ultrafast recombination and reabsorption of the light elsewhere in the sample until the carriers become nondegenerate everywhere. Although this obviously doesn't remove the carriers from the sample, it has the effect of forcing the carrier distribution to have a uniform density over a large region and thereby reducing refractive index gradients. This causes the beam *deflection* to recover in as little as 4 ps, and suggests other methods of using carrier induced PIRIC's for switching effects.

The PIRIC experiments performed using deflection techniques have also allowed us to obtain reasonable interpretations of other types of experiments dealing with the cooling, expansion, and recombination of high density nonequilibrium plasmas in II-VI semiconductors. In particular there has been much debate about the possible role of nonequilibrium phonons in reducing the carrier energy relaxation time in II-VI semiconductors, an effect which is well-known in general [van Driel, 1979] and in the III-V semiconductors [Vasconcellos, 1980; Pötz, 1983; Shah, 1989] in particular. We show that this effect is insignificant under the conditions of our experiments. We further demonstrate that apparent enhanced diffusion rates [Saito, 1985; Rinker, 1989] of high density carriers in II-VI semiconductors which had been attributed to carrier screening [Junnarkar, 1986] or Fermi pressure in a degenerate plasma [Cornet, 1981; Combescot, 1979] can, in certain cases, be explained in terms of stimulated emission *and reabsorption* of the photons. Earlier Dneprovskii and co-workers [1988,1990] observed sharp decays of carrier populations in excess of the threshold density for stimulated emission. We show below that the combination of stimulated emission and reabsorption leads to an effective ultrafast transport process for carriers.

The investigation of refractive index changes in the II-VI semiconductors CdS, $CdS_{0.75}Se_{0.25}$, and ZnSe is also appropriate with respect to particular applications of these materials. $CdS_xSe_{1-x}$ and ZnSe are well characterized optical materials which are used extensively in the production of semiconductor doped glass filters and for window material at infrared wavelengths. All three materials are known to display large PIRIC's and as such are attractive candidates for all-optical switching devices. Both in bulk crystalline form and in semiconductor doped glass form (quantum dots), the PIRIC's in these materials have been extensively studied using nanosecond and picosecond laser pulses [Gibbs, 1990; Haug, 1988; Wherrett, 1990]. The II-VI semiconductors have also been of interest recently because of their wide energy band gaps. Because the band-gap energy corresponds to radiation with wavelengths in the blue and green these materials are attractive candidates for the construction of laser diodes in this spectral region. Diode lasers in this region are being sought for applications ranging from flat optical displays to high density optical memories. Continued improvements in the design and performance of these devices will in part depend on an increased understanding of the physics of high density (degenerate) carrier distributions in these materials, and in particular on an increased understanding of carrier and lattice relaxation processes under strong excitation conditions which can lead to stimulated emission.

**Band Structure Parameters of CdS, $CdS_{0.75}Se_{0.25}$, and ZnSe**

For convenience in the discussions to follow we present some of the salient properties of the II-VI semiconductors of interest here. Unlike the elemental and III-V semiconductors which bond with a cubic (Zincblende) lattice structure, CdS, $CdS_{0.75}Se_{0.25}$, and ZnSe can bond with either a cubic or hexagonal (Wurtzite) lattice structure [Landolt-Börnstein, 1982]. The experiments described here have all been performed using the more common hexagonal form of these materials. This difference in lattice structure has a direct effect on the electronic band structure, in particular on the structure of the hole bands. As shown in Fig. 2 each of the materials has three hole bands, two of which (A and B) are degenerate at the $\Gamma$ point, and one of which (the split off band) corresponds to higher hole energies and does not play a significant role in the experiments described here. Table 1 lists the room temperature values of the band-gap ($E_g$) and the different effective masses for the energy bands of interest here. The A and B hole bands are elliptical with each characterized by one effective mass along the c-axis ($m^{\parallel}_A$,$m^{\parallel}_B$) and another effective mass along directions perpendicular to the c-axis ($m^{\perp}_A$,$m^{\perp}_B$). For the A band the major axis lies along the c-axis, while for the B band it is the minor axis which lies along the c-axis. Values of the heavy ($m_{hh}$) and light hole ($m_{lh}$) masses are also given as is the split-off hole energy where these values are known.

The effective mass values in CdS, $CdS_{0.75}Se_{0.25}$, and ZnSe are not particularly small which explains why these materials have not attracted attention from the electronic device community. Despite this they do offer a convenient band structure with which to investigate semiconductor electronic behaviour. For example, none of these materials has side valleys in either the valence or conduction bands up to energies of at least 3 eV. The presence of side valleys at low energies in GaAs has complicated the study of fundamental relaxation phenomena such as carrier cooling. The II-VI materials are also significantly more polar than the III-V materials. As a result there is a much stronger coupling between the carriers and the LO phonons in the II-VI materials and hence more rapid carrier cooling.

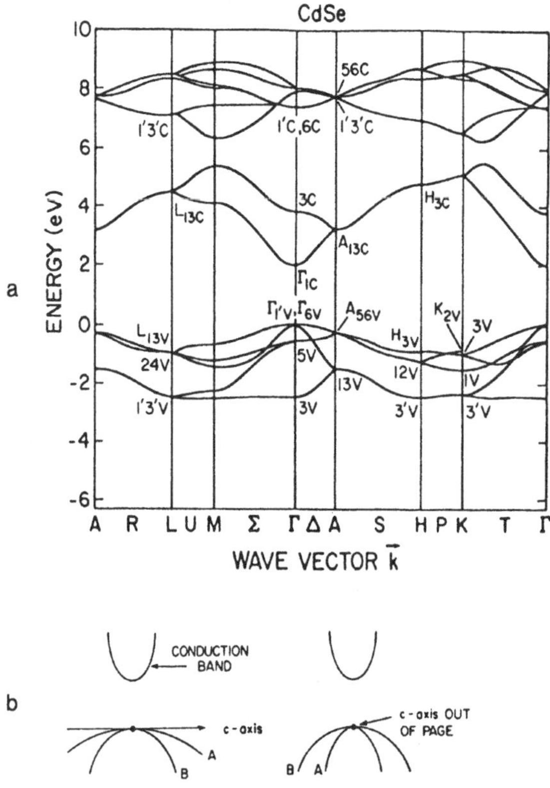


**Figure 2.** a) Schematic diagram of the band structure of a hexagonal II-VI semiconductor (CdSe) [Kobayashi, 1983]. b) Simplified representation of the energy bands showing the relative curvatures of the A and B valence bands in two directions.

The remainder of this article is organized as follows. In the next section we discuss the fundamental sources of PIRIC's in semiconductors in terms of virtual, real carrier and lattice heating effects. The theoretical foundations of the deflection and diffraction techniques are offered as well. We then discuss the experimental deflection and diffraction techniques used to probe refractive changes on a femtosecond and picosecond time-scale in II-VI semiconductors. This is followed by a presentation and discussion of experimental results following which conclusions are drawn.

## THEORETICAL BACKGROUND

In this section we consider how the real and virtual mechanisms considered above contribute to the PIRIC's when below-band-gap radiation interacts with semiconductors, in general, and the II-VI semiconductors, in particular. A discussion of these effects, especially for high intensity laser beams, involves both concepts in nonlinear optics as well

**Table 1.** Band structure parameters of select II-VI semiconductors[a].

| PARAMETER | CdS | CdSe | $CdS_{.75}Se_{.25}$ | ZnSe |
|---|---|---|---|---|
| $m_e / m_0$ | 0.21 | 0.112 | 0.186 | 0.15 |
| $m_A^{\parallel} / m_0$ | 2.8 | 1.2 | 2.4 | - |
| $m_A^{\perp} / m_0$ | 0.68 | 0.45 | 0.62 | - |
| $m_B^{\parallel} / m_0$ | 0.81 | 0.37 | 0.70 | - |
| $m_B^{\perp} / m_0$ | 1.0 | 0.90 | 0.98 | - |
| $m_{hh} / m_0$ | - | - | - | 1.25 |
| $m_{lh} / m_0$ | - | - | - | 0.22 |
| Band gap (eV) (300 K) | 2.42 | 1.72 | 2.125 | 2.67 |
| split-off hole energy (eV) | 0.56 | 0.53 | 0.55 | 0.45 |
| Kane energy parameter (eV) | 21 | 21 | 21 | 24.2 |

a) All entries in this table are from Landolt-Börnstein [1982] except for the Kane energy parameter [Van Stryland, 1985]; $m_0$ is the free electron mass.

as nonequilibrium kinetics. We therefore begin by outlining the salient features of these two areas as they relate to the experiments of interest here.

**Fundamentals of Optical Pulse-Semiconductor Interactions**

When optical radiation interacts with a semiconductor the optical response can be described by the superposition of the responses to the individual monochromatic components which make up the beam. Although the pulses we consider later are as short as 100 fs, this is still long compared to the sub-femtosecond period of the 620 nm electromagnetic radiation. To a good approximation then we are justified in considering the interaction to be quasi-monochromatic at a single, centre frequency $\omega$ and describe the incident electric field by $\vec{E}(\omega)$. The response to the field can be described in terms of an induced polarization density. Because the effects we are interested in relate to induced refractive index changes experienced by the beam itself, we are only interested in contributions to the polarization density at the original frequency, viz., $\vec{P}(\omega)$. The first two non-zero (linear and nonlinear) induced responses are given by [Shen, 1985]:

$$\vec{P}(\omega) = \varepsilon_0 \{ \vec{\chi}^{(1)}(-\omega,\omega) \cdot \vec{E}(\omega) + \vec{\chi}^{(3)}(-\omega,\omega,-\omega,\omega) \cdot \vec{E}(\omega)\vec{E}(-\omega)\vec{E}(\omega) \} \quad (1)$$

where $\varepsilon_0$ is the dielectric permitivity of the vacuum. For the materials considered here the anisotropies in the linear and nonlinear susceptibilities are small and of little interest to us so we will ignore the tensor character of the response. The product $\vec{E}(\omega)\vec{E}(-\omega)$ is related to the instantaneous intensity of the beam given by

$$I(\omega) = \frac{1}{2} c\, n_0(\omega) \varepsilon_0 E(\omega) E(-\omega) \quad (2)$$

with $n_0(\omega)$ (henceforth written simply as $n_0$) being the quiescent refractive index experienced by the beam and c is the speed of light. The scalar polarization density can then be rewritten as:

$$P(\omega) = \varepsilon_0 \left[ \chi^{(1)}(-\omega,\omega) + \frac{2}{c n_0 \varepsilon_0} \chi^{(3)}(-\omega,\omega,-\omega,\omega) I(\omega) \right] E(\omega) \quad (3)$$

$$= \varepsilon_0 \chi^{(1)}_{\mathrm{eff}}(-\omega,\omega) E(\omega) \quad (4)$$

where the effective susceptibility, $\chi^{(1)}_{\mathrm{eff}}(-\omega,\omega)$, is, in general, a complex number. From Eq. 3, and a little algebra, the effective susceptibility leads to a complex refractive index, $\hat{n}(\omega) = n + i\kappa$, whose real part is given by :

$$n(\omega) = n_0(\omega) + n_2(\omega) I(\omega) \quad (5)$$

The quantity $n_2(\omega)$ is often referred to as the nonlinear refractive index. From the fact that the imaginary part $\kappa$ is related to the absorption coefficient $\alpha$ through $\alpha = 4\pi\kappa\lambda^{-1}$, the absorption coefficient can be written as:

$$\alpha(\omega) = \alpha_0(\omega) + \beta(\omega) I(\omega) \quad (6)$$

where $\alpha_0(\omega)$ (henceforth referred to as $\alpha_0$) is the low intensity (single photon) absorption coefficient while $\beta(\omega)$ is the nonlinear absorption coefficient which can be identified with two-photon absorption (2PA).

In the case of the sub-band gap radiation, and in the absence of impurities or defects, $\alpha_0(\omega)$ is initially virtually zero and absorption is dominated by interband 2PA. However, two photon absorption leads to the production of carriers of density N which can in turn influence the value of $n_0$ and $\alpha_0$ through free carrier absorption (FCA). The change in the refractive index can therefore occur directly through $n_2(\omega)$ (part of which, as will be seen later, results from virtual excitations) while a second contribution can occur through two-photon induced real carriers which can then modify $n_0$. The total change in refractive index, $\Delta n$, can be written as the sum of real (r) and virtual (v) effects so that:

$$\Delta n = (\Delta n)_v + (\Delta n)_r \tag{7}$$

where

$$(\Delta n)_v = n_2 I. \tag{8}$$

In a similar fashion, the contributions to $\Delta n$ from real excitations can be expressed in terms of the magnitudes of quantities associated with the real excitations. As shown below, the free excitations considered here are related to free carriers of density N and nonequilibrium phonons of density $N_p$ so that $(\Delta n)_r$ can be written as :

$$(\Delta n)_r = \gamma N + \zeta N_p \tag{9}$$

where the proportionality constant $\gamma$ depends on band structure, temperature of the plasma, etc. while the constant $\zeta$ depends on the types of phonons excited.

Because the parameter $\gamma$ depends on details of the plasma, but also because the plasma itself depends on nonlinear optical effects for its existence, it is worth reviewing some elements of optical absorption and evolution of photoexcited plasmas before we delve further into the theoretical aspects of the contributions to PIRIC's from real or virtual excitations.

In general, linear and two-photon absorption lead to a reduction of the beam irradiance with propagation distance, z, in the sample. The equation governing the irradiance at depth z, I(z) is given by :

$$\frac{dI(z)}{dz} = -(\alpha_0 + \beta I(z))I(z). \tag{10}$$

For sub-band-gap radiation only two photon absorption leads to carrier generation and one electron-hole pair is generated for two absorbed photons (each of photon energy $\hbar\omega$). The local carrier generation rate, dN/dt, can then be written as:

$$\frac{dN}{dt} = \frac{1}{2\hbar\omega}\beta I^2(z). \tag{11}$$

Short optical pulses can be often be modeled using a Gaussian function to describe the radial profile and a $\text{sech}^2$ function to describe the temporal (t) profile. In this case the incident irradiance on a semiconductor can be expressed as:

$$I(t) = I_0 \exp(-2r^2 / r_0^2) \mathrm{sech}^2(1.76t / \tau_p), \tag{12}$$

where r is the radial distance from the centre of the beam, $r_0$ is the $e^{-2}$ radius, $\tau_p$ is the temporal pulse width at half maximum irradiance, and $I_0$ is the peak irradiance. From Eqs. 10 and 12 the irradiance distribution within the sample can be written:

$$I(r,z,t) = \frac{I_0 T \exp(-2r^2 / r_0^2) \mathrm{sech}^2(1.76t / \tau_p)}{1 + Tz\beta I_0 \exp(-2r^2 / r_0^2) \mathrm{sech}^2(1.76t / \tau_p)}. \tag{13}$$

where T is the linear transmission across the surface interface and z represents the depth below the sample surface. This expression can be integrated after combining it with Eq. 11 to give the total carrier density generated by a single pulse:

$$N(r,z) = \frac{\beta}{2\hbar\omega} \int_0^\infty dt \left[ \frac{I_0 T \exp(-2r^2 / r_0^2) \mathrm{sech}^2(1.76t / \Delta t)}{1 + \beta T z I_0 \exp(-2r^2 / r_0^2) \mathrm{sech}^2(1.76t / \Delta t)} \right]^2. \tag{14}$$

This illustrates that for small values of $\beta I_0$ the local carrier density scales quadratically with the incident peak irradiance. At the sample surface one obtains an equation of the form

$$N(r,0) = \delta I_0^2 \exp(-4r^2 / r_0^2). \tag{15}$$

Values of $\delta$ are given in Table 2 for the three semiconductors of interest here in the case of incident 620 nm pulses with pulsewidth of 120 fs.

When carrier densities become large, free carrier absorption may also contribute significantly to the attenuation of an incident beam, but not to carrier generation. The energy lost from the radiation field due to FCA, however leads only to carrier heating. In many cases FCA is described in terms of a cross section, $\sigma_{FCA}$, so that $\alpha_0 = \sigma_{FCA} N$. Although we know of no measurements of $\sigma_{FCA}$ which have been made at $\lambda$ = 620 nm in $CdS_{0.75}Se_{0.25}$ or ZnSe, values of $\sigma_{FCA} = 10^{-18}$ cm$^2$ ($2.8 \times 10^{-18}$ cm$^2$ ) have been reported for ZnSe ($CdS_{0.9}Se_{0.1}$) at 705 (532) nm. If this additional absorption mechanism is included in Eq. 10 then the local irradiance and carrier density are calculated by numerically solving the set of coupled equations 10 and 11. Of particular interest is the ratio of energy absorption due to 2PA (which leads to carrier production) to total energy absorption (2PA + FCA). Analytically it can be shown that the ratio decreases as the product $I_0 \sigma_{FCA} \tau_p$ increases. The fraction of energy absorption associated with 2PA has been calculated in the near surface region using $\sigma_{FCA} = 2 \times 10^{-18}$ cm$^2$ and $\tau_p$ = 120 fs. For light incident with peak irradiances $I_0$ = 100, 250, and 500 GWcm$^{-2}$ the fraction equals 0.98, 0.96, and 0.93 respectively. This shows that for the range of irradiances investigated here free carrier absorption can be ignored as an energy loss mechanism.

The electron-hole pairs are generated with an ionization energy of $E_g$ and total kinetic energy of $(2\hbar\omega - E_g)$ per pair. If one assumes that the electron and hole bands are parabolic and the surfaces of constant energy are spherical, one can describe the electron and hole effective masses by energy-independent scalars, $m_e$ and $m_h$ respectively (the actual situation is more complicated than this as the previous section showed). Simple considerations of energy and momentum conservation establish that the initial excess electron kinetic energy is given by $(2\hbar\omega - E_g)\mu_r/m_e$ and the hole excess energy is $(2\hbar\omega - E_g)\mu_r/m_h$ where $\mu_r = (m_e^{-1} + m_h^{-1})^{-1}$ is the reduced electron-hole mass. Fig. 3 illustrates [van Driel, 1987] how the

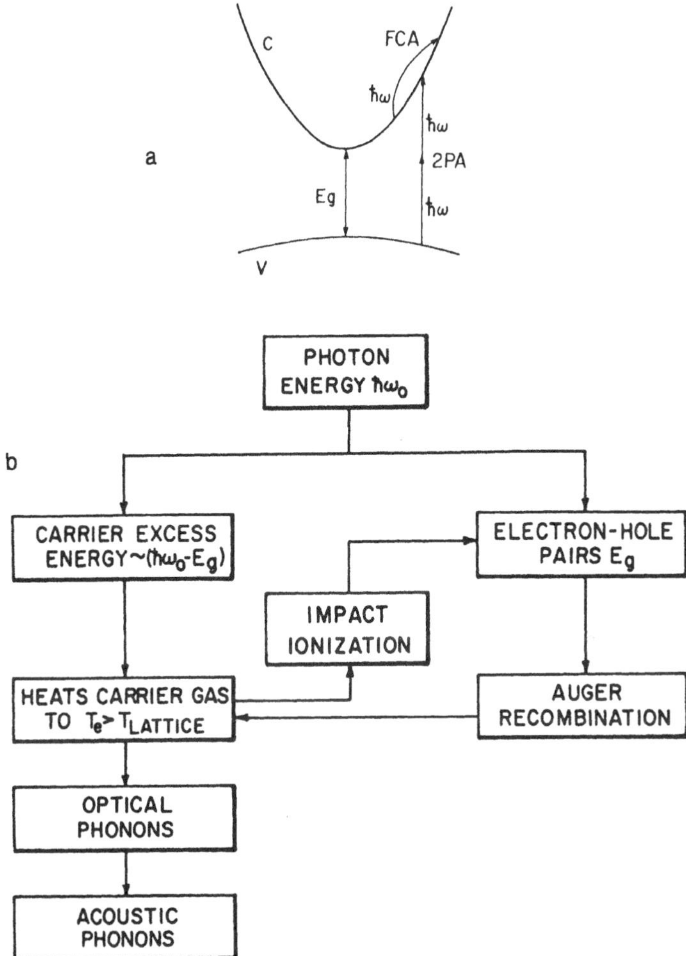


**Figure 3.** a) Schematic diagram of optical excitation of a semiconductor. b) Schematic diagram of flow of photoexcited carrier energy (assuming negligible free carrier absorption).

carrier kinetic energy and carrier density evolve as the semiconductor returns to equilibrium following optical pulse excitation with effective photon energy $\hbar\omega_0 = 2\hbar\omega$. The carrier energy rapidly thermalizes to allow the electrons (holes) to form Fermi-Dirac distributions characterized by temperature $T_e$ ($T_h$) and Fermi level $F_e$ ($F_h$).

$$f_e(E_e) = [1 + \exp\{\beta_e(E_e - F_e)\}]^{-1} \tag{16a}$$

$$f_h(E_h) = [1 + \exp\{\beta_h(E_h - F_h)\}]^{-1} \tag{16b}$$

with $\beta_e=(k_B T_e)^{-1}$ and $\beta_h=(k_B T_h)^{-1}$. The carrier density can then be written as

$$N = N_c \mathcal{F}_{1/2}(\beta_e \eta_e) = N_v \mathcal{F}_{1/2}(\beta_h \eta_h) \tag{17}$$

where $\eta_e$ ($\eta_h$) is the Fermi level of the electrons (holes) relative to its band edge, $N_c$ and $N_v$ are the effective density of states in the conduction and valence band respectively and $F_j$ is the Fermi-Dirac Integral of order j. The temperatures and Fermi levels for the electrons and holes are determined by solving the systems of equations consisting of Eqs. 17 and the following equations (one each for electrons and holes) in which the initial energy of each carrier relative to the relevant band edge is equated to the average carrier energy within the thermalized distribution (weighted by the density of states):

$$\frac{\mu_r}{m_e}\left[2\hbar\omega - E_g\right] = \frac{3}{2} k_B T_e N_c \mathcal{F}_{3/2}(\beta_e \eta_e), \tag{18a}$$

$$\frac{\mu_r}{m_h}\left[2\hbar\omega - E_g\right] = \frac{3}{2} k_B T_h N_v \mathcal{F}_{3/2}(\beta_h \eta_h). \tag{18b}$$

It should be noted that, to be exact, each hole with a distinct effective mass must be treated as a separate carrier species. It can be shown that for low density plasmas this calculational scheme leads to the well known result for Boltzmann statistics that the initial temperature of each carrier species is equal to $(2/3)k_B^{-1}$ multiplied by the initial energy of that species.

Following thermalization the electrons and holes evolve so as to reach a common carrier temperature through collisions with each other. Because the effective mass of the electrons is much smaller than that of the holes, the electrons initially have a much higher temperature than that of the holes but as will be seen below lose much of their energy to the holes on a picosecond time scale. However, on a time scale of a few picoseconds the carriers in polar semiconductors also cool by emission of longitudinal optic (LO) phonons through the Fröhlich interaction [Seeger, 1982]. At high excitation it is possible that the optical phonon distribution is driven far from equilibrium (hot phonon effect) so that the carrier cooling rate is inhibited [van Driel, 1979, Vasconcelos, 1980; Pötz, 1983]. If this occurs then both carriers and optic phonons decay at a rate determined by the optical phonon lifetime, typically 5 ps. The optic phonon modes and subsequently excited acoustic modes come to equilibrium with the remaining modes in the lattice through anharmonic lattice interactions which occur on a time scale of several ps to several ns depending on the particular modes.

Finally the carrier density is reduced through various recombination processes. In our experiments the carrier densities are typically less than $5\text{x}10^{19}$ cm$^{-3}$ so that (spontaneous and stimulated) radiative recombination will dominate with the carriers recombining at a rate,

$$\frac{\partial N}{\partial t} = -BN^2 \tag{19}$$

where B is approximately $10^{-11}$ cm$^{-3}$ [Dneprovskii, 1988] for spontaneous emission but has a much higher value and complicated dependence on $T_e$, $T_h$, $F_e$, and $F_h$ in the case of stimulated emission. Note that because the emitted photons have an energy higher than that of the band-gap, they can be reabsorbed leading to carrier generation in other regions of the sample, in some cases not more than a fraction of a micron from where the initial recombination event occurred. This photon recycling process [Chandresekhar, 1950; Dumke, 1957; Epifanov, 1976; Tsarenkov, 1979] leads to an effective transport of carriers. It is also possible that emitted photons can be reabsorbed by free carriers leading to a heating of the plasma. This would have the effect of altering Fermi levels and changing the recombination kinetics. The point is that all these processes are inter-related and the evolution of a spatially inhomogeneous plasma is very complicated, even without having to consider normal ambipolar diffusion processes.

**Contributions to Refractive Index Changes**

We are now in a position to consider the contributions to the refractive index change induced by a high intensity laser beam. Theoretical expressions for refractive index changes are generally more difficult to derive from first principles than expressions for absorption coeffcient changes. Optical refraction, however, is always accompanied by optical absorption. As a result most calculations of PIRIC's are performed by first calculating the strength of the related absorptive nonlinearity. The results are then transformed using the Kramers-Kronig relations which mathematically link the absorption coefficient and index of refraction.

The Kramers-Kronig analysis of the linear optical response of a material relates the refractive index, n(ω), and the absorption coefficient, α(ω):

$$n(\omega) = 1 + \frac{c}{\pi} P \int_0^{\infty} d\omega' \, \frac{\alpha(\omega')}{\omega'^2 - \omega^2} \tag{20}$$

where the symbol $P$ denotes the principal part of the integral. In order to determine n(ω) for a specific frequency, ω, it is necessary to know α(ω) for all frequencies between zero and infinity. However, the weighting factor in the denominator makes the calculation of n most sensitive to resonances in a range close to the frequency ω. It follows from Eq. 20 that any effect which produces a change in the absorption coefficient, Δα(ω), will also produce an accompanying change in the refractive index, Δn(ω). From an extension of Eq. 20 the relation between the two quantities can be expressed as:

$$\Delta n(\omega) = \frac{c}{\pi} P \int_0^{\infty} d\omega' \, \frac{\Delta\alpha(\omega')}{\omega'^2 - \omega^2} \tag{21}$$

Changes in the absorption coefficient can be due to various perturbations including free carriers, phonons, or strong electromagnetic fields. In what follows, the Kramers-Kronig relations will be applied to the calculation of refractive index changes due to such perturbations. Bound electronic effects will be discussed first. This will be followed by a discussion of free electron effects, and finally lattice heating effects. All of these will be discussed for optical frequencies where the corresponding photon energy lies below the energy of the semiconductor energy band gap which is the regime of experimental interest later.

Absorption changes which are associated with alteration of the bound electron distribution occur only in the presence of a light field. If a material is illuminated by monochromatic radiation at a frequency ω then the induced absorption change at a frequency ω', due to the presence of a light field can be expressed to lowest order as:

$$\Delta\alpha(\omega) = \theta(\omega,\omega')\, I(\omega') \tag{22}$$

where the coefficient β(ω,ω') depends on the nature of the absorption change. When Eqs. 21 and 22 are combined one obtains that the change in refractive index brought about by virtual electronic effects is of the form of Eq. 8.

Sheik-Bahae et al. [1991] have shown that when $\hbar\omega < E_g$ the two most important contributions to Δα are from 2PA and the quadratic (or optical) Stark effect (QSE), while effects due to the linear Stark effect (LSE) and the electronic Raman effect (RAM) are of a lesser significance. In the general case of nondegenerate 2PA an optical transition involves two photons of unequal energy. Phenomenologically, nondegenerate 2PA can be described as a change in the absorption coefficient at one frequency, ω, due to the presence of a light field at a second frequency, ω', so that Eq. 22 becomes

$$\Delta\alpha(\omega) = \beta(\omega,\omega')I(\omega') \tag{23}$$

where β(ω,ω') is the nondegenerate 2PA coefficient and is proportional to the imaginary part of the third order optical susceptibility, Im $\chi^{(3)}(-\omega; \omega,\omega',-\omega')$. Expressions for β(ω,ω') have been derived by Sheik-Bahae et al. [1988] for direct band gap semiconductors using second order perturbation theory within an effective mass approximation. General expressions for the degenerate 2PA show that β (in $\mathrm{cmGW}^{-1}$) can be written:

$$\beta = (3.1 \pm 0.5) \times 10^3 \frac{E_p F_2(2\hbar\omega / E_g)}{n_0^2 E_g^3}, \tag{24}$$

where $E_p$ is the Kane energy parameter in eV (see Table 1) and $F_2$ is a function which depends on the details of the band structure. For semiconductors with parabolic band structures the function $F_2$ can be written:

$$F_2(x) = \frac{(x-1)^{3/2}}{x^5}. \tag{25}$$

The degenerate two-photon absorption coefficient for CdS, $CdS_{0.75}Se_{0.25}$ and ZnSe are displayed in Table 2.

The general expression for the non-degnerate two photon absorption coefficient can be used to deduce a general scaling behaviour for $n_2$ as a function of $\hbar\omega / E_g$. It was found that the value of $n_2$ scales with $E_g^{-4} n_0^{-2}$. It is possible to obtain a dimensionless dispersion function, $G_2(\hbar\omega/E_g)$ for $n_2$ if the latter is divided by $6\pi x 10^8 c^{-1} E_g^{-4} n_0^{-2} E_p^{1/2}$. The dispersion function $G_2$ is plotted in Fig. 4. For the combination of materials and photon energy used here, $\hbar\omega/E_g > 0.75$. In this regime the results indicate that $n_2$ should be less than zero, and that the magnitude of $n_2$ should increase as the photon energy approaches the band gap energy.

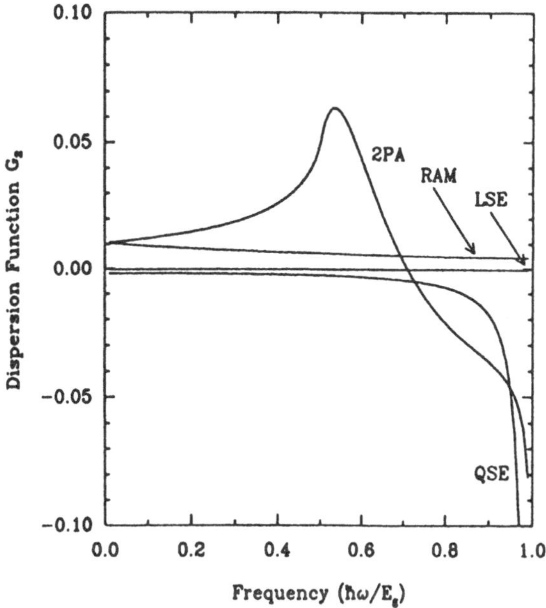


**Figure 4.** Variation of various contributions to nonlinear refractive index with photon energy.
M. Sheik-Bahae *et al., Dispersion of Bound Electronic Nonlinear Refraction in Solids*, IEEE J. Quan. Elec. QE-27, 1296 (1991).

The effect of the quadratic (or optical) Stark effect is to blue shift the electronic band gap in the presence of intense optical radiation. The shift can be calculated by considering the coupling of the valence and conduction band states by the non-secular part of the radiation Hamiltonian. To first order in the intensity of the exciting radiation it is found that the induced absorption changes for photon energies less than the energy band gap scale linearly with the intensity. This is identical to the scaling displayed by 2PA-induced changes. It follows that the quadratic Stark effect contribution to $n_2$ can be calculated through the Kramers-Kronig transformation. The results are displayed in Fig. 4.

Since the absorption changes induced by the quadratic Stark effect are greatest in the immediate vicinity of the band edge, the greatest changes in the refractive index occurs for photons with energies close to the band gap energy. As for the 2PA contribution in the near band edge regime, the quadratic Stark effect contribution to $n_2$ is negative and increases sharply in magnitude as the photon energy approaches the band gap energy.

**Table 2.** Optical properties for select II-VI semiconductors

| Parameter | CdS | CdSe | $CdS_{.75}Se_{.25}$ | ZnSe |
|---|---|---|---|---|
| $n_0$ ($\lambda$=620 nm) [a] | 2.5 | 2.13 | 2.41 | 2.56 |
| $dn_0/dT_L$ ($K^{-1}$) ($\lambda$=620 nm) [a] | $1.1x10^{-4}$ | - | $2.4x10^{-4}$ | $1.6x10^{-4}$ |
| $n_2$ ($cm^2GW^{-1}$) | $-5.2x10^{-5}$ | - | $-2.6x10^{-4}$ | $-1.8x10^{-5}$ |
| $\gamma_{FCA}$ ($cm^3$) | $-4.0x10^{-22}$ | - | $-4.5x10^{-22}$ | $-5.1x10^{-22}$ |
| $\gamma_{BF}$ ($cm^3$) (T=300 K) [b] | $-8.5x10^{-22}$ | - | $-3.6x10^{-21}$ | $-9.4x10^{-22}$ |
| $\gamma_{BF}$ ($cm^3$) (T=300 K) [c] | $-1.3x10^{-21}$ | - | $-4.4x10^{-21}$ | $-7.4x10^{-22}$ |
| $\gamma_{BF}$ ($cm^3$) (high T) [d] | $-3.1x10^{-22}$ | - | $-5.2x10^{-22}$ | - |
| $\gamma_{BF}$ ($cm^3$) (nonthermal) | $-7.8x10^{-23}$ | - | $-8.7x10^{-23}$ | - |
| $\gamma_{LH}$ ($cm^3$) | $+1.6x10^{-23}$ | - | $+4.2x10^{-23}$ | $+1.8x10^{-23}$ |
| $\beta$ (2PA) ($cmGW^{-1}$) | 6.8 | - | 9.0 | 5.7 |
| $\sigma_{FCA}$ ($cm^2$) | ~$10^{-18}$ to $10^{-17}$ | - | ~$10^{-18}$ to $10^{-17}$ | ~$10^{-18}$ to $10^{-17}$ |
| $\delta$ ($cmGW^{-2}$) [e] | $6.1x10^{-14}$ | - | $8.1x10^{-14}$ | $5.1x10^{-14}$ |

a) from Landolt-Börnstein [1982].
b) after Auston et al.[1978].
c) full band filling calculation; performed for a carrier density of $10^{17}$ $cm^{-3}$.
d) CdS: $T_e$=10,400 K, $T_h$=2,200 K; $CdS_{0.75}Se_{0.25}$: $T_e$=12,500 K, $T_h$=2,500 K.
e) when multiplied by the peak irradiance for a 120 fs pulse this gives the surface carrier density.

The two other processes which contribute to $n_2$, the linear Stark effect and the electronic Raman effect, can largely be ignored in the frequency regime of interest here. Linear Stark effect contributions arise from a reduction in the oscillator strength due to the coupling of the conduction band to itself (or of the valence band to itself) by the radiation field. Electronic Raman contributions arise from two photon transitions where the difference in the energies of the two photons is equal to the energy of the electronic transition (a photon at one energy is absorbed while a photon at a second energy is created). The contributions of these two processes to $n_2$ are also plotted in Fig. 4.

We have used the formulas which were employed to generate the plots in Fig. 4 to calculate $n_2$ for CdS, $CdS_{0.75}Se_{0.25}$, and ZnSe for the photon energy (2 eV) of interest later. The results are listed in Table 2. It should be noted that the values are greatest when the energy band gap is close to the photon energy of the exciting radiation. The general variation of $n_2$ with photon energy in a wide range of semiconductors is shown in Fig. 5.

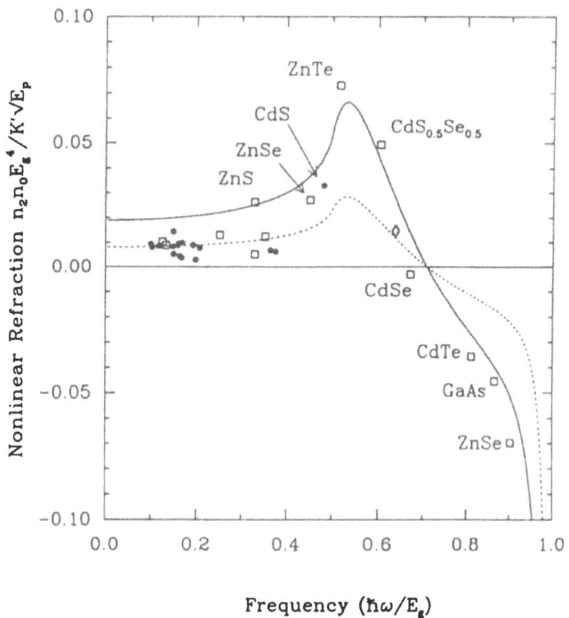


**Figure 5.** Variation of $n_2$ with photon energy in various semiconductors.
M. Sheik-Bahae *et al., Dispersion of Bound Electronic Nonlinear Refraction in Solids,* IEEE J. Quan. Elec. QE-27, 1296 (1991).

Free carrier contributions to the refractive index change, $(\Delta n)_r$, can also be related to absorption changes associated with the presence of excited electrons and holes. Such absorption changes include those due to band filling (BF), free carrier absorption (Drude effects), band gap renormalization (BGR) and plasma screening of Coulomb interactions (e.g., excitonic effects). Band filling nonlinearities are associated with the bleaching of optical interband transitions due to the filling of electron states in the conduction band and hole states in the valence band. Free carrier absorption, as mentioned briefly above, is a process by which electrons and holes make intraband transitions in the conduction and valence bands respectively with the participation of both a photon and a phonon. BGR is a many-body effect whereby the energy band gap is reduced by the presence of free carriers and thus causes changes in optical absorption as a result of shifting energy levels. Plasma screening effects are mainly associated with excitonic contributions to the refractive index; as the plasma density increases the Coulomb interaction which binds the electron-hole pair is reduced which results in a reduction in the oscillator strength of the excitonic transition and enhancement of oscillator strength for the continuum states.

For the experiments described later the most important free carrier contribution to the refractive nonlinearity is from band filling in which originally vacant conduction (or valence) band states become occupied by photoexcited electrons (or holes). Since occupied states cannot participate in interband optical transitions it follows that band filling is accompanied by a reduction in the magnitude of the absorption coefficient, and hence also by a change in the refractive index. At a qualitative level one can treat the semiconductor as a collection of simple Lorentz oscillators in which one oscillator exists for each allowed interband transition, and in which the refractive index is to be evaluated for optical frequencies which lie below the minimum oscillator frequency. Since each oscillator contributes positively to the net refractive index in this regime, and since the contributions of oscillators corresponding to blocked (bleached) transitions vanish, it follows that band filling must contribute negatively to the net refractive index (i.e. $\Delta n_{BF} < 0$). Also, since the magnitude of the refractive index at a particular optical frequency is most affected by oscillators with resonances corresponding to nearby frequencies, it follows that the magnitude of $\Delta n_{BF}$ for frequencies corresponding to photon energies less than the band gap energy will be most affected by the occupancy of the band edge states. The magnitude of $\Delta n_{BF}$ should also increase as the optical frequency approaches the band edge frequency.

In order to quantitatively predict the magnitude of $\Delta n_{BF}$ a more sophisticated treatment than that offered in the previous paragraph is required. The approach which is usually taken is to derive an expression for the change in the absorption coefficient due to band filling, and then to integrate this expression using Eq. 16. The analysis begins by considering the expression for the absorption coefficient corresponding to k-conserving interband transitions for a single pair of conduction and valence bands in a crystalline semiconductor which can be treated within the effective mass approximation [Johnson, 1967]:

$$\alpha(\omega) = \frac{2^{5/2}e^2\mu_r^{3/2}\langle p\rangle^2}{m_0^2 n_0 c\hbar^3\omega}(\hbar\omega - E_g)^{1/2}[1 - f_e(\hbar\omega) - f_h(\hbar\omega)] \tag{26}$$

where $\langle p\rangle$ is the momentum matrix element averaged over all interband transitions, and $f_e(\hbar\omega)$ and $f_h(\hbar\omega)$ are the electron and hole occupancy factors for the states coupled by photon energy $\hbar\omega$. In determining the reduced density of states effective mass, one must recall that for an elliptical hole band, $m_h = (m_{h\parallel}m_{h\perp}^2)^{1/3}$ is the density of states effective hole mass where $m_{h\parallel}$ and $m_{h\perp}$ are the effective hole masses for crystal momenta parallel and perpendicular to the c-axis respectively. For ZnSe the hole bands are spherical and so $m_{h\parallel} = m_{h\perp}$; see Table 1. For a system with multiple hole bands the total absorption coefficient can be evaluated by adding the values calculated individually for each pair of conduction and valence bands. The effects of bandfilling on the above expression for the absorption coefficient are expressed through the occupancy factors, $f_e(\hbar\omega)$ and $f_h(\hbar\omega)$. For the wide band gap II-VI semiconductors the occupancy factors prior to photoexcitation can be assumed to be negligible. From Eq. 21 it follows that the change in the absorption coefficient due to the presence of photoexcited carriers can be written:

$$\Delta\alpha(\omega) = -\frac{2^{5/2} e^2 \mu_r^{3/2} \langle p \rangle^2}{m_0^2 n_0 c \hbar^3 \omega} (\hbar\omega - E_g)^{1/2} [f_e(\hbar\omega) + f_h(\hbar\omega)] \tag{27}$$

The calculation of $\Delta\alpha(\omega)$, and hence also $\Delta n(\omega)$, is thus reduced to a calculation of the occupancy factors for specific scenarios. The simplest case to deal with is that in which the carriers are in quasi-equilibrium at the band edge (i.e. in the limit of a two-level atom). This case has been considered by Auston et al.[1978] who have derived an analytic expression for $\Delta n_{BF}$ in which the magnitude of $\Delta n(\omega)$ is predicted to scale linearly with the density of photoexcited carriers:

$$\Delta n(\omega) = -\frac{2\pi N e^2}{n_0 m_r \omega^2} \frac{\omega^2}{\omega^2 - \omega_g^2} \tag{28}$$

In the derivation of this expression the average momentum matrix element in Eqs. 26 and 27 has been replaced by an approximate value $\hbar\omega_g m_0^2 / 4m_r$ derived from $k \cdot p$ perturbation theory; $m_r$ is the reduced optical effective mass (equal to $\mu_r$ if the bands are spherical) for an electron-hole pair, and $\hbar\omega_g = E_g$ is the band-gap energy. Eq. 21 was derived only for a single pair of conduction and valence bands and so must be evaluated individually for each combination of hole and conduction band. Also, it should be noted that although Eq. 28 can be evaluated for frequencies $\hbar\omega > \hbar\omega_g$ and $\hbar\omega < \hbar\omega_g$, it cannot be used to predict the magnitude $\Delta n_{BF}$ as the ratio $\omega / \omega_g$ approaches unity since damping effects have been neglected. From the values for physical constants listed in Table 1, and for $\hbar\omega = 2$ eV, the values of $\Delta n$ per carrier (or $\gamma_{BF}$) for CdS, $CdS_{0.75}Se_{0.25}$, and ZnSe have been calculated and are presented in Table 2. As expected, the values of $\gamma_{BF}$ are all negative and increase in magnitude as $\hbar\omega$ approaches $\hbar\omega_g$.

For more realistic carrier densities and/or for carrier distributions characterized by a non-zero temperature, the Fermi-Dirac distribution functions must be used in Eqs. 26 and 27. Since only near-band-edge states contribute significantly to the band-filling contribution to $\Delta n$ the split off bands, which are located between 0.45 eV and 0.55 eV below the valence band edges for the materials considered here can be ignored so that only the A and B hole bands (or the light and heavy hole bands in the case of ZnSe) need be considered. The only remaining unknown in Eqs. 26 and 27 are the square of the average momentum matrix element $\langle p \rangle$. As with the model of Auston et al., an analysis based on $k \cdot p$ perturbation theory is usually used whereby $\langle p \rangle$ is replaced by the Kane momentum parameter, $P = h\langle p \rangle / m_0$, which can in turn be related to the nearly material-independent Kane energy parameter, $E_p$. While Auston et al. used an analytical expression for $E_p$, often a value is derived from fits to band structure measurements making the approach somewhat more empirical in nature. However, when compared with experimental data for CdS and CdSe (no detailed data was found for ZnSe), the predicted interband absorption coefficient was found to exceed the measured values by a factor of approximately 2. Because of this discrepancy the calculations of $\Delta n_{BF}$ performed for this work have been made using the experimentally measured absorption coefficient. The predicted values of $\gamma_{BF}$ are listed in Table 2 for a carrier density of $10^{17}$ $cm^{-3}$ and a plasma temperature equal to 300 K. Calculations performed for a carrier density of $10^{17}$ $cm^{-3}$ indicate that $\gamma_{BF}$ is constant (i.e. the refractive index change scales linearly with carrier density).

As indicated above, similar calculations of $\gamma_{BF}$ have been performed and reported previously, but only for low density plasmas where occupancy factors can be modeled using Maxwell-Boltzmann distribution functions. Since highly degenerate carrier distributions are produced in the experiments considered here, and since full Fermi-Dirac distribution functions have been incorporated into the model, calculations of $\gamma_{BF}$ were performed as a function of carrier density for densities between $10^{17}$ cm$^{-3}$ and $10^{21}$ cm$^{-3}$. The results are plotted in Fig.6 (again, for 300 K) for all three materials. The value of $\gamma_{BF}$ is constant for low plasma densities (i.e. the magnitude of $\Delta n_{BF}$ scales linearly with the plasma density), but falls rapidly for carrier densities in excess of $5 \times 10^{19}$ cm$^{-3}$. It follows that in the low to moderate density regime the magnitude of the band filling contribution to $\Delta n$ is a good probe of plasma density. From equation 27 it is also seen that the contribution to the PIRIC depends on the details of the carrier distribution and, for thermal distributions, the carrier temperature. In an idealized picture, prior to thermalization the electrons and holes occupy the states which are optically coupled by the exciting radiation (a transition of 4 eV for the case of interest here), and are therefore far removed from the band edges. These states occupy an energy range equal to the that of the exciting radiation, although the occupation number depends critically on the level of excitation. For calculational ease it can be assumed that only states over a much narrower energy range are occupied, but that within this range all of the states are fully occupied. The width is chosen such that it is sufficiently large to account for all of the photoexcited carriers (this band-width is substantially less than that of the laser pulses for the lower carrier densities investigated in the experiments; 1.2 meV width in the conduction band of CdS for $N=10^{18}$ cm$^{-3}$ versus a pulse band-width of 18 meV). The refractive index changes per carrier can be calculated analytically using Eqs. 21 and 27, and the results are listed for CdS and $CdS_{0.75}Se_{0.25}$ in Table 2 for a carrier density of $10^{17}$ cm$^{-3}$ (the calculation was performed only for these materials since high resolution data for short time delays was obtained only for these materials). One sees that $\gamma_{BF}$ is much smaller for nonthermal distributions than for thermalized distributions, a result consistent with the fact that the number of occupied states near the band edge is significantly reduced for the nonthermal case.

The next stage in the relaxation process for which the carrier distribution functions can be expressed analytically is that following thermalization but prior to significant carrier cooling. Here the electron and hole occupation factors follow the form of those in Eq. 16. The calculated initial temperatures for a plasma density of $10^{18}$ cm$^{-3}$ are listed in the foot-notes to Table 2, while the results of calculations of $\gamma_{BF}$ performed for these temperatures are listed in the same table. The band filling contribution to $\Delta n$ has also been calculated for combinations of electron and hole temperatures between the initial values and 300 K. It is found that the results are primarily sensitive to the electron temperature. For example, in CdS with a fixed electron temperature $T_e$=10,400 K (corresponding to the initial energy expected with two photon excitation at $\lambda$ = 620 nm), $\Delta n$ increases by only 6% when a hole temperature $T_h$=300 K is used rather than the initial hole temperature $T_h$=2,220 K, whereas $\Delta n$ increases by a factor of almost 4 when the electron and hole temperatures decrease from their initial values to room temperature. It follows that to a good level of approximation the temperature dependence of the refractive index can be treated by fixing the hole temperature at 300 K and only considering the changes in the electron temperature. The results are plotted in Fig. 7 for CdS and $CdS_{0.75}Se_{0.25}$. The results for ZnSe are not considered since it was not possible experimentally to obtain carrier cooling data with low noise levels in ZnSe.

Although band filling effects dominate the refractive index change, contributions related to changes in the intraband absorption play a significant role particularly at high plasma temperatures where the interband contributions decrease in magnitude. Here the dominant contribution to $\Delta\alpha$ is due to FCA which, like band filling, scales linearly (to a

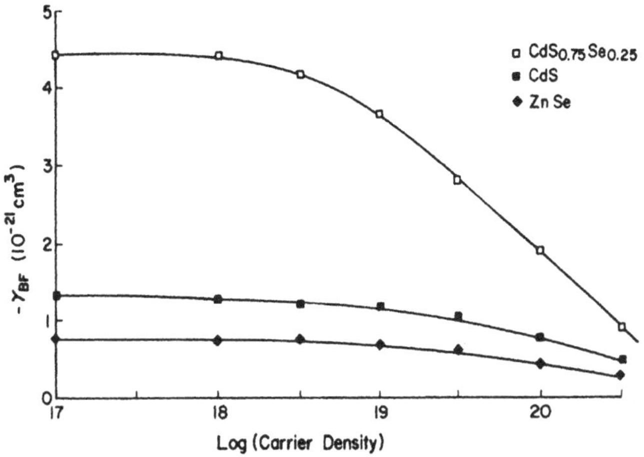


**Figure 6.** Calculated value of $\gamma_{BF}$ due to band filling by a cold-plasma in various II-VI semiconductors.

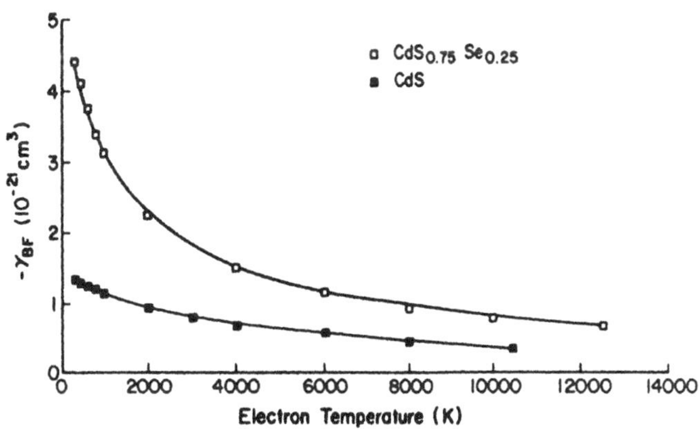


**Figure 7.** Calculated value of $\gamma_{BF}$ versus electron temperature when the hole temperature is 300 K.

first approximation) with carrier density as indicated earlier. The calculation of the magnitude of $\gamma_{FCA}$ can be treated in a similar manner to the calculation of $\gamma_{BF}$ (i.e. one derives an expression for FCA from first principles and performs the integration in Eq. 21), however detailed expressions for FCA with the lattice temperature and carrier temperature dependences included are in general difficult to develop. Instead, the problem can be treated approximately within the context of a simple Drude model whereby the contribution of a specific carrier species to $\Delta n$ can be expressed as follows:

$$\Delta n = -\frac{2\pi N e^2}{\omega^2 m_c n_0^2}. \tag{29}$$

where $m_c$ is the particular carrier effective mass. It follows that $\gamma_{FCA}$ can be written:

$$\gamma_{FCA} = -\frac{2\pi e^2}{\omega^2 m n_0^2}. \tag{30}$$

This effective Drude contribution for each carrier species has been calculated using the above expression (the electron and two hole bands are considered separately). The total Drude contribution to the refractive index is listed in Table 2.

The effect of bandgap renormalization has been investigated using the same formalism as was outlined for the band filling problem. BGR occurs when carrier densities become large enough that exchange and correlation self-energies become significant in relation to the magnitude of the energy band gap. The electron and hole energies are reduced (at a qualitative level one can understand this in terms of the energy that the carriers gain by avoiding each other spatially) which in turn leads to a reduction in the band gap energy and hence to an increase in the refractive index for $\omega < \omega_g$. Simple expressions for the BGR in terms of carrier density are not possible for this many-body effect. Most model calculations are developed for zero temperature and predict a rigid shift in the bandedge which scales as $N^{1/3}$. A more sophisticated analysis performed by Zimmermann [1988] predicts a bandedge shift given by the following expression:

$$\Delta E_g(N, T_c) = -\frac{4.64(Na_B^3)^{1/2} R_H}{\left[Na_B^3 + (0.107 k_B T_c / R_H)^2\right]^{1/4}}, \tag{31}$$

where $a_B$ is the Bohr excitonic radius, $R_H$ is the exciton Rydberg energy, and $T_c$ is the carrier temperature. This expression predicts band gap reductions of 45 meV for $10^{18}$ cm$^{-3}$, 90 meV for $10^{19}$ cm$^{-3}$, and 160 meV for $10^{20}$ cm$^{-3}$ in CdS and $CdS_{0.75}Se_{0.25}$, and approximately 5% lower values for ZnSe, all at 300 K. These predicted shifts are greater than the blue shifts of the band edge due to band filling. The predictions of this model are in sharp contrast to experimental results which indicate that in materials like CdS the effects of BGR dominate over band filling only for low plasma temperatures [Said, 1992]. In particular, for N= $2 \times 10^{18}$ cm$^{-3}$ it was found that the absorption edge was blue shifted by 20 meV which is roughly what one would expect for this carrier density if only band filling were operative. Nonetheless, calculations in which BGR, band filling, and Drude contributions are considered have been performed. The results show that positive values of $\gamma_{BGR}$ are predicted for all but the very highest plasma densities. These results are not consistent with the experimental results of others [Said, 1992] and as a result are not considered to be of consequence for the experiments reported here.

Another many body effect associated with the presence of free carriers is plasma screening of the Coulomb interaction between electrons and holes (i.e. excitonic phenomena). Excitons in general have large oscillator strengths and hence contribute significantly to interband absorption. Even where bound excitons do not exist (e.g. at high temperatures) or where photon energies are not in resonance with excitonic transitions electron-hole correlations remain which contribute to the oscillator strength (i.e. through Coulomb enhancement of the continuum states). One way in which the strength of exciton related absorption can be modified by the presence of a plasma (and hence contribute to $\Delta n$) is through screening which reduces the strength of the Coulomb interaction and thereby reduces the strength of the interband absorption. The majority of the results presented here, however, are for room temperature where the exciton binding energy is comparable to thermal energies (i.e. excitons cannot exist even without the presence of a plasma), and where excitonic enhancement can be ignored. Rinker et al.[1989] point out that excitonic enhancement is negligible for carrier temperatures $T > 100$ K.

The final process which will be considered is lattice heating. The consideration of lattice heating separately from free carrier effects is somewhat artificial since the increase in lattice energy is due to energy transfer from the carrier system to the phonon system as the carriers cool towards the band edge. Of course, the time evolution of carrier and lattice phonon effects will be different. On a time scale of picoseconds, the phonons (of density $N_p$) generated by cooling or recombining carriers, may not have established an equilibrium distribution characterized by an (elevated) lattice temperature ($T_L$). Nonetheless we will use an effective lattice temperature to discuss the lattice heating effects, since a more detailed model is not justified at this time. As the lattice temperature rises the band gap decreases due to many body effects involving the interaction of carriers and the phonons, and due to changes in the lattice constant as the solid expands. To a first approximation the band gap decreases linearly with temperature change, and since linear shifts in the band gap lead to linear shifts in the refractive index, the lattice heating contribution to $\Delta n$ can be treated as a linear function of initial carrier density (on a time scale of more than picoseconds). The magnitude of band gap shifts with temperature change remain difficult to predict at a theoretical level although experimental measurements are legion. The magnitudes of $\Delta n/\Delta N$ through lattice heating effects can be estimated by calculating $\Delta n/\Delta T_L \times E_L/(\rho C)$ where $\rho$ is the material density, C is the heat capacity, and $E_L$ is the energy per carrier transferred to the lattice. The value of C is calculated from the Dulong and Petit Law to be $2.2 \times 10^{18}$ $eVg^{-1}k_B^{-1}$ for CdS, $CdS_{0.75}Se_{0.25}$, and ZnSe. The values of $dn/dT_L$ (for $\hbar\omega = 2$ eV) are listed in Table 1 for the II-VI materials of interest here while $\rho$ is about 5 $gcm^{-3}$ for the semiconductors. Two values of $\Delta n/\Delta N$ are calculated; the first assumes that the carries cool to the bandedge and hence transfer ($2\hbar\omega - E_g$) worth of energy per carrier to the lattice. The second value assumes that the carriers recombine nonradiatively so that each carrier contributes $2\hbar\omega$ to the lattice. In both cases the lattice heating contribution to $\Delta N$ is at least two orders of magnitude smaller than the bandfilling contribution and hence can be ignored.

**Measurement Techniques**

Measurements of the changes in the refractive index have been made using a beam deflection technique and an induced grating diffraction technique. In this section these will be discussed and expressions will be developed in order to interpret the experimental results that follow. Part of the difficulty in the interpretation of the results is due to the attenuation of the pump light by 2PA which leads to an absorption depth of the light which depends on the irradiance. From Eq. 10 (in the limit of negligible free carrier absorption) this depth is given by $(\beta I_0)^{-1}$ where $I_0$ is the incident irradiance.

The discussion in previous sections has illustrated that many different processes may contribute to changes in the refractive index. Many of these processes can be distinguished on the basis of their temporal characteristics (i.e. instantaneous contributions versus long lived contributions). Determination of the scaling behaviour of PIRIC's with respect to incident irradiance is also useful in identifying underlying mechanisms. In this section the expected scaling behaviour of beam deflection angles with respect to incident irradiance for different refractive nonlinearities will be determined.

One of the techniques we have employed to measure time-dependent refractive index changes is the beam deflection technique indicated in Fig. 8 in which identical but orthogonally polarized pump and probe beams are used. A pump beam induces a temporary or long-lived refractive index change at various locations in the sample. A probe beam which follows with a variable delay time can be deflected due to its propagation through a region with a refractive index gradient.

The optical field in different parts of the probe beam will accumulate different amounts of phase in proportion to the local refractive index as the probe pulse traverses the sample. The normal to the surfaces of constant phase for light entering the sample will differ from the normal to the surfaces of constant phase for light leaving the back surface of the sample and hence the beam will propagate in a different direction. In order to calculate the phase front for the transmitted light a full beam propagation model must be used. Since it is not the aim in this work to extract quantitative results for the magnitudes of the nonlinearities, a less rigorous treatment can be used. A rough prediction of the deflection angle can be made by treating the propagating pulse as a pair of parallel rays which traverse the sample at two separate radial positions relative to the centre of the excited spot. The spatial separation of these rays on the phase front is considered to be approximately the same as the size of the probe spot which is small in comparison to the size of the excited spot. It will be assumed that the refractive index experienced by the ray which propagates closer to the centre of the excited spot, ray "a", can be represented as $n_a(z)$, while that experienced by the ray which propagates farther from the centre of the pump spot, ray "b", can be represented by $n_b(z)$. The difference in phase accumulated by the two rays is equal to

$$\Delta\phi = \frac{2\pi}{\lambda}\int_0^L dz\left[n_b(z) - n_a(z)\right] \tag{32}$$

where $\lambda$ is the vacuum wavelength and L is the sample thickness. If the two rays are assumed to enter the sample normal to the surface at a distance r from the centre of the pump beam, then the beam which leaves the sample will propagate at an angle $\varphi$ relative to the surface normal where $\tan(\varphi) = \Delta\phi / s$, and s is the spatial separation between the two rays. As s approaches zero the deflection angle can then be expressed as :

$$\varphi = \tan^{-1}\left[\int_0^L dz\, \frac{\partial\, \Delta n(z,r)}{\partial\, r}\right]. \tag{33}$$

For small refractive index changes the argument of the $\tan^{-1}$ function will be small and $\varphi$ can be approximated as the argument of this function.

The calculation of beam deflection angles is performed by replacing $\Delta n(z,r)$ in Eq. 33 with expressions corresponding to the time-dependent PIRIC processes discussed previously. For example, the beam deflection angle for instantaneous nonlinearities can be predicted by combining Eqs. 8 and 13, differentiating with respect to r, and performing the integral in Eq. 33. The result is however only accurate for a specific time in the probe pulse. To calculate the beam deflection angle for all time delays this result must be weighted by the instantaneous irradiance, integrated over the probe pulse, and normalized

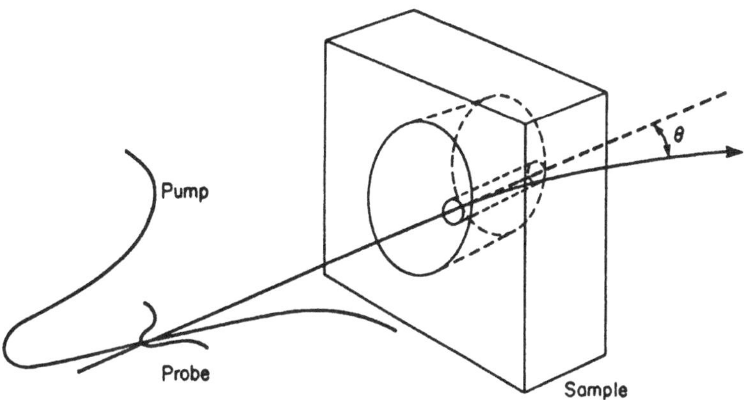


**Figure 8.** Schematic diagram of beam deflection geometry.

by the total energy in the probe pulse. In general one expects that the average deflection experienced by the probe pulse will be a convolution of functions involving probe and pump pulses. For zero time delay one has that the deflection angle is:

$$\varphi(r) = -\frac{4rn_2}{r_0^2\beta}\frac{\int_0^{\infty}dt\left[\frac{\beta I_0 TLe^{-2r^2/r_0^2}\operatorname{sech}^4(1.76t/\tau_p)}{1+\beta TI_0 Le^{-2r^2/r_0^2}\operatorname{sech}^2(1.76t/\tau_p)}\right]^2}{\int_0^{\infty}dt\,\operatorname{sech}^2(1.76t/\tau_p)}. \quad (34)$$

After performing a numerical integration it is found that the deflection angle scales linearly with $I_0$ for low values of $I_0$ ($I_0$ less than approximately 10 $GWcm^{-2}$) and approaches a constant value for high values of $I_0$

The calculation of the beam deflection angle for carrier induced refractive nonlinearities proceeds in a similar fashion. Eq. 14, which expresses the carrier density in terms of the peak irradiance, is combined with Eq. 31 to give:

$$\varphi(r,t) = -\gamma \frac{4\beta L r}{2\hbar\omega r_0^2}\exp(-4r^2/r_0^2)T^2 I_0^2$$

$$\times\int_0^\infty dt\, \mathrm{sech}^4(1.76t/\tau_p)\left[\frac{2+\beta I_0 T e^{-2r^2/r_0^2}\mathrm{sech}^2(1.76t/\tau_p)}{1+\beta T I_0 e^{-2r^2/r_0^2}\mathrm{sech}^2(1.76t/\tau_p)}\right] \quad (35)$$

where $\gamma$ is the total refractive index change per carrier density. The beam deflection angle scales quadratically with $I_0$ for low values of $I_0$ (i.e., < 10 GWcm$^{-2}$), and linearly with $I_0$ for higher values.

Another technique we have used to study PIRIC's is based on time-resolved light scattering from induced refractive index gratings. Induced grating measurements are performed by diffracting a time-delayed (weak) probe pulse from a refractive index grating written by a pair of pump pulses. A schematic diagram of a typical arrangement is shown in Fig. 9.

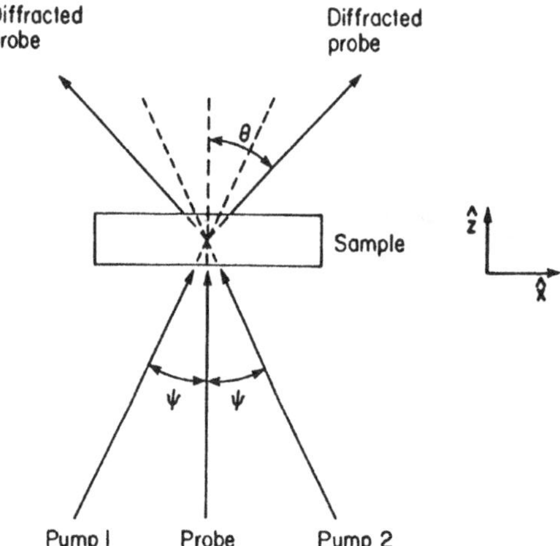


**Figure 9.** Schematic diagram of the beam geometry used for measurements of PIRIC's by the laser-induced grating technique.

For thin gratings (Raman-Nath regime) the diffraction efficiency scales with the square of the magnitude of the refractive index modulation. For the thick gratings produced and investigated here a more detailed analysis is required in which coupled wave equations must be solved. We now consider a formalism with which these calculations can be performed and this will be used to predict the scaling behaviour of the diffraction efficiency.

The formalism follows that described by Solymar and Cooke [1981]. Calculations are performed for transmission geometries, and only first order diffraction is considered. The analysis begins by considering the Helmholtz wave equation for the scattered and unscattered light (the term "scattered" will be used inter-changeably with the term "diffracted"):

$$\nabla^2 E + \tilde{k}^2 E = 0. \qquad (36)$$

In the plane wave approximation E is the superposition of the incident, $P\exp(i\vec{k}^p.\vec{r})$, and scattered, $S\exp(i\vec{k}^s.\vec{r})$, fields associated with the probe radiation ($\vec{k}^p$ and $\vec{k}^s$ are the wave vectors associated with the scattered and unscattered light). The coupling between these two fields is described in terms of the position dependent effective wavevector, $\tilde{k}$, where

$$\tilde{k}^2 = k^2[n_0 + \Delta n_r(\vec{r}) + \Delta n_i(\vec{r})] \qquad (37)$$

with the $\Delta n_r(\vec{r})$and $\Delta n_i(\vec{r})$ terms denoting the pump-induced changes in the real and imaginary parts of the refractive index respectively, and k being the vacuum wavevector. For the purposes of this calculation it is assumed that the surface normal is along $\hat{z}$ and that the induced-grating vector is along $\hat{x}$ so that only the x and z dependencies of S, P, $\Delta n_r(\vec{r})$, and $\Delta n_i(\vec{r})$ need be considered.

For the instantaneous contribution to the diffraction efficiency one can use the changes in the real and imaginary parts of the refractive index as expressed by Eqs. 5 and 6 with an intensity I(x,z) that varies with x and y. The calculation of the diffraction efficiency is performed for a uniform irradiance level so that details pertaining to the temporal shape of the pump and probe pulses do not enter the analysis. This assumption affects the positions of specific features relative to absolute pump irradiances, but does not affect the qualitative features of the scaling behavior. In order to arrive at an expression for I(x,z) it is assumed that each of the pump beams is incident on the surface at an angle $\psi$ relative to the surface normal such that both pump beams lie in the xz plane as illustrated in Fig. 9. If each beam has an intensity $I_0$, the intensity distribution at the sample surface, I(x, z=0), is equal to $I_0$ times the square of the modulus of the sum of the phase factors $\exp(i[k_x x + k_z z])/2$ and $\exp(i[-k_x x + k_z z])$, where $k_x = k\sin\psi$ and $k_z = k\cos\psi$, and can be expressed as,

$$I(x, z=0) = 2I_0[1+\cos(k^g x)], \qquad (38)$$

where $k^g = 2\pi/\Lambda = (4\pi/\lambda)\sin\psi$ is the grating wavevector with $\Lambda$ being the grating spacing. It is assumed that the attenuation of the pump radiation is due to 2PA so that the irradiance at a general point (x, z) can be written as

$$I(x,z) = \frac{2I_0T[(1+\cos(k^g x)]}{1+2\beta I_0 Tz[1+\cos(k^g x)]}. \qquad (39)$$

To generate the wave equation which governs the instantaneous diffraction efficiency this expression is combined with Eqs. 5 and 6, and the expressions for the real and imaginary parts of the refractive index change are inserted in Eq. 37. The resultant equation is simplified by expanding the term $\cos(k_g x)$ in the numerator as $[\exp(ik^g x) + \exp(-ik^g x)]$, and by approximating the denominator as $1+4\beta I_0 Tz$. Moreover, terms are kept only to first order in $n_2$ and $\beta$, and it is assumed that S and P only vary slowly with z so that terms in $d^2P/d^2z$ and $d^2S/d^2z$ can be ignored. These assumptions lead to a set of 3 coupled first order differential equations in $\exp(ik_x{}^S x)$, $\exp(-ik_x{}^S x)$, and $\exp(ik_x{}^P x)$where

$$k_x^s = k_x^p \pm k^g. \tag{40}$$

This relation allows the directions of propagation for the scattered light to be determined. Since the two wavevectors on the right hand side are known it follows that the scattering angle, $\theta$, is given as:

$$\tan\theta = \frac{k_x^g}{\sqrt{k^2 - (k_x^g)^2}} \;. \tag{41}$$

Finally, since the two scattered waves behave identically the scattering problem can be reduced to a set of two coupled differential equations:

$$\begin{aligned} \frac{dP}{dz} &= a_{11}P + 2a_{12}S\exp(-i\Delta kz) \\ \frac{dS}{dz} &= a_{21}S + a_{22}P\exp(i\Delta kz) \end{aligned} \tag{42}$$

where $a_{11}$, $a_{12}$, $a_{21}$, and $a_{22}$ are complex constants which are linear in the coefficients $n_2$ and $\beta$. The factor of 2 which multiplies the second term of the first equation is explicitly included to illustrate that both scattered beams contribute to depletion of the unscattered light. The wave vector $\Delta k$ is defined as :

$$\Delta k = k_x^s - k_x^p. \tag{43}$$

The phase terms $\exp(\pm i\Delta kz)$ enter Eq. 42 since the scattered and unscattered waves do not propagate collinearly, and are significant when the sample thickness, L, does not satisfy the relation $\Delta kL << \pi$. In our experiments the thin grating condition does not apply and the phase terms cannot be ignored.

Eqs. 42 were solved numerically for CdS, $CdS_{0.75}Se_{0.25}$, and ZnSe using the values for the nonlinear optical coefficients listed in Table 2. The results indicate that the diffraction efficiency scales with $I_0^2$ for low irradiances (< 100 $GWcm^{-2}$) and displays a sublinear scaling behaviour for higher irradiances. For the beam deflection scaling behavior discussed above it was found that the change in scaling behavior with increasing irradiance was due to 2PA attenuation of the pump radiation. For diffraction the change in scaling behaviour is due to attenuation of the pump light *plus* 2PA attenuation of the probe light (i.e. for beam deflection measurements it is the deflection angle and not the signal beam fluence which is important).

A similar analysis can be applied to the calculation of the diffraction efficiency for non-zero pump-probe time delays. The real part of the refractive index change is assumed to be dominated by a free carrier nonlinearity such that $\Delta n_r = -(\gamma_{BF} + \gamma_{FCA})N$, while the imaginary part of the refractive index change is assumed to be due to FCA so that $\Delta n_i = -\sigma_{FCA}N(2k)^{-1}$. For the materials and conditions discussed here, however, $\Delta n_i/(2k)$ is 1.5 to 3 orders of magnitude smaller than $\Delta n_r$ and hence can be ignored. The diffraction efficiency was calculated for CdS, $CdS_{0.75}Se_{0.25}$, and ZnSe using the values of the coefficients listed in Table 2 (room temperature, low density limit). The calculations indicate that the diffraction efficiency scales as $I_0^4$ for low irradiance (peak irradiance < 100 $GWcm^{-2}$), and saturates for higher irradiances to a value where 50% of the light initially in the probe beam has been coupled into the scattered beams (i.e. a steady-state is reached between the rate at which light is coupled into the scattered fields and the rate at which light in the scattered fields is coupled back into the incident field). The calculations also indicate that for pump irradiances in the range of 50 $GWcm^{-2}$ to > 500 $GWcm^{-2}$ the carrier induced diffraction efficiency exceeds the diffraction efficiency evaluated at $t = 0$ by at least 2 orders of magnitude.

The discussion thus far has concentrated on the derivation of the equations of motion which describe the dynamics of the scattering of light from the incident probe beam into diffracted beams via a pump induced grating. Laser-induced gratings can be generated and probed in geometries other than the one illustrated in Fig. 9. One geometry which we have found particularly useful is shown in Fig. 10.

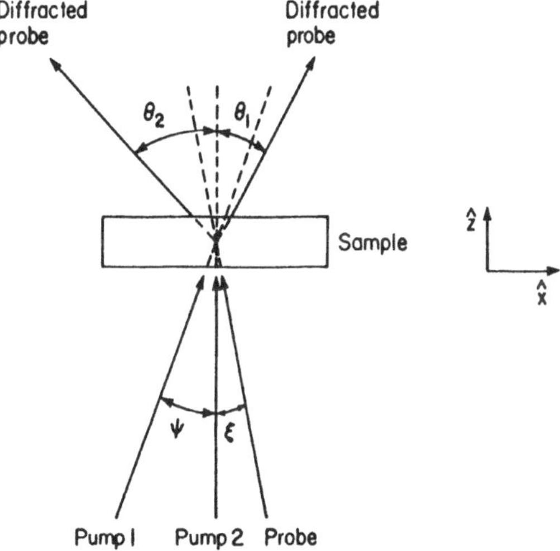


**Figure 10.** Schematic diagram of an alternative beam geometry for the laser-induced grating technique.

A similar analysis to that considered above can be performed for this geometry in which the pump beams are not symetrically located about the probe beam (this geometry is useful for performing experiments in which the grating spacing is varied by altering the angle of incidence of one beam). One pump beam is incident with an angle $\psi$ relative to the surface normal, the second pump beam is normally incident, and the probe beam is incident with an angle $\xi$. The grating wave vector, $k^g$, is equal to $k\sin\psi$ so that the wave vector for the scattered light, $k^s$, is equal to $k\sin\xi \pm k\sin\psi$. It is generally best to work with the beam diffracted furthest from the pump and unscattered beams. It follows that the angle for this diffracted beam, $\theta$, is defined by

$$\tan\theta = \frac{\sin\psi + \sin\xi}{\sqrt{1-(\sin\psi + \sin\xi)^2}}. \quad (44)$$

## EXPERIMENTAL APPARATUS AND TECHNIQUES

In this section the details of the experimental arrangements which were used to study time-resolved refractive index changes and carrier relaxation processes in CdS, $CdS_{0.75}Se_{0.25}$, and ZnSe are described. The optical radiation which was used for all of the experiments was provided by a colliding pulse modelocked (CPM) dye laser. This radiation was amplified by a dye amplifier system pumped by a frequency doubled Nd:YAG laser.

The CPM is a ring dye oscillator [Valdmanis, 1985] which is pumped by a CW argon ion dye laser. The output of the laser consists of a train of modelocked pulses (repetition rate of 100 MHz) with pulse widths which can be varied between 50 fs (FWHM for an assumed $\text{sech}^2$ pulse shape) to more than 200 fs. Average powers of 5 to 10 mW are achieved which correspond to pulse energies of between 50 and 100 pJ. The peak in the spectrum of the laser radiation is between 618 and 620 nm. The pulses are linearly polarized.

The amplifier [Rolland, 1986] consists of 3 Bethune type prism dye cells pumped laterally with frequency doubled pulses from a Q-switched Nd:YAG laser (Quantel Y580) which operates with a repetition rate of 20 Hz. A grating pair (1200 grooves per mm) is used to temporally recompress the amplified pulses. The pulses are linearly polarized in the vertical direction. The energies of the amplified pulses are of the order of 50 μJ and the temporal pulse width is typically 120 fs. The shot-to-shot amplitude fluctuations are typically 10%. Near-Gaussian spot profiles are observed with greater than 90% of the pulse energy in the $e^{-2}$ radius of the beam.

Measurements of the deflection experienced by a time delayed probe pulse due to the influence of a strong pump pulse allow refractive index changes induced by the pump pulse to be investigated. A schematic diagram of the apparatus is illustrated in Fig. 11. The amplified laser pulse is divided by a beam splitter into pump and probe pulses with 6% of the initial energy being directed into the probe arm. The pump pulse passes through a calibrated variable neutral density filter wheel before being directed onto the sample by a steering mirror. A lens (f = 6 cm) is used to focus the pump pulse to a spot size of between 20 and 60 μm ($e^{-2}$ radius) measured at the sample surface. The probe pulse is attenuated by calibrated neutral density filters, passes through a variable delay arm, and is focussed onto the sample (f = 1 cm) to a spot size of <10 μm. The filters in the probe arm were chosen such that the peak irradiance of the of probe light was always less than the peak irradiance associated with the pump light by at least a factor of 3. The variable delay arm consists of an aluminum mirror retroreflector mounted on a translation stage. The translation stage is driven by a stepper motor with a 3.3 μm step resolution (equivalent to a temporal resolution of 22 fs). The pump and probe pulses are both incident on the sample at angles of $10^o$ relative to the surface normal. A half wave plate located in each arm of the setup allowed the polarizations of the pump and probe pulses relative to the sample surface to be adjusted independently. The transmitted probe pulse is recollimated with a f=10 cm lens and then directed onto a two-dimensional diode array which is controlled by Spiricon beam profiler electronics. The beam profiling system allows the beam position to be monitored which in turn allows the magnitude of the deflection to be measured.

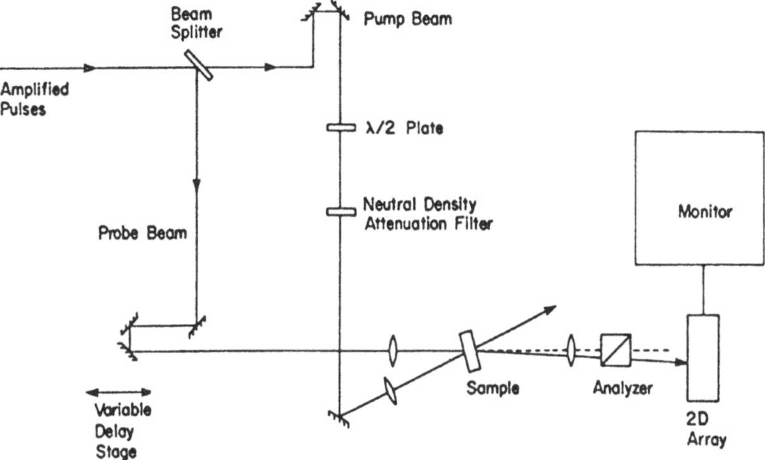


**Figure 11.** Schematic diagram of apparatus used for deflection experiments.

To determine the effect of lattice temperature on PIRIC's, the samples were mounted in a flow-through He cryostat and experiments were performed for temperatures between 77 and 300 K. The temperatures were measured using a silicon diode.

The experimental setup used to measure the scattering efficiencies for the probe pulse diffracted by a grating written by a pair of pump pulses is similar to the setup which was used to perform the beam deflection measurements. A schematic diagram of the diffraction arrangement in illustrated in Fig. 12. The setup is identical with the exception that a secondary beam splitter is located in the pump arm in order to split the pump pulse into two equivalent pulses. The pump pulse passes through a λ/2 plate oriented to rotate the vertically polarized pulse by 45°. The pulse is then directed through a calcite cube polarizing beam splitter which splits the pulse into two equally energetic pulses. The reflected half is reflected by an aluminized retroreflector mounted on a translation stage. The transmitted half is passed through another λ/2 plate which re-orients the polarization so that it lies in the vertical direction. Both pump pulses are then directed through matched (f = 6 cm) lenses onto the sample. For these experiments a λ/2 plate was used to re-orient the polarization of the probe light such that it was orthogonally polarized relative to the pump light. The same range of spot sizes as was used for the deflection experiments was used for the diffraction measurements. Again the peak probe pulse irradiance was always less than the peak pump pulse irradiance by at least a factor of 3.

For the simplest diffraction experiments the radiation probe radiation is incident on the sample along the surface normal, and each pump beam is incident on the sample at an angle of 18° relative to the surface normal. The diffracted light (first order) is collected by a lens (f = 10 cm), passed through an analyzer (to aid in signal discrimination between the pump and probe), and focused onto either a fast photodiode or a photomultiplier tube.

Measurements were also made in which a single pump pulse was used to photoexcite the sample either before or following photoexcitation by the pair of pump pulses used to write the diffraction grating. For these experiments an additional splitter and delay arm was placed in the pump arm. This splitter consisted of another $\lambda/2$ plate followed by a calcite cube polarizing beam splitter. The transmitted light was what was eventually used to write the diffraction grating while the reflected light was passed through a delay arm. The delayed light was then directed into the cube polarizing beam splitter to derive the two pulses used to write the grating at an angle such that the reflected light was coincident but orthogonally polarized with one of the pump pulses used to write the grating structure.

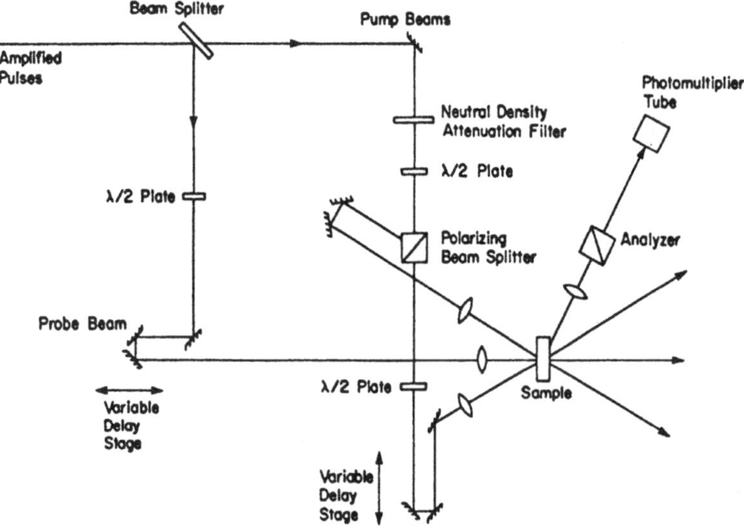


**Figure 12.** Schematic diagram of apparatus used for diffraction experiments.

Since it was suspected that radiative processes might play a role in carrier relaxation through carrier recombination, luminescence yield and luminescence line shape measurements were made for different levels of photoexciation. For these measurements the pump beam was focussed down onto the sample surface to a spot size of between 20 and 60 μm and a collection lens (f = 10 cm) was used to collimate the luminescence emitted from the surface. The luminescence was directed into an optical multichannel analyzer (OMA) system. The OMA system had a wavelength resolution of 0.3 nm.

The CdS and $CdS_{0.75}Se_{0.25}$ samples were obtained from Cleveland Crystals Inc. The material was provided in the form of a 1 mm thick c-plates. These were mechanically polished on both sides to thicknesses of 100 μm by Interoptic Inc. in Ottawa. The samples were not chemically etched following mechanical polishing. The ZnSe was obtained from Dr. R. Bhargava at Philips Research Corp. in New York. The ZnSe was oriented as an a-plate. The sample had been mounted on a sapphire substrate and then mechanically polished to a thickness of 20 mm. No chemical etching was performed on this sample.

## RESULTS AND DISCUSSION

Here we present some of the essential results on time-resolved refractive index changes measured with the beam deflection and induced diffraction grating techniques. The beam deflection measurements were performed in 100 μm thick slabs of CdS and $CdS_{0.75}Se_{0.25}$. The laser induced grating experiments were carried out using these same samples and also in 20 μm thick ZnSe. Experiments were carried out for peak pump irradiances up to 700 $GWcm^{-2}$ and for lattice temperatures between 77 and 300 K.

### Time Resolved Beam Deflection and Induced Diffraction Results

The beam deflection results are treated first. A typical result is displayed in Fig. 13 where the probe pulse deflection angle in $CdS_{0.75}Se_{0.25}$ is plotted versus the pump-probe time delay, t, for a series of different pump pulse (peak surface) irradiance, $I_0$. The radial position of the probe pulse, $r_p$, and the $e^{-2}$ radius of the pump pulse, $r_0$, are both equal to 40 μm (the quoted irradiance is that evaluated at $r_p$ and for these beam deflection experiments will be referred to as $I_{BD}$). The positive deflection angles measured correspond to a (depth averaged) $\Delta n < 0$ in all cases consistent with what is expected theoretically. For the lowest excitation conditions the observed deflection follows the temporal profile of the pump pulse. For larger values of $I_{BD}$ the peak at $t = 0$ is still observed but is followed by a subsequent deflection which peaks at a delay time of 1.5 ps and then decreases on a time scale larger than 200 ps. With increasing excitation the non-instantaneous contribution to $\Delta n$ is seen to completely dominate the instantaneous ones. For $I_{BD} > 25$ $GWcm^{-2}$ the beam deflection still peaks at 1.5 ps, decreases sharply on a time scale as short as 3 ps, and then decays more slowly for times > 100 ps. The data in this high excitation regime is distinctive not only by its fast temporal behavior, but also by the scaling of the deflection for long time delays. In particular the smooth curves drawn through data points corresponding to high values of $I_{BD}$ cross through those corresponding to lower values of $I_{BD}$ so that the maximum deflection at long time delays corresponds to an intermediate excitation level. This behavior will herein be referred to as "curve crossing", and the excitation level which produces the largest beam deflection for long time delays, and which marks the onset of rapid signal decay, will be referred to as the threshold irradiance. This irradiance is dependent on the location of the probe pulse relative to the centre of the pump pulse. For $r_0 = r_p = 40$ μm a threshold value for $I_{BD} = 25$ $GWcm^{-2}$ is measured. For the same pump spot size but with $r_p$=15 μm the threshold corresponds to $I_{BD} = 110$ $GWcm^{-2}$, and for $r_p = 0$ the measured threshold is equal to 120 $GWcm^{-2}$. Qualitatively similar behaviour is seen in CdS for all aspects of the beam deflection data with the exception that the deflections are slightly smaller for the same excitation strength and that the delayed peak in the deflection occurs at $t = 2.0$ ps.

The main features of the data include the instantaneous deflection at t=0, the peak in the non-instantaneous deflection at t=1.5 ps, and the partial recovery in the deflection angle for large values of $I_{BD}$ and time delay greater than 10 ps. Measurements of these deflection

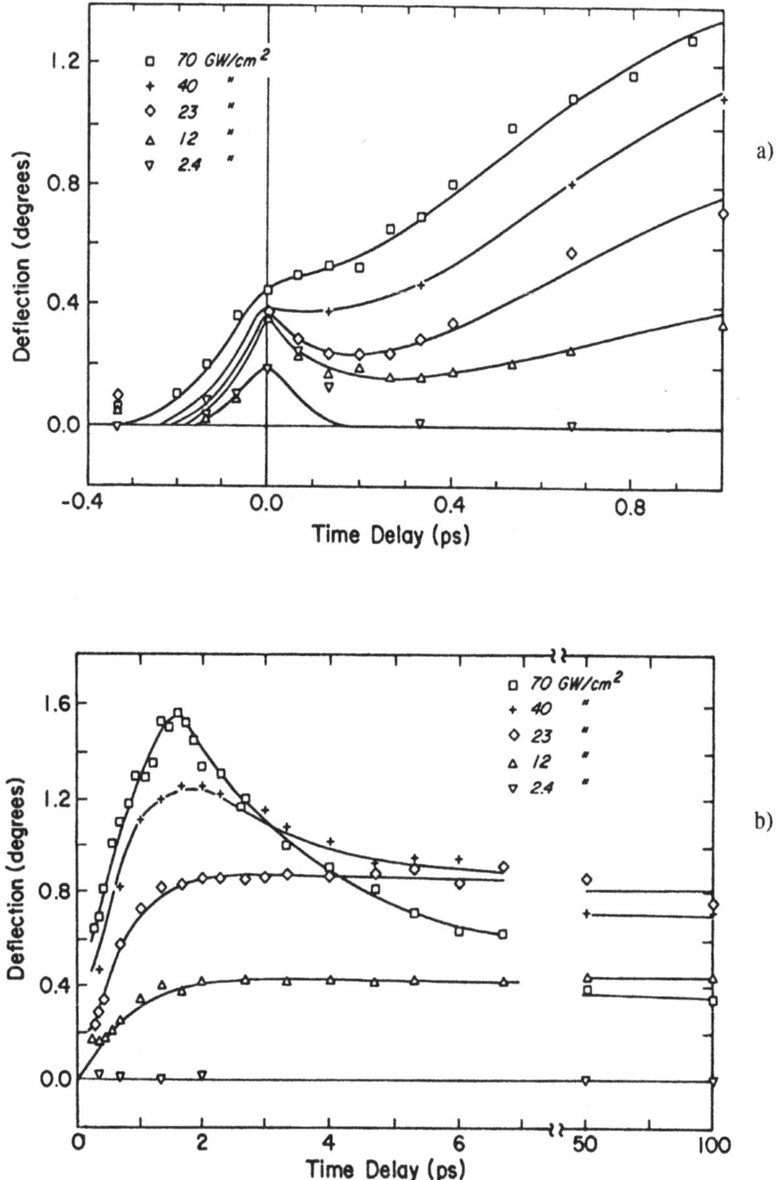


**Figure 13.** Pump beam induced probe deflection in $CdS_{0.75}Se_{0.25}$:a) near zero delay time and b) at longer times.

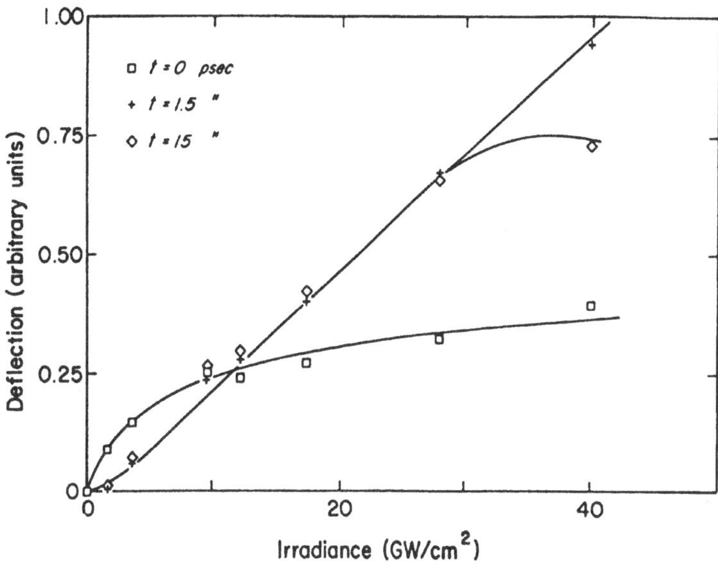


**Figure 14.** Scaling behavior of deflection with irradiance for $CdS_{0.75}Se_{0.25}$ ($r_0$ = 60 mm, $r_p$ = 40 mm).

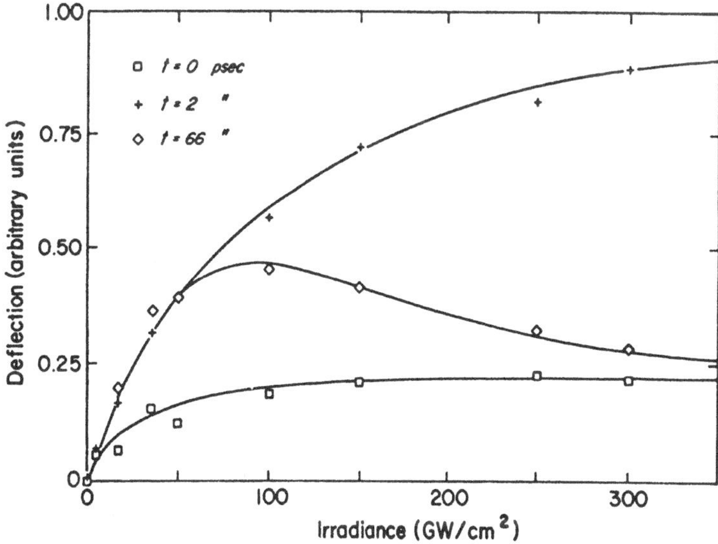


**Figure 15.** Scaling behavior of deflection with irradiance for CdS ($r_0$ = 20 mm, $r_p$ = 20 mm).

signals at t= 0, 1.5 and 10 ps were made as a function of excitation level and are presented in Fig. 14 and Fig. 15 for $CdS_{0.75}Se_{0.25}$ and CdS respectively. It can be seen that in both cases the instantaneous deflection grows linearly with $I_{BD}$ for small values of $I_{BD}$ and saturates for larger values. The magnitude of the deflection at t=1.5 ps grows superlinearly with peak irradiance for small values of irradiance and afterwards grows linearly with irradiance. The deflection magnitude for large time delays is observed to follow that at t = 1.5 ps for values of $I_{BD}$ up to the threshold value of 25 $GWcm^{-2}$, but decreases for large values (curve crossing).

The diffraction grating data displays features similar to those observed using the beam deflection technique. A typical diffraction grating result is displayed in Fig. 16 for CdS where the diffracted signal magnitude (arbitrary units) measured in a 100 μm thick plate of CdS is plotted as a function of time delay for different values of $I_G$. The values of $I_G$ refer to the peak value of the irradiance at the sample surface and evaluated at the peaks in the diffraction grating at the centre of the pump spot; since the two pump spots which produce the gratings have the same irradiance, $I_G$ is four times as large as the peak irradiance from a single pump pulse. Because the measured signal consists of a magnitude only, the sign of the nonlinearity can unfortunately not be deduced from the data. No peak in the diffraction efficiency is observed at t=0 for any excitation strength. Instead the diffraction efficiency grows with increasing time delay beginning with the arrival of the pump pulse. The growth continues until a time delay of 2 ps is reached. For the lower values of $I_G$ the diffraction efficiency decreases slowly beyond t = 2 ps with a time constant greater than 100 ps. For values of $I_G$ greater than 200 to 250 $GWcm^{-2}$ the diffraction efficiency peaks and then displays a rapid decay on a time scale as short as 3 ps. The diffraction data displays a curve crossing behavior similar to that observed in the beam deflection results. Again the onset of the fast decay behavior occurs at an irradiance (which will again be referred to as the threshold irradiance) for which maximum signal amplitudes are measured at long time delays. The threshold irradiances in these experiments were not the same as those obtained from the deflection experiments for reasons to be explained later.

Similar results are obtained in $CdS_{0.75}Se_{0.25}$ and ZnSe as in CdS. For the thin ZnSe sample the signal decay time for large values of $I_G$ can be as short as 4 ps and have a greater depth of recovery than in the thicker samples.

Scaling measurements similar to those made using the beam deflection technique were also performed using the diffraction technique. The results are displayed in Fig. 17 where the logarithm of the peak height (t = 1.5 to 2.5 ps) is plotted versus the logarithm of $I_G$. For low excitation the peak height scales as $[I_G]^{3.5}$, while for high excitation the peak height saturates. Measurements of the absolute diffraction efficiency indicate that in the saturation regime the diffraction efficiency is equal to15% for each of the two diffracted beams.

We will now consider the data in the three main time regimes relative to the theoretical results presented earlier.

**Instantaneous Behavior.** Instantaneous contributions to Δn are associated with virtual excitation of the electronic system and are manifested by a response which is proportional to the instantaneous excitation strength. Although no such response is observed in the diffraction data, the peak at t = 0 in the beam deflection data (Fig. 13) is consistent with such a nonlinearity. From the scaling measurements presented in Figs.13 and 14 the irradiance dependence of the nonlinearity can be deduced. For small excitation strengths the height of the peak at t = 0 grows linearly with $I_0$, while for large excitation strengths the peak height saturates and becomes constant. In the discussion presented earlier it was

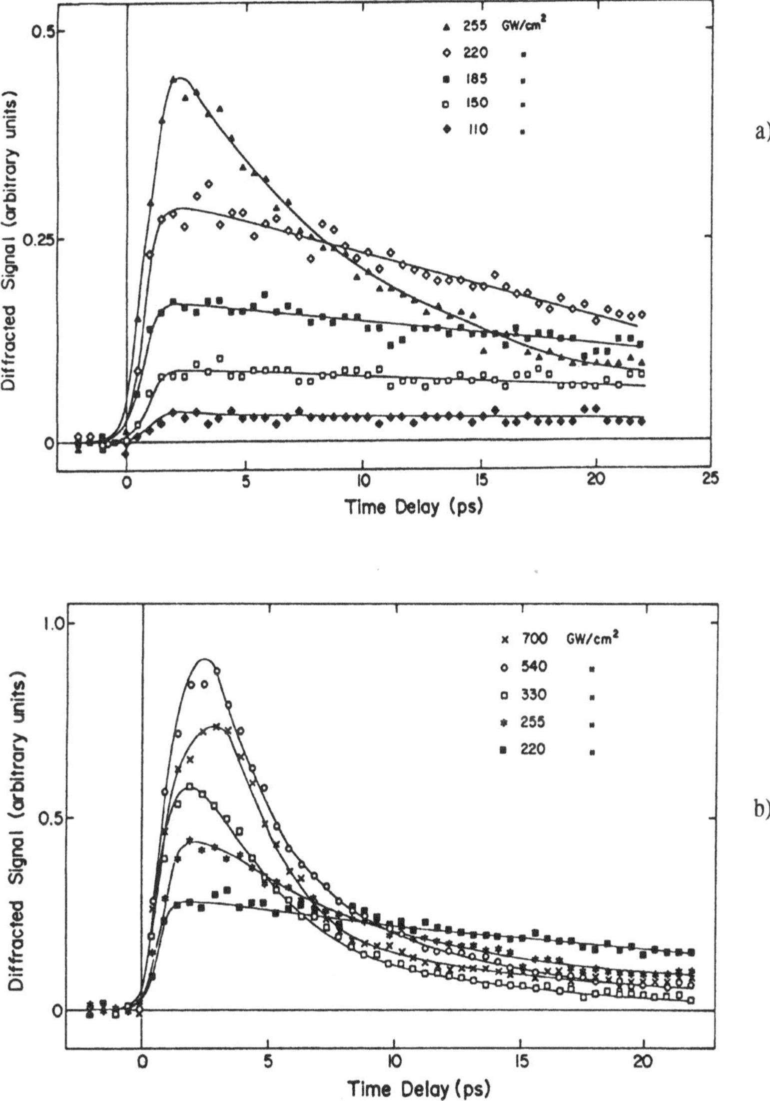


**Figure 16.** Time-resolved diffraction in CdS for a) low values of $I_G$ and b) high values of $I_G$.

shown that this type of scaling behavior follows if $\Delta n$ is linearly proportional to the instantaneous irradiance (and hence linearly proportional to $I_0$). This is consistent with the theoretical expectation that the optical Stark effect and 2PA should dominate the instantaneous nonlinear refractive index, $n_2$, for the combinations of materials and excitation wavelength used. Moreover, both of these effects contribute negatively to $\Delta n$, consistent with the measured changes. In order to extract absolute magnitudes for $n_2$ a full treatment of the beam propagation problem (and improved irradiance measurements) would be required.. Nonetheless the magnitudes of the measured beam deflection angles for $t = 0$ are consistent at the order of magnitude level with the theoretical values of $n_2$ listed in Table 2.

The lack of a peak at $t = 0$ in the diffraction data is expected based on the discussion presented in the theory section. For low irradiances the diffraction efficiency at $t=0$ should scale as $I_0^2$, while for larger irradiances the diffraction efficiency is predicted to scale

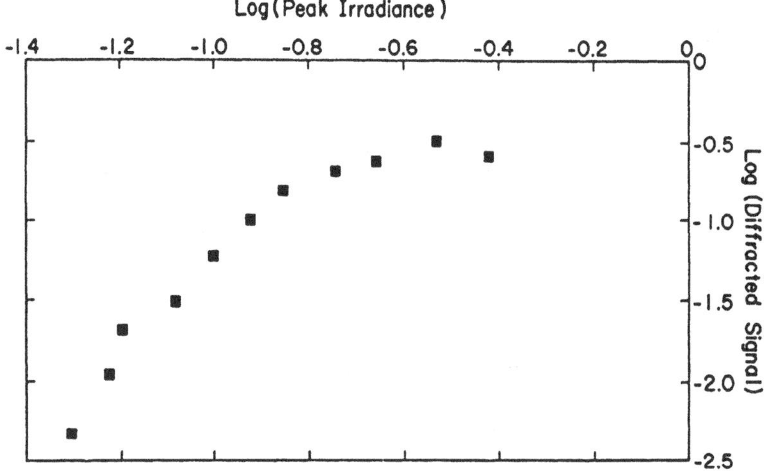


**Figure 17.** Scaling of diffraction intensity with pump irradiance in CdS.

sublinearly with $I_0$. This is due to the 2PA dominated attenuation of the exciting radiation and due to 2PA attenuation of the diffracted light. Using the theoretically predicted refractive nonlinearity values listed in Table 2 the scaling calculations predict that for the range of $I_0$ values used in the diffraction experiments the instantaneous diffraction efficiency should be at least 2 orders of magnitude less than the diffraction efficiency which follows for nonzero time delays.

**Intermediate Time Behavior.** For delay times between 0 and 3 ps both the deflection and diffraction results display changes in the (depth-averaged) refractive index which remain after the pump irradiance has vanished. For time delays between $t = 0$ and 1.5 to 3 ps these refractive index changes increase in magnitude before either saturating or exhibiting the onset of a decay back to zero. The scaling data shows that the height of the peak beam deflection at $t = 1.5$ to 3 ps scales superlinearly with $I_0$ for small values of $I_0$ and linearly with $I_0$ for large values of $I_0$. The magnitude of the peak diffraction efficiency scales as $I_0^{3.5}$. The models presented and discussed in the theory section predict these types of scaling behavior if the local value of $\Delta n$ is proportional to the local carrier density where carrier generation is dominated by 2PA. Moreover the beam deflection data shows that $\Delta n$ is negative. These results suggest that the refractive index change measured between $t = 0$ and 1.5 to 3 ps is due to free carriers with $\Delta n$ being dominated by bandfilling and Drude contributions.

The fact that $\Delta n$ is not reached until a few ps following photoexcitation is due to the thermalization and cooling of the carriers. These processes and their effect upon $\Delta n$ were discussed earlier. The photoexcited carriers are initially located in a narrow band of high energy states coupled by 2PA. Carrier-carrier and carrier-phonon scattering produce thermalized electron and hole distributions which subsequently cool by phonon emission. The thermalization and cooling processes both lead to a redistribution of the carriers in which states close to the band edge become preferentially occupied. Although these processes have no effect on the plasma contribution, the refractive index change due to band filling is highly sensitive to the carrier distribution and increases with thermalization and carrier cooling. Since the pulse width used is 120 fs, and since the carrier densities are larger than $10^{18}$ $cm^{-3}$, it is unlikely that much of the increase in beam deflection strength or diffraction efficiency can be attributed to thermalization (or even that any evidence of nonthermalized distributions exists in the data). Since carrier cooling is expected to occur on a time scale of a few ps, however, most of the signal changes can likely be attributed to this mechanism.

The results of calculations of the refractive index change per carrier for nonthermalized, hot and lattice temperature carrier distributions were presented earlier. For low to moderate carrier densities ( $< 10^{19}$ $cm^{-3}$) the refractive index change per carrier for $CdS_{0.75}Se_{0.25}$ (CdS) should increase by a factor of 1.8 (1.5) as the carriers thermalize and by a further factor of 5 (2.5) as they cool from their initial temperature to the lattice temperature. The ratios are largest for $CdS_{0.75}Se_{0.25}$ and smallest for ZnSe since the Drude contribution is a smaller fraction of the total refractive index change per carrier for $CdS_{0.75}Se_{0.25}$ than for ZnSe. Experimentally the changes in signal magnitudes are as follows. For $CdS_{0.75}Se_{0.25}$ (CdS) the diffraction efficiency changes by a factor of 4 (4) between $t = 0$ and 0.1 ps, and by a factor of 20 (8) between $t = 0.1$ ps and $t = 2$ ps. For $CdS_{0.75}Se_{0.25}$ the beam deflection magnitude changes by a factor of 4 between $t = 0.1$ ps (extrapolated) and $t = 1.5$ ps (the free carrier contribution to beam deflection at $t = 0$ is masked by the instantaneous contribution). No results are quoted for ZnSe since the lower signal sizes recorded on this sample make estimates of the signal sizes in the region about $t = 0$ difficult. For the purposes of comparison between experiment and theory it is assumed that half of the total carrier generation has occurred by $t = 0$, and that carrier generation is complete by $t = 0.1$ ps. If it is assumed that the beam deflection magnitude scales linearly with $\Delta n$ and that the diffraction efficiency scales quadratically with $\Delta n$ then the experimental results indicate that $\Delta n$ increases by a factor of 4.2 (2.8) between $t = 0.1$ ps and $t = 2$ ps. These increases in $\Delta n$ are in close agreement with the theoretically predicted increases which accompany the cooling of the carriers from their initial temperatures to room temperature. It follows that the changes in the beam deflection angles and diffraction efficiencies between $t = 0.1$ ps and $t = 2$ ps can be attributed to carrier cooling, and that the cooling time is of the order of 1 ps. For the theoretically predicted increases in $\Delta n$ which accompany carrier thermalization the diffraction efficiency should increase by a factor of 13 (9). Since the measured changes in diffraction efficiency between $t = 0$ and $t = 0.1$ ps are smaller, it can be concluded that carrier thermalization must take place within 120 fs and that significant carrier cooling does not occur during excitation.

The measured cooling times of 1.5 to 3 ps are consistent with those reported for these materials in the literature [Auston, 1978]. Moreover the calculations described in the theory section predict the electron temperature to decrease at a rate of roughly 10,000 $Kps^{-1}$ until temperatures close to the lattice temperature are reached. This too is consistent with cooling times of between 1 and 3 ps. To investigate the cooling behaviour in more detail diffraction efficiencies were measured in CdS for time delays between t = 0 and t = 5 ps. A range of different excitation levels were used producing peak carrier densities between $6x10^{18}$ $cm^{-3}$ and $1.5x10^{20}$ $cm^{-3}$. The data however display little dependence on the excitation level indicating that the cooling rate is not highly affected by hot phonons or screening of the Fröhlich interaction.[Kocevar, 1985] This is in marked contrast to materials like GaAs where such processes become important for carrier densities of the order of $10^{18}$ $cm^{-3}$. It is interesting to note however that the Fröhlich interaction predicts that the carrier temperature should decrease linearly with time. A detailed analysis of our data shows that the functional dependence is close to exponential with time, and in any event, that there is a period of rapid cooling followed by a reduced cooling rate over the 2-3 ps time. We believe that this may be due to the electrons initially cooling by transferring their energy to holes. When the two distributions reach approximately the same temperature they are thought to cool by the phonon-mediated processes.

**Long time Behavior.** The data plotted in Figs. 13 and 16 show that for small values of $I_0$ the refractive index change saturates and decays slowly with time. For larger values of $I_0$ the measured signal magnitudes display a partial recovery which can be described by a decay time as short as 4 ps. For the deflection data the threshold is dependent on the location of the probe pulse and varies from $I_{BD}$ = 25 $GWcm^{-2}$ for $r_p = r_0$ = 40 μm to $I_{BD}$ = 120 $GWcm^{-2}$ for $r_p$ = 0 and $r_0$ = 40 μm (both for CdS and $CdS_{0.75}Se_{0.25}$). For the diffraction data the threshold value of $I_G$ is of the order of 240 $GWcm^{-2}$ for CdS and $CdS_{0.75}Se_{0.25}$, and is closer to 200 $GWcm^{-2}$ for ZnSe. We will now consider possible reasons for the ultrafast recovery is observed and why it occurs at a threshold that varies with the technique used.

Possible reasons for the decrease in the beam deflection magnitude or diffracted signal strength are first considered. The beam deflection magnitude and diffraction efficiency are both proportional to the strength of the local radial gradient in Δn. Decreases in either the beam deflection magnitude or diffraction efficiency may be due to decreases in the (spatially averaged) value of Δn or to decreases in the local radial gradient in Δn (e.g. with no accompanying change in the average local value of Δn). Since the signal magnitudes measured between t = 0 and t = 2 to 3 ps have been interpreted in terms of a free carrier effect it would be consistent to assume that the measured signal decays may in part be due to carrier recombination (decrease in average magnitude of Δn) and/or carrier diffusion (decreases in the magnitudes of the transverse gradients). Finally, it is possible that Δn is not due to free carriers only and that the signal decreases are due to a process which contributes positively to Δn and which continues to grow in strength after carrier cooling has ceased.

Carrier recombination will be considered first. Carrier recombination should be dominated by radiative and trapping processes for most of the carrier densities investigated in the beam deflection experiments. Auger processes may contribute for the very highest excitation strengths. Using a value of $10^{-10}$ $cm^3sec^{-1}$ [Dneprovskii, 1988] for the radiative recombination coefficient it follows that for $I_0$=25 GWcm$^{-2}$ (i.e. threshold for the onset of decay in the beam deflection data; peak carrier density N=5x10$^{17}$ cm$^{-3}$) the decay time should be of the order of 10 ns. Decay times for recombination via traps are highly dependent on sample quality, but usually fall in the ns regime. No reliable measurements of the Auger coefficient have been reported for the CdSSe family, however most estimates fall in the range of $10^{-29}$ cm$^6$sec$^{-1}$ [Dneprovskii, 1990]. Using this value a decay time of the order of a few hundred ns is calculated for $I_0$ = 25 GWcm$^{-2}$. Since the measured decay time from Fig. 7 is of the order of a few ps it follows that none of these recombination processes can credibly account for the observed decay. Stimulated emission recombination has been used to explain rapid decays of carrier density in samples where the carrier distribution is degenerate. However, the peak carrier density at the location of the probe spot at the threshold irradiance for the onset of decay is an order of magnitude below the critical density for efficient local stimulated emission recombination. The threshold irradiance for the diffraction data are higher ( > 200 GWcm$^{-2}$; peak carrier density between 2x10$^{19}$ and 4x10$^{19}$ cm$^{-3}$) and so the predicted recombination times are closer to the measured values. Using the same recombination coefficients as above, both the radiative and Auger recombination times are predicted to be of the order of 100 to 500 ps, still considerably longer than the measured 4 to 20 ps times. The carrier densities, however, are sufficiently larger than the critical densities for stimulated emission that stimulated emission recombination may play a role. As discussed earlier, recombination times of the order of a few ps are consistent with stimulated emission recombination processes. Carrier recombination, then, might explain the rapid signal decays observed using the diffraction technique, but cannot explain the signal decays observed using the beam deflection technique.

Carrier diffusion has been used to explain observations of rapid plasma expansions in II-VI semiconductors upon high excitation above some threshold [Cornet, 1981, Junnarkar, 1986; Rudolph, 1990; van Lap, 1991]. In these cases diffusion coefficients up to 10$^5$ times larger than the low density value (5 cm$^2$s$^{-1}$) have been invoked. The signal decays displayed in our beam deflection and diffraction measurements have been considered in terms of this enhanced diffusion process. The beam deflection results are considered first. The deflection strength at a time t is proportional to the radial gradient of the carrier density evaluated at the location of the probe pulse. If the time dependence of the carrier density is governed by diffusion with ambipolar diffusion coefficient D, it is simple to show that the local density varies according to:

$$N(r,z,t) = N(0,z,0)\frac{r_0^3}{(16Dt + r_0^2)^{3/2}}\exp[-\frac{4r^2}{(16Dt + r_0^2)}], \tag{45}$$

For the combination of spot size and probe spot location used in the measurement of the data in Fig. 13 it is calculated that the ambipolar diffusion coefficient, D, must exceed 5x10$^5$ cm$^2$s$^{-1}$ in order to explain the factor of 2 reduction in the deflection strength within the first 5 ps. This value is consistent with those which have been reported in the literature, but the plasma density threshold for the onset of the rapid signal decays is considerably lower than the reported threshold values.

A similar analysis can be applied to the diffraction grating results. In this case it is assumed that the diffusive decay of the grating structure can be described as follows:

$$N(r,z,t) = \frac{N(0,z,0)}{2}\{1+\exp(-\frac{4D\pi^2 t}{\Lambda^2})\cos(2\pi x/\Lambda)\} \tag{46}$$

where N(0,z,0) is the carrier density at peaks in the grating structure at a depth z below the surface, and Λ is the grating period. Since the data displayed in Fig. 16 corresponds to a grating period of 1 μm it follows that a 4 ps decay time can be described by a diffusion coefficient of the order of 60 $cm^2s^{-1}$. This value is closer to the accepted value of D for low density plasmas, but is 4 orders of magnitude less than the value of D which describes the decay of the beam deflection data.

In an attempt to determine the extent to which diffusion plays a role in the signal decays, time resolved measurements of the diffraction efficiency were performed using three different grating spacings. The data are displayed in Fig. 18 where the diffraction efficiency is plotted versus time delay for grating spacings between $\Lambda = 1.2$ and 2.9 μm. The data was obtained using a (constant) value of $I_G$ larger than the threshold for the onset of rapid decays. If diffusion is responsible for the fast decays observed in Fig.18 then the time constants which describe the decays should vary by a factor of 6 according to Eq. 44 . Since there is no measurable difference in the decay times it appears that diffusion does not play a significant role. This is consistent with the fact that similar decay times are observed using both the beam deflection and grating diffraction techniques despite the fact that the carrier density gradients vary by a factor of $10^5$.

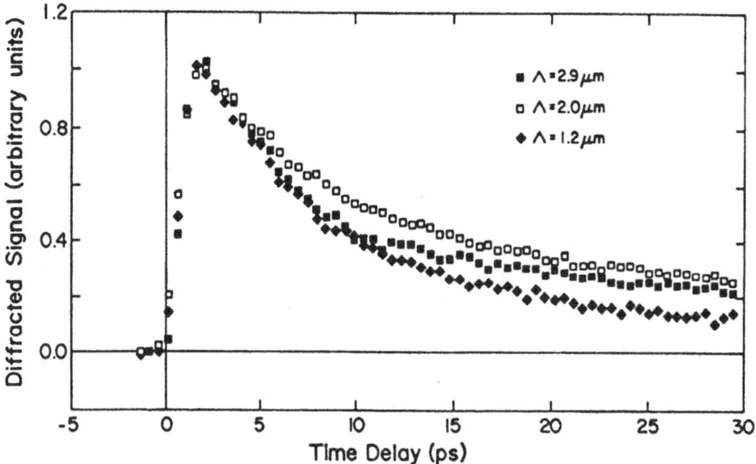


**Figure 18.** Grating decay for gratings with different periods.

In considering carrier recombination and carrier diffusion the assumption has been made that free carrier PIRIC's dominate Δn for long time delays. Other PIRIC's, however, may contribute to Δn, and if these have positive signs and display either a threshold behavior or scale superlinearly with carrier density, then they may in part account for the observed decreases in the beam deflection magnitude and diffraction efficiency for $t > 3$ ps. Lattice heating and band gap renormalization are two processes with positive contributions to Δn. Lattice heating occurs as the carriers cool and transfer their energy to the phonon bath. Phonon relaxation occurs more slowly than carrier cooling and if the lattice contribution to $\Delta n$ increases in magnitude as the phonons thermalize internally then their contribution to $\Delta n$ might not be maximized until many ps after carrier cooling has finished (i.e. at $t > 3$ ps). The calculations which were outlined in the theory section, however, indicate that the magnitude of the lattice heating contribution to $\Delta n$ should be at least 2 orders of magnitude less than the contribution of the free carrier contribution and hence be negligible. Moreover, lattice heating effects should scale linearly with carrier density and are not predicted to display a threshold behaviour. Band gap renormalization, in which the presence of free carriers leads to a reduction in the energy of the band gap, has the potential to produce sizable contributions to $\Delta n$. It is unlikely that it is responsible for the observed signal decays, however, since band gap renormalization occurs nearly instantaneously with carrier generation, and since it is predicted to scale sublinearly with carrier density.

From the above discussion it appears that only recombination by stimulated emission [Fox, 1989; Dubard, 1987] can adequately account for the short time constants observed for above threshold excitation at $t > 3$ ps. Even this recombination, however, cannot explain the signal decays observed in the beam deflection experiments performed with probe spot away from the centre of the pump spot. One aspect of the data not yet considered is the crossing of the curves in the decay region. It is easily shown that if the local rate of carrier loss through recombination is proportional to the local carrier density alone then curve crossing cannot occur. Processes such as radiative, Auger, and trap-dominated recombination can thus be eliminated (it is possible for the curves to cross if the high excitation generates a new recombination channel which remains open after the local carrier density has fallen below the initial value produced by the threshold value of irradiance; no such process has been reported in the literature and it is unlikely to exist). Diffusion and enhanced diffusion can likewise be eliminated if the local diffusion coefficient is proportional to the local carrier density. However, a process such as stimulated emission recombination coupled with some nonlocal process for carrier transfer might explain the rapid signal decays, the threshold behaviour, and the curve crossing.

One process which we have proposed and which is consistent with the data is a photon assisted carrier transfer process [Fox, 1992b; Fox, 1992c]. The process we consider is the stimulated emission analog of spontaneous emission processes which have been used to deal with such disparate phenomena as transport of radiation in stellar atmospheres [Chandreskkhar, 1950] to carrier transport in semiconductors themselves [Dumke, 1957, Epifanov, 1976, Tsarenkov, 1979]. Through the mechanism we consider carriers are moved from regions of high to low density when the conditions for stimulated emission are met. The light generated by stimulated emission recombination is only weakly reabsorbed (e.g. through free carrier absorption) within the region excited above the gain threshold. The light instead propagates until it either escapes the sample or is reabsorbed in a surrounding region characterized by a nondegenerate carrier distribution thereby increasing the local carrier density. By this process a carrier may be lost from a highly excited part of the distribution and redeposited many microns away nearly instantaneously. This nonlocal process can be shown to be consistent with the measured time scales, the threshold behaviour, and the curve crossing in our experiments.

The threshold behaviour will be considered first. At room temperature the electron system in $CdS_{0.75}Se_{0.25}$ (CdS, ZnSe) is degenerate for plasma densities greater than $1.4x10^{18}$ $cm^{-3}$ ($1.8x10^{18}$ $cm^{-3}$, $1.1x10^{18}$ $cm^{-3}$), and gain for bandedge radiation is achieved for plasma densities larger than $5x10^{18}$ $cm^{-3}$ ($6x10^{18}$ $cm^{-3}$, $4x10^{18}$ $cm^{-3}$). For the grating diffraction measurements rapid decays at $t > 3$ ps are observed for peak carrier densities of the order of $3 \times 10^{19}$ cm$^{-3}$. For the beam diffraction measurements the irradiance threshold for decays corresponds to a carrier density of the order of $10^{19}$ cm$^{-3}$ when the probe spot is located at the centre of the pump spot. When the probe spot is located away from the centre of the pump spot (see, for example, Fig. 8) the carrier density at the location of the probe spot might not exceed the gain threshold density although the density at the centre of the pump spot does. Rapid signal decays, then, are accompanied by stimulated emission over some region of the excited volume.

To verify that these stimulated emission processes do take place, the front-surface luminescence yield was measured as a function of pump irradiance. The results for the $CdS_{0.75}Se_{0.25}$ sample are presented in Fig. 19 and show a transition from superlinear to sublinear behavior for the luminescence for $I_0$ about 100 GWcm$^{-2}$. The lineshape of the luminescence was not observed to vary with the level of excitation. The sublinear scaling of the yield at higher irradiances is indicative of a stimulated process which does not emit

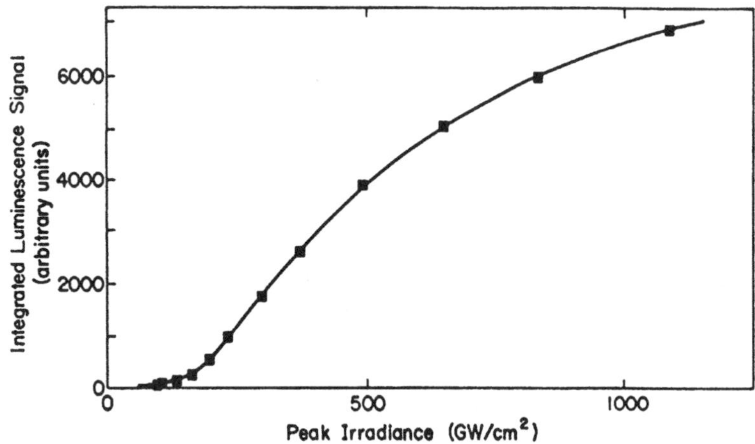


**Figure 19.** Integrated band-edge luminescence versus peak irradiance in CdS. The solid curve is a guide to the eye.

along the surface normal (the measurement direction) and hence must emit preferentially along directions parallel to the surface. In addition, the induced grating experiments were repeated for different lattice temperatures between 300 and 80 K. For lower temperatures the gain threshold carrier density drops (reaching roughly 1/6 of its room temperature value at $T = 80$ K). In general, our results show a decrease in both the threshold irradiance and the decay time with decreasing temperature, both consistent with stimulated emission. The observed irradiance threshold behavior, then, is consistent with stimulated emission recombination.

Stimulated emission recombination alone, however, cannot explain the behaviour displayed in Figs. 13 and 16. In order to explain the observed curve crossing it is necessary to consider the changes which occur in the spatial distribution of the carriers. The steady state spatial changes induced in deflection experiments are considered first.

For the data presented in Fig. 13 decays are observed for peak pump irradiances $I_{BD} > 25$ GWcm$^{-2}$ evaluated at $r_p = r_0 = 40$ µm. The corresponding value of $I_0$ evaluated at the centre of the pump pulse is equal to 185 GWcm$^{-2}$ which corresponds to an initial plasma density at the centre of the spot equal to $2.8 \times 10^{19}$ cm$^{-3}$. The gain volume extends to a depth of 10 µm and has a radial extent of 26 µm. Since the gain volume has its greatest extent in the radial direction most of the carrier redistribution will occur in the plane of the surface. Light produced by radiative recombination is amplified as it traverses the central region of the excited volume, and is absorbed in the region beyond the boundary between gain and absorption. The absorption coefficient climbs rapidly for carrier densities below the gain threshold so that most of the light will be reabsorbed within distances of the order of 1 µm beyond the boundary between regions of gain and absorption. It follows that the carrier distribution produced by the stimulated emission process will consist of a disk of radius R where the carrier density within the disk is equal to the gain threshold density, $N_g$, and where carrier number is conserved, i.e.

$$\int_0^R 2\pi r N(r) dr = \pi R^2 N_g \; . \tag{47}$$

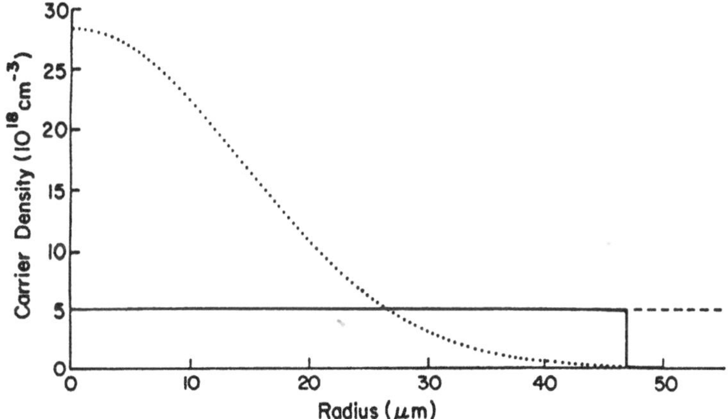


**Figure 20** Radial carrier density profile at the surface of $CdS_{0.75}Se_{0.25}$ sample for $I_0 = 185$ GWcm$^{-2}$ and $r_p = r_0 = 40$ mm. The dotted line corresponds to the density profile before stimulated emission, the solid line shows the distribution immediately after transfer, and the dashed line corresponds to the gain threshold density.

Fig. 20 shows the carrier density radial profile before and after the carrier transfer process for the $CdS_{0.75}Se_{0.25}$ sample for $I_0 = 185$ GWcm$^{-2}$.

For the excitation conditions quoted for Fig. 13 the radius R is calculated to be 47 µm. Since the radial carrier density gradient is zero within the disk (of density equal to threshold density for gain) it follows that wherever the radius R exceeds the position radius of the probe spot there will be no contribution to beam deflection. The maximum beam deflection will occur for an excitation strength just below that for which the radius R at the surface is equal to the probe spot position radius. For higher irradiances the depth from which no contribution to beam deflection is made increases. If the local radial gradient evaluated at the position of the probe is integrated over the sample thickness it is found that the beam deflection magnitude should decrease as the excitation strength increases (i.e. the data should exhibit curve crossing) provided that the radial position of the probe, $r_p$, satisfies $r_p < 0.8r_0$. As the probe spot position is moved closer to the centre of the excitation region the threshold irradiance decreases and the depth of the recovery increases. Since for a 100 μm thick sample the carrier density at the back surface never reaches the gain threshold density the beam deflection magnitude never vanishes completely.

Of course this analysis is extremely simplistic and gives only a qualitative understanding of the data. In a more sophisticated analysis the probe beam would be treated not as a simple ray but rather as a beam with non-zero radial extent, and a full beam propagation model would be constructed. Finally, in the above analysis it was assumed that the absorption depth beyond the boundary of the gain region approaches zero for the radiation emitted in the stimulated process. In fact the radiation may travel significantly beyond this boundary and produce smoother density profiles if bandgap renormalization is significant (i.e. the peak gain will occur for wavelengths closer to or even below the bandedge evaluated in the less excited regions).

The analysis of the diffraction experiments can be treated in a similar fashion. Here the gain volume consists not of a disk but of long submicron wide ribbons separated by the grating spacing, Λ. Stimulated emission, and hence carrier redistribution, will occur preferentially along directions parallel to these ribbons. Since the probe beam is located at the centre of the pump-induced carrier distribution it follows that the carriers redistributed by the stimulated emission process will be deposited in regions not sampled by the probe radiation. The magnitude of the refractive index grating at the location of the probe will decrease to a uniform value (with the peak value of $\Delta n$ equal to the value corresponding to the gain threshold carrier density) in the near surface region, and remain unchanged for depths beyond the extent of the original gain volume as illustrated in Fig. 21. Calculations of the diffraction efficiency have been performed for the steady state distribution produced by the stimulated emission process by integrating the coupled equations for the propagation of scattered light in a thick grating (see Theory section). The calculations indicate that the diffraction efficiency should fall rapidly to values a factor ~4 smaller than the peak diffraction efficiency for $I_0$~120 $GWcm^{-2}$, and to a factor of ~70 smaller for the very highest values of $I_0$. These results can be understood in terms of the phase matching aspects of the thick grating problem. If the grating modulation as a function of depth is constant then the net contribution from a depth equal to an integral multiple of the inverse of the wavevector mismatch will be zero. The observation of non-zero effective diffraction efficiencies in the data is due to the fact that the grating modulation decreases with increasing depth (i.e. 2PA attenuation leads to a carrier density distribution which decreases with increasing depth). It follows that negligible contribution to the scattered signal will be made by the region between the surface and the maximum gain depth, the same region responsible for the largest fraction of the effective diffraction efficiency prior to stimulated emission. The experimental results don't fall off as sharply with increasing excitation as predicted by the model, but are in qualitative agreement with the model results. Finally, the fact that the threshold excitation strength required to induce decays in the diffraction efficiency is larger than that required to induce decays in the beam deflection magnitudes is consistent with the model of stimulated emission recombination since gain can only occur along one dimension in the case of diffraction rather than along two dimensions in the case of deflection.

The final consideration in the analysis of the model of stimulated emission recombination and carrier diffusion is the temporal evolution of the measured beam deflection magnitudes and diffraction efficiencies. In order to calculate these quantities a model is required in which local carrier densities and photon fluxes can be determined as a function of time. The evolution of highly excited carrier distributions in which stimulated emission dominates carrier recombination have been considered theoretically [Goebel, 1977; Kalafati, 1991]. Although these models are quite sophisticated with such processes as carrier cooling and intervalley scattering being considered, they are all performed for spatially homogeneous carrier distributions (and hence light field distributions) as might be found in a semiconductor laser, and are thus local in nature. In order to calculate beam deflection

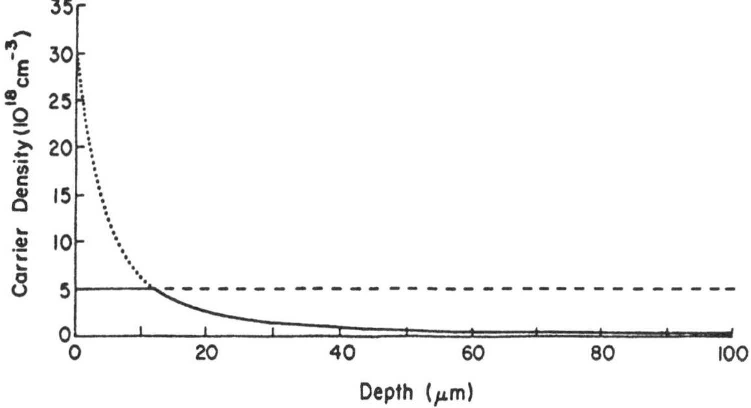


**Figure 21.** Depth carrier density profile in CdS for $I_G$ = 220 $GWcm^{-2}$. The dotted line corresponds to the carrier density profile prior to carrier transfer, the dashed line corresponds to the gain threshold density, and the solid line corresponds to the carrier density following carrier transfer.

magnitudes and diffraction efficiencies the non-uniform nature of the spatial carrier distributions cannot be ignored. In turn, the non-uniform spatial nature of the radiation field must also be considered making the problem a highly nonlocal one (i.e. the strength of the radiation field at a particular position, time, and in a particular direction is highly dependent on past values of the carrier and light field distributions in the vicinity). Nonetheless it is possible at a qualitative level to understand the processes which determine the temporal characteristics. The initial electron (hole) distributions are described by temperatures of the order of 10,000 K (3,000 K). Gain, and hence stimulated emission, cannot occur until a degenerate distribution is created by carrier cooling (the electron distribution for an arbitrarily chosen density $N=5x10^{19}$ $cm^{-3}$ is degenerate for temperatures less than 2,800 K, 3,150 K, and 3,900 K for CdS, $CdS_{0.75}Se_{0.25}$, and ZnSe respectively; the corresponding temperatures for holes are 730, 730, and 450 K). This is consistent with the observation that significant signal decays do not begin until t ~ 3 ps, the time associated with carrier cooling. If no stimulated emission occurs until the carriers have fully cooled then peak gains in excess of $3x10^4$ $cm^{-1}$ would result and gain depletion would take place nearly instantaneously (see the discussion below). The processes of carrier cooling and gain depletion are highly coupled, however, so that the temporal dependence of the decay in the gain is determined by a balance between continued carrier cooling (which increases the gain), stimulated emission (which reduces the gain), carrier heating from free carrier absorption of the emitted light and from the preferential recombination of near-band-edge carriers (which decreases the gain), and the evolving size and shape of the gain region.

Although it is not possible to predict a theoretical decay time based on simple arguments, the 4 to 20 ps times measured are of the same order of magnitude as those predicted using the local models discussed above. The times are also consistent with the temporal widths of light pulses produced by simple gain switching in semiconductor lasers.

## CONCLUSIONS

We have tried to offer a summary of theoretical and experimental results illustrating the various effects that contribute to refractive index changes in semiconductors. The techniques mentioned offer fundamental insight into several microscopic mechanisms which contribute to the changes in the refractive index, both through virtual carrier and real carrier effects. In the process of conducting the experiments we have also found a new ultrafast photon induced carrier transport process which causes a rapid decay in switching behavior by removing the gradient of the carrier density as opposed to the carriers. This process may also account for reports of ultrafast diffusion processes in the literature.

## ACKNOWLEDGEMENTS

We gratefully acknowledge support of the Natural Sciences and Engineering Research Council of Canada and the Premier of Ontario's Technology Fund. The technical help of Mr. G. Rawlings with the experiments is appreciated.

## REFERENCES

Auston, D.H., McAfee, S., Shank, C.V., Ippen, E.P., Teschke, O., 1978, Picosecond spectroscopy of semiconductors, *Solid State Electron.* 21:147.

Anderson, K.K., Lagasse, M.J.,. Wang, C.A, Fujimoto, J.G., Haus, H.A., 1990, Femtosecond dynamics of the nonlinear index near the band edge in AlGaAs waveguides, *Appl. Phys. Lett.* 56:1834.

Chandresekhar, S., 1950, "Radiative Transfer," Dover, New York.

Combescot, M., 1979, Hydrodynamics of an electron-hole plasma created by a pulse, *Solid State Commun.* 30:81.

Cornet, A., Pugnet, M.,. Collet, J,. Amand, T, Brousseau, M., 1981, Spatial expansion of hot electron-hole plasma at high density in CdSe, *J. de Phys.* C7:471.

Cotter, D., Ironside, C.N., Ainslie, B.J., Girdlestone, H.P., 1989, Picosecond pump-probe interferometric measurement of optical nonlinearity in semiconductor-doped fibers, *Opt. Lett.* 14:317.

Ding, Y.J., Guo, C.L. Swartzlander, G.A., Jr., Khurghin, J.B., Kaplan, A.E., 1990, Spectral measurement of the nonlinear refractive index in ZnSe using self-bending of a pulsed laser beam, *Opt. Lett.* 15:1431.

Dneprovskii, V.S., Klimov, V.I. , Novikov, M.G., 1988, Recombination dynamics of an electron-hole plasma in cadmium sulfide, *Sov. Phys. Solid State* 30:1694.

Dneprovskii, V.S., Efros, A.L., Ekimov, A.I., Klimov, V.I., Kudriavstev, I.A., Novikov, M.G., 1990, Spontaneous and stimulated collapse of high density electron-hole system in CdSe, *Solid State Commun.* 74:555.

Downer, M.C. and Shank, C.V., 1986, Ultrafast heating of silicon sapphire by femtosecond optical pulses, *Phys. Rev. Lett.* 56:761.

Dubard, J., Oudar, J.L., Alexandre, F., Hulin, D., Orszag, A., 1987, Ultrafast absorption recovery due to stimulated emission in GaAs/AlGaAs Multiple quantum wells, *Appl. Phys. Lett.* 50:821.

Dumke, W.P., 1957, Spontaneous radiative recombination in semiconductors, *Phys. Rev.* 105:139.

Eichler, H.J., Gunter, P., Pohl, E.W., 1986, "Laser Induced Dynamic Gratings," Springer-Verlag.

Epifanov, M.S., Galkin, G.N., Bobrova, E.A., Vavilov, V.S., Sabanova, L.D., 1976, Photon transfer of excitation of nonequilibrium carriers in gallium arsenide, *Fiz. & Tekh. Poluprovodn.* 10:889 (*Sov. Phys. Semicond.* 10:526).

Fork, R. L., Greene , B.I., Shank, C.V., 1981, Generation of optical pulses shorter than 0.1 psec by colliding pulse mode locking, *Appl. Phys. Lett.* 38:671.

Fox, A. M., Manning, R.J., Miller, A., 1989, Picosecond relaxation mechanisms in highly excited GaInAsP, *J. Appl. Phys.* 65:4287.

Fox, E.C., Canto-Said, E.J., van Driel, H.M., 1991, Femtosecond time-resolved refractive index changes in $CdS_{0.75}Se_{0.25}$ and CdS, *Appl. Phys. Lett.* 59:1878.

Fox, E.C., Canto-Said, E.J., van Driel, H.M., 1992a, Separation of bound and free carrier contributions to the refractive index change induced in II-Vi semiconductors by femtosecond pulses, *Semicond. Sci. Technol.* 7 B183.

Fox, E.C., Canto-Said, E.J., van Driel, H.M., 1992b, Femtosecond time-resolved refractive index changes in CdSSe, SPIE meeting on "Ultrafast Phenomena in Semiconductors and Superconductors," Sommerset, N.J., U.S.A.

Fox, E.C. and van Driel, H.M., 1992c, Ultrafast carrier recombination and plasma expansion via stimulated emission in II-VI semiconductors, *Phys. Rev. B*, in press.

Friberg, S.W. and Smith, P.W., 1987, Nonlinear optical glasses for ultrafast optical switches, *I.E.E.E. J. Quantum Electron.* QE-23:2089.

Gibbs, H.M., 1985, "Optical Bistability: Controlling Light with Light," Academic Press, New York.

Gibbs, H.M., Khitrova, G., Peyghambarian, N., 1990, "Nonlinear Photonics," Springer Verlag, Berlin.

Goebel, E.O., Hildebrand, O., Lohnert, K., 1977, Wavelength dependence of gain saturation in GaAs lasers, *I.E.E.E. J. Quantum Electron.* QE-13:848.

Haug, H., 1988, "Optical Nonlinearities and Instabilities in Semiconductors," Academic Press, San Diego.

Johnson, E.J., 1967, in "Semiconductors and Semimetals," vol. 3, ed. Willardson, R.K. and Beer, A.C., Academic Press, London.

Junnarkar, M.R. and Alfano, R.R., 1986, Photogenerated high-density electron-hole plasma energy relaxation and experimental evidence for rapid expansion of the electron-hole plasma in CdSe, *Phys. Rev. B* 34:7045.

Kalafati, Y. D. and Kokin, V.A., 1991, Picosecond relaxation processes in a semiconductor laser excited by a powerful ultrashort light pulse, *Sov. Phys. J.E.T.P.* 72:1003.

Kobayashi, A., Sankey, O.F., Volz , S.M., Dow, J.M., 1983, Semiempirical tight-binding band structures of wurtzite semiconductors: AlN, CdS, CdSe, ZnS, and ZnO, *Phys. Rev. B* 28:935.

Kocevar, P., 1985, Hot phonon dynamics, *Physica* 134 B+C:155.

Kressel, H. and Butler, J.K., 1977, "Semiconductor Lasers and Heterojunctions," Academic Press, New York.

LaGasse, M.J., Anderson, K.K., Haus, H.A., Fujimoto, J.G., 1989, Femtosecond all-optical switching in AlGaAs waveguides using a time division interferometer, *Appl. Phys. Lett.* 54:2068.

Landot and Börnstein, 1982, "Numerical Data and Functional Relationships in Science and Technology, New Series," vol. 17 & 22, Springer-Verlag.

Majumder, F.A., Swoboda, H.-E., Kempf, K., Klingshirn, C., 1985, Electron-hole plasma expansion in the direct-band-gap semiconductors CdS and CdSe, *Phys. Rev. B* 32:2407.

Miller, A., Miller, D.A.B., Smith, S. D., 1981, Dynamic non-linear optical processes in semiconductors, *Adv. in Physics* 30:697.

Pötz, W. and Kocevar, P., 1983, Cooling of highly photoexcited electron-hole plasma in polar semiconductors and semiconductor quantum wells: a balance-equation approach, *Phys. Rev. B* 82:7040.

Puls, J., Rudolph, W., Henneberger, F., Lap, D., 1988, Femtosecond studies of room temperature optical nonlinearities in wide-gap II-VI semiconductors, *Phys. Stat. Sol. (b)* 150:419.

Pugnet, M., Collet, J., Cornet, A., 1981, Cooling of hot electron-hole plasmas in the presence screened electron-phonon interactions, *Solid State Commun.* 38:531.

Rinker, M.,. Swoboda, H.-E, Majumder, F.A., Klingshirn, C., 1989, Diffusive and thermal properities of the electron-hole plasma in CdS and CdSe, *Solid State Commun.* 69:887.

Rolland, C. and Corkum, P.B., 1986, Amplification of 70 fs pulses in a high repetition rate XeCl pumped dye laser amplifier, *Opt. Commun.* 59:64.

Rudolph, W., Puls, J., Henneberger, F., Lap, D., 1990, Femtosecond studies of transient nonlinearities in wide-gap II-VI semiconductor compounds, *Phys. Stat. Sol. (b)* 159:49.

Said, A.A., Sheik-Bahae, M., Hagan, D.J., Wei, T.H., Wang, J., Young, J., Van Stryland, E.W., 1992, Determination of bound and free-carrier nonlinearities in ZnSe, GaAs, CdTe, and ZnTe, *J. Opt. Soc. Am. B* 9:405.

Saito, H. and Göbel, E.O., 1985, Picosecond spectroscopy of highly excited Cds, *Phys. Rev. B* 31:2360.

Seeger, K., 1982, "Semiconductor Physics, an Introduction," Springer Verlag, Berlin.

Shah, J., 1989, Photoexcited hot carriers: from CW to 6 fs in 20 years, *Solid State Electron.* 32:1051.

Shank, C.V., Auston, D.H., Ippen, E.P., Teschke, O., 1978, Picosecond time resolved reflectivity of direct gap semiconductors, *Solid State Commun.* 26:567.

Sheik-Bahae, M., Said, A.A., Van Stryland, E.W., 1989, High-sensitivity, single-beam $n_2$ measurements, *Opt. Lett.* 14:955.

Sheik-bahae, M., Hutchings, D.C., Hagan, D.J., Van Stryland, E.W., 1991, Dispersion of bound electronic nonlinear refraction in solids, *I.E.E.E. J. Quantum Electron.* QE-27:1296.

Shen, Y.R., 1984, "Principles of Nonlinear Optics," John Wiley & Sons, Toronto.

Solymar, L. and Cooke, D.J., 1981, "Volume Holography and Volume Gratings," Academic Press, New York.

Stegeman, G.I. and Stolen, R.H., 1988, "Nonlinear Guided Wave Phenomena," special issue of *J. Opt. Soc. Am. B* 5:264-574.

Tsarenkov, G.V., 1979, Drift of recombination in a variable gap semiconductor, *Sov. Phys. Semicond.* 13:641.

Valdmanis, J.A., Fork, R.L., Gordon, J.P., 1985, Generation of optical pulses as short as 27 femtoseconds directly from a laser balancing self-phase modulation group velocity dispersion, saturable absorption, and saturable gain, *Opt. Lett.* 10:131.

van Driel, H.M., 1979, Influence of hot phonons on energy relaxation of high-density carriers in germanium, *Phys. Rev. B* 19:5928.

van Driel, H.M., 1987, Kinetics of high-density plasma generated in Si by 1.06- and 0.53-mm picosecond laser pulses, *Phys. Rev. B* 35:8166.

van Lap, D., Peschel, U., Ponath, H.E., Rudolph, W., 1991, Investigation of carrier temperature relaxation with femtosecond transient grating experiments in $CdS_xSe_{1-x}$ semiconductors, Inst. Phys. Conf. Ser. No. 126: Section V, 357, presented at Int. Symp. on Ultrafast Processes in Spectroscopy, Bayreuth.

Van Stryland, E.W., Vanherzeele, H., Woodall, M.A., Soileau, M.J., Smirl, A.L., Guha, S., Boggess, T.F., 1985, Two photon absorption, nonlinear refraction, and optical limiting in semiconductors, *Opt. Eng.* 24:613.

Vasconcellos, A. and Luzzi, R., 1980, Coupled electron-hole plasma-phonon system in far-from-equilibrium semiconductors, *Phys. Rev. B.* 22:6355.

Wherrett, B.S., 1988, Nonlinear Refraction for CW Optical Bistability in "Optical Nonlinearities and Instabilities in Semiconductors," ed. Haug, H., Academic Press, San Diego.

Wherrett, B.S., Darzi, A.K., Chow, Y.T., McGuckin, B.T., Van Stryland, E.W., 1990, Ultrafast thermal refractive nonlinearities in bistable interference filters, *J. Opt. Soc. B* 7:215.

Zimmermann, R., 1988, Nonlinear optics and the Mott transition in semiconductors, *Phys. Stat. Sol.* (b) 146:371.

# TUNNELING OF ELECTRONS AND HOLES IN ASYMMETRIC DOUBLE QUANTUM WELLS

Jagdeep Shah, Karl Leo*, D. Y. Oberli**, and T. C. Damen

AT&T Bell Laboratories
Holmdel, N. J. 07733

*ABSTRACT*

*The quantum mechanical phenomenon of tunneling is of interest from fundamental as well as device points of views. Semiconductor microstructures, and in particular asymmetric double quantum well structures (a-DQWS), provide ideal systems for investigating tunneling phenomena, especially when an electric field is applied to bring various electronic levels into resonance. We first discuss various resonant and nonresonant tunneling processes for electrons and holes, including both coherent and incoherent resonant tunneling, and then review recent extensive investigations of resonant and non-resonant tunneling rates of electrons and holes in asymmetric double quantum well structures using time resolved spectroscopy. We also show that cw and ultrashort pulse experiments probe different physics and that it is difficult to obtain quantitative information about resonances and resonant tunneling rates from cw experiments.*

## I. INTRODUCTION

Tunneling is a quantum mechanical phenomenon that is of interest in many branches of physics. Recent advances in the growth of high quality quantum nanostructures, consisting of thin, epitaxial layers of semiconductors with different bandgaps, have provided a versatile system for investigating the fundamental physics of tunneling. Besides this fundamental interest, tunneling plays an important role in perpendicular transport of carriers in superlattices. Such transport has many interesting properties as first discussed by Easki and Tsu. [1] Tunneling also plays a fundamental role in resonant tunneling diodes (RTD) that have been shown recently to operate at

*Ultrashort Processes in Condensed Matter*, Edited by
W.E. Bron, Plenum Press, New York, 1993

frequencies as high as 460 GHz. [2] These factors make investigation of tunneling in semiconductor microstructures a very active field of current research.

Perpendicular transport in superlattices and tunneling in RTDs were first investigated by steady-state measurement of current-voltage characteristics. Such static investigations have given considerable information and many groups continue to pursue this approach. In contrast to such static investigations, the investigation of the dynamics of these phenomena has been undertaken only in recent years. The high frequency characteristics of RTD were first investigated by Sollner and coworkers [3] whereas the transient photocurrent response of perpendicular transport in superlattices and multiple-quantum-well structures (MQWS) was investigated by Minot et al., [4] Tarucha et al. [5, 6], and by Schneider et al. [7-10] in many different experiments.

The time resolution in such experiments can be improved considerably by using electro-optic sampling techniques instead of measuring current directly. Such measurements have indeed been performed, for example by Whittaker et al. [11] on resonant tunneling diodes. However, the elaborate device preparation effort needed to reduce the RC time constant limitation has discouraged such effort. The best techniques to obtain time resolution in the sub-picosecond and femtosecond regimes are *all-optical techniques*, such as excite-and-probe spectroscopy, transient four-wave-mixing (FWM) techniques or luminescence spectroscopy, whose time resolution is limited primarily by the laser pulsewidth. Such techniques allow one to investigate as-grown wafers with minimum processing and are not limited by RC considerations.

In double barrier structures used for resonant tunneling diodes, all-optical techniques were first used by Tsuchiya et al. [12] who optically excited the quantum well and determined the escape rate of electrons from the quantum well in a double barrier structure by measuring the decay of quantum well luminescence. Escape times in double barrier structures have also been investigated by Norris et al., [13] who investigated the dependence on electric field, and Jackson et al. [14]. It should be mentioned that cw spectroscopy has also been used to determine parameters relevant to the operation of double barrier diodes, in particular the charge accumulation in the well. [15-19]

All-optical techniques have also been used to determine perpendicular transport of carriers across a large distance ($\approx 1\ \mu m$). The screening of the applied electric field by the space charge generated by the transport of electrons and holes was used to determine the velocity overshoot [20] in bulk GaAs and to determine the tunneling-dominated transport of electrons across a multi-quantum-well sample [21] Luminescence spectroscopy has been used to investigate tunneling-dominated transport of photoexcited carriers [22] and Bloch transport of carriers in a superlattice. [23-25] These studies provide valuable insights into the nature of perpendicular transport, but they are influenced by tunneling as well as other physical processes. Therefore, for developing a quantitative understanding of tunneling, it is better to investigate *isolated structures* such that only tunneling contributes to the measured effect.

An asymmetric double quantum well structure (a-DQWS), consisting of a wide well (WW) and a narrow well (NW) separated by a barrier provides an ideal isolated structure. Tunneling in these structures is between two quasi-2D systems. In these systems, the term tunneling is used to describe two different physical situations. In a strict sense, one must speak of the energy levels of the entire double well structure rather than the levels of an individual well. Therefore, the wavefunction of each level has a component in both wells and transitions of interest in the absence of any resonance between electronic levels for the same in-plane wavevector requires a momentum conserving process and are inter-subband transitions between levels of the entire structure. However, these transitions can be thought of as tunneling transitions if the carriers are initially localized primarily in one well and become localized primarily in the other well after the transition. We will refer to such transitions as *non-resonant tunneling*. When the electronic states in the two wells at the same in-plane wavevector are in resonance, tunneling interaction between them will lead to a splitting of these levels. We will refer to tunneling under these conditions as *resonant* or *coherent tunneling*. This distinction between the two different meanings of the word tunneling must be kept in mind. In general, we will use the term tunneling to describe both situations, but we will use the terms resonant (coherent) or non-resonant tunneling explicitly if there is a chance of confusion.

The first time resolved experiments in a-DQWS were performed on isolated a-DQWS by Tada et al. [26] with 300 ps time resolution and on non-isolated coupled quantum wells by Tsuchiya et al. [27] with a time resolution of $\approx$50ps. Norris et al. [28] first reported on the dynamics of charge transfer luminescence (electrons and holes in separate wells). The first quantitative determination of the electron tunneling rates and the demonstration that electronic resonance strongly enhances these rates were reported by Oberli et al. [29-31] These authors were also the first to investigate the effect of barrier thickness on tunneling rates [29] and to observe a threshold in the tunneling rate at the optical phonon energy, [29, 32] demonstrating the importance of optical phonon-assisted tunneling processes.

Many other studies of tunneling in a-DQWS have since been reported, primarily using time resolved luminescence spectroscopy. These studies include InGaAs/InP system, [33] pressure dependence in GaAs, [34] use of luminescence correlation spectroscopy to measure tunneling rates, [35] electron and hole tunneling in GaAs, [36] and resonant and non-resonant behavior as a function of well thickness [37, 38] and electric field. [39-41] Tunneling between two quantum wells has also been used to demonstrate fast recovery of excitonic absorption. [42, 43] In addition, resonant and non-resonant tunneling of holes has been investigated by Leo et al. [44, 45] who demonstrated the importance of collisions and relaxation on tunneling. There have also been studies of cw luminescence from coupled quantum wells; [46, 47] however, such studies provide only limited information about the dynamics of tunneling, as discussed later.

There have also been a number of theoretical studies related to tunneling in a-DQWS. One of the first was by Weil and Vinter [48] who investigated phonon-assisted transfer rates in a-DQWS. A discussion of coherent oscillations in a-DQWS has been given by Luryi [49, 50] and assisted relaxation processes have been discussed by Ferreira and Bastard [51] and Bastard et al. [52] Monte Carlo simulations of tunneling in a- DQWS have been carried out by Lary et al. [53] It should also be mentioned that general aspects

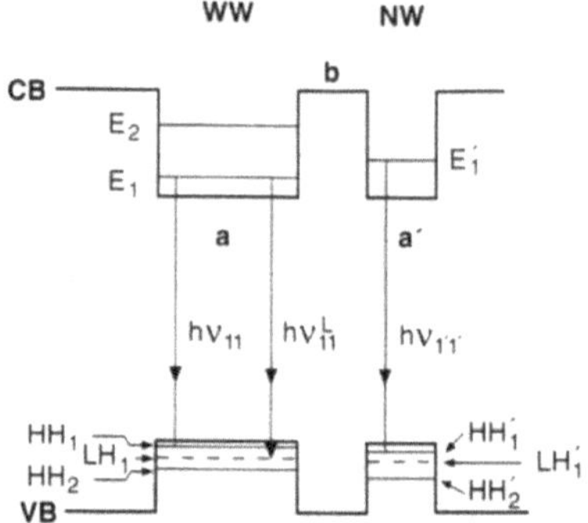


**Fig. 1** A schematic diagram of an asymmetric double quantum well structure (a-DQWS) showing the potential profile for the valence and the conduction bands, various energy levels and optical transitions as discussed in the text.

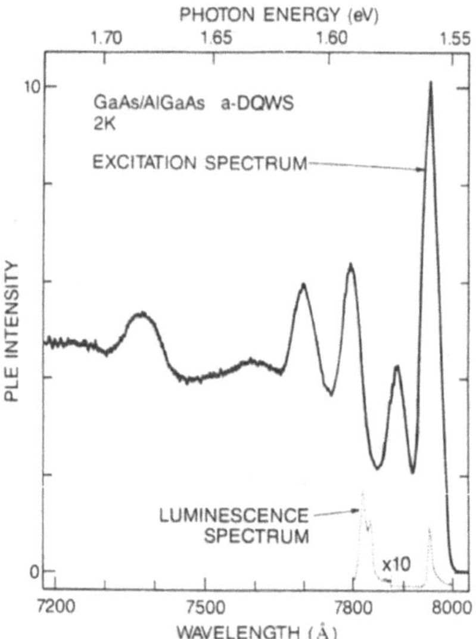


**Fig. 2** Photoluminescence (PL) and Photoluminescence Excitation (PLE) spectra of an a-DQWS. The PL spectrum shows the WW and the NW luminescence peaks. The two peaks in the NW PL probably arise from regions of different well widths within the excitation spot. The PLE shows the lowest HH and LH transitions in each well and some higher transition.

of tunneling in presence of scattering have been discussed theoretically by Stone and Lee [54], Leggett et al. [55], Wingreen et al. [56], Büttiker, [57] and Gelfand et al. [58]

This brief introduction makes it clear that tunneling in semiconductor microstructures has received considerable attention in the last few years. The purpose of this chapter is to provide a general discussion of various tunneling processes and then to present a review of the extensive time-resolved spectroscopy measurements of tunneling in a-DQWS. The emphasis in on new concepts, and their illustration by various experiments, and no attempt is made to present a complete survey of the literature. The organization of the chapter is as follows. In Sec. II, we discuss some basic concepts related to tunneling in a-DQWS. We first discuss how optical spectroscopy allows a direct determination of the tunneling rates in such structures. By applying an external electric field, one can tune the separation between various electronic levels, and isolate and investigate specific tunneling processes. For non-resonant tunneling, we discuss the dependence on barrier thickness and the role of optical phonons. For resonant tunneling, we first consider the near-ideal case of weak collisions/relaxation. We consider how coherent oscillations of a wavepacket comes about near resonance and then discuss how different considerations apply to cw and ultrashort pulse experiments. The final part of Sec. II discusses the effects of collisions and relaxation on coherent tunneling process. In Sec. III, we discuss some experimental details. The results for non-resonant as well as resonant tunneling of electrons and holes are presented and discussed in Sec. IV. Observation of coherent oscillations of an electronic wavepacket is also discussed here. Sec. V presents a summary.

## II. BASIC CONCEPTS

In this section, we discuss some basic concepts related to resonant and non-resonant tunneling in asymmetric double quantum well structures.

### II.A Tunneling in Asymmetric Double Quantum Well Structures

An asymmetric double quantum well structure (a-DQWS) is shown schematically in Fig. 1. It consists of two quantum wells, a wide well (WW) of width a and a narrow well (NW) of width $a'$, separated by a barrier of width b. All quantities related to the NW have a prime ($'$) associated with them. We will refer to samples by a notation of the form $a_L/b/a_R$ where all the thicknesses are in Å, and $a_L$ and $a_R$ refer to the thicknesses of the left and the right well respectively. The right well is grown first and hence is closer to the $n^+$ substrate. The electron, heavy hole and light hole confinement energies are denoted by $E_i$, $HH_i$ and $LH_i$ respectively for the WW and $E_i'$, $HH_i'$ and $LH_i'$ respectively for the NW. Transition energies between various levels are denoted by $h\nu_{ij}$ where i and j are the level indices for the initial and final states respectively (both i and j can be primed quantities). Transitions energies involving light holes will be denoted by $h\nu_{ij}^L$.

Different quantum confinement energies for electrons and holes in each well lead to different interband transition energies for the two wells, giving a distinct spectral signature to each well. Typical photoluminescence (PL) and Photoluminescence Excitation (PLE) spectra for a 63/50/90 a-DQWS are shown in Fig. 2a and Fig. 2b respectively. There are two luminescence peaks, the lower energy peak (at photon energy $h\nu_{11}$) corresponding to the heavy-hole exciton transition in the wide well (WW) and the high energy peak (at energy $h\nu_{1'1'}$ ) corresponding to the heavy-hole exciton transition in the narrow well (NW). The double peak in the NW feature is probably a result of different well-widths in different regions of the sample. The PLE spectrum shows the heavy and light hole excitonic transitions in each well.

In an a-DQWS, each spectral feature corresponds to a specific well (WW or NW) so that the dynamics of carriers and excitons in each well can be independently investigated by studying the dynamics of different spectral features, using absorption, reflection, or luminescence spectroscopies. Time resolved luminescence provides a particularly simple and direct technique for such studies. For large values of b, the luminescence in each well decays with the usual recombination rate $1/\tau_{rec}$. With decreasing b, tunneling from one well (say the NW) to the other well (WW) becomes important and increases the decay rate for the NW luminescence. In this case the tunneling rate $1/\tau_t$ is the difference between the measured decay rate $1/\tau_d^{NW}$ and the usual recombination rate in the absence of tunneling:

$$1/\tau_t = 1/\tau_d^{NW} - 1/\tau_{rec}^{NW} \tag{1}$$

The recombination rates for the WW and the NW are approximately the same if the well thicknesses are not too different.[60] Since there is no tunneling out of the WW, the recombination rate in the WW is approximately the same as the decay rate of the WW. In this case, the tunneling rate is the difference in the decay rates of the two wells. If the well thicknesses are very different, then one must, of course, compare the decay rates of the NW in the absence and presence of tunneling.

Several points need to be stressed. First, we stress again that the term tunneling is used for both resonant (coherent) as well as non-resonant tunneling as discussed earlier. Second, when we talk of tunneling times, we mean the *inverse of the tunneling rate* as defined above and *not the traversal time* through a barrier which has been discussed with considerable controversy in the literature.[61-64]

Another point that should be discussed at the outset is that optical excitation creates both electrons and holes in a-DQWS and the optical transitions investigated in such studies (typically made at liquid He temperatures) correspond to excitonic transitions rather than free carrier transitions. One generally assumes that since the binding energy of the exciton is small compared to the heights of the barriers, the tunneling rates for the electron or the hole are not influenced significantly by the excitonic effects. However, this is an assumption that needs to be examined in more detail since recent calculations shows that excitonic effects are substantial for certain cases.[52, 65]

Finally, it has been shown recently[66, 67] that for non-resonant excitation (which creates free carriers rather than excitons), it may take several hundred picoseconds to form the $K \approx 0$ excitons, the only excitons that can directly couple to photons. This is true for high quality samples showing no Stokes shift between PL and PLE spectra,[67]

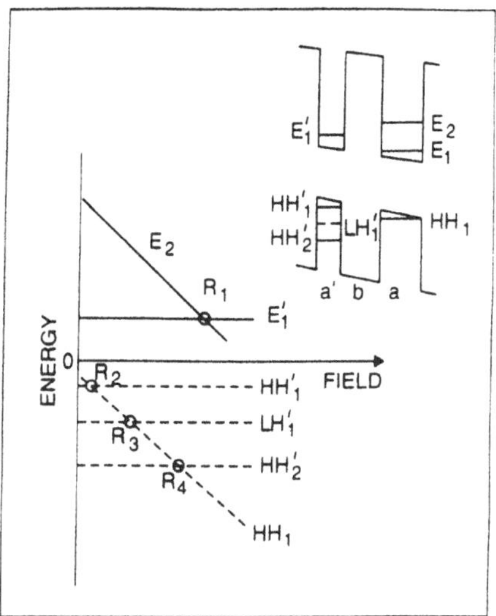


**Fig. 3** A schematic of the energy levels of an a-DQWS in an electric field. The electrostatic potential is taken to be zero at the center of the left well (the NW in this case). Various resonances are indicated by $R_i$.

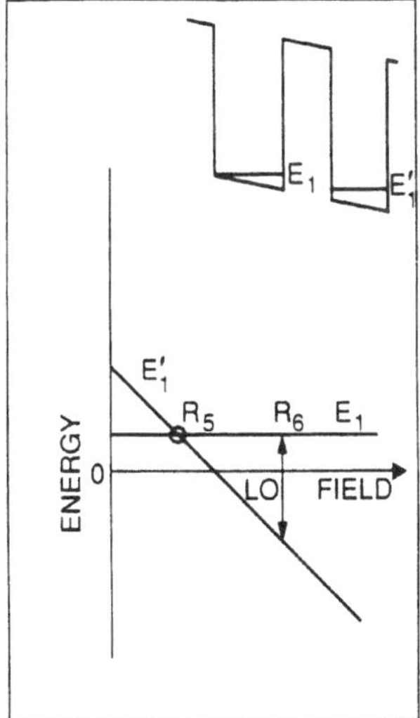


**Fig. 4** A schematic of the energy levels of an a-DQWS in an electric field. The electrostatic potential is taken to be zero at the center of the left well (the WW in this case). Various resonances are indicated by $R_i$.

and also if localization effects are important.[68-70] Since most optical studies investigate excitons, this implies that the increase in the luminescence or the time evolution of the bleaching signal in the destination well is a less reliable measure of the tunneling rate than the decrease in the luminescence or the bleaching signal in the initial well. One may try to circumvent this problem by choosing a higher temperature, [36] or by using doped wells.[37, 38] However, it may be best to rely only on the decay of the signal from the initial well for determining tunneling rates.

**II.B a-DQWS in Electric Field**

With a-DQWS under flat-band conditions, one can measure tunneling rates from the lowest electron or hole level in the NW to the corresponding level in the WW as a function of barrier width b, [30, 31] or compare samples with varying well widths to investigate resonance between some of the electronic levels.[37, 38] The real strength of the a-DQWS system, however, stems from the fact that by applying an appropriate electric field, one can vary the relative positions of electron and hole levels and thereby isolate and selectively investigate different non-resonant tunneling processes (e.g., electrons vs. hole tunneling or phonon vs. impurity assisted tunneling ). Furthermore, one can also investigate resonant tunneling, i.e., tunneling when two electronic levels are brought into resonance. Such electric fields may be applied by imbedding the a-DQWS in the intrinsic region of a p-i-n structure or by using semi-transparent Schottky contacts and using a reverse biased Schottky diode. In both cases, it is important to apply only such bias that there is no significant current flow. This restriction makes it clear that different structures have to be designed to investigate different phenomena, e. g. the WW may be on the left or the right, the well thicknesses and the barrier thickness may be different, etc.. By using appropriate sample structures and biases, one can selectively investigate various electron or hole resonant tunneling processes.

When discussing the effects of electric field, we will adopt the convention that the zero of the electrostatic potential is taken at the center of the left well. Application of reverse bias will therefore move the levels of the right well lower with respect to the levels of the left well. We will neglect the changes in the confinement energies with electric field (the Quantum Confined Stark Effect) for the qualitative discussion in this section. Using these conventions, Fig. 3 and Fig. 4 schematically illustrate how various electron and hole energy levels move with respect to each other for two generic structures, one in which the WW is on the left and the other in which the NW is on the left. Two different classes of processes can occur: when no electronic levels are in resonance, a tunneling transition requires a momentum-conserving process such as scattering with a phonon or a defect as discussed above. For this case (non-resonant tunneling), tunneling can be properly thought of as an inter-subband transition. The second case (resonant tunneling) is that of electronic resonance when tunneling can be a coherent quantum mechanical process without any need for the participation of a momentum conserving process. Various resonances are indicated as $R_i$ in Fig. 3 and Fig. 4 and the special case of non-resonant tunneling when optical phonon assisted process becomes allowed is also indicated as $R_6$ in Fig. 4. A detailed discussion of these processes follows.

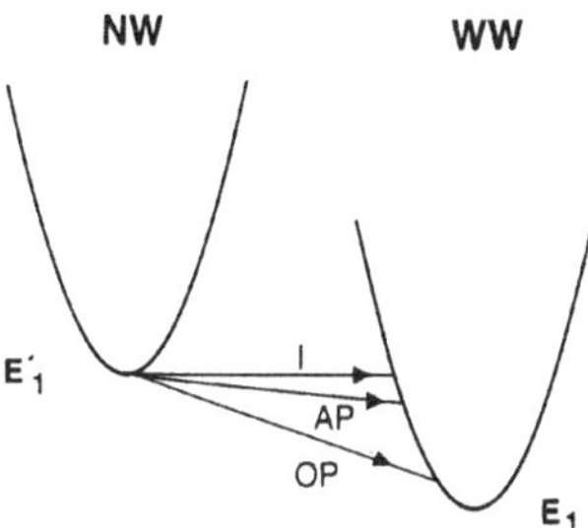


**Fig. 5** A schematic of the band diagrams for the two wells showing that an electron at the bottom of the two-dimensional band corresponding to $E_1'$ in the NW needs an energy and momentum conserving scattering process for tunneling to the $E_1$ level in the WW. I, AP and OP denote impurity-assisted, acoustic phonon-assisted and optical phonon-assisted processes respectively.

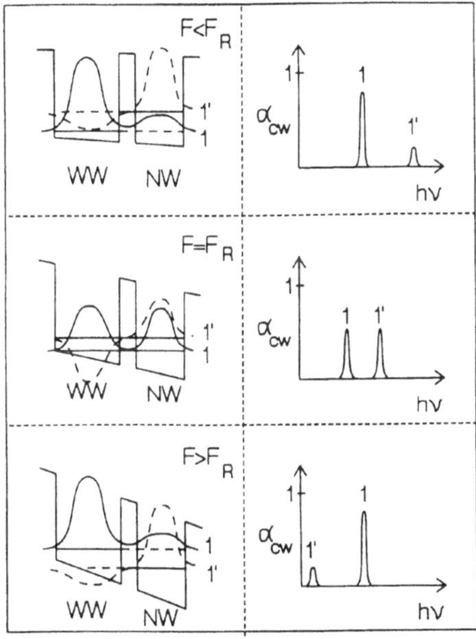


**Fig. 6** Left panel: a schematic of the conduction band potential diagrams, energy levels and wavefunctions for three different electric fields near the $R_5$ resonance; right panel: a schematic of the cw absorption spectrum associated with the $HH_1$ level near resonance $R_5$.

Various tunneling processes can be investigated by measuring the decay of luminescence intensity or some other optical property as a function of electric field. The changes in the decay rates for an isolated quantum well as a function of applied electric field have been investigated extensively.[71, 72] It was shown that for thin wells ( < 100Å ) and fields smaller than $10^5$ V/cm, the effect of field on decay rates is negligible. Even when there is an effect, it is to decrease the decay rate, whereas participation of a new tunneling process at resonance is expected to increase the decay rate. Therefore, the variation of the decay rate of a single quantum well with an electric field is not an important factor in most tunneling measurements.

### II.C Non-Resonant Tunneling

To be specific, we consider the case of electron tunneling, although similar considerations also apply to hole tunneling. Under flat-band conditions, the lowest electron level in the NW is above that in the WW (Fig. 5). Therefore, the $k_{//} = 0$ state of the lowest NW level is at the same energy as some $k_{//} \neq 0$ state of the lowest WW level, where $k_{//}$ is the wavevector parallel to the interface plane. Since the momentum parallel to the interface planes must be conserved,[73] a transition between the two states requires a momentum conserving scattering process, e.g. scattering by impurities or interface defects (elastic collisions), scattering by acoustic phonons (nearly elastic) or scattering by optical phonons (inelastic). This is schematically illustrated in Fig. 5.

For any of these non-resonant processes, the tunneling rates depend on the square of the overlap between the initial and the final wavefunctions. The wavefunction overlap is expected to decrease exponentially with increasing b. It is important to emphasize that, for a given b and given height of the barrier, the tunneling rate also depends on the *well widths* because the effective barrier heights and hence the penetrations of the wavefunctions change with the confinement energies. Another factor that enters into consideration is the fact that application of an electric field distorts the wavefunctions and affects the tunneling rate for a given barrier thickness.

For tunneling involving scattering of optical phonons, only the emission of phonons is important at low temperatures. Therefore the energy separation $\Delta E$ must exceed $\hbar\omega_{LO}$ for this process to be important. As shown schematically in Fig. 4, the energy separation $\Delta E$ can be varied through $\hbar\omega_{LO}$ by applying an electric field. This provides the clearest method for determining the role of optical phonons in tunneling in a-DQWS and comparing its importance with other assisted processes such as impurity and acoustic phonon scattering. It should be noted here that one of the important questions in this regard is the nature of optical phonon modes in such structures. Experiments at $R_6$ threshold can be potentially useful in providing information about these modes.

### II.D Resonant Tunneling: General Concepts in the Ideal Case

While the case of non-resonant tunneling can be analyzed in a straightforward manner, tunneling of carriers when electronic levels are brought into resonance brings many new quantum mechanical features. In this section, we discuss some general features of resonant tunneling for the ideal or near-ideal case of no or weak collision/relaxation. The information that one can obtain depends strongly on whether one considers cw or time resolved experiments. These two cases are examined in detail in the next two sections. Finally, the last section discusses the effects of strong collisions and relaxation on resonant tunneling.

We first consider resonance $R_5$ between the lowest electron levels $E_1$ and $E_1'$ (Fig. 4). It is well known that there is an anti-crossing of the levels at the resonance. Such anti-crossing behavior has been observed in a number of experiments.[65, 74-80] There are important changes in the wavefunctions and the absorption spectra close to resonance, as schematically illustrated in Fig. 6. The wavefunctions are localized in one well or the other far away from the resonance but become progressively more delocalized as one approaches resonance. This behavior of the two electron wavefunctions $\Psi_1$ and $\Psi_1'$ is schematically shown by solid and dashed curves respectively in the left panel of Fig. 6. The nature of the excitonic absorption spectrum also changes dramatically as one varies the field through this resonance. A schematic of the cw absorption spectrum for the exciton involving the first heavy hole level ($HH_1$) in the WW and electron levels $E_1$ and $E_1'$ is also shown in Fig. 6. Away from resonance, $\Psi_1'$ is localized primarily in the NW with little overlap with the wavefunction of $HH_1$ in the WW. Therefore the cw absorption spectrum is dominated by the spatially direct absorption in the WW, with a weak peak at higher energy due to spatially indirect transition. At resonance, the two electron wavefunctions are completely delocalized with the result that the cw absorption spectrum from $HH_1$ in the WW has two nearly equal peaks, with peak heights smaller than the that of the strong peak away from resonance. Finally, for fields much greater than resonance, the wavefunctions are once again localized in their respective wells and the absorption spectrum is dominated by a strong, spatially direct transition at the same energy as before. A small peak due to spatially indirect absorption is expected at lower energies as shown in Fig. 6.

We have discussed the behavior for the lowest electron state in each well but it should be clear that qualitatively similar behavior is expected for resonance between any two electron levels, two heavy hole levels and two light hole levels. The case of resonance between a heavy and a light hole level is more complicated. Hybridization between the heavy and light hole wavefunctions occurs only for $k_{//} \neq 0$; therefore, heavy and light hole levels can cross each other at $k_{//} = 0$ , but there is an anti-crossing leading to a minimum level separation at $F_R$ for $k_{//} \neq 0$ . The degree of hybridization determines the extent of anti-crossing; however, in the ideal case *the delocalization of the wavefunction is complete* at resonance, independent of the minimum level separation, so long as there is some hybridization.

While these considerations are quite clear and well understood, there seems to be considerable confusion in the literature concerning how they affect various experiments. We emphasize that considerations for cw or long pulse experiments where the exciting laser has a spectral width smaller than the separation $\Delta E$ between the levels under consideration are very different from the considerations for the case when excitation is with a short pulse laser whose spectral width is larger than $\Delta E$ . We consider these two cases in detail in the next two sections, where the discussion is limited to the near-ideal case of collision and relaxation rates smaller than the resonant tunneling rates. The

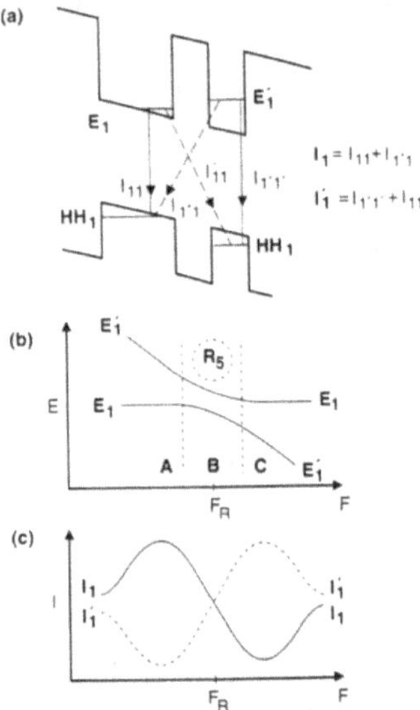


**Fig. 7** A closer look at resonance $R_5$: (a) a schematic of the band potentials, energy levels, transition energies and intensities, (b) a schematic of anti-crossing of levels $E_1$ and $E_1'$ at resonance, and (c) a schematic of the cw luminescence intensities as a function of electric field.

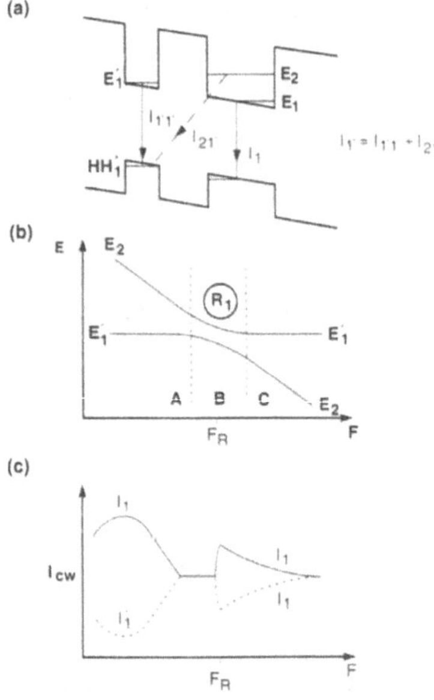


**Fig. 8** A closer look at resonance $R_1$: (a) a schematic of the band potentials, energy levels, transition energies and intensities, (b) a schematic of anti-crossing of levels $E_1'$ and $E_2$ at resonance, and (c) a schematic of the cw luminescence intensities as a function of electric field.

case of strong collisions/relaxation brings into picture many novel features which are discussed in Sec.II.G.

**II.E Effects of Resonance on cw Experiments**

The quantum mechanical tunneling interaction between the two wells leads to a splitting of levels and a complete delocalilzation of wavefunctions at resonance. Although the delocalization is incomplete away from resonance, there may be still be a substantial probability of the wavefunction in both wells. For excitation with a near-monochromatic light source (e. g. a cw or a long-pulse laser), one always excites the eigenstates of the system. Therefore, even if the excitation energy is such that the hole is excited only in the WW, the excited electron wavefunction may still have a large probability in the NW. In this case, the excitation process *directly creates* electrons in both wells and no additional transitions are required to populate the NW with electrons.

To illustrate what is expected for cw experiments near electronic resonances, we discuss the behavior of cw luminescence for the case of electron resonances $R_1$ (Fig. 3) and resonance $R_5$ (Fig. 4). Similar considerations apply near hole resonances and also for other optical experiments. In order to keep the discussion simple, we assume that there is no other resonance near this field, i.e. only two electron levels are near resonance, and that hole tunneling is not important. Even with these restrictions, it is quite clear that the behavior of the cw luminescence intensity as the field tunes the two electron levels through a resonance is determined by such factors as non-resonant tunneling time and radiative and non-radiative decay times in each well. A rate equation analysis is necessary for a quantitative understanding and such an analysis has indeed been reported.[47] However, some qualitative conclusions can be drawn from simple arguments. These conclusions are quite illuminating and we discuss them below for the two resonance cases mentioned above.

Consider first what happens to the cw luminescence near resonance $R_5$ between the two lowest electron levels as shown in Fig. 4. This is illustrated in more detail in Fig. 7, with (a) showing the a-DQWS and the nomenclature, and (b) showing the anti-crossing behavior. Consider an excitation energy such that only the lowest subbands of both wells are excited. At any field, the luminescence spectrum at low temperatures will in general show four spectral features at energies and intensities indicated in Fig. 7. These correspond to spatially direct and indirect transitions to the highest hole levels in the two wells. The total intensity of the spatially direct ($I_{11}$) and indirect ($I_{1'1}$) transitions associated with $HH_1$ level in the WW is denoted by $I_1$ , and the total intensity of the spatially direct ($I_{1'1'}$) and indirect ($I_{11'}$) transitions associated with $HH_1'$ level of the NW is denoted by $I_1'$ .

We consider three different field regions: in region A (C) the field is so low (high) that the energy level separation is large, there is no significant delocalization of the wavefunctions, and only spatially direct transitions are significant. Region B is the resonance region in which there is significant delocalization of the wavefunctions. We limit the field to values such that the energy level separation is less than the optical phonon energy so that impurity or defect-assisted or acoustic phonon-assisted transitions dominate. For such transitions, Ferreira and Bastard [51] have shown that the non-resonant tunneling rate is maximum when the level separation is minimum and decreases as one moves away from resonance. In the following, we assume that the non-resonant tunneling rate is larger than the recombination rate for small energy level separations, but becomes smaller than the recombination rate for large separations.

Consider first what happens at exact resonance ( $F = F_R$ ). At this field, the wavefunctions are completely delocalized so that the two intensities $I_1$ and $I_1'$ are nearly equal, *regardless of thermalization of population between the two electron levels*. In other words, even if all the electrons are in the lower state, the intensities $I_1$ and $I_1'$ are nearly equal because the wavefunction is delocalized. As one moves away from the resonant field, the relative population of the two levels, which is determined by non-resonant tunneling between the levels, also becomes a factor in determining the two luminescence intensities. Therefore, as one moves away from the resonance, the intensity corresponding to the direct transition of the lower electron level increases and that for the direct transition of the higher electron level decreases. For further increases in level separations (region A and C), non-resonant tunneling rate approaches and then becomes smaller than the recombination rate with the result that the intensity of the lower (higher) electron level reaches a maximum (minimum). Finally, for large separations when there is no delocalization and negligible non-resonant tunneling, the two intensities approach each other. The resulting qualitative behavior is illustrated in Fig. 7c.

The ratio $I_1'/I_1$ has a maximum and a minimum as a function of field, in qualitative accord with rate equation analysis.[47, 52] However, in contrast to these analyses, our qualitative arguments show that the maximum does *not* occur at the resonant field. In fact, the ratio is *nearly unity* at the resonant field, independent of any non-resonant tunneling rates and thermalization processes because *it is determined primarily by the delocalization of the wavefunctions*. Therefore, such experiments provide no information about resonant tunneling rates and can not determine the precise positions of resonances. Also, it is difficult to obtain quantitative information about non-resonant tunneling from the variation of intensities with field because several different factors enter into determining such variations.

The behavior of cw luminescence near resonance $R_1$ (Fig. 3), corresponding to a resonance between the lowest electron level in the NW with the second electron level in the WW, is schematically illustrated in Fig. 8. In the following discussion, we assume that $E_2 - E_1$ is much larger than the optical phonon energy so that the relaxation rate of electrons from $E_2$ to $E_1$ in an isolated WW is much faster than the normal recombination rate. Assume further that $E_1' - E_1$ under flat band conditions is also larger than the optical phonon energy so that, at low fields, non-resonant tunneling is dominated by optical phonon-assisted scattering processes. This non-resonant tunneling rate decreases as the separation between $E_1'$ and $E_1$ increases with increasing electric field in field region A where $F < F_R$ (see the discussion in Sec. IV-B). Therefore, $I_1$ dominates at low fields but $I_1$ and $I_1'$ approach each other as the field increases in region A. This is illustrated schematically in Fig. 8(c).

As the field approaches $F_R$ (region B), the wavefunction of the lowest electron level in the NW begins to have a substantial amplitude in the WW as a result of resonance between $E_1'$ and $E_2$. This increases the overlap between $\Psi_1'$ and $\Psi_1$, the wavefunction of the lowest electron level in the NW and the WW respectively, and provides an additional channel of decay for electrons in the NW. Since this decay rate is largest at the resonance, and the region of resonant coupling is rather small, $I_1$ ( $I_1'$) rapidly reaches a maximum (minimum) at the resonant field.

With increasing field beyond resonance (region C), the wavefunctions localize just as rapidly, but the intensities may change slowly because impurity, defect and acoustic phonon-assisted non-resonant tunneling from $E_1'$ to $E_2$ keeps the intensity of the NW luminescence low. With further increase in the field, this non-resonant tunneling rate decreases and the two intensities once again approach each other. This behavior is illustrated in Fig. 8c. Note that the intensity ratio $I_1'/I_1$ shows a minimum at the resonant field.

It is important to note that although extrema in intensities are expected at $F_R$ , these intensities are completely controlled by the amplitudes of $\Psi_1'$ in the WW, i.e. *by delocalization of the wavefunction* and the inter-subband transition rate in the WW. In particular, the intensity at resonance is *independent of the barrier thickness b*, so long as the picture of complete delocalization of wavefunction is valid. However, the intensities away from $F_R$ do depend on b because the *non-resonant tunneling* (i.e. the tunneling away from $F_R$ ) depends exponentially on b. Therefore, such cw measurements provide no information about tunneling rates at resonance. Note also that the ideal picture of weak or no collisions/relaxation breaks down for large values of b and one has to bring in the considerations of Sec. II.G below.

These arguments lead to the conclusion that cw luminescence provides no information about tunneling rates at resonance. Similar arguments can be made for other cw spectroscopies. This conclusion is a direct result of the fact that one creates and probes the stationary eigenstates of the system in cw experiments.

**II.F Time Resolved Experiments Near Resonance**

Considerations for ultrashort pulse experiments are very different because such an excitation creates a *linear superposition* of the eigenstates within the spectrum of the laser rather than an eigenstate of the system as in cw experiments.

Consider the case of resonance $R_5$ (Fig. 4) between the two lowest electron eigenstates $E_1$ and $E_1'$ . When this system is excited at the $h\nu_{11}$ resonance by an ultrashort laser pulse whose spectral width is larger than $\Delta E$ , the separation between the eigenstates, a *linear superposition of these eigenstates* is created. Since holes are excited only in the WW under these conditions, the linear superposition is such as to create an electron wavepacket in the WW. In contrast to the cw case, *time dependent* Schrödinger equation must be invoked to understand the behavior of this wavepacket. The time evolution of each of the eigenfunctions constituting the wavepacket is slightly different because the eigenstates have slightly different energy[81] As a consequence, the amplitude of the wavepacket in each well oscillates from a maximum to a minimum as a function of time. For the WW, the maximum amplitude of the wavepacket as a function of time is unity and the minimum depends on the field, approaching zero at resonance. For the NW, the minimum amplitude of the wavepacket is always zero and the maximum depends on the field, approaching unity at resonance. The period of oscillations is inversely proportional to $\Delta E$ and is therefore largest at resonance and decreases away from resonance whereas the amplitude of oscillations is maximum at resonance and decreases away from resonance. This behavior is illustrated in Fig. 9.

This may be mathematically expressed as follows. The wavefunction of this electronic wavepacket $\Psi(z,t)$ is given by:[81]

$$\Psi(z,t) = \Psi_1(z,t)e^{-iE_1 t/\hbar} + \Psi_1'(z,t)e^{-iE_1' t/\hbar} \tag{2}$$

and the probability density $P(z,t) = |\Psi(z,t)|^2$ is given by

$$P(z,t) = |\Psi_1(z)|^2 + |\Psi_1(z)|^2 + 2\mathrm{Re}\Psi_1{}^*\Psi_1' e^{-i(E_1-E_1')t/\hbar} \tag{3}$$

Therefore, the probability density oscillates in time between two extreme values $(|\Psi_1| - |\Psi_1'|)^2$ and $(|\Psi_1| + |\Psi_1'|)^2$ with a *period* $\tau_{coh}$ given by

$$\tau_{coh} = h/\Delta E \tag{4}$$

where $\Delta E$ is the energy separation $|E_1 - E_1'|$.

In an ideal case where there are no collisions and or relaxations, these *coherent oscillations* of the electronic wavepacket will go on forever. In a real system, however, collisions and or relaxations damp these oscillations. In the case where collision and relaxation rates are much smaller than the coherent oscillation frequency ($\nu_{coh} = 1/\tau_{coh}$), (the near-ideal case we have been discussing) the frequency of coherent oscillations is not significantly affected by the collisions/relaxations. However, the amplitude of the oscillations is damped with increasing time after the creation of the wavepacket. The qualitative behavior of $P_{WW}(t)$ and $P_{NW}(t)$, the probability amplitudes in the WW and the NW respectively, is schematically illustrated on the left side of Fig. 9 for three electric fields. The qualitative behavior of $|\Psi_1(t)|^2$ and $|\Psi_1'(t)|^2$, the probability densities of being in the eigenstates with energies $E_1$ and $E_1'$ respectively, are schematically illustrated on the right side of Fig. 9 for the three representative electric fields. The damping of the oscillations is determined by the rate at which the coherence of the polarization at the frequency corresponding to $\Delta E$ is destroyed. This dephasing can be brought about by any collision that randomly changes the phase of the wavefunctions. Note that such collisions do not have to transfer particles between the two states although such relaxation processes between the eigenstates clearly put an upper limit on how long the dephasing time can be. On the other hand, the probability of being in the higher eigenstate decays with the rate of relaxation between the two levels and is not influenced by the dephasing rate. The relaxation rate, in general, is expected to be smaller than the dephasing rate of the polarization.

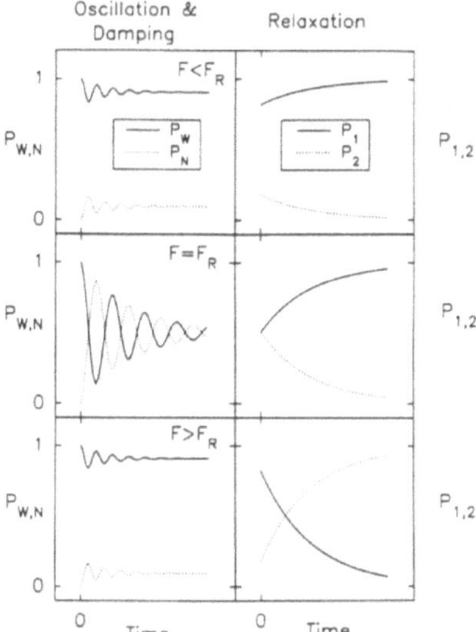


**Fig. 9** Effect of coherent oscillations on the probability of being in the WW or the NW is shown schematically in the left panel for three different electric fields. The amplitude as well as the period of the oscillations change depending on the level separations at different fields. The damping of the oscillations is due to dephasing. The right panel illustrates how the occupation probabilities of the two eigenstates varies with time for three different fields. The time constant of the decay in the right panel corresponds to the relaxation from the upper level to the lower level. The dephasing times responsible for the damping of the coherent oscillations (left panel) can be much shorter than the energy relaxation times ponsible for the decay in the right panel.

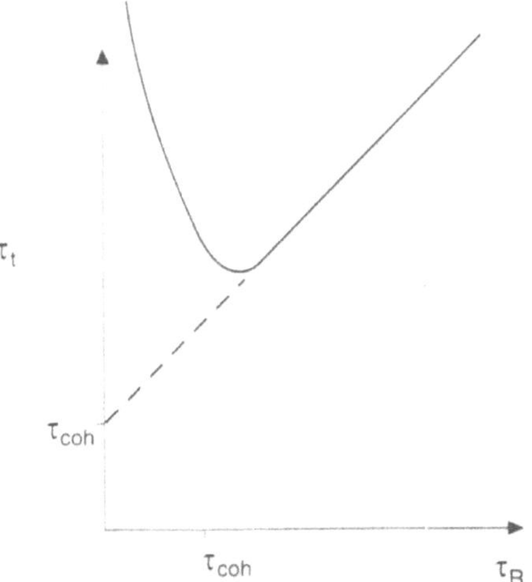


**Fig. 10** Resonant tunneling times in the sequential picture (dashed line) and the unified picture of tunneling (solid line) at resonance. The sequential picture predicts a monotonic decrease of the tunneling time as the relaxation time decreases. The unified picture of coherent tunneling including the effects of collisions and relaxation predicts a non-monotonic behavior of the tunneling time (see the text).

The coherent oscillations of the wavepacket is expected to give rise to oscillatory behavior in the optical properties of the WW as well as the NW. Such discussion in terms of occupation probabilities has been given.[49, 50] It should be noted, however, that considerations of the coherent aspects of the phenomena are essential in obtaining a proper understanding of the optical properties at ultrashort times.[59]

It is clear from this discussion that the behavior in time resolved experiments is quite different from that in cw experiments. In particular, time resolved luminescence measurement at the WW luminescence energy for the case discussed above is expected to reveal an oscillatory behavior with a period $\tau_{coh}$. Similar oscillations are expected for other time resolved measurements such as pump-and-probe and transient four-wave-mixing experiments. Therefore, in contrast to the cw case, such a measurement will *directly determine the coherent tunneling time*. It should be pointed out that although the variation of cw luminescence intensity with electric field does not provide any information about the tunneling rate, the spectroscopic information about the splitting $\Delta E$ can be used to deduce $\tau_{coh}$. However, such information may be difficult to extract because of strong effects of inhomogeneous broadening. Also, information about the *dynamics* of various relaxation processes contained in time resolved experiments is not available in cw experiments.

We now turn our attention to resonance $R_1$(Fig. 3) between $E_1'$ and $E_2$ . We consider in this section the case when collisions/relaxations are slow compared to the coherent tunneling rate. The case of strong collisions and relaxations is considered in the next section. We only consider the case of exact resonance. One new feature that enters into the picture in this case is that the wavepacket has a substantial overlap with $\Psi_1$ when it is in the WW but not while it is in the NW. The other new feature is the possibility of relaxation to a state that is lower than the states involved in resonance.

The behavior of time resolved luminescence of the NW depends on the relative importance of various dephasing and relaxation times. For dephasing time shorter than the inter-subband relaxation time at resonance, the oscillatory luminescence from the NW will be damped with the dephasing time constant and will settle to half its initial value in a short time. This luminescence intensity then decays to zero with a time constant approximately equal to twice the time constant of $E_2$ to $E_1$ inter-subband relaxation away from resonance; i.e. when the wavefunction $\Psi_2$ is localized in the WW. We have assumed in this qualitative discussion that $\tau_{21}$ and $\tau_{1'1}$ are the same at resonance. If the inter-subband relaxation time is short compared to other dephasing times, then the time constant $2\tau_{21}$ determines both the damping of the oscillatory luminescence and the decay of the luminescence intensity. In either case, the damping of oscillations and time constants for luminescence decay are determined by inter-subband relaxation rates and other dephasing rates and *not* by coherent tunneling rates. The long term decay of luminescence is determined only by the inter-subband relaxation rate. Only the period of oscillatory luminescence is determined by the coherent tunneling rate and depends strongly on the barrier thickness. We emphasize once again that these conclusions hold only in the case of *weak* collisions and relaxations, where collision/relaxation rates are smaller than the coherent oscillation frequency.

It is instructive to obtain the dependence of $\tau_{coh}$ on barrier thickness b. The best means of calculating this involves making a numerical calculation of the energy splitting $\Delta E$ (e. g. by using the transfer matrix method) and using Eq. 4 to determine $\tau_{coh}$. However, a useful estimate for $\tau_{coh}$ may be obtained using the following expression:[82]

$$\tau_{coh} = \frac{\lambda(m_e/m_0)\exp(2\pi b/\lambda)}{\hbar} \quad (5)$$

$$\tau_{coh}(s) = 8.6\times10^{-17}\lambda(\text{Å})b(\text{Å})(m_e/m_0)\exp(2\pi b/\lambda) \quad (6)$$

where

$$\lambda = \frac{h}{[2m_B(V_0-E)]^{1/2}} \quad (7)$$

$$\lambda(\text{Å}) = \frac{12.3}{[(m_B/m_0)(V_0(eV)-E_i(eV))]^{1/2}} \quad (8)$$

where $m_e$ and $m_B$ are the electron masses in the well and the barrier respectively, $V_0$ is the barrier height and E is the confinement energy of the electron.

It is also instructive to note that the escape time of an electron from a quantum well surrounded by barriers of thickness b on each side (i. e. a double barrier structure) varies as $\exp(4\pi b/\lambda)$. The square of the overlap between the WW and the NW wavefunctions in an a-DQWS away from resonance is also expected to show approximately the same dependence on barrier thickness. Comparing this with Eq. (5) leads to the conclusion that the dependence of $\tau_{coh}$ on b is weaker than the two cases discussed above. Numerical calculations show that $\tau_{coh}$ is also considerably smaller than the escape time as well as non-resonant tunneling times in a-DQWS.[82]

We have presented the discussion as if there are only two discrete states. In reality, there is a quasi-continuous distribution of states as one moves away from $k_{//}$ = 0. Excitons also have a dispersion in the direction parallel to the interface. This complicates the discussion because the laser spectrum will encompass several $k_{//}$ states in a given subband. One may ignore these complications in the first consideration of the problem because, for every $k_{//}$, the two subbands contain a pair of states separated by $\Delta E$. However, detailed discussion of these considerations will become necessary as our understanding of the problem increases.

### II.G Resonant Tunneling in Presence of Strong Collisions

The concept of coherent oscillations of an electronic wavepacket discussed in the preceding sections breaks down when collision/relaxation rates become comparable to or larger than the frequency of coherent oscillations in the absence of collision/relaxation. Under this condition, there are no coherent oscillations because strong collisions/relaxation destroys the coupling between the two states. The two states become independent, non-interacting states which can cross each other, and the wavefunction of each state is localized in either the WW or the NW; i.e. there is no delocalization.

In order to be specific, we consider the case of resonance $R_1$ in Fig. 3 because it specifically corresponds to some experiments. Although there are no coherent oscillations, an electron created in $E_1'$ has a finite probability to resonantly tunnel to the WW and end up in $E_1$. If this probability is sufficiently large, one can measure it by, for example, comparing the decay rates of the NW luminescence at resonance and away from resonance. In contrast to the non-resonant case, where this transfer rate is simply given by the inter-subband transition rate between $E_1'$ and $E_1$ , the situation in the present case is considerably more complicated.

The general case of collision/relaxation rates comparable to or larger than the ideal coherent oscillation frequency is best treated by a density matrix formalism. The more general problem of resonant tunneling in presence of dissipation has been treated by Leggett and coworkers and has been reviewed by Leggett et al. [55] However, it was argued [44] that the theory is particularly simple in the absence of scattering processes (collisions). Then the wave vector parallel to the quantum wells $k_{//}$ is a good constant, and transitions occur only between states of the same $k_{//}$ . For this case one needs to consider only the effect of relaxation between $E_2$ and $E_1$ and it can be shown that the equations of motion for the state amplitudes are the same as those for coupled damped harmonic oscillators. In this case the effect of relaxation is to broaden the resonance by the relaxation rate ($1/\tau_R$ ) between $E_2$ and $E_1$. If the relaxation rate is much larger than the coherent oscillation frequency ($1/\tau_R >> 2\pi/\tau_{coh}$), it can be shown that the tunneling rate at resonance ($1/\tau_t$) is given by

$$1/\tau_t \cong (1/\tau_{coh})(4\pi^2\tau_R/\tau_{coh}) \tag{9}$$

$$\tau_t \cong \tau_{coh}(\tau_{coh}/4\pi^2\tau_R) \tag{10}$$

In the limit of strong relaxation, the tunneling rate at resonance *decreases* as the inter-subband relaxation rate *increases*, because the rapid damping prevents the buildup of state amplitude in the WW. Therefore, one expects to see a non-monotonic behavior of the effective tunneling rate as a function of the relaxation rate. This result, which is *counter-intuitive* in a sequential picture of the coherent tunneling and relaxation processes, is a consequence of the loss of coherence due to relaxation. This non-monotonic behavior is schematically illustrated in Fig. 10 where the tunneling time goes through a minimum when the coherent tunneling time and the relaxation times are approximately equal. For relaxation times much longer than the coherent tunneling time, the tunneling time will be twice the relaxation time as discussed above. For relaxation time much shorter than the coherent tunneling time, the tunneling time will be larger than the coherent tunneling time by a factor proportional to the ratio of the relaxation time to the coherent tunneling time, as in Eq. 10.

In a real physical system, there are additional scattering processes (e.g., carrier-carrier-scattering) destroying the phase coherence between the eigenstates of energies $E_1'$ and $E_2$ . A full density matrix formalism is required in this case. It can be shown that the width of tunneling resonance in energy increases from $\hbar/\tau_R$ to $(\hbar/\tau_R + (2\times\hbar)/\tau_2)$, where $1/\tau_2$ is the dephasing rate, or the additional decay rate of the phase coherence due to scattering processes (collisions). It is noteworthy that in either

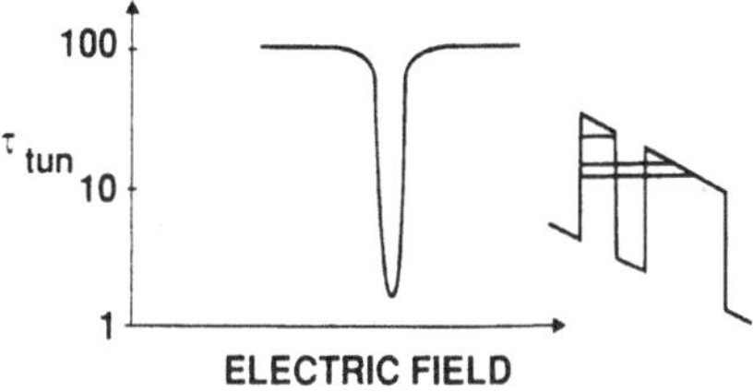


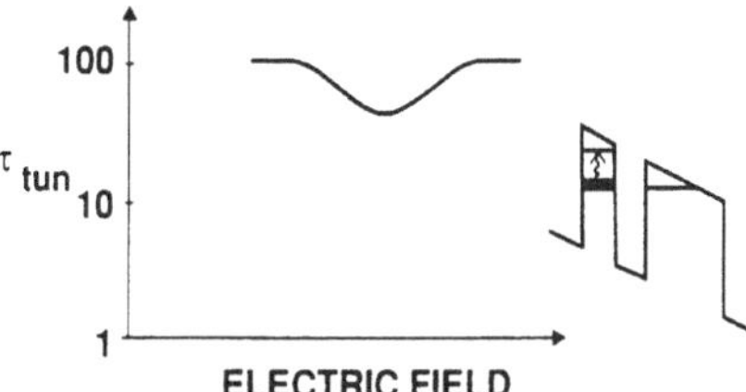


**Fig. 11** A schematic diagram showing that the tunneling times are expected to exhibit a strong and narrow resonance for the near-ideal case of weak or no collisions (top figure), and a weak and broad resonance for the case of strong collisions/relaxation (bottom figure). The area of the resonance is expected to be conserved (see the text).

| | | | | |
|---|---|---|---|---|
| $p^+$ | GaAs | | 2000 Å | |
| $p^+$ | AlGaAs | | 600 Å | |
| i | AlGaAs | | 600 Å | |
| i | GaAs | | $a_1$ | ×8 |
| i | AlGaAs | | $b_1$ | ×8 |
| i | GaAs | | $a_2$ | ×8 |
| i | AlGaAs | | 150 Å | ×8 |
| i | AlGaAs | | 2000 Å | |
| $n^+$ | AlGaAs | | 5000 Å | |
| $n^+$ | GaAs | BUFFER | 2000 Å | |
| $n^+$ | GaAs | SUBSTRATE | | |

**Fig. 12** A typical sample structure in which multiple periods of the a-DQWS are imbedded in the i region of a $p^+$ - i - $n^+$ diode.

case, the integral of the tunneling decay rate with respect to the energy detuning is independent of the width ($\hbar/\tau_R$ or $\hbar(1/\tau_R + 2/\tau_2)$) of the resonance and is given by

$$\text{Area} \cong h\pi^2(1/\tau_{coh}^2) \tag{11}$$

$$\text{Area (meV/ps)} \cong 40.8/(\tau_{coh}\text{ (ps)})^2 \tag{12}$$

The broadening of resonance can also be simply understood in terms of lifetime broadening of the level $E_2$ in the case of relaxation without any other collisions. Such a lifetime broadening reduces the available density of final states at any given detuning between the levels and hence reduces the tunneling rate. However, this does not change the area under the resonance curve. This is schematically illustrated in Fig. 11 for the near ideal case and the case of strong collisions/relaxation. We note that the effect of collisions on the transmission through a double barrier structure has been considered by Stone and Lee.[54] They reached similar conclusions concerning broadening of the transmission resonance but did not discuss the effect on tunneling rates.

One further point of interest concerns the effects of homogeneous versus inhomogeneous broadening. It should be obvious that the considerations of broadening of resonance discussed above applies only when the intrinsic homogeneous broadening due collision/relaxation is larger than any inhomogeneous broadening. If the sample quality is poor, the inhomogeneous broadening may dominate and broaden the resonance even if collision/relaxation effects are small. In a real system, both broadening mechanisms may contribute and may be difficult to separate.

To summarize this section, these considerations show that the sequential picture of tunneling followed by relaxation is not valid in the case of strong relaxation and collisions and that one must use a unified picture. Furthermore, Eq. 9 shows that in this limit the tunneling rate is proportional to $(1/\tau_{coh})^2$ which from Eq. 5 is proportional to $\exp(-4\pi b/\lambda)$. This dependence on the barrier thickness is much stronger than for the case of ideal coherent tunneling and is the same as expected for the escape time of a carrier from a barrier of thickness b or the non-resonant tunneling time ( see the discussion following Eq. 7 ).

## III. EXPERIMENTAL TECHNIQUES

We discuss in this section experimental details such as sample structures, lasers and techniques for measuring tunneling rates using optical spectroscopy.

### III.A Samples

Two types of samples structures have been generally used in these experiments: either a $p^+$-i-$n^+$ structure, or i-$n^+$ structure. Electric fields were applied by reverse biasing the diode in the first case and by depositing and reverse biasing a Schottky contact in the second case. Multiple (usually 8 or 10 ) periods of asymmetric double quantum wells were grown in the i-region of the structure in each case. Each period was isolated from the next by a thick (usually 250 Å ) barrier. This was sufficient to prevent tunneling between the neighboring units in nearly all cases. A typical $p^+$-i-$n^+$ structure is shown in Fig. 12. With direct bandgap barriers in the GaAs/AlGaAs

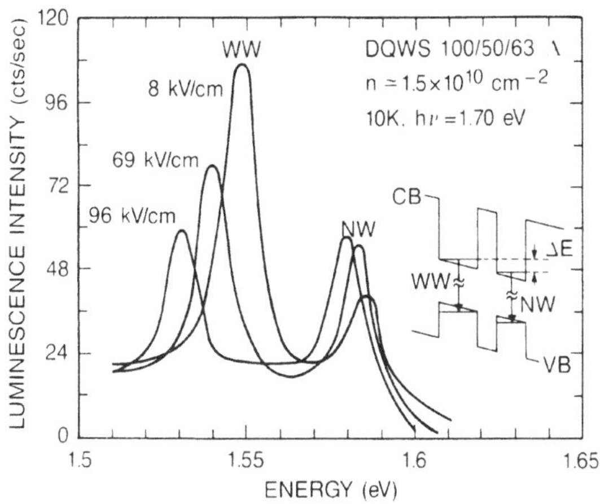


**Fig. 13** Typical time-integrated luminescence spectra at various fields showing shifts in the luminescence peaks as a results of the Quantum Confined Stark Effect (QCSE).

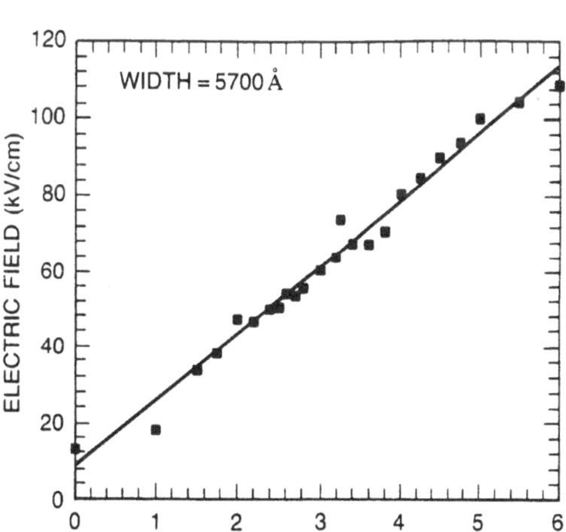


**Fig. 14** Electric field in the sample (deduced from comparing the spectral positions of luminescence peaks with those calculated from QCSE) as a function of the applied voltage.

system, the height of the barriers is limited to about 300 meV in the conduction band and about 150 meV in the valence band. The effective barrier height decreases with applied electric field and this may lead to a significant escape of carriers out of the DQWS at high fields. Sample structures are generally
designed to minimize this effect.

The quality of the structures is generally assessed by a number of different techniques, including PL (photoluminescence), PLE (photoluminescence excitation) and PC (photocurrent) spectroscopy as well as X-ray diffraction and TEM studies. [83] These characterization studies allows selection of high quality microstructures and a determination of their parameters.

In an ideal p-i-n or a Schottky diode, the i region is insulating and the applied electric field under reverse bias conditions is constant. In a real structure, unintentional doping in the i-region leads to non-uniform electric field in the sample. C-V measurements allows one to determine the background impurity level in the i region of the structures, and select samples expected to give good (better than 10 %) field uniformity in the i region.

The internal field is best determined by using a direct technique. It is well known that the confinement energies of electrons and holes in a quantum well decreases with increasing electric field perpendicular to the interfaces [84]. This quantum confined Stark effect (QCSE) is well understood and serves as a calibration of the internal electric field in the i region of the sample. The procedure is to measure the interband transition energy using any of the standard optical techniques (PL, PC, PLE, absorption) and deduce the internal electric field by comparing the measured variation of this energy with the electric field with the calculated variation. Fig. 13 shows typical PL spectra of a 100/50/63 a-DQWS at three different electric fields. The WW luminescence shifts more than the NW luminescence because the QCSE is larger for wider wells. By comparing the WW peak position with the calculated position, one can obtain an electric field for each bias. Fig. 14 shows that the electric field varies linearly with bias voltage in this case, as expected for a good quality sample. The electric fields determined by this procedure are quite accurate for fields greater than about 20 kV/cm.

In addition to measuring the fields by the shift of optical transition energies, the *broadening* of the spectral features gives a good indication of the uniformity of the field. These measurements show that the field uniformity for many of the samples for which results are discussed below is better than 5 % across the multiple periods of the structure. So long as the voltage drop between the two wells in the first period and that in the last period differ by less than the inhomogeneous broadening of the spectral feature under investigation, the use of multiple period structures is advantageous because it enhances the signal.

Most of QCSE measurements are performed either in cw spectroscopy or time integrated spectroscopy. Clearly, the time dependence of the electric field can not be determined from such measurements. Note, however, that a significant variation of the field with time will be reflected in a broadening of the spectral features. Also, time resolved optical spectroscopy can be performed if it is essential to determine the temporal variation of the field.

For the luminescence experiments discussed below, the p-i-n diodes were defined on the samples by mesas. First, an ohmic contact was formed on the backside by electrolytic deposition on AuSn, followed by a ramp annealing at 450°C. Then, $200\mu m \times 200\mu m$ mesas were defined on the front side by standard photolithographic processes and a wet etching step. The mesas are contacted by $100\mu m \times 50\,\mu m$ Au contact pads.

For the FWM and pump-probe experiments, the substrate has to be removed to allow experiments in transmission configuration. All these measurements were performed with i-$n^+$ samples with Schottky contacts. In a first step, an ohmic contact similar to the pin-diodes was applied to the backside. Then, a semi-transparent Schottky contact (10 Å Cr, 60 Å Au) with a diameter of about $500\mu m$ was evaporated on the front surface. The samples were then glued with epoxy on a sapphire disk, and the substrate was removed by standard selective wet etching.

The structures so prepared were mounted on the copper cold finger of an optical cryostat using a sapphire disk if necessary. The lowest sample temperature in this arrangement was $\approx$10K.

**III.B Lasers**

All measurements discussed below were performed with synchronously pumped dye lasers in two different configurations. In the first configuration, an IR dye laser was pumped with the compressed, doubled output of a modelocked Nd:YAG laser, whereas in the second configuration, an IR dye laser was pumped with the output of a Rh6G laser pumped with the doubled output of a modelocked Nd:YLF laser. With the IR dye Styryl 8 or IR 751, the IR laser operated with pulsewidths between 300 fs and 500 fs, was tunable from 710 to 810 nm and had an average power of 150 mW.

**III.C Luminescence Spectroscopy**

Most of the data were obtained with the luminescence upconversion technique which has been discussed in detail earlier. [85] In this technique, the laser beam is divided into two; one of the beams excites the sample in the cryostat and the luminescence excited by this beam is collected and focussed on a nonlinear crystal. The second beam is suitably delayed and is also focussed on the nonlinear crystal. Sum frequency photons are generated by angle-tuned, phase-matched process only during the time that the delayed laser pulse is incident on the crystal. Therefore, the upconversion technique acts as an ultrafast gate with time resolution determined by the laser pulsewidth and the group velocity dispersion in the crystal. Time resolution of about 60 fs has been demonstrated. [86]

These measurements were performed at 10 K unless otherwise stated. The laser excitation energy was such that only the wells were excited ($h\nu_{laser} < E_g$(barrier)). In some cases the laser energy was selected to excite only the WW. Note that only insignificant amount of light was absorbed in the layers above the i-region and that the carriers created in the GaAs buffer layer and the substrate were prevented from entering the i-region by the wide AlGaAs layers between the DQWS and the buffer layer. The excitation density was in the range of 1 - $2\times10^{10}$ cm$^{-2}$ unless stated otherwise.

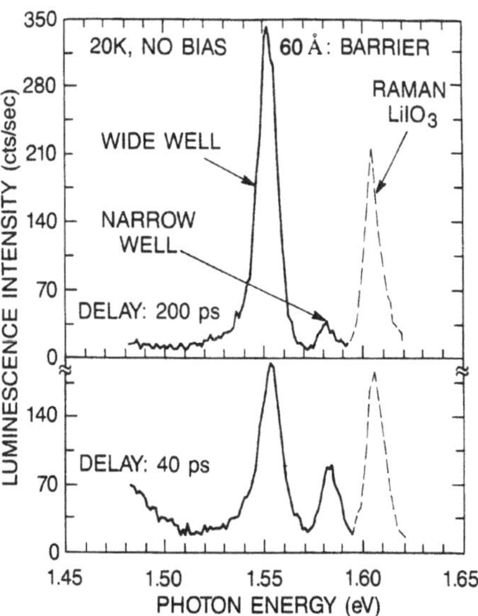


**Fig. 15** Typical time resolved luminescence spectra from a 63/60/90 a-DQWS for two different delays; 20 K, no bias. The dashed features result from Raman scattering in the nonlinear crystal $LiIO_3$.

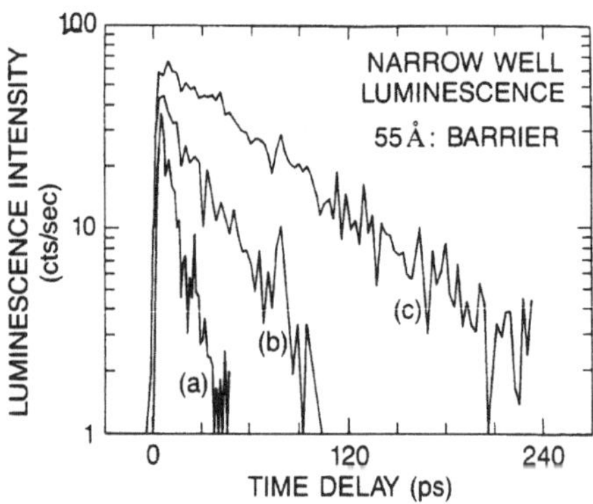


**Fig. 16** Typical time evolution of the luminescence of the NW in a 63/50/90 a-DQWS for three different electric fields, showing the near-exponential decay and the variation of the decay with electric field. The change in the decay time results from tunneling from the NW to the WW. (a) 82 kV/cm, (b) 65 kV/cm and (c) 59 kV/cm.

The measured decay times were independent of density for at least a factor of two increase or decrease in the density. It is essential to check this independence on the excitation density because tunneling of electrons and holes tend to produce a space charge field that opposes the applied electric field. The independence of the tunneling time on density makes sure that the space charge field is small and does not unduly influence the measurement.

Fig. 15 shows time resolved luminescence spectra from a 63/50/100 a-DQWS for two different time delays. Each spectrum shows two features; the one at low energy is due to the WW and the other due to the NW. We see that the NW luminescence intensity decreases between 40 and 200 ps whereas that for the WW increases in this time. This shows that carriers excited in the NW have a shorter decay time. This reduced decay time can be shown to be a result of tunneling. Fig. 16 shows the time evolution of the NW luminescence for three different electric fields. There is a strong dependence of the decay time on the field, which in this case is due to a resonance of the type $R_1$. These two figures illustrate typical spectral and temporal information that can be obtained from time resolved luminescence experiments. The spectral feature marked "Raman $LiIO_3$ " is an artifact arising from Raman scattering in the nonlinear crystal.

### III.D Excite-and-Probe And DFWM Techniques

The powerful techniques of excite-and-probe spectroscopy and degenerate four-wave-mixing (DFWM) have also been used to investigate tunneling in these structures. A schematic diagram of these techniques is shown in Fig. 17. Once again, the laser beam is divided into two beams, and a delay is introduced in beam # 2. Both beams are then focussed to overlap on the sample. In the excite-and-probe transmission spectroscopy, the second beam is much weaker than the first beam and acts as a probe. The differential transmission of the probe beam (i.e. the change in the transmission with and without the pump beam ) is measured as a function of the delay between the two beams.

In the DFWM experiments, the two beams are of the same intensity. If the macroscopic polarization created by the first pulse (beam # 1 ) is still present when the delayed pulse from the second beam arrives, a polarization grating is created. The second beam is diffracted by this grating and the experiment measures the total diffracted energy as a function of the time delay between the two beams. The strength of the signal varies with the time delay and provides information about the dephasing rates for the polarization induced by the laser. Since this diffracted signal propagates along a different direction compared to the incident beams (Fig. 17), DFWM experiments provide a background-free signal.

### III.E Relative Merits of Various Techniques

Luminescence spectroscopy also provides a background-free signal and has been used extensively to study tunneling [82] and transport [24] using the "marker technique". Most optical measurements on tunneling have been performed using luminescence

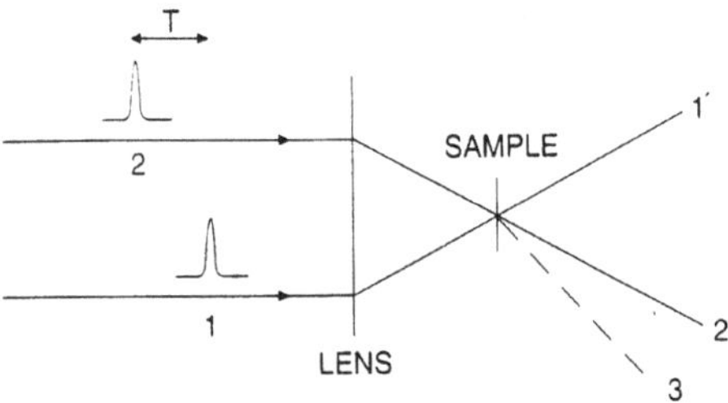


**Fig. 17** A schematic diagram of the experimental arrangement for the excite-and-probe and self-diffracted transient four-wave-mixing experiments. The pulse in beam 2 is delayed by time T compared to the pulse in beam 1. In the excite-and-probe transmission measurements, the probe beam 2 is weaker than the pump beam 1 and the change in the intensity of 2′ induced by the pump is measured as a function of T. In the FWM experiment, the direction of the diffracted beam 3 is determined by phase-matching conditions, and the intensity of beam 3 is measured as a function of T.

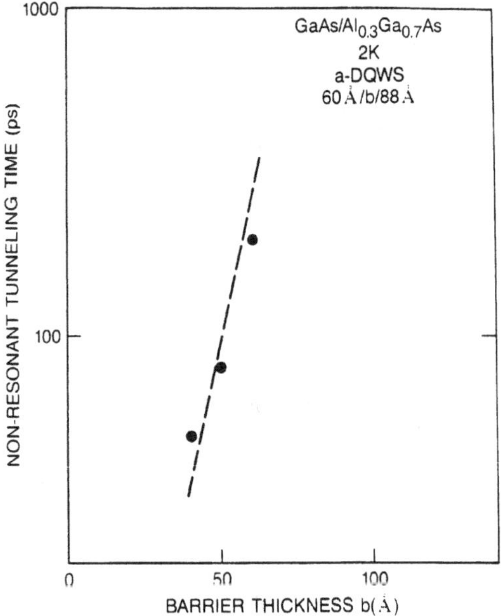


**Fig. 18** Dependence of the measured tunneling time on barrier thickness b for a-DQWS 60/b/88. Data obtained by measuring luminescence decay time of the NW for three different samples.

techniques because of the simplicity and versatility of the technique and the ease of interpretation. One difficulty with the luminescence measurements arises from the fact that scattered light prevents one from measuring *resonantly excited luminescence* in the usual upconversion technique. Although one can get close to resonance by using a double spectrometer for analyzing the upconverted photons, by using type II crystal for upconversion or by exciting with one laser and upconverting with a laser of different wavelength, [87] exact resonant excitation is very difficult to achieve. Therefore, the relaxation of the photoexcited carriers or excitons from the state in which they are created to the state in which they are probed must be considered in the interpretation of the data. This is true even if one measures the decay of carriers from the initial well rather than the arrival of carriers in the target well, as discussed in Sec. II. This problem can be overcome only by developing a true resonant time resolved luminescence technique with femtosecond time resolution. Note that streak camera allows investigation of time resolved resonant luminescence for samples with good surface quality, but its time resolution for measuring weak luminescence signbals is several picoseconds, about *two orders of magnitude* worse than what is demonstrated for the luminescence upconversion technique.

The major strength of the excite-and-probe and DFWM techniques is that they can be performed with *resonant* excitation so that one does not need to worry about relaxation phenomena. The excite-and-probe transmission signals have a coherent as well as an incoherent component [88] The DFWM signal as described above has only a coherent component. The coherent signals decay with rates determined by various dephasing processes which are much faster than the carrier relaxation processes. Therefore, the coherent signals are not influenced by these slow relaxation processes and provide much better time resolution than the *non-resonant* luminescence upconversion experiments whose time resolution may be affected by the slower carrier relaxation processes. An additional advantage of the FWM techniques is that they can be performed with densities which are one to two orders of magnitude lower than the luminescence upconversion experiments. Therefore, one need not worry about space charge fields induced by tunneling.

## IV. RESULTS AND DISCUSSION

In this section, we present some experimental results on tunneling in a-DQWS, discuss them in light of the basic concepts discussed above. We first discuss measurements of nonresonant tunneling rates for electrons and then consider resonant tunneling. We then compare the unified model of tunneling in presence of strong collisions/relaxation with the experimental results. Finally, we present results on tunneling obtained by FWM and then discuss observation of coherent oscillations.

### IV.A Non-Resonant Tunneling: Dependence on Barrier Thickness

Oberli et al. [29-32] investigated 63/b/100 a-DQWS for three different values of b (b = 40, 50 and 60 Å). The excitation wavelength was such that electrons were excited in both $E_1$ and $E_1'$ but not in $E_2$ . No bias was applied to the structure and the built-in field was ≈20kV/cm. The energy separation between $E_1'$ and $E_1$ under these conditions was larger than the optical phonon energy. The tunneling rate from the NW to the WW was obtained by subtracting the WW decay rate from the total decay rate of the NW (Eq. (1)).

The tunneling times ( = 1/(tunneling rate)) for the NW are plotted in Fig. 18 as a function of b. The plot shows that the tunneling times depend exponentially on the barrier thickness. This is to be expected because the wavefunction overlap integral varies exponentially with b. Other groups [37-41] have extended these data and shown that the exponential dependence on b is valid for a larger range of b.

**IV.B Non-Resonant Tunneling: Effect of Optical Phonons**

The exponential dependence of tunneling rate on the barrier thickness b is expected to be valid for a large number of tunneling mechanisms. Therefore, a measurement of non-resonant tunneling times can not provide direct information on the importance of various processes which contribute to tunneling (Fig. 5). In order to distinguish between different mechanisms we have varied the separation between the initial and the final level through an optical phonon energy. [29, 30, 89] The results provide a *direct experimental determination* of the dominant tunneling mechanism.

The experiment was performed on a 100/50/63 a-DQWS. Application of reverse bias brought $E_1'$ below $E_1$ allowing electrons in the WW to tunnel to the NW. This tunneling rate was determined by measuring the decay rates for the WW luminescence at various electric fields. The results were independent of whether only the WW or both the WW and the NW were excited by the laser pulse. The decay time of the WW luminescence as a function of an applied electric field is shown in Fig. 19. The decay time increases slightly with electric field at low fields but then drops abruptly, goes through a minimum and then increases gradually. A possible explanation for the initial increase in the decay time is a change in tunneling rates as levels $E_1'$ and $E_1$ cross each other; another possible explanation is an increase in the recombination time because of the delocalization of the wavefunctions near resonance. We concentrate here on the abrupt drop and the subsequent increase in the decay time.

At very low fields, the decay time is close to the normal recombination time for thick barriers showing that tunneling is extremely weak for small $\Delta E$ ( = $E_1$ - $E_1'$ ). The drop in tunneling time occurs at a field when $\Delta E$ approaches $\hbar\omega_{LO}$ , the optical phonon energy (36 meV in GaAs ). This large decrease in the tunneling time close to the optical phonon energy is a result of the onset of non-resonant tunneling involving emission of optical phonons. It shows that *optical phonon assisted tunneling is considerably stronger than impurity, defect and acoustic phonon assisted tunneling* which is allowed for $\Delta E$ smaller than $\hbar\omega_{LO}$ .

The tunneling time then goes through a minimum when the energy separation is 47 meV and increases for larger $\Delta E$ . This increase is a result of two factors: the minimum wavevector of phonon that must be emitted to conserve energy and momentum conditions increases with increasing electric field. Since the Fröhlich Hamiltonian varies as 1/q this leads to an increase in the tunneling time. The other

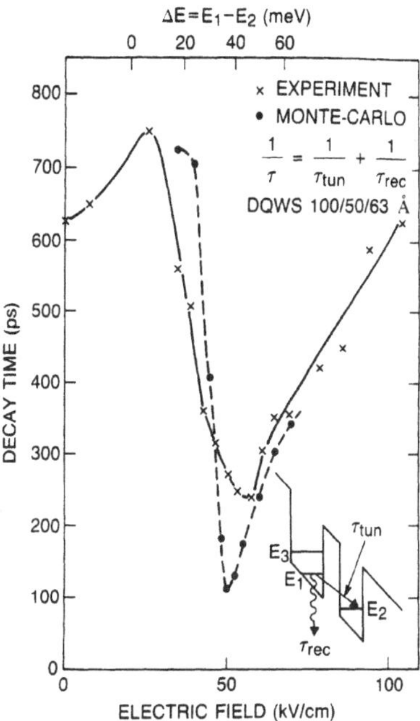


**Fig. 19** The measured dependence of the decay time of the WW luminescence for a 100/50/63 a-DQWS as a function of the applied electric field. The separation $\Delta E$ between $E_1$ and $E_2$ is shown on the scale at the top. The solid line is to guide the eye. The drop in the decay time beginning at 35 kV/cm is due to the onset of optical phonon-assisted tunneling. The solid dots are a result of the Monte Carlo simulation of the problem by Lary and Goodnick.[90]

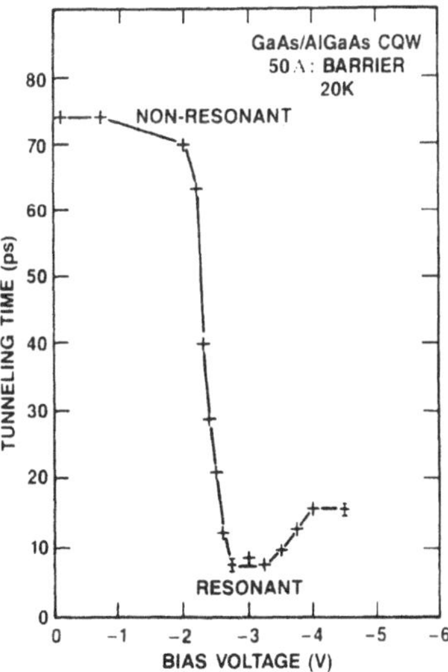


**Fig. 20** Tunneling times of electrons from the NW to the WW in a 63/50/88 a-DQWS as the electric field is tuned through resonance $R_1$ shown in Fig. 3. The system time resolution is 0.5 ps, considerably shorter than the tunneling time of $\approx$ 7 ps measured at resonance.

factor contributing to an increase in tunneling time is the dependence of the overlap integral on the applied electric field. [90]

The tunneling times for the conditions of these experiments have been calculated using Monte Carlo simulations by Lary and Goodnick. [53] Their calculation of tunneling time is compared with the data for the decay rates in Fig. 19 by assuming that the recombination time is 750 ps and assuming that phonons correspond to bulk phonon modes in GaAs. The calculation reproduces the experimental trends quite nicely but quantitative agreement is lacking in two areas. First, the experimental minimum occurs for $\Delta E$ = 47 meV, considerably larger than the 36 meV LO phonon energy in GaAs. This energy is in fact the energy of the AlAs like phonon in AlGaAs barrier. This suggests that confined phonon modes as well as interface modes in the entire structure may play an important role in the tunneling process. Once these modes are correctly incorporated in the calculation, the second disagreement, namely the value of the tunneling time close to the minimum, may also be reduced. Efforts are under way to incorporate the correct phonons in the calculation. [91] Other effects such as inhomogeneous broadening and inhomogeneous electric fields may also contribute to the discrepancy and should be considered for obtaining a better quantitative fit to the data.

We conclude this section by noting that this approach allows a direct comparison of tunneling with and without phonon participation in the same structure, and it directly shows that optical phonon assisted tunneling dominates over defect or acoustic phonon assisted tunneling in good quality samples. This approach is also capable of providing detailed information about the nature of phonons involved in tunneling. Of course, a pre-requisite to the success of this approach is small impurity/defect concentration because with large concentrations of impurities/defects, the change in the tunneling rate at the phonon resonance may be too small to be observed.

This concludes our discussion of non-resonant tunneling. In the next sections, we discuss some results on tunneling rates for electrons and holes when two electronic levels are brought into resonance by applying an electric field. As discussed in Sec. III.E, time resolved luminescence studies with *non-resonant* excitation are influenced by slow relaxation to the states emitting light. Therefore, we first discuss time resolved luminescence studies near resonances such as $R_1$, $R_3$, and $R_4$ and then discuss results obtained by FWM techniques when the resonance is between the two lowest electronic levels (resonances $R_2$ and $R_5$ in Fig. 3 and Fig. 4 respectively).

**IV.C Resonant Tunneling of Electrons**

The first measurement of tunneling rates near electronic resonances was performed by Oberli et al. [29-31] for the case of electron resonance $R_1$ (Fig. 3) in a 63/50/90 a-DQWS. The decay times of the NW luminescence were measured as a function of the applied electric field. These results are plotted in Fig. 20. At low fields, the separation between $E_1'$ and $E_1$ is already larger than the optical phonon energy. Therefore the decay is dominated by tunneling and the decay time directly gives the tunneling time. With increasing field, the tunneling time shows a sharp decrease, followed by a broad minimum and then an increase. This *non-monotonic* behavior

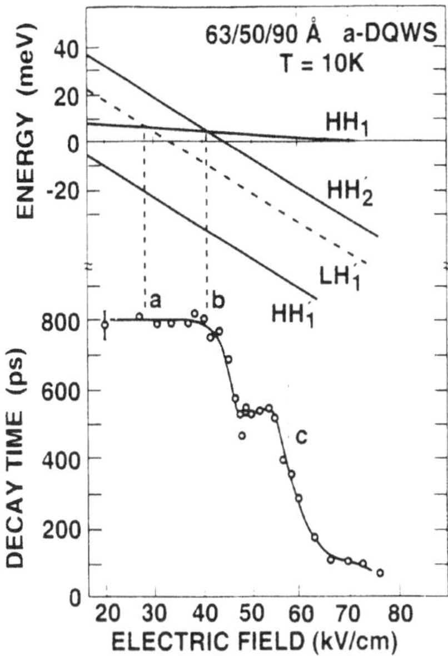


**Fig. 21** The measured decay time of the WW luminescence in a 63/50/90 a-DQWS. The top of the figure shows various hole energy levels with the center of the WW taken as the zero of the electrostatic potential. The resonance between the $HH_1$ level of the WW with the $LH_1$ level of the NW does not affect the decay time of the WW, indicating that there is no significant tunneling of holes under this condition. The decrease in the decay time beginning at 40 kV/cm results from the resonance between two HH levels as indicated. The drop in the decay time at fields larger than 55 kV/cm results from electron tunneling from the WW to the NW in the next period to the right.

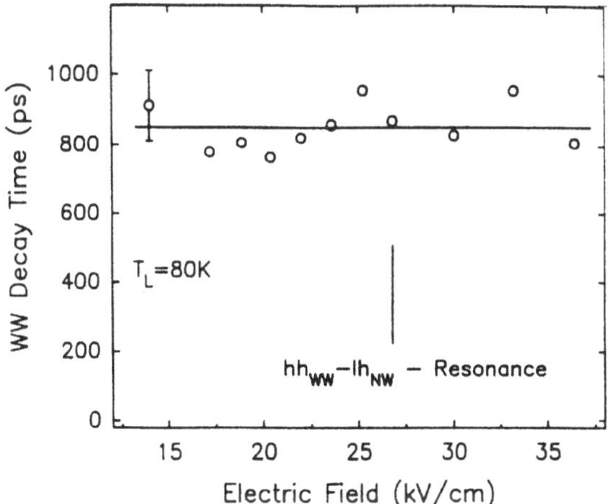


**Fig. 22** The measured decay time of the WW luminescence as a function of field spanning the region of the HH-LH resonance discussed in Fig. 21. The measurements were made at 80 K to increase the occupation of higher $k_{//}$ states in the WW.

directly shows the resonant nature of the phenomenon under investigation. These results provided the first experimental evidence of the sharp decrease in tunneling time at a resonance.

The resonant field (i. e. the field at which tunneling time is minimum) was found to be slightly smaller at lower excitation density, showing the presence of a small space charge field effect. However, the value of tunneling time at resonance did not vary with the excitation density. Also, the resonant field agrees with the calculated resonant field within 10 %, confirming that electron resonance is responsible for the observed dramatic change in the tunneling times. Finally, the tunneling time at large fields did not increase to the value at low fields, possibly because of tunneling of electrons out of the DQWS.

Another interesting observation is that the minimum value of the tunneling time ($\approx$7ps) is considerably larger than the value of $\approx$1ps expected for the coherent oscillation period (Sec. II.D) in this structure and about an order of magnitude larger than the time resolution of the experiment. A number of explanations were put forward to explain this large discrepancy. These included *extrinsic* possibilities, such as inhomogeneously broadened luminescence linewidths and inhomogeneous field distribution. The explanations also included *intrinsic* possibilities such as coherence breaking collisions and relaxation processes, including those involving optical phonons as well as rate limiting inter-subband relaxation. We postpone a discussion of this topic after the data on hole tunneling are discussed in the next section.

**IV.D Resonant Tunneling of Holes**

The first study of tunneling rates for holes near resonance between hole levels was performed by Leo et al. [44] on a 63/50/100 a-DQWS. Only the WW was photoexcited and the decay times of the WW were measured as the electric field was tuned through the resonances $R_3$ and $R_4$ in Fig. 3.

The measured decay times for the WW are plotted as a function of the electric field in Fig. 21. The decay time remains nearly constant at 800 ps, close to the expected recombination time for the WW, at low fields showing that tunneling is not important in this region. With further increase in the field, there is a sharp drop in the decay time, then a plateau or a slight increase, and finally a gradual decrease extending to the highest field investigated. As discussed earlier, [44] the last decrease results from the tunneling of *electrons* from the WW through the wide barrier into the NW of the next period on the right. Therefore, at high fields, the DQWS are no longer isolated. We concentrate on the two interesting aspect of the data at fields below 50 kV/cm.

First, the resonance $R_3$ between $HH_1$ and $LH_1'$ is expected to occur at 20kV/cm. From the data we see that there is no change in the decay time at this resonance. Since the mixing between the heavy and light hole bands is expected to increase as one increases $k_{//}$, these measurements were repeated at 80K to see if occupation of higher $k_{//}$ has any influence. The results for 80K are shown in Fig. 22. Once again, there is no change in the decay time near the expected resonance. From the experimental uncertainty in determining the decay time, it was estimated that the tunneling time at this resonance is longer than 8000 ps. This conclusion was valid up to a density of $1x10^{11}cm^{-2}$ for which the Fermi wavevector at 0 K is $(125Å)^{-1}$, within a factor of 2 of the inverse well thickness for both the NW and the WW. There has been considerable discussion of the role of heavy hole-light hole coupling in connection with the double barrier diodes and there have been suggestions that this coupling may play an important role in such structures.[92-94] Similarly, strong effects resulting from such coupling have been predicted and used to explain cw luminescence data in a-DQWS.[47, 51, 52] These results indicate that this effect is not universally strong. In a recent experiment using the same concept of a-DQWS in electric field, but with different parameters for the well thicknesses, a resonance is indeed observed close to the expected field.[40, 41, 95, 96] Since the coupling depends on the thicknesses of the wells, it is possible that these effects are structure dependent. A systematic investigation of HH-LH resonance has not been reported to our knowledge.

We now discuss the more positive result, namely the reduction in the decay time at 40 kV/cm followed by a slight increase. From the calculated energy levels on the top of the figure, this decrease in the decay occurs at the $HH_1$ - $HH_2'$ resonance (of the type $R_4$ in Fig. 3.) Subtracting the decay rate at low fields (1/800 ps) from the rate at the minimum (1/500 ps) gives a tunneling time of 1330 ps at this resonance. The coherent oscillation time $\tau_{coh}$ calculated numerically for this structure is 56 ps, corresponding to an energy splitting of 0.074 meV. Therefore, the measured tunneling time at resonance is approximately *45 times longer* than $\tau_{coh}/2$.

We can rule out several trivial possibilities very simply. The estimated full-width-half-maximum (FWHM) of the resonance curve is 6kV/cm, considerably larger than the 2kV/cm inhomogeneity in the electric field estimated from the QCSE linewidth data. This FWHM of the measured curve implies a resonance linewidth of 6-7meV which is considerably larger than the inhomogeneous broadening of about 2 meV for the luminescence lines. Therefore, these extrinsic effects can be ruled out as a cause for the increase in the tunneling time. By measuring several interband transition energies for the NW in PC spectra at finite fields, it was shown that the $HH_2'$ - $HH_1$ energy separation is 39 meV, larger than the TO and LO phonon energies in GaAs. Therefore, the inter-subband relaxation is expected to be much faster than the measured tunneling rate and does not act as a rate limiting mechanism in this measurement. Therefore, the final possibility mentioned in the work on resonant electron tunneling by Oberli et al.[32], namely departure from the simple theory as a result of coherence breaking collisions and relaxations is the likely explanation for the inordinately large tunneling times observed. A quantitative understanding of this process has been developed as we discussed in Sec. II.G. In the next section we make a quantitative comparison of the model with the results discussed above.

### IV.E Resonant Tunneling with Strong Collisions and Relaxation

If we assume that the measured resonance width of 6 meV is due to a rapid inter-subband relaxation, then the relaxation time $\tau_R$ is 0.1 ps, a value quite consistent with the expected inter-subband rates for TO and LO assisted phonon transitions for holes. Inserting a value of 56 ps for $\tau_{coh}$ and the above value for $\tau_R$ in Eq. (8), we obtain a tunneling time of about 800 ps, a factor of 1.7 smaller than the experimentally determined value. Similarly, the area under the resonance curve is given by the product $(\pi/2)$(FWHM)(maximum tunneling rate) which in this case is 7.2$\mu$eV/ps and compares favorably with the value of 13$\mu$eV/ps expected from Eq. (12). There is thus an agreement within a factor of two in both cases. This is quite good considering that the original discrepancy was a factor of 45, that there is some uncertainty in the determination of the area and the width of resonance, and that there must be some contribution from inhomogeneous broadening..

We now consider how this model applies to the case of electron resonant tunneling data by Oberli et al. discussed above and more recent data of others. The FWHM of the resonance in Fig. 3 of Oberli et al. is 1.4 V which corresponds to 25 kV/cm and a resonant width of 30 meV. If we attribute the entire width to lifetime broadening, the scattering time would be 0.022 ps, considerably shorter than expected for inter-subband optical phonon scattering. This implies that a substantial part of the broadening, and hence a substantial part of the observed reduction in the tunneling rate at resonance, is caused by other factors such as inhomogeneous broadening and non-uniform electric field. However, the area under the resonance curve may still be conserved. The area under the experimental curve is $\approx$6.5meV/ps which is about a factor of 3.5 smaller than the value of 22 meV/ps expected from Eq. (12). While this discrepancy is still large, it is considerably smaller than the discrepancy between the measured value of 7.5 ps and $\tau_{coh}/2$ = 0.67 ps. It is clear, however, that the model of strong collisions/relaxation needs to be extended to include the effect of inhomogeneities in samples where such effects are substantial.

Two other sets of resonant tunneling data have appeared in the literature since the original work of Oberli et al. Deveaud et al. [37, 38] have investigated the effect of resonance by measuring tunneling times in three sets of samples: In the first set, they kept a′ and b constant but varied a to go through the the resonance between $E_1'$ and $E_2$. In the second set, they kept a and a′ constant to maintain the resonance between $E_1'$ and $E_2$, but varied the barrier thickness b; this is the "on resonance" set. Finally, they investigated an "off-resonance " set in which the well thicknesses were fixed such that $E_1'$ and $E_2$ were not in resonance and varied the barrier thickness b. It should be mentioned that one can not be certain of being at precise resonance in this approach.

For the first set, Deveaud et al. [37, 38] observed a broad resonance as the WW thickness a is varied through the expected resonance. For the "off-resonance" set, they observed an exponential dependence of tunneling time on the barrier thickness. This is in agreement with the theoretical expectations based on the wavefunction overlap and the data presented in Sec. IV.A. For the "on-resonance" set, they obtained a very interesting set of data. The measured tunneling time remained constant at 2 ps for b < 50 Å but increased exponentially with b for b > 50 Å . The authors have interpreted these results as follows: for b < 50 Å , they propose that the near-ideal coherent tunneling model discussed in Sec. II.F for resonance $R_1$ applies and that the measured time is twice the inter-subband scattering time for an isolated well. However, the measured time is approximately a factor of 3 larger than expected for this model (assuming that the inter-subband transition time is 0.3 ps). Therefore, the applicability of this model to the data must be scrutinized in more detail. Some of the factors that need to be considered are the influence of intervalley scattering, the influence of various phonon modes on the inter-subband scattering times, the variability of samples and the difficulty of determining when an exact resonance is achieved by this technique. Electric field measurements [29, 31, 39, 40, 41] which continuously tune through a resonance, show a minimum value different from the 2 ps value obtained in these experiments; also the minimum value shows a strong dependence on barrier thickness b. The influence of strong collisions/scattering on the model must also be examined.

For b > 50 Å, Deveaud et al. [37, 38] observed that the measured tunneling time depends exponentially on the barrier thickness. They attributed this behavior to difficulty in achieving exact resonance. However, one may also want to consider the model of strong collisions/relaxation which broadens the resonance and leads to an exponential dependence of the tunneling time on b ( $\tau_t$ prop $\exp(4\pi b/\lambda)$), as can be seen through Eq. (9) and Eq. (5). This behavior is the same as that expected for the non-resonant tunneling time and the experimental data are consistent with this expectation. Similarly the data for 60 and 75 Å barriers, should also be compared with the model of strong collisions discussed above.

Alexander et al. [39, 40] have extended the measurements of Oberli et al. [29, 30, 89] and measured tunneling times near the $R_1$ resonance with an applied electric field for samples with three different barrier thicknesses b. They observed a well defined minimum in the tunneling times at certain electric fields for samples with barrier thicknesses of 60 Å and 80 Å. From the observation that the measured tunneling time at the minimum was a function of b and different from twice the inter-subband phonon scattering time, they concluded that the coherent model is not applicable and a *sequential* model must be invoked. From our discussion in Sec. II.G (Eq. 8), it is evident that a strong dependence on barrier thickness is predicted by the unified model of coherent tunneling in presence of strong collisions/relaxation proposed by us. Such a model also gives a minimum tunneling time which is much longer than the inter-subband phonon scattering time. There is therefore no need to invoke the sequential model of two quantum wells coupled by a strong resonant scattering process, as proposed by Alexander et al. [39]

It is, in fact, interesting to see how the results by Alexander et al. [39, 40] and Nido et al. [41] compare with the predictions of the coherent tunneling model including strong collisions and relaxation. A comparison with the data for 60 Å barrier is not possible because the data at resonance are limited by the system time resolution. For the sample with b = 80 Å, the measured time at resonance is 50 ps as compared to $\tau_{coh} = 11.2$ ps. If we attribute the entire width of the resonance curve ($\approx$ 9 kV/cm or 14 meV ) to collision/relaxation broadening, it would imply a collision/relaxation time of approximately 50 fs. Inserting this into Eq. 8 would lead to a tunneling time of 55 ps, in close agreement with the experiment. This agreement may be fortuitous because inhomogeneous broadening certainly plays a role as can be inferred from the luminescence linewidths in these samples and the fact that the required phonon scattering time of 50 fs in above calculation is too short. Therefore the effects of strong collisions/relaxation as well as the effects of inhomogeneous broadening must be considered more carefully to quantitatively explain these data.

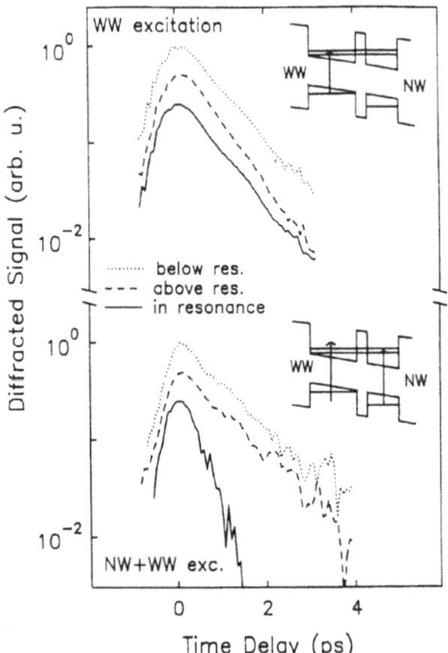


**Fig. 23** Top: Transient FWM signal from a 67/48/42 a-DQWS for three different electric fields for excitation in resonance with the lowest interband transition of the WW. Bottom: same as above, but with excitation in resonance with the lowest interband transition of the NW which creates free carriers in the WW in addition to excitons in the NW.

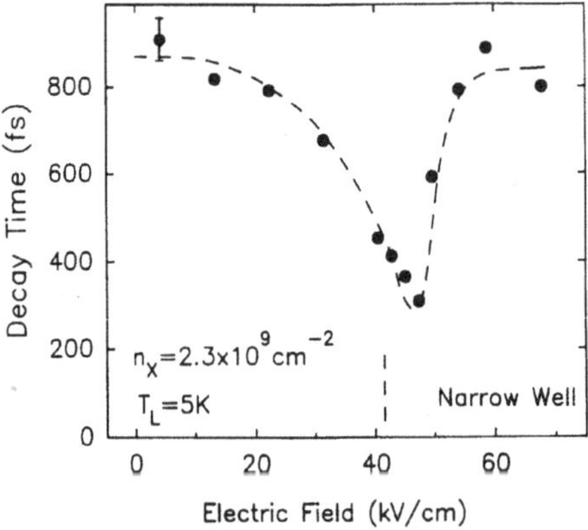


**Fig. 24** Dependence of the decay time of the FWM signal on electric field for excitation in resonance with the lowest interband transition of the NW. The expected value of the resonant field $F_R$ is indicated by a dashed vertical line near the bottom axis.

We conclude from this discussion that some of the data in the literature can be explained primarily on the basis of the unified model of tunneling in presence of strong collisions/relaxation. Some other data, presumably on samples of lower quality, requires consideration of inhomogeneous broadening of the line in addition to this unified model. In any case, it is clear that the model of sequential tunneling is not applicable in most experimental situations.

One may also ask under what conditions the coherent oscillations predicted by the ideal or near-ideal model of coherent tunneling can be observed. Since collisions and relaxation must be reduced, it seems logical to look for such coherent oscillations at the $R_5$ resonance rather than the $R_1$ like resonances we have been considering so far. Also, the sample quality must be extremely good so that inhomogeneous broadening effects are small. Leo et al. [59] have been able to satisfy these conditions and have indeed observed these oscillations at $R_5$ resonance. However, we postpone a discussion of these results until Sec. IV. G.

**IV.F Tunneling Probed by FWM Experiments**

As discussed in Sec. III.E, transient four-wave-mixing experiments can be performed with resonant excitation and hence can provide time resolution better than that obtained by non-resonant luminescence experiments. We discuss here an experiment first reported by Leo et al. [97]

These authors investigated a 67/48/42 a-DQWS with $Al_{0.37}Ga_{0.63}As$ barriers. Self-diffracted, transient FWM signals were recorded in the forward direction (transmission geometry) as a function of the time delay between the two beams, as discussed in Sec. III.D. Reverse bias was applied through a semi-transparent Schottky contact and resonance behavior near $R_5$ was investigated. Fig. 23a shows the transient signals for three electric fields when the laser was in resonance with the lowest interband transition in the WW. We see that the transient signal decays with the same time constant regardless of the electric field. Fig. 23b shows similar data when the laser is tuned to the lowest interband transition of the NW so that it excites excitons in the NW and free electron hole pairs in the WW. Under this condition, the decay of the transient FWM signal shows a pronounced dependence on the field. In fact, the decay time as well as the diffracted intensity at $\tau = 0$ show a distinct minimum close to the expected resonance, as shown in Fig. 24 and Fig. 25 respectively.

The reason for this minimum is the following: for excitation in resonance with the NW, free electron hole pairs are created in the WW. Away from resonance, the wavefunctions are localized so that the measured dephasing time represents the dephasing of the wavefunction in the NW. The free carriers in the WW have no effect on this dephasing time. However, at resonance, a wavepacket is created which is expected to oscillate with an estimated period of $\tau_{coh}$ = 0.5ps. While this wavepacket is in the WW, it is dephased more efficiently by the presence of free carriers in the WW. Therefore, at resonance, the measured decay time is considerably shorter. A more detailed discussion of the various times involved is given by Leo et al. [97, 98] We note

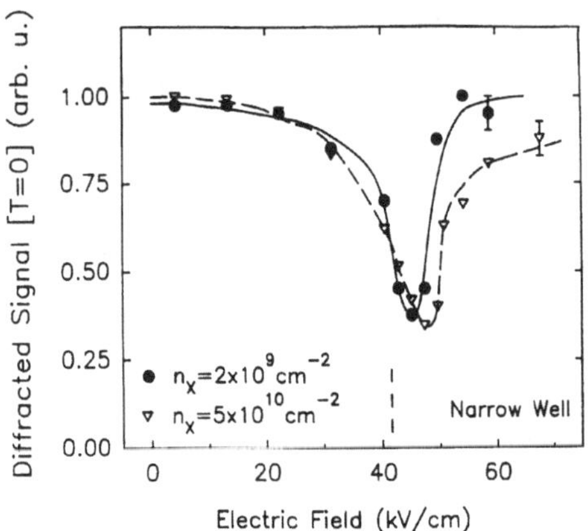


**Fig. 25** Dependence of the peak intensity of the FWM signal (i.e. intensity for delay T = 0 ) on electric field for excitation in resonance with the lowest interband transition of the NW. The expected value of the resonant field $F_R$ is indicated by a dashed vertical line near the bottom axis. The resonance shifts slightly due to reduced space charge effect and is sharper for the lower excitation density.

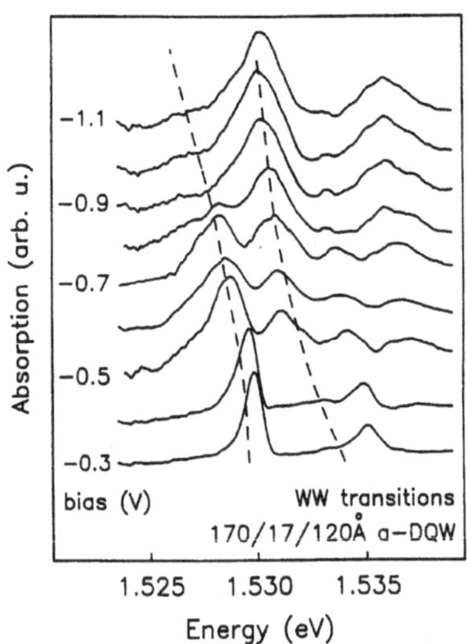


**Fig. 26** The measured cw absorption spectrum for a 170/20/120 a-DQWS at 5 K in the region of the lowest HH and LH transitions in the WW, for various electric fields close to resonance. The two well defined transitions at -0.3 V correspond to the lowest HH and LH excitons of the WW. The new features appearing on the high energy side of these features correspond to the spatially indirect transitions at low fields. These new transitions become stronger and dominate after anti-crossing at resonance. Note that the features at high fields are considerably broader than those at low fields, probably because an additional channel of scattering into lower states is available and also because inhomogeneities in the field lead to

here that these results imply that the wavepacket is oscillating as expected but no oscillatory structure is observed, probably because the sample did not meet the extremely high standards required for the observation of the coherent wavepacket oscillations. A different sample structure does indeed show more direct evidence for such oscillations as discussed in the next section.

**IV.G Coherent Oscillations of an Electronic Wavepacket**

The major problem with the structure used in the above experiment was that the inhomogeneous broadening of the spectral features was comparable to the minimum energy separation at the anti-crossing. This was a result of the relatively narrow well-widths (42 Å and 67 Å ) of the a-DQWS. It is well known that the widths of the spectral features decrease as the well-width is increased, because the fractional fluctuations in the well-widths is smaller when the well-width is larger. The new structure was a 170/17/120 a-DQWS with $Al_{0.35}Ga_{0.65}As$ barriers. The barrier thickness was reduced considerably to compensate for the increase in the effective barrier heights because of the reduction in the confinement energies. The calculated minimum energy separation between the two lowest electron levels at resonance was ≈3meV leading to $\tau_{coh} = 1.3$ ps.

Fig. 26 shows the cw absorption spectrum of the sample near the lowest interband transition energy of the WW for various electric fields near the electronic resonance $R_5$ obtained by Leo et al. [59] At -0.3 V, below the resonance, there are two peaks due to the transitions to the $E_1$ level from the $HH_1$ and and $LH_1$ levels. With increasing reverse bias, the $HH_1$ transition moves to lower energies and a peak develops on its high energy side. This new feature arises from the absorption between $HH_1$ and the higher of the two coupled electron states (anti-symmetric wavefunction). This new transition slowly gains strength and then dominates the absorption spectrum as the spatially indirect transition vanishes. The resonance can be identified as the field at which the ratio of the absorption coefficients is $(a/a')2$. This occurs at ≈-0.6 V. The separation between the peaks at resonance is measured to be 2.6 meV without considering the overlap between the peaks and 2.7 meV considering the overlap of two Gaussian peaks. A proper deconvolution will yield somewhat larger value. The agreement with the expected value of 3 meV is quite good. We note that the measured behavior of the cw absorption spectrum in Fig. 26 is similar to the behavior qualitatively expected for the ideal picture and shown in Fig. 6.

Leo et al. [59] have performed pump-and-probe transmission and transient FWM experiments at various electric fields for this sample as well as a 140/25/120 a-DQWS. This latter structure has a larger splitting at resonance so that the coherent oscillation period is about 800 fs. We discuss here the results on this improved sample [99]. The laser was tuned to the $HH_1$ -$E_1$ resonance of the WW. The bandwidth of the laser was approximately 4 meV, sufficient to encompass the two coupled levels near resonance but not so large as to excite too many free carriers in the bands. The latter condition is necessary to avoid reducing the dephasing time due to more efficient collisions with free carriers. The differential transmission of a weak probe beam (i.e. the change in its transmission with and without the presence of a strong pump ) as a function of the delay between the pump and the probe is plotted on a linear scale in Fig. 27. At resonance, the differential transmission signal shows well defined peaks. For higher fields, the period of the oscillation decreases and the amplitude diminishes. In all cases, the non-oscillating differential transmission signal persists for times much longer than the oscillation period. As discussed earlier by Leo et al. [59], this long lasting signal is a result of the incoherent population effects and will not be discussed further.

These results can be viewed as confirming the idea of coherent oscillations of the electronic wavepacket. In a qualitative way, one can think that the bleaching of the WW excitonic absorption is larger when the wavepacket is in the WW than when it is in the NW. At longer times, the excitation relaxes to the lower of the two coupled states so that the bleaching persists but there are no oscillations. The decrease in the period and the amplitude of the oscillations is consistent with the ideal picture developed in Sec. II.E. The pump-and-probe signal is made of of a coherent part and an incoherent part. The oscillations are present in the coherent part.

Transient FWM signal far away from resonance is expected to decay with a time constant related to the dephasing time of the WW exciton. Near the resonant field, the FWM signal is also expected to show oscillatory behavior with an oscillation period related to the separation between the two electronic levels. Our results for the same sample as above are shown in Fig. 28 for three different fields. The most pronounced oscillation are present at the resonant field and the general behavior with change in the field is similar to that of the pump-and-probe signal in Fig. 27.

As discussed in Sec. II.E on resonant tunneling in the near-ideal case, simple quantum mechanics leads one to expect an oscillatory motion of the electronic wavepacket under the conditions of the experiment. The observed cw absorption spectrum, and the time evolution of the differential transmission and the FWM signals and their dependence on the electric field lead to the conclusion that the oscillatory motion of the wavepacket is taking place. This represents the first observation of the coherent oscillation of an electronic wavefunction in a solid. Previously, the oscillatory motion of a wavepacket in molecules [100-102] and in atoms [103, 104] was investigated experimentally as well as theoretically. In the case of the molecule, the configuration coordinate space is quite complicated. For atoms, many closely spaced Rydberg states are involved. Our case is considerably simpler with only one spatial coordinate. Furthermore, it offers the unique possibility of tuning the resonance with the application of an electric field.

The period of oscillation can also be obtained from the cw spectroscopy by measuring the splitting of the levels at resonance, provided inhomogeneous broadening effects are not strong. However, the transient differential absorption and FWM measurements have the potential of providing much more information concerning the *dynamics* of these oscillations. Schmitt-Rink and Schäfer [105] have developed a theory for analyzing these effects. The theory in its simplest form was outlined in an earlier publication.[59] These authors have shown that the differential transmission signal and the FWM signal are damped by different dephasing mechanisms. The former is influenced by the dephasing of polarization at the frequency corresponding to the splitting of the

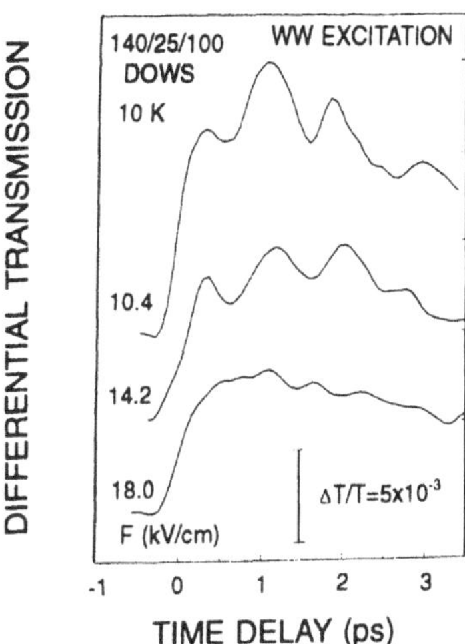


**Fig. 27** Differential transmission signal for a a-DQWS as a function of time delay T for various fields. The oscillations correspond to the coherent oscillations of an electronic wavepacket.

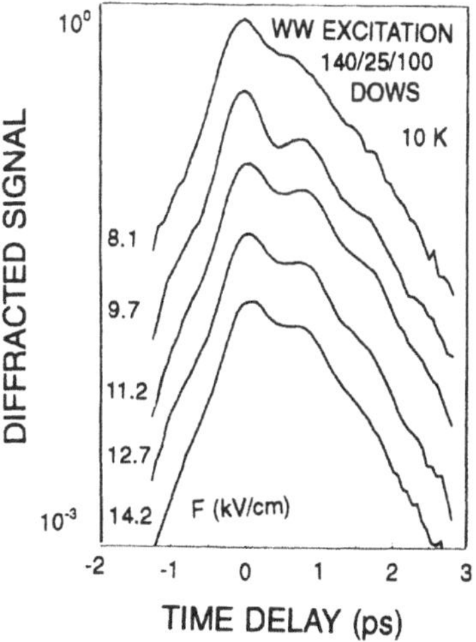


**Fig. 28** Transient FWM signal for several different electric fields for the same sample as in Fig. 27. Note the logarithmic vertical scale.

electron levels whereas the latter is influenced by the dephasing of the polarizations connecting the ground state to the two excited states. There is thus a wealth of information to be obtained from a detailed analysis of the dynamic behavior. Clearly, this is just the beginning of an exciting new direction.

We conclude this discussion with two comments. We have discussed the oscillation of the wavepacket in terms of electrons. However, the optical excitation creates both electrons and holes so the correct description of the process is in terms of the oscillation of an excitonic wavepacket, from a spatially direct exciton wavepacket to a spatially indirect exciton wavepacket. It turns out that including the exciton effects does not change the splitting at resonance to the first order and hence the period of oscillation, or the tunneling time, is not affected. However, the excitons do make a difference in a more subtle way. It has been shown [65, 106] that, since the binding energy of an indirect exciton far from resonance is smaller than that of a direct exciton far from resonance, the resonant field is larger for resonance involving the HH1 level (the WW hole) compared to that involving the $HH_1'$ lcvel (the NW hole). This effect has been experimentally verified [65, 99, 106]. Finally, Roskos et al. [107] have shown very recently that this oscillating wavepacket emits coherent THz radiation at the oscillating frequency, providing unequivocal evidence for the oscillating wavepacket, and opening up further possibilities of novel experiments.

## V. SUMMARY

Tunneling is a quantum mechanical process that is of fundamental as well as applied interest. Semiconductor microstructures provide an ideal system for investigating various tunneling phenomena. In this chapter, we have discussed basic concepts related to nonresonant as well as resonant tunneling in asymmetric double quantum well structures and discussed recent time resolved experiments as illustrations of these concepts. Considerable insight into the nature of tunneling has been obtained using these time resolved optical spectroscopy.

## VI. ACKNOWLEDGEMENTS

It is a pleasure to acknowledge fruitful collaboration and discussions with F. Capasso, D. Chemla, J. E. Cunningham, E. Göbel, K. Köhler, J. M. Kuo, S. Luryi, D. A. B. Miller, S. Schmitt-Rink, W. Schäfer and C. W. Tu.

## REFERENCES

* Present Address: Institut für Halbleitertechnik, RWTH Aachen, Aachen, Germany
**Present Address: Walter Schottky Institute, Garching, Germany

1. L. Esaki, and R. Tsu, IBM J. Res. Dev. **14**, 61 (1970).
2. E. R. Brown, T. C. L. G. Sollner, C. D. Parker, W. D. Goodhue and C. L. Chen, Appl. Phys. Lett. **55**, 1777-1779 (1989).
3. T. C. L. G. Sollner, W. D. Goodhue, P. E. Tannenwald, C. D. Parker and D. D. Peck, Appl. Phys. Lett. **43**, 588-590 (1983).
4. C. Minot, H. Le Person, F. Alexandre and J. F. Palmier, Physica **134B**, 514-518 (1985).
5. S. Tarucha, K. Ploog and K. v. Klitzing, Phys. Rev. B **36**, 4558-4560 (1987).

6. Seigo Tarucha, and Klaus Ploog, Phys. Rev. B **39**, 5353-5360 (1989).
7. H. Schneider, K. v. Klitzing and K. Ploog, Superlattices and Microstructures **5**, 383-386 (1989).
8. Harald Schneider, Holger T. Grahn, Klaus v. Klitzing and Klaus Ploog, Phys. Rev. B **40**, 10040-10043 (1989).
9. H. Schneider, W. W. Rühle, K. v. Klitzing and K. Ploog, Appl. Phys. Letters **54**, 2656-2658 (1989).
10. H. Schneider, K. v. Klitzing and K. Ploog, Europhys. Letters **8**, 575-577 (1989).
11. J. F. Whittaker, G. A. Mourou, T. C. G. Sollner and W. D. Goodhue, Appl. Phys. Letters **53**, 385 (1988).
12. M. Tsuchiya, T. Matsusue and H. Sakaki, Phys. Rev. Letters **59**, 2356-2359 (1987).
13. T. B. Norris, X. J. Song, W. J. Schaff, L. F. Eastman, G. Wicks and G. A. Mourou, Appl. Phys. Letters **54**, 60-62 (1989).
14. M. K. Jackson, M. B. Johnson, D. H. Chow, T. C. McGill and C. W. Nieh, Appl. Phys. Letters **54**, 552-555 (1989).
15. J. F. Young, B. M. Wood, G. C. Aers, R. L. S. Devine, H. C. Liu, D. Landheer, M. Buchannan, A. J. SpringThorpe, and P. Mandeville, Phys. Rev. Letters **60**, 2085-2088 (1988).
16. W. R. Frensley, M. A. Reed and J. H. Luscombe, Phys. Rev. Letters **62**, 1207-1207 (1989).
17. N. Vodjdani, E. Costard, F. Chevoir, D. Thomas, D. Cote, P. Bois, and S. Delaitre, in "OSA Proceedings on Picosecond Electronics and Optoelectronics", edited by T. C. L. G. Sollner and D. M. Bloom, Optical Society of America; 251-253 (1989).
18. I. Bar-Joseph, T. K. Woodward, D. S. Chemla, D. Sivco and A. Y. Cho, Phys. Rev. B **41**, 3264-3267 (1990).
19. Hisao Yoshimura, Joel N. Schulman and Hiroyuki Sakaki, Phys. Rev. Letters **64**, 2422-2425 (1990).
20. C. V. Shank, R. L. Fork, B. I. Greene, F. K. Reinhart and R. A. Logan, Appl. Phys. Letters **38**, 104 (1981).
21. G. Livescu, D. A. B. Miller, T. Sizer, D. J. Burrows, J. E. Cunningham, A. C. Gossard and J. H. English, Appl. Phys. Letters **54**, 748 (1989).
22. Yasuki Masumoto, Seigo Tarucha and Hiroshi Okamoto, Phys. Rev. **B 33**, 5961-5964 (1986).
23. B. Deveaud, Jagdeep Shah, T. C. Damen, B. Lambert and A. Regreny, Phys. Rev. Lett. **58**, 2582-2585 (1987).
24. B. Deveaud, Jagdeep Shah, T. C. Damen, B. Lambert, A. Chomette and A. Regreny, IEEE J. of Quan. Electronics **QE-24**, 1641-1651 (1988).
25. B. Lambert, F. Clerot, B. Deveaud, A. Chomette, G. Talalaeff, A. Regreny and B. Sermage, J. of Lum. (1990).
26. Tetsuya Tada, Atsushi Yamaguchi, Toshyuki Ninomiya, Hisao Uchiki, Takayoshi Kobayashi and Takafumi Yao, J. Appl. Phys. **63**, 5491-5494 (1988).
27. M. Tsuchiya, T. Matsusue and H. Sakaki, in "Proceeding of the Sixth International Conference on Ultrafast Phenomena", edited by T. Yajima, K. Yoshihara, C. B. Harris and S. Shionoya, Springer-Verlag, Berlin; 304-306 (1988).
28. T. B. Norris, N. Vodjani, B. Vinter, and C. Weisbuch, Phys. Rev. B **40**, 1074 (1989).
29. D. Y. Oberli, Jagdeep Shah, T. C. Damen, C. W. Tu and D. A. B. Miller, Optical Society of America, Technical Digest Series **10**, 272-275 (1989).
30. D. Y. Oberli, Jagdeep Shah, B. Deveaud and T. C. Damen, in "Proceedings of the Topical Meeting on Picosecond Electronics and Optoelectronics", edited by T. C. L. G. Sollner, (1989).
31. D. Y. Oberli, Jagdeep Shah, T. C. Damen, T. Y. Chang, C, W, Tu, D. A. B. Miller, J. E. Henry, R. F. Kopf, N. Sauer and A. E. DiGiovanni, Phys. Rev. B **40**, 3028-3031 (1989).
32. D. Y. Oberli, Jagdeep Shah, T. C. Damen, R. F. Kopf, J. M. Kuo, and J. E. Henry, in "OSA Proceedings on Picosecond Electronics and Optoelectronics", edited by T. C. L. G. Sollner and D. M. Bloom, 111-114 (1989).
33. M. G. W. Alexander, Nido, M.; Rühle, W.W.; Sauer, R.; Ploog, K.; Köhler, K.; Tsang, W.T. , Sol. St. Electron. **32**, 1621-1625 (1989 ).
34. M. G. W. Alexander, Nido, M.; Reimann, K.; Rühle, W.W.; Köhler, K. , Appl. Phys. Lett. **55**, 2517-2519 (1989 ).
35. N. Sawaki, R. A. Höpfel, E. Gornik, and H. Kano, Appl. Phys. Letters **55**, 1996-1998 (1989).
36. M. Nido, M. G. W. Alexander, W. W. Rühle, T. Schweizer and K. Köhler, Appl. Phys. Letters **55**, 355-357 (1989).
37. B. Deveaud, F. Clerot, A. Chomette, A. Regreny, R. Ferreira and G. Bastard, Europhys. Letters **11**, 367-369 (1990).
38. B. Deveaud, A. Chomette, F. Clerot, P. Auvray, A. Regreny, R. Fereira and G. Bastard, Phys. Rev. B to be published (1990).

39. M. G. W. Alexander, Nido, M.; Rühle, W.W.; Köhler, K. , Phys. Rev. B **41**, 12295-12298 (1990 ).
40. M. G. W. Alexander, M. Nido, W. W. Rühle and K. Köhler, in "Proceedings of NATO workshop", (1990).
41. M. Nido, M. G. W. Alexander, W. W. Rühle and K. Köhler, in "Proceedings of SPIE Conference, Den Haag, Netherlands, March 10, 1990", (1990).
42. Atsushi Tackeuchi, Tsugo Inata, Shunichi Muto and Eizo Miyauchi, Jap. J. Appl. Phys. L750-L753 (1989).
43. B. Deveaud, B. Lambert, A. Chomette, F. Clerot, A. Regreny, Jagdeep Shah, T. C. Damen and B. Sermage, in "Optical Switching in Low-Dimensional Systems", edited by H. Haug and L. Banyai, Plenum Press, London ; (1989).
44. K. Leo, Jagdeep Shah, J. P. Gordon, T. C. Damen, D. A. B. Miller, C. W. Tu and J. E. Cunningham, Phys. Rev. B **42**, 7065-7068 (1990).
45. K. Leo, Jagdeep Shah, J. P. Gordon, T. C. Damen, D. A. B. Miller, C. W. Tu, J. E. Cunningham and J. E. Henry, **1283**, in "Proceedings of SPIE Conference on Quantum-Well and Superlattice Physics III", edited by G. Dohler, E. Koteles and J. N. Schulman, SPIE; 35-44 (1990).
46. R. Sauer, K. Thonke and W. T. Tsang, Phys. Rev. Letters **61**, 609-612 (1988).
47. H. W. Liu, R. Ferreira, G. Bastard, C. Delalande, J. F. Palmier, and B. Etienne, Appl. Phys. Letters **54**, 2082-2084 (1989).
48. T. Weil, and B., Vinter, J. Appl. Phys. **60**, 3227 (1986).
49. S. Luryi, Sol. St. Commun. **65**, 787-789 (1988).
50. S. Luryi, Superlattices and Microstructures **5**, 375-382 (1989).
51. R. Ferreira, and G. Bastard, Phys. Rev. B **40**, 1074-1086 (1989).
52. G. Bastard, C. Delalande, R. Ferreira and H. W. Liu, J. Lumin. **44**, 247-263 (1989).
53. Jennifer Lary, S. M. Goodnick, P. Lugli, Jagdeep Shah and D. Y. Oberli, Sol. St. Electron. **32**, 1283-1287 (1989).
54. A. D. Stone, and P. A. Lee, Phys. Rev. Letters **54**, 1196-1199 (1985).
55. A. J. Leggett, Chakravarty, S.; Dorsey, A.T.; Fisher, P.A.; Garg, A.; Zwerger, W. , Rev. Mod. Phys. (USA) vol. 59, no.1, pp.: 1-85 **59**, 1-85 (1987 ).
56. N. S. Wingreen, K. W. Jacobsen and J. W. Wilkins, Phys. Rev. Letters **61**, 1396-1399 (1988).
57. M. Buttiker, IBM J. Res. Develop. **32**, 63-75 (1988).
58. B. Y. Gelfand, S. Schmitt-Rink and A. F. J. Levi, Phys. Rev. Letters **62**, 1683-1686 (1989).
59. K. Leo, Jagdeep Shah, Ernst O. Göbel, T. C. Damen, Stefan Schmitt-Rink, Wilfried Schäfer and Klaus Köhler, Phys. Rev. Letters **66**, 201-204 (1991).
60. J. Feldmann, G. Peter, E. O. Göbel, P. Dawson, K. Moore, C. Foxon and R. J. Elliott, Phys. Rev. Letters **59**, 2337-2340 (1987).
61. S. Collins, D. Lowe and J. R. Barker, J. Phys. C: Solid State Phys. **20**, 6213-6232 (1987).
62. S. Collins, D. Lowe and J. R. Barker, J. Phys. C: Solid State Phys. **20**, 6233-6243 (1987).
63. M. Buttiker, and R. Landauer, Physica Scripta **32**, 429 (1985).
64. E. H. Hauge, and J. A. Stovneng, Rev. Mod. Phys. **61**, 917-936 (1989).
65. A. M. Fox, D. A. B. Miller, G. Livescu, J. E. Cunningham, J. E. Henry and W. Y. Jan, Phys. Rev. B **42**, 1841-1844 (1990).
66. T. C. Damen, Jagdeep Shah, D. Y. Oberli, D. S. Chemla, J. E. Cunningham and J. M. Kuo, Phys. Rev. B **42**, 7434-7438 (1990).
67. T. C. Damen, Jagdeep Shah, D. Y. Oberli, D. S. Chemla and J. E. Cunningham, J. of Lumin. **45**, 181-185 (1990).
68. Jun-chi Kusano, Yusaburo Segawa, Yoshinobu Aoyagi, Susumu Namba and Hiroshi Okamoto, Phys. Rev. B **40**, 1685-1691 (1989).
69. H. Stolz, D. Schwarze, W. von der Osten and G. Weimann, Superlattices and Microstructures **6**, 271-282 (1989).
70. T. Amand, F. Lephay, S. Valloggia, F. Voillot, M. Brousseau and A. Regreny, Superlattices and Microstructures **6**, 323-328 (1989).
71. H. J. Polland, L. Schultheis, J. Kuhl, E. O. Göbel and C. W. Tu, Phys. Rev. Letters **55**, 2610- 2613 (1985).
72. G. Bastard, E. E. Mendez, L. L. Chang and L. Esaki, Phys. Rev. B **28**, 3241-3245 (1983).
73. J. Smolnier, W. Demmerle, G. Berthold, E. Gornik, G. Weimann and W. Schlapp, Phys. Rev. Lett. **63**, 2116-2119 (1989).
74. H. Kawai, J. Kaneko and N. Watanabe, J. Appl. Phys. **58**, 1263 (1985).
75. Tomofumi Furuta, Kazuhiko Hirakawa, Junji Yoshino and Hiroyuki Sakaki, Jap. J. Appl. Phys. **25**, L151-L154 (1986).

76. M. N. Islam, R. L. Hillman,m D. A. B. Miller, D. S. Chemla, A. C. Gossard and J. H. English, Appl. Phys. Letters **50**, 1098-1100 (1987).
77. Y. J. Chen, E. Koteles, B. S. Elman and C. A. Armineto, Phys. Rev. B **36**, 4562-4566 (1987).
78. H. Q. Le, J. J. Zayhowski and W. D. Goodhue, Appl. Phys. Letters **50**, 1518-1520 (1987).
79. J. E. Golub, P. F. Liao, D. J. Eilenberger, J. P. Harbison, L. T. Florez and Yehiam Prior, Appl. Phys. Letters **53**, 2584-2586 (1988).
80. Y. Tokuda, K. Kamamoto, N. Tsukada, and T. Nakayama, Appl. Phys. Letters **54**, 1232-1234 (1989).
81. A. Messiah, in "Quantum Mechanics", North-Holland Publishing Company, Amsterdam; (1961).
82. Jagdeep Shah, in "Proceedings of NATO Advanced Research Workshop, Venice 1989", edited by G. Fasol. A. Fasolino and P. Lugli, Plenum Press; 535-560 (1989).
83. Footnote1, (1991).
84. D. A. B. Miller, D. S. Chemla, T. C. Damen, A. C. Gossard, W. Wiegmann, T. H., Wood and C. A. Burrus, Phys. Rev. B **32**, 1043-1060 (1985).
85. Jagdeep Shah, IEEE J. Quan. Electronics **24**, 276-288 (1988).
86. T. C. Damen, and Jagdeep Shah, Appl. Phys. Lett. **52**, 1291-1293 (1988).
87. M. R. Freeman, D. D. Awschalom and J. M. Hong, Appl. Phys. Letters **57**, 704-706 (1990).
88. Y. R. Shen, John Wiley and Sons, New York; (1984).
89. D. Y. Oberli, Jagdeep Shah, and T. C. Damen, Phys. Rev. B **40**, 1323-1324 (1989).
90. D. Y. Oberli, Jagdeep Shah, T. C. Damen, J. M. Kuo, J. E. Henry, Jennifer Lary and Stephen M. Goodnick, Appl. Phys. Letters **56**, 1239-1241 (1990).
91. S. M. Goodnick, and P. Lugli, private communication (1990).
92. E. T. Yu, M. K. Jackson and T. C. McGill, Appl. Phys. Lett. **55**, 744-746 (1989).
93. R. Wessel, and M. Altarelli, Phys. Rev. B **39**, 12802-12807 (1989).
94. C. Y-P Chao, and S-L Chuang, Phys. Rev. B **to be published**, (1991).
95. M. Nido, Alexander, M. G. W.; Rühle, W. W., Phys. Rev. B **43**, 1839-1842 (1991).
96. T. B. Norris, Vodjdani, N; Vinter, B; Costard, E. and Bockenhoff, E., Phys. Rev. B**43**, 1867 (1991).
97. K. Leo, T. C. Damen, J. Shah, E. Göbel and K. Köhler, Appl. Phys. Lett. **57**, 19 (1990).
98. K. Leo, Jagdeep Shah, Ernst O. Göbel, T. C. Damen, K. Köhler and Peter Ganser, Superlattice and Microstructures **7**, 427-432 (1990).
99. K. Leo, Jagdeep Shah, E. O. Goebel, T. C. Damen, S. Schmitt-Rink, W. Schaefer and K. Koehler, OSA Proceedings on Picosecond Electronics and Optoelectronics, Vol. 9, edited by T. C. L. G. Sollner and J. Shah, Optical Society of America, Washington, DC (1991); pp. 204-209.
100. H. L. Fragnito, J. Y. Bigot, C. H. Brito-Cruz, R. L. Fork and C. V. Shank, Chem. Phys. Lett. **160**, 101-104 (1989).
101. P. C. Becker, H. L. Fragnito, J. Y. Bigot, C. H. Brito-Cruz, R. L. Fork and C. V. Shank, Phys. Rev. Letters **63**, 505-508 (1989).
102. W. T. Pollard, C. H. Brito-Cruz, C. V. Shank and R. A. Mathies, J. Chem. Phys. **90**, 199 (1989).
103. J. A. Yeazell, and C. R. Stroud, Phys. Rev. Lett. **60**, 1491-1497 (1988).
104. A. ten Wolde, L. D. Noordam, A. Legendijk and H. B. van den Heuvell, Phys. Rev. Lett. **61**, 2099-2102 (1988).
105. S. Schmitt-Rink, and W. Schäfer, private communications (1990).
106. G. Bastard, Delalande, C.; Ferreira, R.; Liu, H.U. , J. Lumin. **44**, 247-263 (1989).
107. H. G. Roskos, M. C. Nuss, J. Shah, K. Leo, D. A. B. Miller, A. M. Fox, S. Schmitt-Rink and K. Köhler, Phys. Rev. Letters **68**, 2216-2219 (1992).

# ULTRASHORT EXCITATIONS IN SEMICONDUCTORS

Walter E. Bron

Department of Physics
University of California, Irvine
Irvine, California 92717 USA

## I. INTRODUCTION

In keeping with the goals of the Advanced Study Institute, the following is a tutorial review of recent research accomplishments in the field of ultrashort excitations, and the subsequent return of the excited material to thermal equilibrium. The designation of "ultrashort" in the present context refers, for the most part, to events which last of the order of femtoseconds to picoseconds. Electrical circuits are inherently several orders of magnitude slower. Thus, observation of ultrashort events, directly in this time domain, had to await the development of ultrashort optical pulses. Even though readily useful laser sources are now only a decade or so old, the literature on the application of these laser systems is very large. No attempt is made here to cover a major part of the pertinent literature. I choose instead a subset of experiments, which illustrate the richness of the scientific results which have been obtained so far. Specifically, I trace the ultrashort dynamics in semiconductors of nonequilibrium distributions of optical phonons and polaritons in contact with thermal baths, with impurities, with electronic carriers, or with each other. The discussion here includes the experimental apparatus whenever it is unique in some way, and I include discussions of the theoretical basis behind the experimentation, whenever available.

The origin of this line of research lies in the early experiments by von Gutfeld and Nethercot[1] which involved the generation of "heat pulses" and the determination of their transport through various materials. These early experiments had a number of drawbacks as compared to their modern counterparts; namely, the inability to determine the spectral distribution of the constituent phonons composing the heat pulse, and temporal resolution limited for the most part to nano- to microseconds. I have, from time to time, reviewed the progress in this field.[2] The current review starts with an equally important experiment reported by von der Linde, *et al.*

*Ultrashort Processes in Condensed Matter*, Edited by
W.E. Bron, Plenum Press, New York, 1993

on the generation of optically excited longitudinal optical (LO) phonons and their subsequent decay as measured directly in the picosecond time domain. They found a decay rate of 7ps for LO phonons in GaAs held at 5 K.

Before it becomes possible to discuss the significance of these results, it is necessary to step aside for a moment, and discuss the theoretical basis behind most of the experimentation.

## II. THEORETICAL BASIS FOR THE EXPERIMENTAL TECHNIQUES

The experiment by von der Linde, cited above, was one of the first to apply time-resolved spontaneous anti-Stokes Raman scattering to the study of nonequilibrium incoherent optical phonons and their interaction with photoexcited hot electrons and holes. This method was soon superseded by the technique referred to as time-resolved coherent anti-Stokes Raman scattering (TR-CARS)[3] which brought to this line of research all the advantages of coherent excitation and detection and the versatility of multi-wave mixing. In the absence of multi-wave mixing of laser beams, the excitation of the medium occurs only if the incident laser beam is directly resonantly absorbed or is a part of two-photon (or higher order) absorption. Even more interesting is the excitation by mixing two, preferably synchronized, phase coherent, laser beams. The mixing produces electromagnetic waves in the medium which can oscillate at the *sum* and *difference* frequencies. Electromagnetic beams at very low (difference) frequencies can excite phonons and polaritons, and at high (sum) frequencies which can interact with electron-hole pairs and plasma in semiconductors and insulators. These techniques are part of the general category of multi-wave scattering.[3]

As we shall see, many of these techniques involve Raman active excitations. According to Placzek[4], in a Raman active medium the important terms in the optical polarizability, $\alpha$, can be written as

$$\alpha = \alpha_0 + (\partial\alpha/\partial Q_v)Q_v + \cdots, \tag{1}$$

in which $Q_v$ is the coordinate corresponding to some excitation of the solid, and $\alpha_0$ refers to a static polarizability if one is present.

An electromagnetic field, $E$, can interact with the polarization. The pertinent term in the interaction Hamiltonian, $H_1$, corresponding to the lowest-order Raman process, is

$$H_1 = \frac{1}{2}\frac{\partial\alpha}{\partial Q_v}Q_v E^2. \tag{2}$$

It follows that as a result a force,

$$F = -\frac{\partial H_1}{\partial Q_v} = \frac{1}{2}\frac{\partial\alpha}{\partial Q_v}E^2, \tag{3}$$

acts on the medium. Moreover, under the action of both the excitation and the electromagnetic field, a polarization, $P$, is induced, namely,

$$P = -\frac{\partial H_1}{\partial E} = N\frac{\partial\alpha}{\partial Q_v}Q_v E, \tag{4}$$

in which $N$ is the number density of the excited modes. One of the hallmarks of the experimentation to be discussed below is that the laser induced excitation of the solid continues to decay, at some rate, $\Gamma$, even after the laser excitation is turned off. We arrive, therefore, at a picture of the excitation of the solid as a series of damped, driven oscillators, $Q_v$, of mass $m$, and write an equation of motion

$$\ddot{Q}_v + 2\Gamma\dot{Q}_v + \omega_v^2 Q_v = \frac{1}{2m}\frac{\partial\alpha}{\partial Q_v}E^2. \tag{5}$$

As we shall discover below, under coherent excitation, $\langle Q\rangle$ becomes the coherent amplitude, $1/\Gamma$ is related to the dephasing time $T_2$, and $\omega_v$ is the resonance frequency of the oscillator.

In the actual experiment, an electromagnetic field is chosen which contains two frequency components, $\omega_\ell$ and $\omega_s$, produced by two well-defined synchronously pumped coherent lasers, which produce output beams propagating with wavevector $\vec{k}_\ell$ and $\vec{k}_s$. The em field amplitude due to these beams is

$$E(x,t) = \frac{1}{2}E_\ell \exp[i(\omega_\ell t - \vec{k}_\ell\cdot\vec{x})] + \frac{1}{2}E_s \exp[i(\omega_s t - \vec{k}_s\cdot\vec{x})] + c.c., \tag{6}$$

in which $\vec{x}$ is the position vector and $t$ is the temporal coordinate. It is often the case that the available ultrashort pulsed lasers have output frequencies limited to near, or in, the visible spectrum. Clearly if the goal is to produce an effective excitation in the far infrared, then $\omega << \omega_\ell, \omega_s$. In contrast, excitations in the ultraviolet require that $\omega >> \omega_\ell, \omega_s$.

In order to solve Eq. 5 for $Q_v$, we use the trial function,

$$Q_v = \frac{1}{2}q_v \exp(i\omega t) + c.c., \tag{7}$$

and pick out the terms which are consistent with the inequalities in $\omega_\ell, \omega_s$ and $\omega$, i.e.,

$$(\omega_v^2 - \omega^2 - i2\Gamma\omega)q_v \exp(i\omega t) = \frac{1}{2m}\frac{\partial\alpha}{\partial Q_v}E_\ell E_s \exp[i(\omega_\ell \pm \omega_s)t - (\vec{k}_\ell - \vec{k}_s)\cdot\vec{x}]. \tag{8}$$

Note that the temporal terms in Eq. 8 requires that $\omega = \omega_\ell \pm \omega_s$. If $\omega_\ell = \omega_s$ and $\vec{k}_\ell = \vec{k}_s$, the result is the well known case of two-photon absorption (TPA) to which we shall return from time to time. For the present purposes we limit discussion to the case $\omega_\ell \neq \omega_s$, but $\omega_\ell - \omega_s = \omega_v$ with $\omega_v$ corresponding to Raman active phonon (polariton) modes.

Solving Eq. 8 for $Q_v$ yields

$$Q_v = \frac{1}{2m}\frac{\partial\alpha/\partial Q_v}{[\omega_v^2 - (\omega_\ell - \omega_s)^2 + i2\Gamma(\omega_\ell - \omega_s)]}E_\ell E_s \exp[-i(\omega_\ell - \omega_s)t - (\vec{k}_\ell - \vec{k}_s)\cdot\vec{x}]. \tag{9}$$

It is clear from Eq. 9 that a strong resonant excitation can occur if $\omega_\ell - \omega_s$ equals the frequency of a Raman active mode, $\omega_v$, providing that the phase-matching condition $\vec{k}_v = \vec{k}_\ell - \vec{k}_s$ can be met.

It is most important to note that the range of frequencies and wavevectors allowed for the excited mode is highly limited by the wavevectors of the incident beams, the degree of focus, and the uncertainty of $\omega_\ell$ and $\omega_s$, etc. But these effects are very small. Thus, unlike incoherent spontaneous Raman or stimulated Raman scattering, the phase space available to the excitation is severely limited. Consequently, the occupation probability of the excited states may be quite large even with modest laser intensities. The resultant "coherent phonon state" does indeed resemble most closely a classical harmonic oscillator[5] and is the basis for Eq. 5. In real media there always exists the possibility that the coherent state somehow scatters from some component of the media. For example, a vibrational mode may scatter at crystal surfaces, bulk imperfections, electronic carriers, or through phonon-phonon interactions, etc. During the scattering process the wavevector of a component of the coherent state may *change*. The excitation is said to "dephase" with a characteristic rate of $2/T_2$.

The discussion so far has neglected to explicitly include the fact that the laser output is not cw but rather consists of trains of ultrashort duration pulses. If the durations of the exciting laser pulses $\Delta t_\ell$ and $\Delta t_s$ are short compared to the relaxation time of the coherent excitation and to the time between successive laser pulses, then the excitation can be observed to decay, through a coherent Raman interaction between the excitation and a time delayed third (pulsed) "probe" laser beam.

Specifically, the presence of a coherent excited state causes an oscillatory polarizability at frequency $\omega_v$, as indicated in Eqs. 1-5 which can, in turn, interact with a synchronized and coherent third laser beam at frequency $\omega_3$ to yield a non-linear polarization $P^{NL}$

$$P^{NL} = \chi^{(3)} E_\ell E_s E_3, \tag{10}$$

where $\chi^{(3)}$ is the total third-order non-linear susceptibility which, in general, contains contributions from the response of the electronic, $\chi_E^{(3)}$, and vibrational $\chi_R^{(3)}$ components. The polarization acts as a source term in Maxwell equations (see, e.g., Armstrong *et al.*[6] and Maker and Terhune[7]) to produce a strong coherent output beam with $\omega_4 = \omega_3 + (\omega_\ell - \omega_s)$ provided that $\vec{k}_4 = \vec{k}_\ell - \vec{k}_s + \vec{k}_3$ can be satisfied. It is usual to set $\omega_3 = \omega_\ell$ by taking a small component of the incident "$\ell$" laser beam. Therefore, in practice, $\omega_4 = 2\omega_\ell - \omega_s$. The corresponding energy and wavevector diagrams are shown in Figs. 1 and 2. It is clear from the figure that the interaction leads to coherent anti-Stokes Raman scattering (CARS). Accordingly, we identify $\omega_{AS}$ with $\omega_4$ and $\vec{k}_{AS}$ with $\vec{k}_4$, and $\Gamma \propto 2/T_2$.

Figure 2 illustrates yet another property of the CARS technique. In this figure, the resonant coherent excitation is taken to be a lattice vibration (phonon, polariton) with energy $\hbar\omega_v = \hbar\omega_\ell - \hbar\omega_s$, and wavevector $\vec{q}_v$. It is, however, not necessary that the coherent excitation be resonantly excited. As Fig. 1 clearly illustrates, four wave mixing can be satisfied by transitions to virtual states. Polarization of bound electrons due to the mixing of $\omega_\ell$ and $\omega_s$ is a case in point. This excitation is always present whether $\omega_\ell - \omega_s = \omega_v$ or not.

In an illustrative case, to be discussed in some detail below, the nonresonant component is represented by the polarization of bound electrons and the resonant component is represented by the generation, and subsequent decay, of longitudinal optical (LO) phonons in simple compound semiconductors such as GaP. If we add to this picture the coherent, synchronous, time delayed, probe beam, then the total nonlinear polarization of Eq. 10 becomes

$$P^{NL} = \chi^{(3)} E_\ell E_s E_p = (\chi_E^{(3)} + \chi_R^{(3)}) E_\ell E_s E_p \tag{11}$$

It follows from the above discussion and Eqs. 1, 3, 4, and 5, that the intensity of the TR-CARS signal, as a function of the delay time between the pump and the probe laser pulses, can be written as[8]

$$I(\Delta t) = AS(\Delta k) \int_{-\infty}^{+\infty} dt |E_p(t - \Delta t)[NR_a Q(t) + 3\chi_E^{(3)} E_\ell(t) E_s^*(t)]|^2, \tag{12}$$

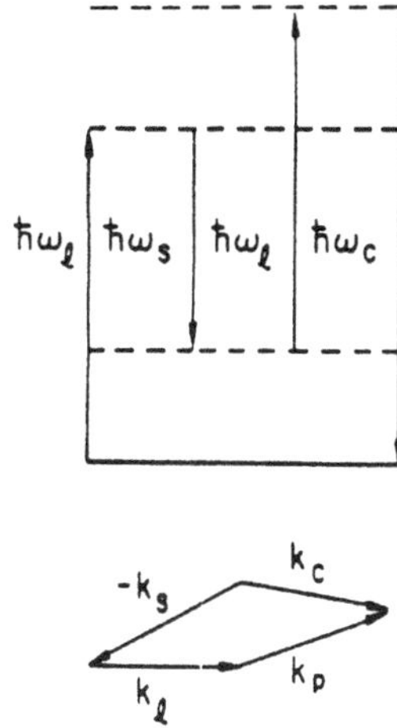


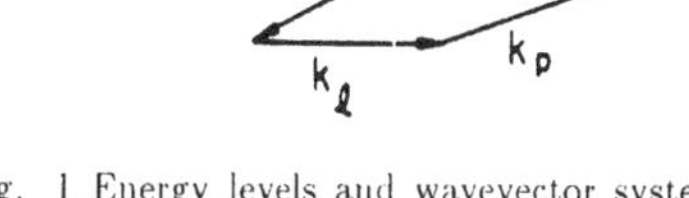


Fig. 1 Energy levels and wavevector system for a non-resonant "virtual" Raman polarization of bound carriers.

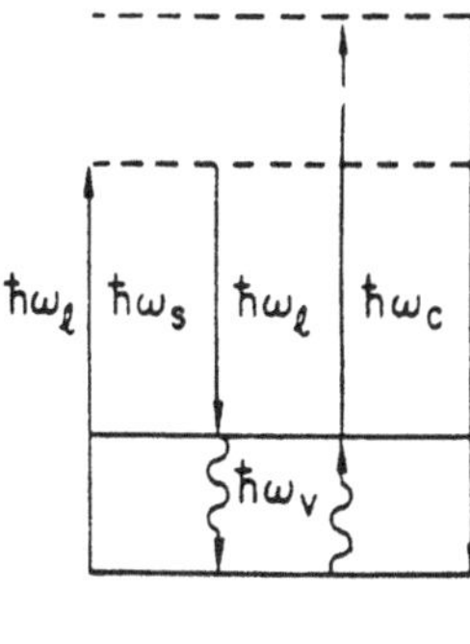


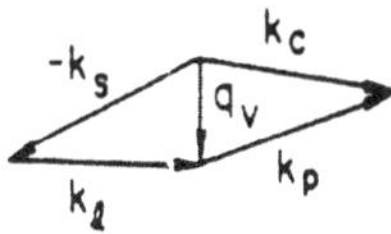


Fig. 2 Energy levels and wavevector system for a resonant excitation of a low frequency Raman active vibrational mode. In both Fig. 1 and 2 energy and momentum (k-vector) conservation is obeyed. Both cases are called coherent anti-Stokes Raman scattering (CARS).

where

$$S(\Delta k) = \frac{\sin^2(\Delta kL/2)}{(\Delta kL/2)^2}, \tag{13}$$

and

$$\Delta k = k_4 - [k_\ell + k_p - k_s], \tag{14}$$

and

$$\ddot{Q} + 2\Gamma\dot{Q} + \omega_v^2 Q = \frac{1}{2m}\frac{\partial\alpha}{\partial Q_v}E^2 = R_a m^{-1}E_\ell(t)E_s^*(t) - \frac{4\pi}{\epsilon_x}\chi^{(2)}E_\ell(t)E_s^*(t). \tag{15}$$

In these expressions $R_a$ is the Raman tensor which is related to $\chi_R^{(2)}$, and the coherent amplitude $Q$ of the excited phonon packet is given by Eq. 15. The wavevector mismatch is given by $\Delta k$; it is set experimentally to approximately zero (hence, $S(\Delta k) \approx 1$). Finally, $\Delta t$ is the temporal delay between the pulses of the probe laser, $p$, and those of the lasers $\ell$ and $s$; $m$ is the reduced lattice mass, $N$ is the number of primitive cells per unit volume and $A = 2\pi\omega_4^2 L^2/c\epsilon_4$. In the expression for $A$, $c$ is the speed of light, $L$ is the effective length within the medium over which spatial overlap of the three-wave mixing exists, and $\epsilon_4$ is the dielectric constant of the medium at $\omega_4$.

That brings to a close a brief discussion of the components of a theoretical basis which describes the experimental observables; namely, the response of condensed matter which has been coherently excited by ultrashort duration optical pulses. The application of the experimental techniques to the excitation and dephasing of various physical phenomena is the major remaining task of this review. However, before proceeding to the physics of optical excitations in solids, it is necessary to at least introduce the very special laser system used to carry out the experimentation.

## III. EXPERIMENTAL BACKGROUND

With but two exceptions the experimental techniques of importance here involve anti-Stokes Raman scattering using pulsed (or in one case cw) laser sources. These sources are applied to produce ultrashort incoherent or coherent pulses with durations in the picosecond and/or femtosecond regime. Again the literature which has accrued on various applicable laser sources is very large, and even a casual perusal of current scientific journals in the field finds ever more new and more sophisticated methods in every new issues of the journals. I, therefore, limit the following to a review of mostly that apparatus developed to accomplish the research in the illustrative set of experiments.

The basic experimental apparatus necessary to conduct TR-CARS experiments is illustrated in Fig. 3. Mode-locking is a technique to phase order the various oscillating modes of a laser. This technique together with inhomogeneous broadening of certain gain media, permit the formation of orderly pulse trains of very short duration pulses.[9] The mode locked laser, shown as part of Fig. 3, is actively mode locked using an acousto-optic resonator. The duration of the output pulses is about 100 ps. The two tunable dye lasers, are themselves passively mode locked using cavity lengths which are matched to that of the "mode-locked laser". The output of the Nd:YAG mode-locked laser is doubled to 532 nm with a KTP crystal and is split in half so as to synchronously pump the two dye lasers. The pulse repetition rate is 76 MHz, and pulse duration of the output of the dye lasers is $\sim$ 7ps with an average output power of $\sim$ 200 mW. The reduction of the output pulse duration of the dye laser stems from the larger spectral width of the dye. The output beams of the two dye lasers are tuned such that $\omega_\ell - \omega_s$ is equal to some Raman active excitation of the solid in which the laser beams propagate. Of course, in order to achieve mixing, the pulses which had been separated at the entrance of the two dye lasers, must be reunited such that the two pulses achieve maximum spatial and temporal overlap.

In a number of experiments the output power of the pulse laser needs to be increased by several orders of magnitude. For these purposes a regenerative amplifier (RA) was added to the basic laser system.[10] We decided to use a cw-pumped Nd:YAG regenerative amplifier which amplifies the dual beams of the synchronously pumped dye laser system to an energy of 12 $\mu$J per pulse. A schematic of the laser system used in this case is shown in Fig. 4. The system starts with the cw mode-locked Nd:YAG laser mentioned above. A small portion of the undoubled output (1.06 $\mu$m) of the Nd:YAG laser is used to seed a cw Nd:YAG regenerative amplifier, which, in turn, produces pulses of 1.8-mJ energy at 1.06 $\mu$m at a repetition rate of 1 kHz. A detailed sketch of the regenerative amplifier is shown in Fig. 5. The input and output of the amplifier are switched by a $LiNbO_3$ Pockels cell controlled by a two-pulse cell driver (Medox Electro Optic model DR 85-A) synchronized to the mode locker. Feedback from the regenerative amplifier to the primary Nd:YAG laser is minimized by an acousto-optic deflector. To prevent laser-induced damage to the $LiNbO_3$ Pockels cell, we inserted a 1:3 beam expander in the cavity of the regenerative amplifier. The amplifier output is doubled with a beta barium borate (BBO) crystal and reaches an energy per pulse of 0.8 mJ at 532 nm (45% conversion efficiency).

The doubled output of the amplifier is used to longitudinally pump dual, two-stage, dye amplifier chains using cuvettes with flowing dyes of Rhodamine 590 at a concentration of $10^{-4}$ M in water (see the right hand side of Fig. 4). Saturable absorber jets with linear absorption of about 99% have been placed between the amplifier stages to reduce the background, consisting of the unamplified 76-MHz pulse train and the amplified spontaneous emission of the first stages, and also to sharpen the leading edge of the pulses. In this manner an amplification factor of $6 \times 10^4$ is reached in each of the two dye laser output beams. Thus the total incident beam intensity available in the solid for two photon excitation reaches an amplification of the order of $4 \times 10^9$ times that available from unamplified laser beams.

For a number of experiments within (and without) the illustrative set, it is necessary to obtain trains of 50 to 100fs duration pulses. For this purpose we have constructed a synchronized femtosecond laser which can be substituted for one of the

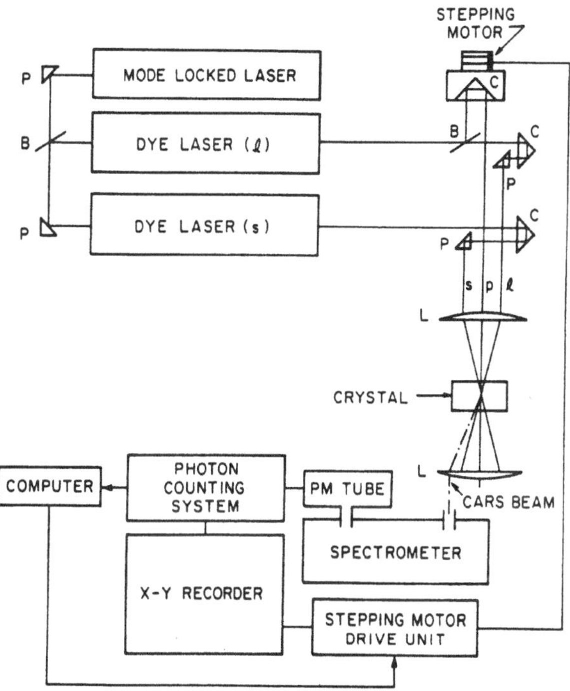


Fig. 3. Basic apparatus for TRCARS four-wave mixing of the incident laser beams in the overlap volume inside some solid. B denotes a beam splitter, C a retroprism, L a lens, and P a turning prism.

synchronized dye lasers discussed above. It should be understood that multi-wave mixing reaches its maximum output only when the maximum of all pulses overlap temporally and spatially. Therefore as we mix one pico-and one-femtosecond pulse, nonlinear excitation of the medium occurs only as long as the shortest duration laser pulse is present. Thus, for the generation of femtosecond excitation, one femtosecond laser is substituted for one of the picosecond dye lasers.

The femtosecond dye laser is synchronously pumped by the doubled output of the Nd:YAG mode-locked master oscillator. The femtosecond laser consists of a standard six-mirror linear cavity dye laser, shown in Fig. 6. The laser contains

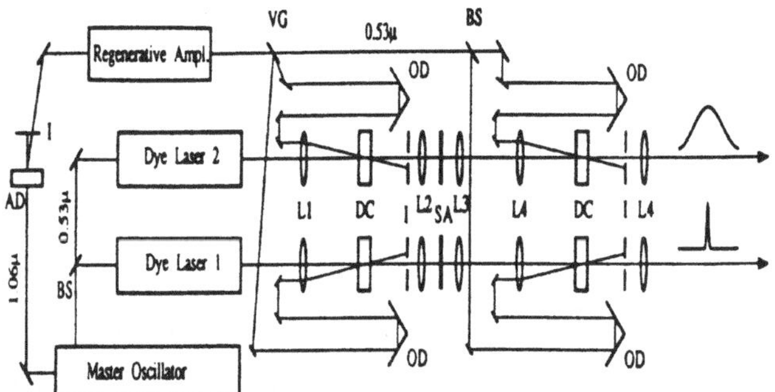


Figure 4. Sketch of a dual, synchronous, tunable amplifiable ultrashort pulsed laser system. The master oscillator (ML) is a mode-locked Nd:YAG laser, DL are two tunable dye lasers, RA is a regenerative Nd:YAG amplifier, AD is an acousto-optic deflector, BS are beam splitters, L are lenses, DC are flow through dye laser quvettes, SA saturable absorber, OD optical delays, I are irises and VG glass wedges

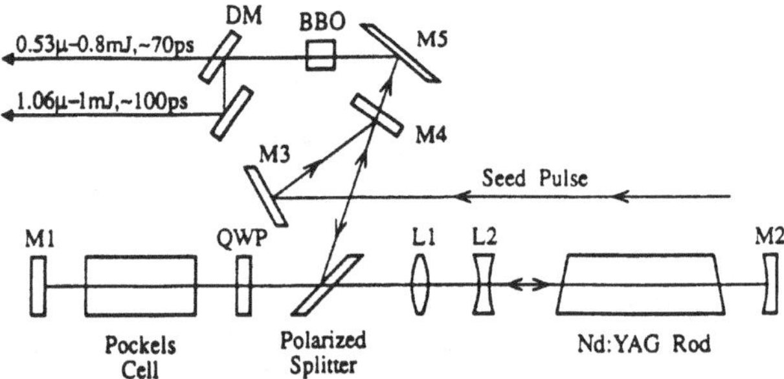


Figure 5. Regenerative amplifier: L stands for lenses, M for mirrors, QWP quarter wave plate, BBO is a beta barium borate doubling crystal. For more detail see ref. 10.

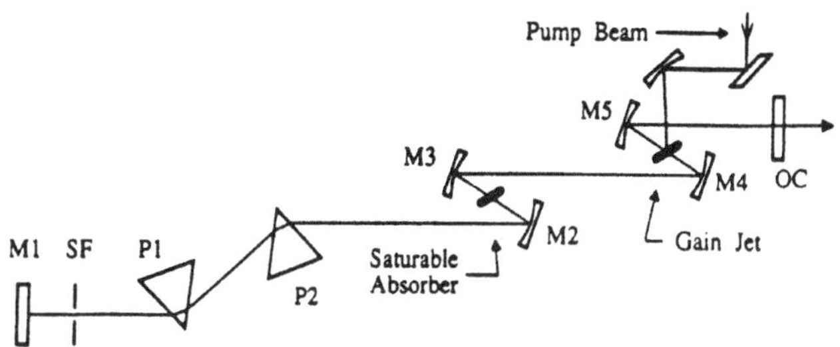


Figure 6. An essentially linear femtosecond laser containing a dye jet gain material and a saturable absorber. M stands for mirrors, OC is output coupler, P for prisms for negative group velocity dispersion. For further details see ref. 10.

only single stack mirrors plus Brewster-angle prisms to provide intracavity control of the group velocity dispersion (GVD). At optimum GVD compensation, the laser delivers 60-80 fs pulses at approximately 630 nm wavelength using Rhodamine 590 tetrafluoroborate gain, and DODCI absorber, dyes. The output power is 50 mW and the amplitude fluctuation of the output pulses is less than 8 percent. As with the case of the picosecond lasers discussed above, the output of the femtosecond laser can be synchronously amplified with the output of the regenerative amplifier.

The femtosecond laser pulses undergo significant temporal broadening when passing through the amplifier chain due to the positive GVD inherent in the lenses and cuvette walls. The pulse duration is 130 fs after passing the unpumped amplifier chains, while the duration of the amplified pulse is 180 fs. Instead of using an external pair of prisms for pulse shape restoration, shorter amplified pulses were achieved using the intracavity prism pair to generate negatively dispersed pulses. With the overcompensated cavity, the laser generated pulses of 235 fs duration. After passing the unpumped chain, the pulses were compressed, due to the glass in the chain which is a source of positive GVD, and leads to an output pulse duration of 70 fs. This concludes the discussion of the elementary basis and the experimental apparatus which form the backbone of the research to be described below. Any deviation of the experimental apparatus, or expansion of the theoretical basis, are covered directly in the pertinent sections of this review.

## IV. CARRIER LIFETIMES AND OPTICAL PHONON DECAY

As noted in the introduction, the first experiment along these lines was performed on GaAs by von der Linde *et al.*,[11] who used time-resolved spontaneous Raman scattering to study the dynamics of non-equilibrium *incoherent* optical phonons generated during the decay of photoexcited hot electrons and holes.

Electron-hole pairs (e-h pairs) were generated with pulses from dye lasers with autocorrelation duration of 2.5 ps. The "excess" energy of the electron-hole pairs, above the minimum of the direct gap, is roughly 17 times that of the energy of the zone centered LO phonon in GaAs, namely, 36.5 meV.

A number of sign-posts exist in our knowledge of the dynamics of such excitations in solids. For example, it is known that highly excited electron-hole pairs, in polar semiconductors, decay into lower energy states via the emission of low wavevector, (LO) phonons primarily through the so-called "Fröhlich Interaction".[12] The decay

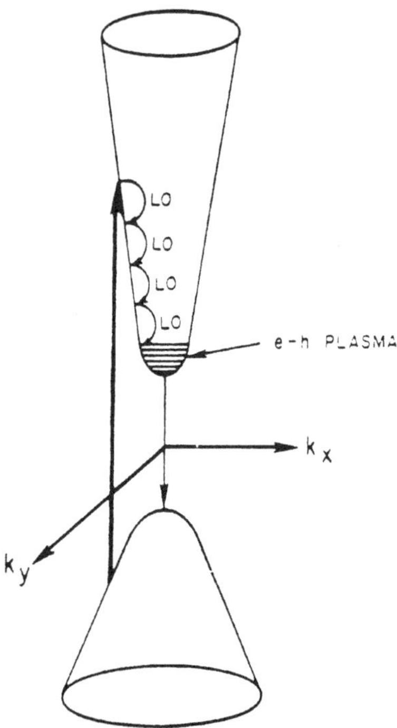


Figure 7. Schematic of the excitation and subsequent decay of a hypothetical direct gap semiconductor with a simple parabolic band structure.

proceeds through a cascade of carrier-phonon interactions terminating in the creation of an electron-hole plasma (gas or liquid) at some minimum of the conduction band. Fig. 7 is an illustration of the excitation and decay in a hypothetical direct gap semiconductor with simple parabolic bands. There are a number of auxiliary features associated with the dynamics of the decay process. For typical polar semiconductors the electron effective mass is approximately $m^* = 0.25m_e$. This means that the curvature of a simple parabolic conduction band is such that the LO phonon which can interact with an electron in the cascade must possess a wavevector, $\vec{q}$, of the order of $10^5$ cm$^{-1}$ and that this value needs to change as the cascade proceeds. This conclusion holds true for electrons with energies of up to 20 times the LO phonon energies above the band minimum. Accordingly, measurements of the decay by time resolved incoherent Raman scattering, or by two-, or three-wave mixing experiments, such as the CARS method described above, all need to be performed in the backward scattering geometry if phonons with such relatively high wavevectors are to be detected. It has also been predicted(12) that the average time interval between successive steps in the cascade is of the order of 100 to 200 femtoseconds which is shorter than the carrier-carrier scattering time in semiconductors with carrier concentrations of less than $10^{17}$ cm$^{-3}$.

The interaction of phonons with an electron-hole plasma, which is formed at the end of the cascade, has been studied in some detail as described in Section IX below. In an earlier work it was found experimentally(13) in GaP that the LO phonon frequency increases as the plasma concentration exceeds above $0.5\times10^{18}$ cm$^{-3}$, but that little effect is observed below that concentration and that, curiously, no change in the phonon lifetime is observed up to concentrations of about $5\times10^{18}$ cm$^{-3}$.

As will be discussed in some detail in later sections, in polar semiconductors, such as GaP, the LO phonons themselves decay into acoustic phonons in times of the order of picoseconds and that the dominant decay mechanism, at least at low temperatures, is three-phonon anharmonic processes.

It should already be apparent to the reader that the ultrashort dynamics of electrons and phonons in these materials exhibits multifaceted and interesting properties. In the following sections, I shall address a number of these facets in more detail. The subsequent decay of the excited electron-hole pair described above and as illustrated in Fig. 7, was monitored by a probe pulse (of the same laser frequency) which preceded or followed the pump pulse. The time delayed anti-Stokes Raman scattering off the LO phonon distribution is, accordingly, a function of the temporal evolution of the LO phonons (with $q \sim 10^5$ $cm^{-1}$). The experimentally measured anti-Stokes Raman signal is illustrated in Fig. 8. The zero of the delay time is fixed at the maximum of the pump pulse. The time independent background of approximately 70 counts per second is the result of carrier excitation by the probe pulse, and the weak signal at about 28 counts per second results from residual stray laser light and detector dark current. The rise of the signal above $\sim$ 70 counts per second indicates the growth of the LO phonon as the result of the decay of the photoexcited electron-hole pairs. The rise time of the signal of approximately 6 ps is related to the cascade of the e-h pairs and their decay into the LO phonons. But, since the temporal resolution in this experiment was only 2.5 ps, no further attempt was made to use this information to determine the electron-phonon interaction time.

The decrease in the anti-Stokes signal is a function of the decay of the LO phonons into acoustic phonons; a point which will be substantiated to a considerable degree in later sections of this review. As shown in Fig. 9, the decay follows an exponential function, $\exp(-\Delta t/\tau')$, with $\tau' = 7 \pm 1$ ps when the sample is held at 77°K. The solid curve in Fig. 9 is a convolution of the temporal evolution of the number of phonons and the evolution of the probe pulse, whereas the dashed curve is the laser autocorrelation signal. The experiment, described above, was repeated by Kash, *et al.* [14]through incoherent Raman scattering measurements in the spectral domain but, more importantly carried out with femtosecond duration laser pulses after various delays relative to the pump pulse. In this experiment the laser pulse duration was only 600fs. The pump and probe pulses were polarized perpendicularly to each other as was the LO phonon Raman (backward) scattering excited by the pump and the probe pulses. However, the pump pulse was observed to also produce a spectrally broad, but unpolarized emission band. This emission, plus the signal from stray laser light and detector dark current, were compensated for by taking difference measurements between spectra taken with the probe delay time, $\Delta t$, fixed at various values less that obtained when $\Delta t < 0$ ps.

Due to the better temporal resolution available to Kash, *et al.*[14] compared to that available to von der Linde, *et al.*[11] it becomes possible to resolve the electron cascade in more detail. It is found that the total cascade consists of 12 steps each of approximately 165fs long; in very good agreement with the predictions. The phonon decay time in GaAs near 80K was found to be 7.5 ps, in good agreement with the earlier results of von der Linde, *et al.*

The observed decay time, $\tau$, is related only remotely to the fundamental phonon lifetime. As should be apparent from Fig. 7, the e-h pair decay produces an ensemble of optical phonons with a spread of $\vec{q}$. The decay of such an ensemble may well differ from the decay of a well defined coherent phonon state as first reported by Kuhl and Bron.[15]

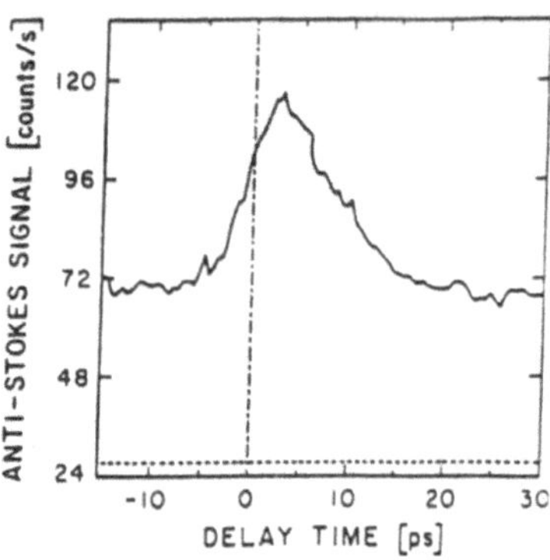


Figure 8. Time resolved incoherent anti-Stokes Raman scattering intensity from an optically pumped GaAs sample.

These relatively simple experiments yield considerable understanding of the dynamics of the optical excitation and the subsequent decay of the excitation. A number of facets of these dynamics are now clear. In polar semiconductors, hot carrier decay proceeds via a cascade of emission of LO phonons with each step of the cascade lasting of the order of 150 fs. The LO phonons produced possess wavevectors of the order of $10^5$ cm$^{-1}$ and decay in GaAs at 77K in about 7 ps.

## V. COHERENT EXCITATION OF LO PHONONS AND THEIR SUBSEQUENT DEPHASING

My first illustration of coherent multiwave generation and dephasing of optical phonons, concerns the dynamics of coherent LO phonon distributions in GaP (other compound semiconductors are cited in the literature). In this example, two laser pulse trains $(\ell, s)$ are tuned so that $\omega_\ell - \omega_s$ is equal to that of the near zone-center LO phonon (or polariton) in compound semiconductors. The excitation yields a coherent phonon state with wavevector $\vec{q}_v$. The excitation takes place over periods of a few picoseconds,[16] or hundreds of femtoseconds.[17] After the laser pulse ceases, the excitation is observed to decay (more correctly dephase). The coherent state is monitored with a time delayed probe pulsed laser train.

The subsequent temporal variation of the coherent LO phonon packet is measured with the TR-CARS technique. For this purpose a small part of the $\omega_\ell$ beam is extracted and delayed by an optical delay line, so that pulses from it appear at the overlap volume at any delayed time, $\Delta t$, of between 0 and 0.3 ns, with a resolution of fractions of picoseconds. As described above, the delayed beam scatters from that component of the LO phonon state which remains coherent after $\Delta t$, producing thereby a coherent beam propagating along $\vec{k}_4$.

Figure 10 is an example of the typical temporal variation of the TRCARS signal intensity, $I(\Delta t)$, for near zone-center LO phonons with $\omega_v$ = 403 cm$^{-1}$ and at an ambient temperature of 5 K for a GaP crystal with a carrier concentration of about $10^{16}$ cm$^{-3}$. The component of the signal from $\approx$ - 40 to $\approx$ +20 ps is primarily the

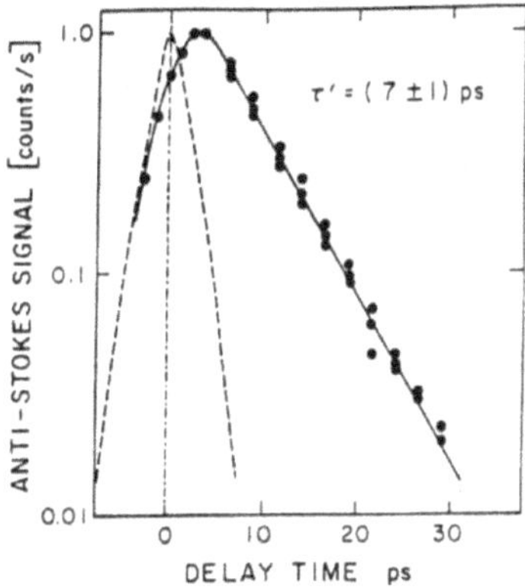


Figure 9. Logarithm of the incoherent anti-Stokes Raman scattering intensity and the laser auto correlation signal (dashed line). The signal for positive delay times decreases exponentially at a time of 7± 1ps.

result of non-linear excitation of the bound electronic system, i.e., through $\chi_E^{(3)}$ as described in section II. The full shape of this signal (dashed curve) can be obtained by slightly detuning $\omega_\ell - \omega_s$ away from $\omega_v$.

The signal beyond 20 ps is a measure of the dephasing of the coherent intensity $\langle Q \rangle^2 \propto \exp[-2t/T_2]$, where $\langle Q \rangle$ is the coherent amplitude of the excited LO mode[18]. The signal intensity is seen from Fig. 10 to decay exponentially over several orders of magnitude with a time constant $\tau = T_2/2$. Here $T_2/2$ is the traditional "dephasing time". The temperature dependence of $\tau$, as observed over a range from 5 to 300 K, is shown in Fig. 11 (solid circles). It is generally accepted that the linewidth $\delta v$ (in $cm^{-1}$) of the same Raman mode, $\omega_v$, as observed from spontaneous Raman scattering is related to $T_2$ by $\delta v = (\pi c T_2)^{-1}$ provided that the Raman line is homogeneously broadened[18]. Linewidth measurements of the LO phonon (at $k \approx 0$) have been

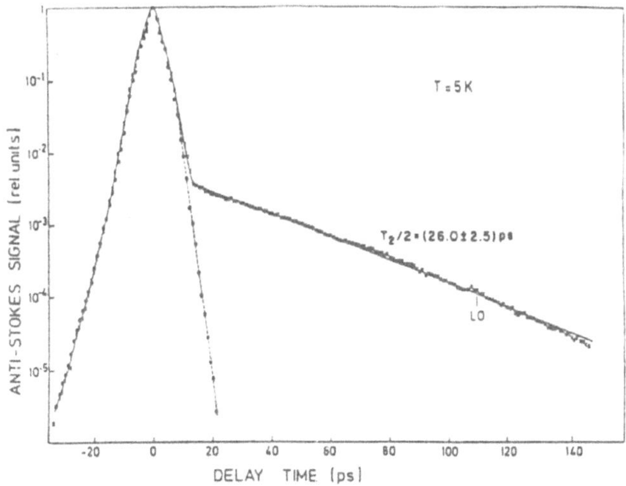


Figure 10. Coherent anti-Stokes Raman scattering signal as a function of the delay time. These results were obtained for a very high purity sample of GaP held at 5K. The measured dephasing time is 26.0 ± 2.5ps.

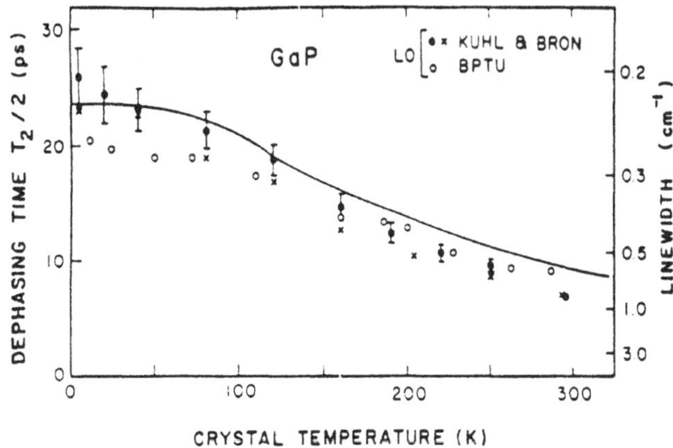


Figure 11. Temperature dependence of the dephasing time and the bandwidth of LO phonons in GaP. The solid line is a theoretical comparison to the data.

reported[19] through high-resolution Fabry-Pérot interferometry. The corresponding dephasing times are designated as BPTU in Fig. 11 (open circles). We have repeated these measurements using our samples and a standard double monochromator providing 0.15 cm$^{-1}$ resolution. Phonon dephasing times are obtained by deconvoluting the spectral function of the instrument from the measured Raman spectrum assuming a Gaussian profile for the spectrometer function and a Lorentzian profile for the Raman line. The results are displayed as crosses in Fig. 11.

The time $T_2$ includes contributions from all dephasing phenomena, such as elastic scattering from surfaces, from impurities, from imperfections, and from electronic carriers, etc., and also includes phonon-phonon scattering which leads to phonon population decay. The population decay time is conventionally referred to as $T_1$. Since $T_2$ contains $T_1$ it follows that $T_2 \leq T_1$.

In the absence of any detailed knowledge of the strength of the various scattering processes it is not possible to calculate the magnitude of $T_2$, although it is possible to predict theoretically its temperature dependence. Contributions to the temperature dependence of $T_2$ can arise from scattering from thermally activated carriers and from various depopulation processes. We concentrate here on the latter processes. At low ambient temperature, such that $\hbar\omega_v \gg k_B T$, the spontaneous three-phonon decay of $k \approx 0$ optical phonons into two acoustic phonons (plus the corresponding recombination process) should dominate all other phonon-phonon interactions.

There are three possible three-particle interactions:

(i) a LO phonon decays into two acoustic phonons;

(ii) two acoustic phonons recombine to form a LO phonon; and

(iii) a LO phonon combines with an acoustic phonon to form an optical phonon.

Channel (iii) is not active since it would produce a phonon of higher energy than the LO phonon. This is not possible in GaP to within three-phonon processes.

The governing equation of motion for a zone centered LO phonon, with resonant frequency $\omega_{LO}$, nonlinearly coupled to two acoustic phonons $gj$ and $g'j'$, and being driven by a pulsed laser with temporal profile, $F(t)$ can be written as

$$\ddot{A}_{LO}(t) + \omega_{LO}^2 A_{LO}(t) + \omega_{LO}\Sigma_{gjj'} V_3 \begin{pmatrix} 0 & q & q' \\ LO & j & j \end{pmatrix} A_{gj}(t) A_{g'j}(t) = F(t) \tag{16}$$

in which $V_3$ is the (cubic) anharmonic coupling coefficient, $A_{LO}$ is the LO phonon normal coordinate and $A_{gj}$ is that of the acoustic phonons. Here $g$ refers to the mode's wavevector, and $j$ to the mode's dispersion branch.

Eq. 16 can be rewritten in terms of the self energy, $\tilde{\Pi}_{LO}(\tau)$,

$$\ddot{A}_{LO}(t) + \omega_0^2 A_{LO}(t) + 2\omega_{LO} \int_{-\infty}^{\infty} \Pi_{LO}(t-t') A_{LO}(t') dt' = F(t) \tag{17}$$

with the Fourier Transform of $\tilde{\Pi}_{LO}(\tau)$ being:

$$\tilde{\Pi}_{LO}(\omega) = \Delta_{LO}(\omega) + i\Gamma_{LO}(\omega) \tag{18}$$

and $\Delta_{LO}(\omega)$ is the Kramers-Kronig transform of $\Gamma_{LO}(\omega)$. This is the well known result that the anharmonicity leads to a change of the LO phonon frequency ($\Delta_{LO}$) and a broadening of its spectral profile ($\Gamma_{LO}$).

The solution of the governing equation of motion for the damping rate is found to be

$$\Gamma_{LO}(\omega) = \frac{\pi}{2} \sum_{\vec{q}} \left| V_3 \begin{pmatrix} 0 & \vec{q} & -\vec{q} \\ LO & j & j' \end{pmatrix} \right|^2 \delta(\omega - \omega_{\vec{q}j} - \omega_{\vec{q}j'})[n_{\vec{q}j} + n_{\vec{q}j'} + 1]. \tag{19}$$

In Eq. 19 $\delta(\omega)$ is the joint density of states and the $n_{qi}$ are the occupation numbers of the acoustic phonon states into which LO phonon decays. Equation 19 may be simplified to

$$\Gamma_{LO}(\omega) = \Gamma_{LO}^0 [n_{qj} + n_{q'j'} + 1] \tag{20}$$

where $\Gamma_{LO}^0$ is the damping rate of the LO phonons at T=0 K.

The topography of the dispersion relations for GaP[21] suggest that the zone-center LO phonon can only decay into two LA phonons each with energy $\hbar\omega_{LO}/2$ and wavevector $\vec{q}$ and $-\vec{q}$ so to maintain energy and momentum conservation. As we shall see in section VI on polariton decay, this restriction to the "half energy" LA phonons is not the general case. It is, however, the simplest illustrative case. Symbolically, we write the decay process as

$$LO(\omega, 0) \rightarrow LA(\omega/2, q) + LA(\omega/2, -q). \tag{21}$$

Thus in Eq. 19 $\omega_{\vec{q}j} = \omega_{\vec{q}j'}$ and under thermal equilibrium

$$n_{qj} = n_{q'j'} = n_{\vec{q},j}^T \tag{22}$$

where $n_{\vec{q}j}^T$ is determined from Bose-Einstein statistics to be

$$n_{\vec{q}j}(\omega, T) = [\exp(\frac{\hbar\omega}{k_B T}) - 1]^{-1}. \tag{23}$$

Expression 23, with its magnitude fit to the low-temperature data, appears as the solid line in Fig. 11. Comparison with the data shows indeed that the low-temperature dependence of $T_2$ is dominated by three-phonon anharmonic interactions, i.e., by $T_1$.

Beyond this temperature additional processes, including higher order phonon-phonon interactions and scattering from thermally activated carriers, must also be considered. We conclude, therefore, that it is possible to measure the temperature dependence of the dephasing time of optical phonons and to account for the observation in terms of Bose-Einstein statistics.

Finally we note, that in these experiments the coherent optical phonons were purposely generated well within the bulk of the sample. As a consequence of the low group velocity of optical phonons, it becomes possible to neglect boundary scattering during the limited experimental observation time ($\sim$ 200 ps). It is also possible to neglect impurity scattering. It is easy to show that elastic scattering at point imperfections is, unlike anharmonic decay, essentially independent of temperature. Thus, the effective dephasing time $T_2/2$ in the presence of both elastic and anharmonic decay becomes

$$\frac{2}{T_2} = \frac{1}{T_{imp}} + \frac{(1 + n' + n'')}{\tau_0}, \tag{24}$$

where $T_{imp}$ is the dephasing time due to elastic scattering at point imperfections. The best fit to the data of Fig. 11 results in values for $T_{imp} = \infty$ and $\tau_0 = 24.6$ ps. Thus, impurity scattering does not play a role in optical phonon dephasing for samples with impurity content of $\leq 10^{16}$ cm$^{-3}$. It will be further demonstrated in Section X that in high purity samples, and low ambient temperature, that scattering with electronic carriers is also weak. Thus it turns out that, for high-purity samples and laser average power of $\sim$10 mW, the main source of optical phonon dephasing is anharmonic decay.

## VI. COHERENT EXCITATION OF POLARITONS AND THEIR SUBSEQUENT DEPHASING

We shall return to the interaction between optical phonons and electronic carriers in sections IX and X below. Instead we discuss next the interaction of polaritons with thermal phonons, in the same sense as we have discussed the interaction of LO phonons with a thermal bath. Polaritons are excitations of coupled transverse optical (TO) phonons and the polarization from electromagnetic excitations traveling in a medium. The net result of this coupling is illustrated in Fig. 12 which is the dispersion relations for polaritons in GaP. For the smallest values of $\hbar ck$ (a measure of the crystal momentum) the properties of the polariton approach that of pure photons, whereas the properties approach those of TO phonon modes at high values of $\hbar ck$. The most interesting section of the polariton dispersion is clearly in the intermediate region, in which the polaritons can be considered to be a mixture of photons and phonons. We have determined the polariton lifetime in the coupled region in much the same way as the lifetime of LO phonons discussed above.

It should be noted from the outset that the restrictions imposed by the dispersion relation for the lifetime of LO phonons in GaP are somewhat relaxed for polaritons in GaP. Namely, the restriction on LO phonon decay into two LA phonons, each of half the energy of the LO phonon (and wavevector $\vec{k}$ and $-\vec{k}$) is lifted for polariton decay in GaP. Depending on the frequency of the polariton, $\omega_\pi$, energy and momentum may be conserved, during polariton ($\pi$) decay into one LA phonon with wavevector

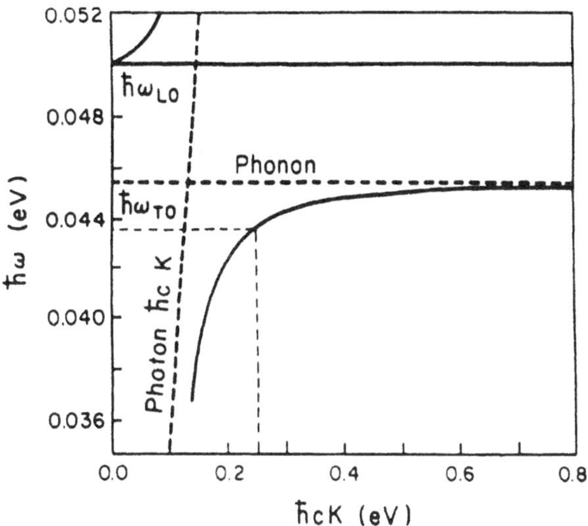


Figure 12. Dispersion relations for photons and LO, TO phonons and polaritons for very small values of the k-vector.

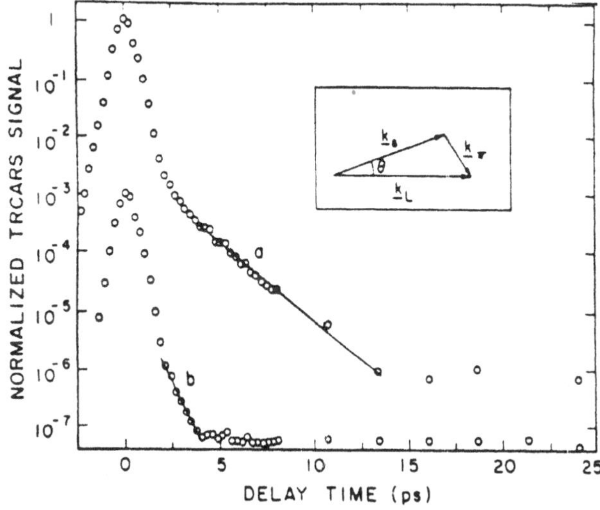


Figure 13. Plot a: Normalized TRCARS signal as a function of the delay of the probe pulse. Plot b: The residual TRCARS signal after the linear portion of plot "a" has been analytically subtracted. The inset depicts wavevector conservation between the $\ell$ and s incident laser beams and the polariton wavevector $k_\pi$. The ambient temperature is $\sim$ 5K.

near the X-point and one TA phonon near the minus X-point of the Brillouin zone, in addition to decay into two LA phonons each with energy $\omega_\pi/2$ and equal, but oppositely signed wavevector.

The experiment used to evaluate the more complicated dephasing of polaritons as compared to LO phonons, is described below. The existence of polariton decay into the TA(X) state has also played a major role in the discovery of anamously long-lived phonons to be discussed in section VIII.

The femtosecond laser system discussed in section III above is adapted for the coherent generation of polaritons, and their lifetime is determined through the TR-CARS technique. In this case, the laser system contains one synchronously pumped tunable picosecond and one synchronously pumped linear cavity femtosecond laser. The femtosecond pulse train is split into equal parts to produce the $\omega_\ell$ and the $\omega_p = \omega_\ell$ pulse trains, and which together with the picosecond $\omega_s$ pulse train, are focused in the bulk of a high purity ($< 10^{16}$ cm$^{-1}$) GaP sample. The focal spot (beam waist) is ~50 $\mu$m in diameter.

A number of factors influence the effective spectral distribution of polaritons generated through CRE. Among these are the forward scattering geometry which limits the wavevector of the coherent polaritons (see insert to Fig. 13), and the narrow spectral bandwidth of the picosecond laser (~ 5 cm$^{-1}$) compared to that of the femtosecond laser (~45 cm$^{-1}$) means that the latter laser dominates the effective bandwidth of the CRE. In this experiment, the femtosecond laser peak frequency was fixed at 624 nm, and that of the picosecond laser is mixed with the femtosecond laser to produce various excitation frequencies. Once the distribution of polariton wavevectors is determined, the corresponding spectral distribution can be obtained through the polariton dispersion relations of GaP.[20] A typical result is illustrated in the insert of Fig. 14. In this example the central polariton wavevector is $q \sim 2 \times 10^3$ cm$^{-1}$, which corresponds to a spectral distribution peaked at ~354 cm$^{-1}$. However, a 120 mm focal length lens is used to focus the $\ell, s$ and $p$ beams into the sample, which results in an angular spread in the wavevector of the laser beams and, hence, in a spread in polariton wavevector of between ~$1.40 \times 10^3$ to ~ $2.6 \times 10^3$ cm$^{-1}$. The corresponding range of polariton frequencies is from ~330 cm$^{-1}$ to ~360 cm$^{-1}$ (see insert to Fig. 14).

In order to avoid the generation of LO phonons ($\omega_{LO}$ ~403 cm$^{-1}$) in addition to the polaritons, the bandwidth of the femtosecond laser was limited to ~45 cm$^{-1}$ which corresponds to a pulse duration of ~300 fs. This is achieved through adjustments of two intracavity prisms in the femtosecond laser, through which it is also possible to produce pulse durations of less than 70 fs (see section III above). The energy per pulse in the output of each laser was ~0.6nJ.

Figure 13 illustrates the temporal dependence of the TR-CARS signal amplitude at an ambient temperature of 5K. The TR-CARS signal shows three different components. The first component from ~0 ps < $\Delta t$ <~2.5 ps is the response of the bound electrons in GaP through the third order nonlinear electronic susceptibility, $\chi_E^{(3)}$, (see section II above). The second component of the signal (~1.5 ps < $\Delta t$ <~ 5ps), which occurs after the CRE is turned off, is nonlinear on a semilog plot, i.e., the dephasing of the polaritons in this time interval does not correspond to a single decay rate. The third component (~5 ps < $\Delta t$ < 12.5 ps) illustrates a clearly exponential decay corresponding to a polariton dephasing time $T_2/2 = 1.72 \pm 0.16$ ps. Analytically subtracting this third component from the total signal, uncovers a second exponential decay with a dephasing time $T_2/2 = 620 \pm 60$ fs. In order to investigate the temperature dependence of the dephasing of the polariton, these measurements are repeated at 50K intervals from 5K to 300K. As a result of the limiting (300 fs) temporal resolution inherent in the measurements it is possible to observe the dephasing of the faster (shorter $T_2/2$) decay channel only up to 200K. Figure 14 displays typical experimental results at 300K. At this temperature the dephasing time of the slower decay component is $T_2/2 = 630 \pm 50$ fs, whereas the dephasing time of the faster component is shorter than the limiting temporal resolution (see Fig. 14). The temperature dependence of $T_2/2$ and that of the polariton dephasing rate (in cm$^{-1}$)[$\Gamma_\pi = (2\pi c T_2/2)^{-1}$], for both decay channels, is indicated

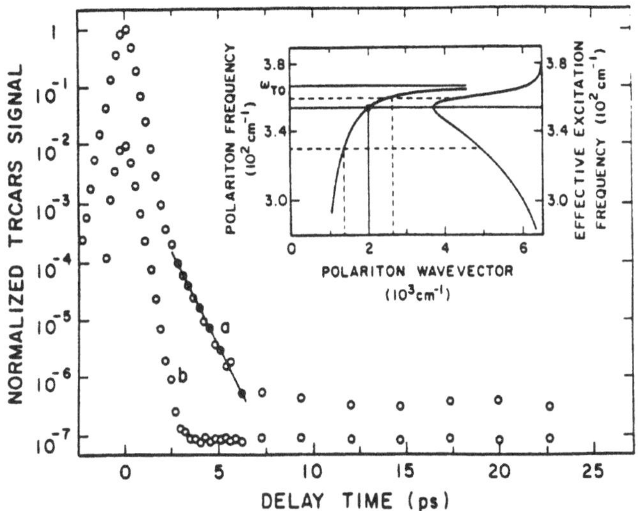


Figure 14. Plot a: Normalized TRCARS as a function of the probe delay. The ambient temperature is 300K. Plot b: Residual signal after the linear part of plot a has been analytically subtracted.

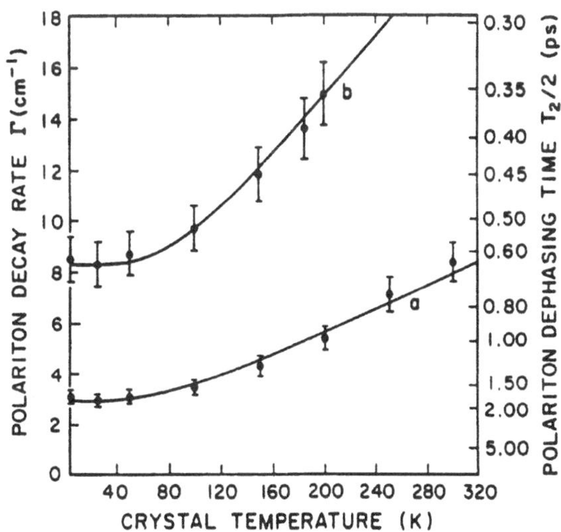


Figure 15. Temperature dependence of the polariton decay rate for the "half-energy" decay channel (b), and the "x-point" decay (a).

in Fig. 15. As noted earlier, at low ambient temperatures, such that the polariton energy $\hbar\omega_\pi >> k_B T$, three particle decay should dominate all other polariton-phonon interactions. We now set out to prove that in the present case the two possible decay channels, mentioned at the beginning of this section, are the source of the more complicated experimental observations. The two possible decay channels are written symbolically as $\pi(\omega_\pi, q \sim 0) \rightleftharpoons LA(\omega_\pi/2, q') + LA(\omega_\pi/2, -q')$ and $\pi(\omega_\pi, q \sim 0) \rightleftharpoons LA(\sim X) + TA(\sim -X)$, in which $\pi$ refers to polaritons, and $X$ refers to the X-point of the Brillouin zone. For simplicity, we refer to the first as the "half-energy" channel, and to the second as the "X-point" channel.

The governing equation of motion is the same as that for LO phonons (see Eq. 19) with $\omega_\pi$ substituted for $\omega_{LO}$. The formal result is the same; namely

$$\Gamma(\omega_\pi) = \frac{\pi}{2} \sum_{\vec{q}} \left| V_3' \begin{pmatrix} 0 & \vec{q} & \vec{q}' \\ \pi & j & j' \end{pmatrix} \right|^2 \delta(\omega - \omega_{\vec{q}j} - \omega_{\vec{q}'j'}) \{ n_{\vec{q}j} + n_{\vec{q}'j'} + 1 \}, \tag{25}$$

We performed two-parameter, least $\chi^2$, fits of the data of Fig. 15 to Eq. 25. With $n_{qj}$ obtained from Eq. 23 above. The two fitting parameters are $\Gamma_0$ and $\omega_1$ with the constraint that $\omega_\pi = \omega_1 + \omega_2$. The fits to the data appear as the solid lines in Fig. 15. For the longer $T_2/2$ (Fig. 15) with $\omega_\pi = 354$ cm$^{-1}$, the best fit is $\omega_1 = (118 \pm 16)$ cm$^{-1}$, $\Gamma_0 = 2.96 \pm 0.4$ cm $^{-1}$, $\omega_2 = (236 \pm 16)$ cm$^{-1}$ with $\chi^2 = 0.9$. This result, when compared to the phonon frequencies at the X-point as obtained from neutron scattering $\{\omega_{TA(X)} = (107 \pm 3)$ cm$^{-1}$ and $\omega_{LA(X)} = (249 \pm 4)$ cm$^{-1}\}$ identifies this as the "X-point" decay channel. Values of $\chi^2 \leq 1$ are obtained for $\omega_\pi$ over a restricted range of $\pm$ 6 cm$^{-1}$ about 354 cm$^{-1}$.

For the shorter $T_2/2$ (Fig. 15) a best fit is obtained with $\omega_\pi = 344$ cm$^{-1}$, $\omega_1 = (172 \pm 26)$ cm$^{-1}$, $\Gamma_0 = (8.2 \pm 0.3)$ cm$^{-1}$, $\omega_2 = (172 \pm 26)$ cm$^{-1}$, and $\chi^2 = 0.5$. For all values of $330 \lesssim \omega_\pi \lesssim 348 cm^{-1}$, $\omega_1 = \omega_2 = \omega_\pi/2$ with $\chi^2 \lesssim 0.8$. This identifies the "half-energy" decay channel. The dispersion relations for GaP[21] permit only excitation of LA phonons for these values of $\omega_\pi/2$. It is obvious from these results that the general case of the decay of excitations need not be limited to single decay channels. On the contrary, the general case has multi-channel decay. However, the analytical tools presented here can be used to ferret out the various contributing decay channels.

## VII. NEW NONEQUILIBRIUM PHONON STATE

So far we have discussed optical excitation of solids for which the concentration of the excitation (so far LO phonons and polaritons), exceeds those at thermal equilibrium by only a small amount. In fact, we explicitly assumed that the excitation lifetime is controlled by its interaction with a thermalized ensemble of phonons. We now turn to an example in which both the concentrations of the excited state (here again LO phonons) and the states into which the excitation decays (here again acoustic phonons) are both considerably above thermal equilibrium, in fact, with concentrations which approach those near and above the damage threshold.

We observe under these conditions evidence for an optically driven nonequilibrium phonon state whose properties differ markedly from the corresponding state at thermal equilibrium. At thermal equilibrium the lifetime of an optical phonon, decaying into two acoustic phonons, decreases and the peak frequency of the optical -phonon spectral distribution shifts to lower values, as the ambient temperature increases.

These effects are a direct consequence of the inherent anharmonic vibrational potential and the properties of thermalized phonon distributions whose spectral distributions are controlled by Bose-Einstein statistics. A theoretical analysis by Bulgadaev and Levinson[22] predicts that these effects are supplanted by others when thermalized phonons are replaced by nonequilibrium phonon distributions. In fact, they predict that a strongly excited narrow bandwidth, LO phonon spectral distribution, in the presence of a highly excited narrow bandwidth acoustic-phonon spectral distribution, will exhibit two (rather than one) decay rates below a critical acoustic-phonon occupation number, $N^*$, and a single decay rate above $N^*$. In addition, below $N^*$ the peak frequency $\omega_0$ of the LO phonon spectral distribution is predicted to remain the value associated with thermal equilibrium, but the spectral distribution splits into two differing peak frequencies above the critical value. A part of these predictions, namely, the observation of two decay rates for the optical-phonon dephasing in the case of acoustic-phonon occupation numbers less than $N^*$ has been observed by us.[23] Some of the other properties of this state are also observed.

The strongly excited nonequilibrium phonon distributions are generated through interactions of a solid with high-power pulsed laser beams described in section III. The equation of motion for this situation is, with only a couple of exceptions, the same as Eq. 16, as is its solution as given by Eq. 19. The exception being simply that the occupation numbers, $n_{\vec{q}j}$ are <u>not</u> those obtained from Bose-Einstein statics and the spectral distribution is no longer Plankian. The actual spectral distribution of the LA phonon is not known at this time. Bulgadaev and Levinson suggest a Lorentzian trial distribution consistent with their model of a new set of coupled acoustic phonon modes. We adopt this assumption here, and write for the new occupation numbers, $n_{\vec{q}j}(\omega)$ as

$$n_{\vec{q}j}(\omega) = n_A \frac{(\Delta\omega/2)^2}{(\omega - \omega_{LO}/2)^2 + (\Delta\omega/2)^2} \tag{26}$$

in which $n_A$ is the occupation number at the peak of the distribution and $\Delta\omega$ is its FWHM. The applicability of this particular distribution is shown by Levinson to be valid on theoretical grounds in the limits of very low and very high optical excitation. As shown below, we observe exponential decay throughout the experiment which is consistent with Lorentzian spectral distributions.

We assume that the intrinsic lifetime of the LA phonons toward decay to even lower energy phonons is long compared to the experimental observation time (~200 ps). Then Eqs. 16-19, and 26 lead to the somewhat unexpected result that the LO phonon excitation exhibits two, rather than a single, exponential decay rate. It is further found that for [$n_A$ less than $N^* = (\alpha - 1)^2/8\alpha$].

$$\Gamma_{1,2} = \frac{1}{2}\Gamma_0[(\alpha + 1) \mp \{(\alpha - 1)^2 - \beta\}^{\frac{1}{2}}], \tag{27}$$

with $\alpha = 2\Delta\omega/\Gamma_0$ and $\beta = 8\alpha n_A$ (typical values for GaP are $\alpha \sim 1$, so that $2\Delta\omega \sim \Gamma_0 \sim 1$ cm$^{-1}$ and $\omega_0 \sim 400$ cm$^{-1}$). Alpha, $\alpha$, turns out to be a crude measure of the deviation from equilibrium phonon distributions. Note that $\Delta\omega$ is related to the spectral width of the nonequilibrium acoustic phonon distribution, whereas $\Gamma_0$ refers to the bandwidth of the equilibrium optical phonon distribution. Thus $\Delta\omega$, and consequently $\alpha$, is expected to increase as the system is driven further into the nonequilibrium state. Note also that as $n_A$ approaches zero, $\Gamma_1$ and $\Gamma_2$ approach $\Gamma_0$ and $2\Delta\omega$, respectively. Thus, the solution with $\Gamma_1 = \Gamma_0$ is the one usually observed in spontaneous Raman measurements at relatively low laser intensities. Bulgadaev and Levinson predict that the strength of the Raman signal associated with $\Gamma_2$, for small $n_A$, is so weak as to be unobservable. This branch of the solution of Eq. 27, however, gains intensity with increasing $n_A$ until $N^*$ is reached, beyond which the intensities are the same and $\Gamma_1 = \Gamma_2$.

Thus, Eq. 27 leads directly to a single decay rate as observed for LO phonons when the acoustic-phonon occupation probabilities are evaluated at thermal equilibrium, i.e., under Bose-Einstein statistics. If, on the other hand, $n_{q,j}(\omega)$ follows a Lorentzian spectral distribution, Eqs. 16-19, and 26 lead directly to two decay rates for the optical phonon. We demonstrate below that, indeed, the two decay rates are observed experimentally.

Before proceeding to the experimental results, it should be noted that the peak laser irradiance (per laser beam) available at the sample must be $\sim 1.3 \times 10^2$ times that normally used (2 $MW/cm^2$) before the second decay process could be observed in our experiment. Note also that the actual energy absorbed by the sample is proportional to the product of the peak irradiance of each of the two "pump" lasers. Thus, considerable energy must be absorbed before the second decay rate becomes observable. This is probably the major reason why the second decay rate has not been previously observed by others.

The amplified picosecond laser system is described in Ref. 10 and reviewed in section III above. In this experiment, however, the irradiance of the pump lasers $\ell$ and $s$ is varied between its "equilibrium" value of ~2 $MW/cm^2$ and a maximum of ~0.7 $GW/cm^2$. A more descriptive independent variable is the peak occupation number $n_A$ of the LA phonons generated by the LO phonon decay (see Eq. 26). For values of $n_A \leq 10^{-2}$ a typical result of the TR-CARS measurement is shown in Fig. 16a. As noted in Ref. 23, the TR-CARS signal intensity has two readily distinguishable components. The first is again attributable to the nonlinear response of bound electrons in the GaP sample. The remaining signal for $\Delta t > 0$ reflects again the dephasing of the LO phonons generated near $\Delta t = 0$. Subtraction of the long-term, exponentially decreasing phonon part of the total signal leaves only the response of bound electrons and a time-independent residue ascribable to various sources of stray background radiation (see Fig. 16b). For values of $n_A \geq 10^{-2}$, however, a second, additional (shorter) dephasing time is readily observed. The faster decay rate is, again, extracted by analytically subtracting the long-term component from the total TR-CARS signal (Figs. 16c and 16d).

Figure 17 (open circles) is a compilation of observed values of $\Gamma_1$ and $\Gamma_2$ (in $cm^{-1}$) for the range of $n_A$ over which two different decay rates are observed. Note that as long as both $\Gamma_1$ and $\Gamma_2$ are separately determined for any one laser excitation level, $\alpha$ may be determined from Eq. 27 since $\Gamma_1 + \Gamma_2 = \Gamma_0(\alpha + 1)$. Moreover, separate evaluations of $n_A$ for fixed values of $\alpha$ lead to $\Gamma_1$ and $\Gamma_2$ as predicted through Eq. 27 (dashed lines). Note that $\alpha$ increases as the laser intensity (and, therefore, $n_A$) increases (see middle part of Fig. 17). A nonlinear dependence of $\alpha$ on $n_A$ was indeed predicted by Levinson. Also note from the top part of Fig. 17 that the calculated value of $N^*$ always exceeds $n_A$. Thus, we are unable in these experiments to reach $N^*$.

We conclude that evidence for the new nonequilibrium phonon state previously proposed by Bulgadaev and Levinson has been observed, and that its properties, for the most part, agree with those predicted. Time-resolved coherent anti-Stokes scattering offers a unique technique for determining the detailed properties of this state. The precise role of the new state as a precursor to laser damage, and in other strongly optically driven phenomena, is being further investigated.

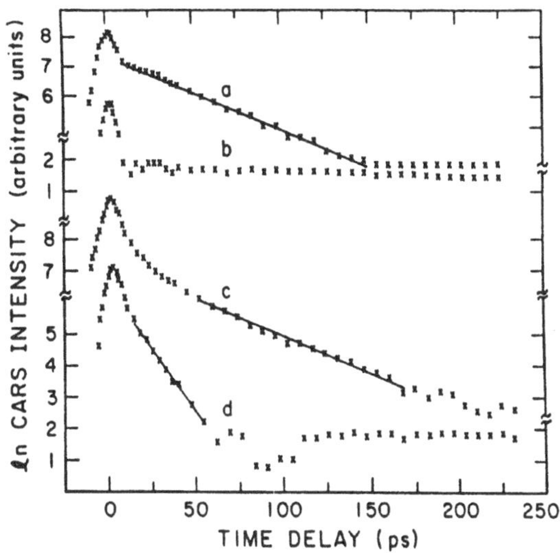


Figure 16. a) and b) TRCARS signal as a function of the probe delay time and with a laser intensity 27 $MW/cm^2$. Traces c and d are for laser intensity of 440 $MW/cm^2$. For details of the method see ref. 15, 16, and 23.

## VIII. DYNAMICS OF ACOUSTIC PHONONS OBSERVED THROUGH VIBRONIC SIDEBAND PHONON SPECTROSCOPY

In the foregoing sections we have identified the decay channels of LO phonons and polaritons in GaP by analyzing the temperature dependence of the dephasing of the excitation. Although the TR-CARS technique used in these analyses is very accurate, it would be helpful to have available an independent experimental observation with which to check the decay channels. The obvious auxillary measurement is the spectral profiles of the acoustic phonon generated through LO phonon or polariton decay.

Unfortunately, the acoustic phonons are not Raman active (in first order) and thus, cannot be detected by TR-CARS. Vibronic sideband phonon spectroscopy (VSPS), which the author developed some time ago for the detection of heat pulses, was adapted to the present experiment in order to monitor the population of polaritons, and the LA and TA acoustic phonons formed as the result of their decay. In VSPS, electronic transitions (zero phonon luminescence), localized at some optically active impurity, are modulated by phonons or polaritons which can couple to the transition. As a result Stokes and anti-Stokes sidebands arise in much the same way as its radio frequency, FM, counterpart. In the presence of phonons an anti-Stokes sideband

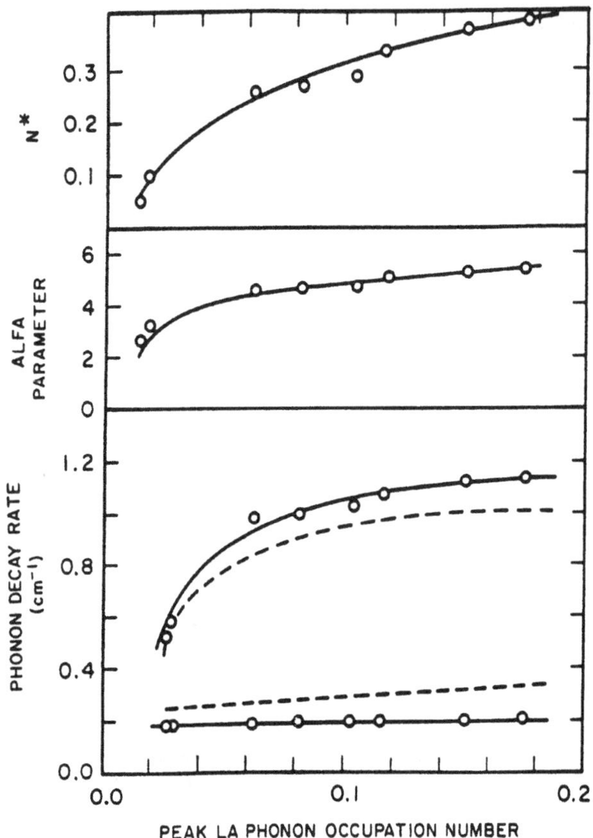


Figure 17. Phonon decay rate, $\alpha$-parameter and $N^*$ as a function of the peak LA phonon occupation number, $n_A$.

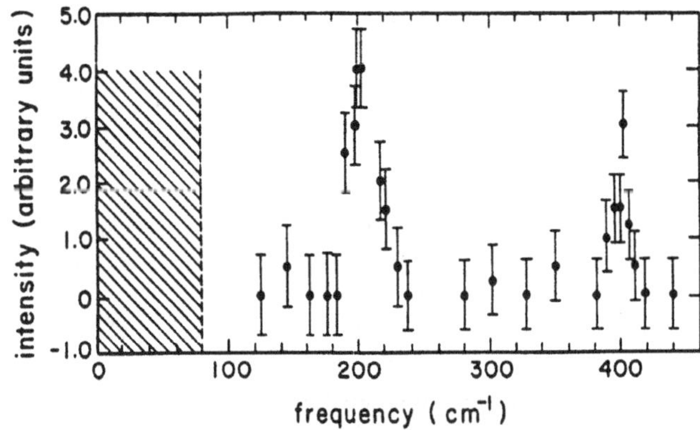


Figure 18 Anti-Stokes vibronic sideband intensity as a function of the phonon frequency.

appears which reflects the phonon spectral distribution. In this way the concentration of polaritons and TA phonons were monitored as a function of a delayable probe laser pulse which triggered the zero phonon luminescence.

The first observations along these lines were reported in reference 24 and demonstrated that indeed, the excitation in GaP of LO phonons at 403 $cm^{-1}$ lead to vibronic sidebands only at 201.5 $cm^{-1}$, i.e. only through the "half energy" decay. These early experiments, shown in Fig. 18, are relatively crude, but nevertheless demonstrate the point. More recent experimentation, at better signal-to-noise ratios, on the decay of polaritons leads to more detailed results. Most importantly, temporal resolution on the nanosecond scale has been accomplished and is incorporated into the VSPS measurements of the polaritons in GaP.

As before, polaritons were coherently generated through CRE. However, in place of the probe laser of TR-CARS, another pulse laser is substituted to excite the vibronic luminescence via VSPS. Specifically, the laser system used in the experiment contains two pulsed nitrogen lasers, one of which, together with two Rodamine 6G tunable dye lasers, is used to produce two laser beams with linewidth of 3 $cm^{-1}$, pulse duration of 5 ns, and 70 $\mu J$ energy per pulse. The two "yellow" output beams are tuned such that their frequency difference equals 354 $cm^{-1}$ or 341 $cm^{-1}$ and the beam directions are oriented to conserve crystal momentum when a polariton is generated through CRE. A second pulsed nitrogen laser is used to pump a dye laser containing coumarin 485 to excite a "green" VSPS zero-phonon luminescence at 535 nm. The source of the luminescence is known to result from an excitation of and subsequent decay of a bound exciton associated with nitrogen impurities in GaP.[25] The luminescence can be delayed electronically over a range of ~100 ns with a resolution of ~5 ns. The diameter of the focal spot of the various laser beams inside the sample was 100 $\mu m$. The output of the VSPS signal is collected and digitally stored in an optical multichannel analyser, OMA. The delay between the "yellow" lasers and the "green" lasers is separately monitored through a digital oscilloscope. The range of detection of the OMA was set to cover the frequency from that of polaritons to that of the TA phonons and to block out the very strong zero-phonon luminescence. The entire system is calibrated using standard reference lines. The sample is held to 4 to 5K, in order to obtain a sideband signal with the best signal to noise ratio. The multiple beam overlap volume should be as near as possible to the entrance surface of the crystal because of the strong absorption at the "green" frequency.

From the earlier work on polariton decay, we expect $\omega_\pi \sim 341$ $cm^{-1}$ polaritons to decay into two LA phonons with frequency $\omega_{LA} = \omega_\pi/2$. This result is indeed observed and is demonstrated in the top panel of Fig. 19 which shows sidebands only at the polariton frequency and at the $\omega_\pi/2$ frequency. In the current context we are more interested in the formation and decay of the TA phonon at the X-point of the Brillouin zone which is formed by the decay of the $\omega_\pi \sim 354$ $cm^{-1}$ polariton. The corresponding experimental results are indicated in Fig. 19 which is a display of the VSPS signal amplitude vs. the phonon (and polariton) frequencies for various delays between the "yellow" pump lasers and the "green" probe laser pulses.

The second highest panel of Fig. 19 corresponds to $\Delta t = 0$ and indeed indicates the presence of $\omega_\pi \sim 354 cm^{-1}$, the LA(X) phonon at $\sim 236$ $cm^{-1}$ and the TA(X) phonon at $\sim 118$ $cm^{-1}$. Note, within a delay of $< 10$ ns, that both the polariton and the LA phonons disappear which is consistent with their expected short lifetimes. All of these observations support the predictions of the earlier results on polariton decay channels.[17]

Of some particular note is the decay of the $\omega_\pi \sim 354$ cm$^{-1}$ polariton into an LA phonon near the X-point of the Brillouin zone and on TA phonons at the minus X-point or vice versa. All of the phonons and polaritons, with one important exception, possess lifetimes in the picosecond or femtosecond regime. The important exception is the TA phonon at (near) the X-point for which Figs. 19 and 20 lead to an observed decay of 66.5 $\pm$ 5 ns. The existence of the long-lived TA phonon has been a source of speculation for nearly three decades. We have now observed a TA phonon at the minus (or positive) X-point which exhibits a lifetime which is 4 to 6 orders of magnitude longer than other phonons and polaritons in GaP. One possible reason for this difference in decay times might be differences in the density of states into which these various phonons and polaritons decay. However calculations of the single and joint density of phonon states[26] are not consistent with sharply different decay times. As an alternative to the above explanation, Ohrbach and Vredevoe[27] have proposed that, in dispersive media, the reason for the anomalously long lifetime is that the TA phonons cannot readily decay by three-particle interactions and simultaneously obey energy and (crystal) momentum conservation. As a consequence, these phonons should possess anomalously long lifetimes.

At the time of Ohrbach and Vredevoe's paper the existence of such long-lived phonons would have verified the simple, but critical, tenet that the phonon branches are in general dispersive and, in the absence of other scattering mechanisms, that the lowest order phonon-phonon anharmonic interaction, plus energy and momentum conservation, govern phonon lifetimes. Although most of these tenets have since been verified by other means, none of these means have unambiguously measured the lifetime of the long-lived phonons. This missing link in phonon physics has attracted, over time, the imagination of a number of investigators. Since long-lived phonons would also be useful as probes of various interactions in condensed matter, the continued hunt for these phonons has, at times, been referred to as the "Holy Grail" of phonon physics.

Thus it has been demonstrated that a long-lived species of TA phonons exists with wavevector near the X-point of GaP. The lifetime of these phonons is 66.5 $\pm$ 5 ns. This lifetime is much longer than that of LO ($\sim$ 26 ps), polaritons ($\sim$ 0.1 ps), and LA phonons. The relatively long life results from restrictions due to energy and momentum conservation.

## IX. INTERACTION OF COHERENT LO PHONONS WITH A TWO-COMPONENT PLASMA

We return now to the TR-CARS technique to determine the effect of a two-component (electron-hole) plasma on the dephasing of near-zone-center LO phonons in GaP. We then compare the results, obtained in this manner, with those obtained for a one-component (electron) plasma through standard incoherent Raman scattering in the frequency domain.

We start with the recently published results on a two-component plasma.[28] In this case, an electron-hole plasma is formed in a pure crystal through the decay of carriers excited after they had been through two-photon absorption (TPA) of incident pulsed laser radiation. This method of photoexcitation may produce a plasma which is nonstationary (i.e., varies in time) during the experimental observation period. It is, therefore, appropriate to conduct time-resolved observations to provide

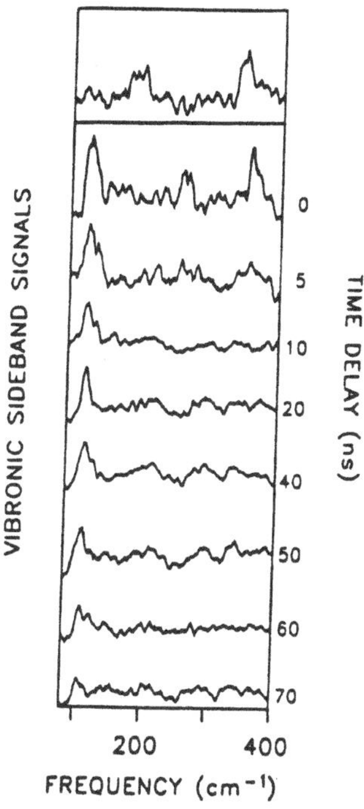


Figure 19. Vibronic sideband signal as a function of particle energy and the delay time of the probe laser pulse. The uppermost panel is the sideband obtained for $\Delta t = 0$ and polariton peak (341 $cm^{-1}$) and the "half energy" peaks. The remaining panels depict the decay of the original distribution of "X-point" formed through the original decay of 354 $cm^{-1}$ polaritons.

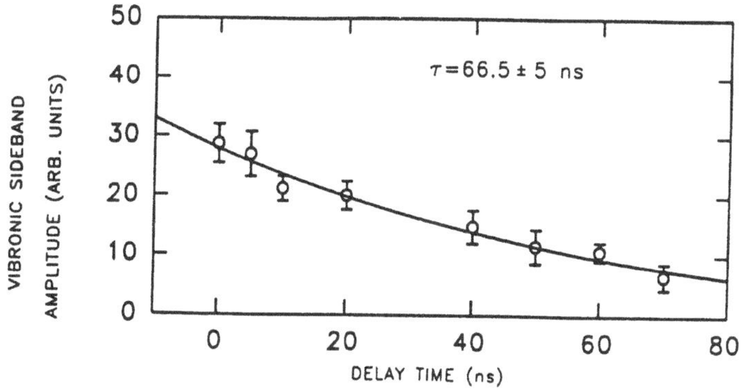


**Figure 20.** Compilation of the vibronic sideband signal following the slow decay of the TA phonons at the X-point.

information on the dynamics of the interaction between the two-component plasma and the phonons. As has been described in section V, VI, and VII above, time-resolved coherent anti-Stokes Raman scattering (TR-CARS) is an outstanding tool for measurements of the dephasing rate of near-zone-center LO phonons in GaP and to determine its dependence on the ambient temperature. We have also shown that the dynamics of the LO phonon dephasing is influenced by the presence of high concentrations of acoustic phonons. However, the dynamics of the LO phonon dephasing in the presence of a two-component plasma has received only limited attention. An understanding of the plasma-LO-phonon interaction is particularly important when the intensity of the incident laser pulse, used to inject the two-component plasma, approaches the (bulk) damage threshold of the material.

Recently, we have applied TR-CARS to the study of the temporal evolution of the interaction of a two-component plasma and LO phonons in GaP. The investigation provided, for the first time, direct observation of the transient dynamics of the plasma-phonon interaction. We observed an increase in the instantaneous dephasing rate of the LO phonons during the first 150 ps of the interaction. The dephasing rate increases as the incident laser irradiance increases (i.e., with increasing plasma density). Furthermore, we observe that the increase in the dephasing rate becomes negligible after 600 ps which suggests plasma diffusion out of the laser interaction volume.

The phonons and the two-component plasma are excited using the dual synchronously amplified picosecond laser system operating at 1 kHz (see section III) with outputs at the frequencies $\omega_\ell$ and $\omega_s(\omega_\ell = 2.137$ eV and $\omega_s = 2.087$ eV). We add to this laser system an additional laser beam used to inject the two-component plasma into the mutual interaction volume of all the laser beams. The difference in output frequencies of the two laser beams $(\omega_\ell - \omega_s)$ is resonantly tuned to the LO phonon frequency (50 meV) and a variably time delayed part of the $\omega_\ell$ beam is used to probe the dephasing of the LO phonons; note that as usual the probe frequency $\omega_p = \omega_\ell$. An additional synchronized laser pulse with frequency $\omega_s$ is used to "inject" a nonstationary electron-hole plasma (NEHP) via two-photon absorption, TPA, $(2\omega_s)$ with an initial excess energy of 1.05 eV above the direct gap of 2.7 eV. (Nonstationary simply means that the effect of the plasma on the phonons varies with time.) The injection laser pulse arrives at the interaction volume at a fixed time after the excitation of the LO phonons. The NEHP density is varied by changing the incident irradiance of the injection laser pulse. The intensities of the $\omega_\ell, \omega_s$, and $\omega_p$ beams are kept sufficiently weak to avoid additional TPA and intense LO phonon generation that could cause nonexponential dephasing of the LO phonons;[23] thus pulse energies for phonon excitation and probing are kept below 1 nJ, whereas the injection pulse energies are varied from 5 to 500 nJ. The four beams are focused into a high purity ($< 10^{16} N$ impurities per cm$^3$) GaP crystal maintained at 5 K. The experiment is carried out by alternately measuring the TR-CARS signal with and without the presence of the NEHP at a set of fixed probe delay times in the interval between - 50 and 200 ps, in order to minimize long-term laser fluctuations.

Figure 21 displays typical experimental results on the intensity of the TR-CARS signal as a function of the probe delay time for three values of the irradiance of the injection laser pulse. The upper curve, which represents the LO phonon dephasing without NEHP injection, clearly shows exponential dephasing. The lower two curves indicate LO phonon dephasing in the presence of the NEHP. The dip at ~26 ps delay is consistent with the timing of the injection pulse, but its origin is not yet

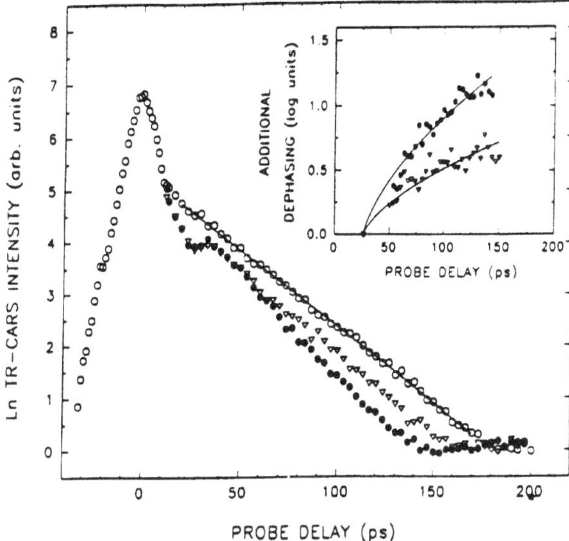


Figure 21. TRCARS signal as a function of probe delay for high purity GaP with no plasma present (open circles) and increasing presence of a nonstationary electron-hole plasma (open triangles and closed circles). The inset depicts the additional dephasing. The solid curves represent the theoretical prediction of the additional dephasing.

understood. An increase in the dephasing rate, due to the presence of the NEHP, follows the injection pulse. The inset of Fig. 21 illustrates the difference between the natural logarithm of the TR-CARS signal observed with the NEHP and that observed without the NEHP as a function of delay time, and indicates the additional dephasing resulting from the plasma-phonon interaction. The additional dephasing, depicted in the inset of Fig. 21, represents a time-dependent interaction between the LO phonons and the NEHP.

We now present a synopsis of a theoretical model which we apply to the experimental observations. The interaction of the LO phonons and the NEHP can be described in terms of a coupled vibrational system consisting of the LO phonons and the electron-hole plasma interacting by way of a local polarization field. The equations of motion of such a system are given by

$$\ddot{W} + \omega_t^2 W + \Gamma \dot{W} = (e^*/M)E, \tag{28}$$

$$m_e \ddot{X}_e = -(1/\tau_e) m_e \dot{X}_e - \eta(\dot{X}_e - \dot{X}_h) + eE, \tag{29}$$

$$m_h \ddot{X}_h = -(1/\tau_h) m_h \dot{X}_h - \eta(\dot{X}_h - \dot{X}_e) - eE, \tag{30}$$

$$E + 4\pi\left(\frac{e^*}{M}E + \frac{e^*}{V_0}W + ne(X_e - X_h)\right) = 0, \tag{31}$$

where $W$ is the amplitude of the relative interatomic lattice displacement; $\omega_t$ is the TO phonon frequency, $\Gamma$ is the spontaneous anharmonic phonon decay rate; $E$ is the local electric field; $M$ and $e^*$ are the reduced mass and effective charge of Ga and P, respectively; $\dot{X}_3, \dot{X}_h$ and $m_e, m_h$ are the drift velocities and effective masses of the electrons and holes, respectively; $\tau_e$ and $\tau_h$ are the momentum relaxation times of the electrons and holes due to scattering by acoustical phonons; and $\eta$ is due to electron-hole Coulomb scattering.

In order to solve Eqs. 28 - 31 we adopt the following assumptions: (a) The plasma frequency $\omega_p << \omega_{LO}$; this corresponds to a low plasma density under which the optical lattice vibrations are only slightly perturbed. The plasma density is estimated in the present experiment as $n \sim 10^{17}$ cm$^{-3}$, thus $\omega_p \sim 10^{13}s^{-1}$ and $\omega_{LO} \sim 10^{14}s^{-1}$; (b) $1/\omega_{LO} << \tau_{e-e}, \tau_{e-h}, \tau_{h-h}$; this indicates slow carrier-carrier scattering compared to the period of an LO phonon oscillation. For electrons and holes in GaP, $\tau_{e-e}, \tau_{e-h}, \tau_{h-h} \sim 10^{-13}s^{-1}$. Under these conditions, the coupled system yields a solution for the temporal evolution of the coherent phonon intensity $|A(t)|^2$ in the presence of the plasma. The result can be expressed in the following form:

$$|A(t)|^2 = |A(0)|^2 \exp[-\Gamma t - \beta \int_0^t \gamma(t')dt'], \tag{32}$$

where $A(0)$ is the coherent phonon amplitude at $t = 0, \beta = (1 - \epsilon_\infty/\epsilon_0)\omega_p/\omega_\ell$ is the coupling constant between the LO phonons and the NEHP, and $\gamma(t)$ is the time-dependent damping of the plasma oscillations. In the absence of a NEHP, the solution reduces to the standard exponential LO phonon dephasing with decay rate $\Gamma$.

In order to arrive at an expression for $\gamma(t)$, it is necessary to determine the temporal dependence of the carrier temperature. After an initial carrier cascade due to the Frölich interaction (<2 ps) (see section IV) the carriers form a plasma at energies below that of the $k \cong 0$, LO phonon and the phonons generated as a result of inter-X-valley electron scattering ("XX scattering"). Starting at this point of the cooling of the plasma, we have calculated the further cooling as the carriers undergo additional LO phonon and $XX$ scattering assisted by carrier-carrier scattering, deformation potential (DA) scattering, and piezoelectric (PA) scattering from acoustical phonons. The result of this calculation appears as the upper left inset to Fig. 22. It can be seen from Fig. 22 that the plasma cools from a starting temperature of approximately 150 K to a temperature of 30 K within the first 25 ps after injection. The initial rapid cooling is due to LO phonon emission and $XX$ scattering. The cooling rate then decreases as DA and PA scattering become the dominant mechanisms of energy relaxation. In order to determine the plasma damping, it is necessary to consider only those carrier interactions which do not conserve current. Assuming a nearly parabolic energy dispersion for the electrons and neglecting contributions due to the anharmonicity of the hole band, the expression for the plasma damping reduces to

$$\gamma(t) = \frac{1}{\tau_{e-h}} = n\sigma v = n(4\pi r_D^2)\left(\frac{3k_BT_c}{m^*}\right)^{\frac{1}{2}}, \tag{33}$$

where $n$ is the plasma density, $\sigma$ is the scattering cross section, $v$ is the thermally controlled drift velocity of the carriers, $T_c(t)$ is the plasma temperature as given in the upper inset to Fig. 22, $r_d(\mathring{A})$ is the Debye screening radius at temperature $T_c$, $k_B$ is the Boltzmann constant, and $m^*$ is the reduced mass for electrons and holes. Substituting Eq. 33 into 32 and using the derived carrier temperature $T_c$ from Fig. 22 yields an expression for the dependence of the additional phonon dephasing on the carrier density $n$. A best fit is then made to the additional dephasing using the theoretical expression to determine the carrier density $n$. The resulting additional dephasing is shown in Fig. 21. The one-parameter fit to the data indicates good agreement with the data and yields $n = 2.1 \times 10^{17}$ and $1.2 \times 10^{17}$ cm$^{-3}$ for the illustrative cases.

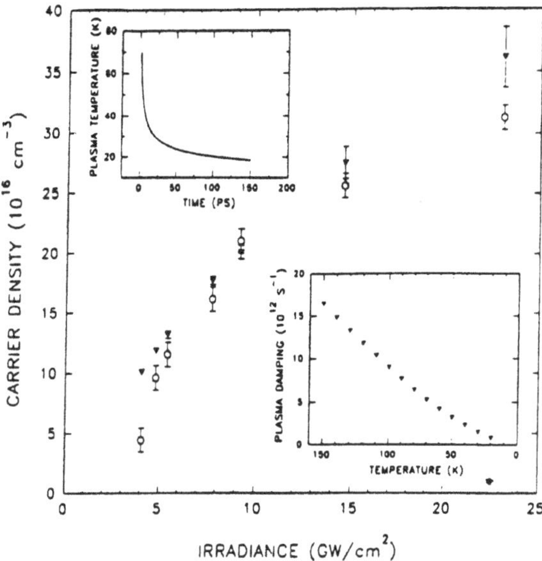


Figure 22. Carrier density as a function of the pump laser irradiance obtained theoretically (open circles) and from two-photon absorption (closed circles). Upper left inset: Calculated plasma temperature as a function of time. Lower right inset: Plasma damping as a function of the ambient temperature.

Whereas the temporal evolution of the additional dephasing indicates good agreement between the theoretical model and the data, it is necessary to determine the reliability of the fitted parameter, i.e., the carrier density $n$. An alternate evaluation of the carrier density $n$ can be obtained through a calculation of the number of carriers generated through TPA in the mutual interaction volume of the three laser beams. Only carriers excited in the interaction volume are available for the interaction, since transport of carriers out of the volume is negligible during the 150-ps interaction time. The main part of Fig. 22 is a comparison between the carrier density obtained from fitting Eq. 32 and that calculated to arise from the TPA.

It is clear that we have observed additional phonon dephasing due to the interaction of lattice vibrations and plasma oscillations. The model is in good agreement with the observed dephasing of the LO phonons, which depends on the plasma density and plasma damping. Moreover, the plasma density as obtained from the model and that calculated independently from TPA are also in good agreement. Since ballistic transport at the thermal velocities does not result in significant carrier transport out of the interaction volume within the first 200 ps of the observation, we eliminate changes in the plasma density as a source of the change in the magnitude of the additional phonon dephasing. The plasma damping results from electron-hole scattering as described in Eqs. 32 and 33. The plasma oscillations are damped by any carrier-scattering event which does not conserve current. The assumption that electron-hole scattering dominates the damping appears valid for the conditions of the experiment, although the deviation seen in Fig. 22 may be due to effects of plasma damping due to hole-hole scattering in a nonparabolic hole band at higher densities. The temporal dependence of the magnitude of the scattering is due to the cooling of the plasma which decreases the plasma velocity and thus the scattering cross section. For the convenience of the reader we include, in the lower inset to Fig. 22, the temperature dependence of the plasma damping as obtained from Eq. 33 and the fitted values of the carrier density.

Carriers which have been transported out of the volume no longer participate in the interaction causing the initial carrier density to decrease as the transport out of the region increases. As a result, the observed additional phonon dephasing is negligible once preinjection is greater than 600 ps, at which time, apparently, all carriers have left the interaction volume. Further analysis of this effect will be presented in a later publication. Kardontchik and Cohen[13] have reported negligible damping of a two-component plasma in GaP as observed through incoherent Raman scattering with laser pulses of 2 ns duration and probe delays also of the order of several nanoseconds. Thus their findings support our observation that the plasma-phonon interaction ceases for delay times greater than of the order of nanoseconds. Prior to this explanation of Kardontchik and Cohen's statement that the plasma was "undamped", we had speculated that the damping of a one-component plasma and that of a two-component plasma is fundamentally different. However, a comparison of the present plasma damping rate and those reported earlier on a one-component plasma indicate no appreciable difference. Comparing high purity samples with carrier concentrations of the order of $10^{17}$ cm$^{-3}$, we find from Figs. 1 and 6 of ref. 1 that for a one-component plasma $\gamma \sim 100$ cm$^{-1}$($\sim 9 \times 10^{12} s^{-1}$), and typical values from the inset of Fig. 22 of the present experiment on two-component plasma are $\gamma \sim 10 \times 10^{12} s^{-1}$ at $T \sim 100$ K.

In conclusion, we have observed, an interaction of an optically induced two-component electron-hole plasma with LO phonons in GaP using the TR-CARS technique. A time-dependent increase in the dephasing rate of the LO phonons is observed to arise from the coupling of the vibrational modes of the plasma and those of the phonons. It is demonstrated, on experimental and theoretical grounds, that the increase of the dephasing rate depends on the density and on the damping of the plasma. The temporal dependence of the increased dephasing arises from a nonstationary plasma temperature, and the diffusion of carriers out of the excitation volume. Plasma damping is readily observable for temporal resolution of the order of picoseconds. The present experimental technique makes it possible to observe, directly in the time domain, the interaction of a plasma with optical phonons.

## X. INTERACTION OF OPTICAL PHONONS WITH A ONE-COMPONENT PLASMA

The following illustrative example of the interaction of optical phonons with a one-component plasma is discussed last, not because it is least interesting, but rather that the experimental technique is different from those which employ TR-CARS or VSPS. Here we return to the spontaneous <u>incoherent</u> Raman scattering described in sections II and IV. Moreover, the information on the interaction is obtained in the spectral domain rather than the temporal domain.[29]

In experimentation in the spectral domain the observables are the bandwidth, peak height, and spectral profile of the incoherent Raman scattering from the LO-phonon-coupled mode. These results are compared to measurements in the temporal domain in which the experimental observable is instead the dephasing time of near-zone-center LO phonons by TR-CARS. It can readily be shown, as long as the incoherent

Raman-scattering spectral function is Lorentzian and the phonon component of the TR-CARS signal decays exponentially, that the bandwidth $\Delta v$ and the dephasing time $T_2$ are simply related by $\Delta v(cm^{-1}) = (\pi c T_2)^{-1}$. Thus, either measurement yields $\Delta v$ and $T_2$, so that the choice of the measuring technique need depend only on the experimental conditions ($c$ is the speed of light).

In forward-scattering incoherent Raman processes, or forward-scattered coherent Raman excitation (CRE), only phonons with wavevector of the order $10^2 - 10^3$ $cm^{-1}$ can be excited and detected in typical compound semiconductors. Moreover, electronic carriers photoexcited high into the conduction band of GaAs, at densities below $5 \times 10^{16}$ $cm^{-3}$, have been shown to lose energy by LO-phonon emission through a multiple-step cascade with a time of $\sim$ 150 fs per cascade step.[14] See also section IV. A few picoseconds are required to form a thermalized electron-hole plasma at some conduction-band minimum. We assume that the temporal factors are essentially the same in GaP as in GaAs (see section IV). GaP differs from GaAs in that GaP is a so-called "indirect-gap" semiconductor, whereas GaAs is a "direct-gap" semiconductor (see Fig. 7). This refers to the fact that in GaAs photoionization from the maximum of the valence (say at the $\Gamma$ point) to a minimum in the conduction band can be reached without a change in the carrier wavevector, whereas that is not the case in GaP, which has lowest-lying conduction-band minima at the $X$-point in the band structure. In the present experiment, carrier excitation occurs through thermal ionization of donor impurity states in which changes in wavevector do not play a role. Nevertheless, once thermally excited the carriers in GaP will most likely form a plasma at equivalent minima in the band structure near the $X$ point of the $\Gamma_1$ band. Such plasmas succumb to carrier-scattering processes which differ from those in, e.g., GaAs, which possesses a single band minimum.

Plasma are damped even in the absence of phonons, the damping arising from, among other factors, carrier-carrier, intervalley, and interband scattering. The plasma damping rate in GaP is known to be high, possibly due to the additional intervalley scattering which may be present in indirect-gap semiconductors. Thus, as a result of phonon-plasma coupling, even modest carrier concentrations in GaP readily cause a decrease in $T_2$ below the 100 fs to two picosecond temporal resolution of the TR-CARS system. This is the basis for choosing incoherent Raman scattering, rather than TR-CARS measurements, except for the case of a very pure sample.

Incoherent Raman scattering was carried out in a standard backward-scattering geometry using a cw krypton laser operating at 647.1 nm and at a power of 100 mW. Samples of GaP, with $n$-type doping of various nominal concentrations (see Table I), were held at ambient temperatures from 5 to 300 K. The temperature dependence

**TABLE I. Identity of samples.**

| Sample | Concentration at 285 K ($cm^{-3}$) |
|---|---|
| a | $< 1.5 \times 10^{16}$ |
| b | $3.6 \times 10^{16}$ |
| c | $1.7 \times 10^{17}$ |
| d | $3.6 \times 10^{17}$ |
| e | $1.3 \times 10^{18}$ |
| f | $1.8 \times 10^{18}$ |

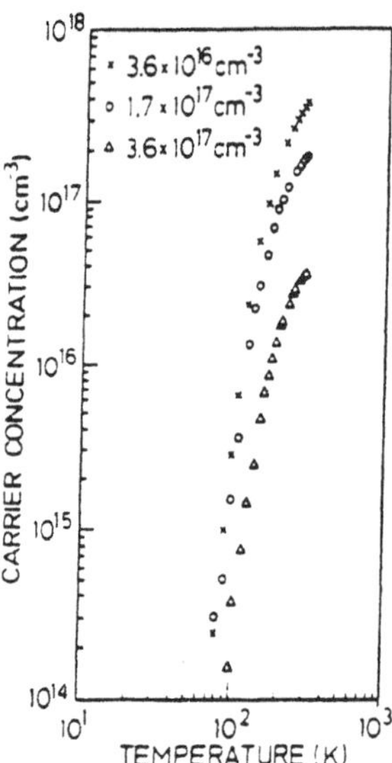


Figure 23. Hall coefficient measurements of the carrier concentration for samples of varying doping levels.

of the Hall coefficient was independently determined for each of the samples of GaP (except sample $a$). Values for the corresponding temperature dependence of the carrier concentration, $n_c(T)$, for three relatively weakly doped samples are shown in Fig. 23. We chose $n_c$ (285 K) as a way of identifying the various samples as indicated in Table I. Samples $a - d$ and $f$ were doped with tellurium, whereas sample $e$ was doped with silicon. Within the spirit of this review article, we drop from further discussion the observation from samples d, e and f. These results may stem from the further presence of electron-hole liquids although no definitive evidence exists to confirm this hypothesis. We refer the reader to reference 29 for further discussion.

Several measurements of the incoherent Raman-scattering intensity were carried out for each sample, and at each ambient temperature, except for the purest sample, for which several TR-CARS measurements were performed. Figure 24 indicates a series of Raman-scattering lines from sample c held at 50, 100, 175, and 250 K. The solid circles represent a subset of all data points taken and the solid lines are Lorentzian spectral distributions fitted to the observed peak frequencies and bandwidth (full width at half maximum (FWHM)). Similar Lorentzian spectral profiles are obtained for all other samples and temperatures.

We now demonstrate that, for $n$-doped GaP the temperature dependence of the bandwidths, peak frequencies, and other parameters of a coupled phonon-plasma system can be accounted for in terms of simple dielectric-response theory without the introduction of fitting parameters. Specifically, we have investigated in some detail the interaction of LO phonons in GaP with a thermally activated (electron) plasma of varying concentration.

Since the pertinent observations in this case are time independent, it is not necessary to construct a governing equation of motion to describe the observations. We treat instead the coupled phonon-plasma system in terms of a dielectric response

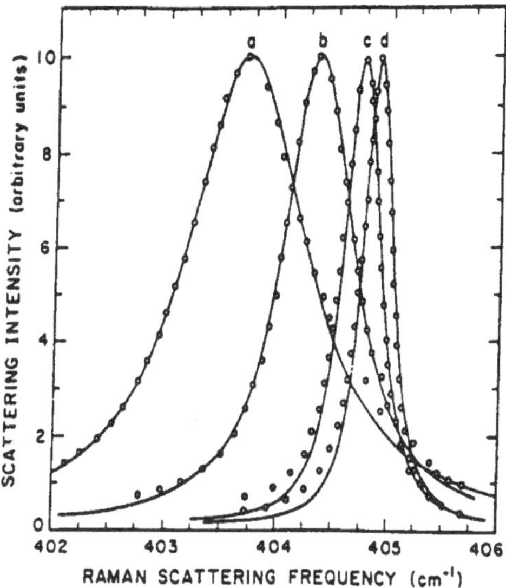


Figure 24. Incoherent Raman spectral profiles for a sample with 1.7 × $10^{17}$ donors. The GaP is held at ambient temperatures of a, 250K; b, 175K; c, 100K; and d, 50K.

function, $\epsilon(\omega)$, given by

$$\epsilon(\omega) = \epsilon_\infty + \frac{\Omega^2}{\omega_T^2 - \omega^2 - i\omega\Gamma} - \frac{\omega_p^2}{\omega(\omega + i\gamma)}. \tag{34}$$

In Eq. (34), $\Omega^2 = \epsilon_\infty(\omega_L^2 - \omega_T^2), \omega_L$ is the LO-phonon frequency, $\omega_T$ is the transverse-optical-(TO)-phonon frequency, $\Gamma$ is the damping rate of the LO phonons in the absence of the plasma, $\gamma$ is the damping rate of the plasma in the absence of the phonons, $\omega_p^2 = \epsilon_\infty\tilde{\omega}_p^2, \epsilon_\infty$ is the static dielectric constant in the presence of the ion-core background, and $\tilde{\omega}_p$ is the plasma frequency in the presence of the ion-core background.

The real part of $\epsilon(\omega)$ set equal to zero leads to the eigenfrequencies of the coupled modes and the imaginary part of $1/\epsilon(\omega)$ leads to a spectral function, $S(\omega)$, analogous to the structure factor of nuclear scattering. In the past, various parameters of Eq. 34 have been neglected. Thus, either $\gamma$ and/or $\Gamma$ are not taken into account depending on their relative magnitude. We have obtained instead a rigorous solution of Eq. 34 which contains $\gamma$ and $\Gamma_{eff}$ with the only requirements that $\Gamma_{eff}(\sim 5cm^{-1}) << \omega_+(\sim 400cm^{-1})$ and that $\omega \cong \omega_L$ (i.e., that $\epsilon(\omega)$ is evaluated only in the neighborhood of the peak of the LO-phonon-coupled mode). The solutions to Eq. 34, with these restrictions, are

$$\begin{aligned} \omega_\pm^2 &= \frac{1}{2}(\omega_L^2 + \tilde{\omega}_p^2 - \gamma^2) \\ &\pm \frac{1}{2}[(\omega_L^2 + \tilde{\omega}_p^2 - \gamma^2)^2 \\ &+ 4(\omega_L^2\gamma^2 + \tilde{\omega}_p^2\gamma\Gamma - \tilde{\omega}_p^2\omega_T^2)]^{\frac{1}{2}} \end{aligned} \tag{35}$$

and

$$S(\omega) = \frac{\Omega^2(\omega_L^2 + \gamma^2)}{\epsilon_\infty^2(\omega_L^2 - \gamma^2 - \tilde{\omega}_p^2)} \frac{1}{4\omega_+} \left( \frac{\Gamma_{eff}}{(\omega_+ - \omega)^2 + (\Gamma_{eff}/2)^2} \right) \tag{36}$$

Here,

$$\Gamma_{eff} \equiv \Gamma + \left( \frac{\tilde{\omega}_p^2 \Omega^2 \gamma}{(\omega_L^2 + \gamma^2 - \tilde{\omega}_p^2)\omega_L^2 \epsilon_\infty} \right) \tag{37}$$

where $\Gamma_{eff}$ is the effective damping rate of the LO-phonon-coupled mode in the presence of the plasma. Here, $\omega_-$ stands for the "plasmalike" coupled mode. Note from Eq. 36 that the predicted spectral function, $S(\omega)$, has a Lorentzian shape as, indeed, observed experimentally, and as demonstrated in the solid lines in in Fig. 24 which are Lorentzian fits to the data.

Thus, there are two equations, (i) the $\omega_+$ component of Eq. (35) and (ii) Eq. (36), for the two unknown quantities $\tilde{\omega}_p$ and $\gamma$. $\Gamma$ is obtained from the previously reported TR-CARS measurements on sample $a$, i.e., on the very-high-purity sample of GaP. In what follows it is assumed that in the "pure" sample the effect of thermally activated carriers is negligibly small. Thus, the curves marked $a$ in Fig. 25 represent the effects on $\Gamma(T)$ and $\omega_L(T)$ solely due to thermal expansion of the lattice and anharmonic dephasing of the LO phonons. As the temperature increases the lattice displacements increase, and the average distance between ions increases; lowering thereby vibrational potential the ions experience, and shifting the peak of the ionic frequency to lower values. Thus, deviations from the parameters which describe the pure sample imply the additional interaction with the thermally activated one-component plasma.

Exact solutions of Eqs. (34) for $\tilde{\omega}_p$ and $\gamma$ as a function of temperature are obtained through an iterative process which starts with an approximate value of $\gamma$ and ends when stable values of $\tilde{\omega}_p$ and $\gamma$ are found. The corresponding values are displayed in Figs. 26 and 27. It follows from summing $\omega_+^2$ and $\omega_-^2$ of Eq. 35 that $\omega_-^2 = \tilde{\omega}_p^2 - \gamma^2$. We find, for the full range of samples and temperatures, that $\gamma \geq \omega_-$; that is, that the Raman scattering peak at $\omega_-$ is either strongly, or as in most cases, overdamped. This accounts for the observation that the scattering peak associated with $\omega_-$ is only barely detectable.

It is now possible to make a number of general observations of the behavior of the coupled phonon-plasma system. First, note that Eqs. (36) and (37) correctly reduce $\Gamma_{eff}$ to $\Gamma$ and $\omega_+$ to $\omega_L$ as $\tilde{\omega}_p$ goes to zero. Also note from Figs. 25a and Eq. (36) that deviations from $\omega_+$ relative to $\omega_L$ vary roughly as $\tilde{\omega}_p^2 - \gamma^2$. Thus, as the carrier concentration (and $\tilde{\omega}_p^2$) increases, $\omega_+$ increases relative to $\omega_L$, but the plasma damping, $\gamma$, tends to reduce this increase. Examination of Eq. (36) further yields, that if $\tilde{\omega}_p = 0$, regardless of the value of $\gamma, \omega_+ = \omega_L$ and $\Gamma_{eff} = \Gamma$. In contrast, if $\gamma = 0, \Gamma_{eff} = \Gamma$ and $\omega_+ \geq \omega_L$ depending on whether $\tilde{\omega}_p \geq 0$. Thus, if, as in Fig. (25) for $T \leq 175$ K, $\Gamma_{eff} = \Gamma$, but, as in Figs. 27, $\omega_+ > \omega_L$ it follows that $\gamma = 0$. This accounts for the various apparent intercepts of $\gamma(t)$ with the temperature axis of Fig. 26.

Figure 27 illustrates the expected trend that $\omega_+$ increases with temperature, i.e., with increasing carrier concentration. However, the intercepts with the temperature axis are grouped into two different values depending on whether or not $n_c$ (285 K) is greater than, or less than, $\sim 10^{18}\ cm^{-3}$. Sample $e$ is doped with silicon, whereas sample $f$ is doped with tellerium. Thus, the dopant species is not a likely cause of this observation. On the other hand, higher dopant concentration may cause defect clusters which may have differing ionization energies. However, we have not pursued

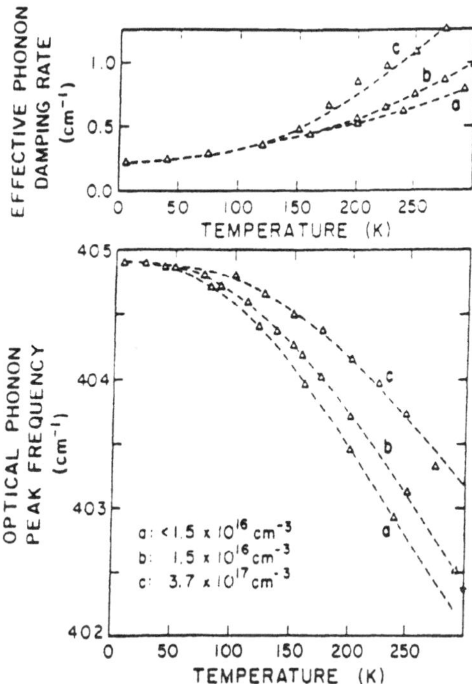


Figure 25. Lower graph: compilation of $\omega_+$ (open triangles) as a function of the ambient temperature. Upper graph: Effective damping rate of LO phonon-coupled mode for samples of varying dopant level.

this point further. It is also not clear why $\gamma(T)$ saturates, and even eventually decreases, with increasing temperature. Although saturation of $\gamma(T)$ may imply depletion of the donor states or the formation of e-h liquids, it is difficult to account for the decrease in $\gamma(T)$ and the lack of a corollary effect on $\tilde{\omega}_p$. It is clear that the dielectric response function given by Eq. (34) adequately accounts for the interaction between the damped LO-phonon modes and strongly damped one-component plasma modes. The derivation adopted here for the parameters of the coupled phonon-plasma system represents a straightforward application of dielectric response theory. Moreover, in the present method, various useful parameters, such as the effective phonon damping

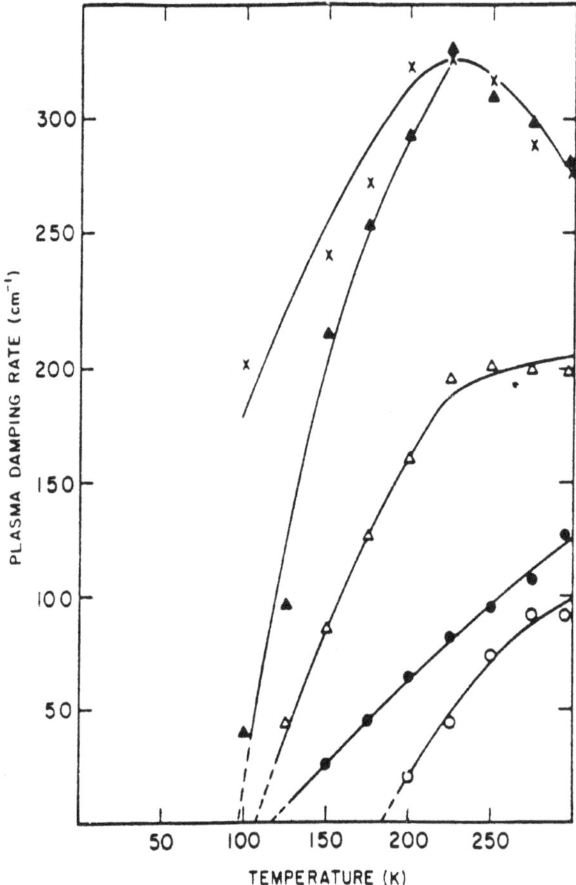


Figure 26. Plasma damping rate ($\gamma$) as a function of the ambient temperature as obtained theoretically (solid lines). See ref. 29 for details.

rate, the plasma damping rate, the plasma frequency, and the dependence of these parameters on temperature and carrier concentration, can be readily recovered.

## XI. SUMMARY REMARKS

I have discussed, in various detail, a set of illustrative investigations on the ultrashort dynamics of various processes in condensed matter. Experimentation is conducted (for the most part) directly in the time domain. For this purpose a unique set of laser systems have been developed that permit temporal resolution down to tens of femtoseconds. Moreover, coherent excitations and coherent detection are brought to bear which permits detection over many orders of magnitude. The experimental arrangement permits investigation over a wide variety of physical phenomena. Among those which have so far been investigated are:

- Dephasing of LO phonons

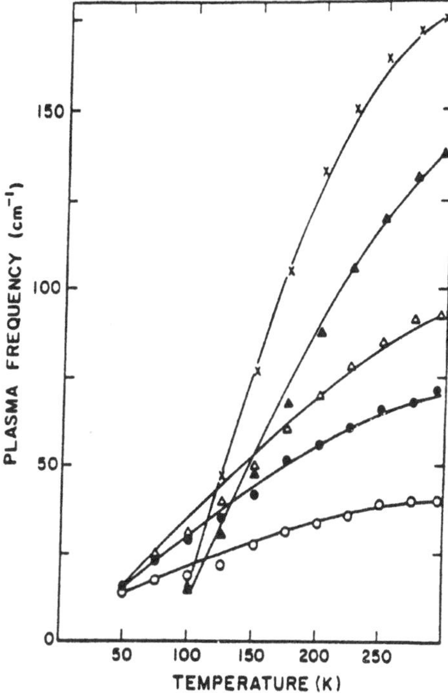


Figure 27 Plasma frequency as a function of ambient temperature and as obtained theoretically.

- Dephasing of polaritons
- Generation and detection of long lived phonons
- Nonlinear coupling between LO and LA phonons
- Dephasing of phonon interacting with one-and two component plasma

Not covered in this review, but just as interesting, is our applications of ultrashort laser pulses to

- Thermomodulation of thin normal metal films
- Ultrashort switching of Pb and high Tc superconducting strip lines
- Carrier dynamics in $C_{60}$ (Buckyballs) thin films
- Optically induced ultrashort laser damage

Many other phenomena await investigation. This field of inquiry is far from complete.

## ACKNOWLEDGMENT

The author acknowledges support through NSF DMR88-3888, NSF DMR 89-13289, and ARO DAAL-03-89-K-0600. He also thanks the many co-workers who have contributed to this research effort.

## REFERENCES


1. R. J. Von Gutfeld and A. H. Nethercot, Phys. Rev. Lett. **12**, 641 (1964).
2. See for example W. E. Bron, Studies of nonequilibrium dynamics in the tune domain, in: "Nonequilibrium Phonon Dynamics", W. E. Bron, ed., Plenum, NY (1985)., W. E. Bron, Phonon generation, transport and detection through electronic states in solids, in: "Nonequilibrium Phonons in Nonmetallic Crystals," W. Eisenmenger and A. A. Kaplyanskii, ed., North Holland, Amsterdam (1986), and W. E. Bron, Phonon transport-Experiement, in: "Dynamical Properties of Solids", G. K. Horton and A. A. Maradudin, ed., Elsevier, Amsterdam (1990).
3. See for example W. Demtroder, "Laser Spectroscopy", Springer, Berlin (1981).
4. G. Placzek, "Handbuch der Radiologie", E. Marx, ed., Akademische Verlagsgesellschaft, Leipzig, (1934).
5. See for example L. L. Schiff, "Quantum Mechanics" 3rd Edition, McGraw Hill, New York (1968). See also R. Loudon, "The Quantum Theory of Light," Clarendon Press, Oxford, 1983 146pp.
6. J. A. Armstrong, N. Bloembergen, J. Ducuing and P. S. Pershan, Phys. Rev. **127**, 1918 (1962).
7. P. D. Maher and R. W. Terhune, Phys. Rev. **137**, A801 (1965).
8. W. Zinth, A. Laubereau and W. Kaiser, Opt. Commun. **49**, 935 (1978).
9. See for example J. T. Verdeyen, "Laser Electronics" 2nd edition, Prentice Hall, Englewood Cliffs, NJ (1989).
10. T. Juhasz, J. Kuhl and W. E. Bron, Opt. Lett. **13**, 577 (1988), T. Juhasz, G. O. Smith, S. M. Mehta, K. Harris and W. E. Bron, IEEE J. Quantum Electronics, **25**, 1704 (1989).
11. D. von der Linde, J. Kuhl and H. Klingenberg, Phys. Rev. Lett. **44**, 1505 (1980).
12. See for example E. M. Conwell and M. O. Vassel, IEEE Trans. Electron. Devices **13**, 22 (1966).
13. J. E. Kardontchik and E. Cohen, Phys. Rev. Lett. **42**, 669 (1979).
14. J. A. Kash, J. C. Tsang and J. M. Hvam, Phys. Rev. Lett. **54**, 2151 (1985).
15. J. Kuhl and W. E. Bron, Sol. State Commun. **49**, 935 (1984).
16. W. E. Bron, J. Kuhl and B. K. Rhee, Phys. Rev. **34**, 6961 (1986).
17. T. Juhasz and W. E. Bron, Phys. Rev. Lett. **63**, 2385 (1989).
18. A. Laubereau and W. Kaiser, Rev. Mod. Phys. **50**, 607 (1978).
19. B. Kh. Bairamov, Yu. E. Kitaev, V. K. Negoduiko and Z. M. Kashkotzev, Fiz. Tver. Tela **16**, 2036 (1974); [Sov. Phys. Solid State **16**, 1323 (1974) and B. Kh. Bairamov, D. A. Parshin, V. V. Toporov and Sh. B. Ubaidullav, Pis'ma Zh. Tekh. Fiz. **5**, 1116 (1979).]; [Sov. Tech. Phys. Lett. **5**, 466 (1979)].

20. S. Ushioda and J. D. Mullen, Sol. State Commun. **11**, 299 (1972).

21. See for example, H. Bilz and W. Kress, "Phonon Dispersion Relations in Insulators", Springer Verlag Berlin (1979).

22. S. E. Bulgadaev and Y. B. Levinson, Pis'ma Zh. Eksp. Teor. Fiz. **19**, 583 (1974), [JETP Lett. **19**, 304 (1974)]; S. E. Bulgadaev and Y. B. Levinson [Zh. Eksp. Teor. Fiz. **67**, 2341 (1974); [Sov. Phys. JETP **40**; 1161 (1974)].

23. W. E. Bron, T. Juhasz and S. Mehta, Phys. Rev. Lett. **62**; 1655 (1989). See also S. Mehta, T. Juhasz and W. E. Bron, Phys. Rev. **B45**, 209 (1992).

24. B. K. Rhee and W. E. Bron, Phys. Rev. **34**, 7107 (1986).

25. D. G. Thomas, M. Gershenson and J. J. Hopfield, Phys. Rev. **150**, 580 (1966).

26. K. Kunc, Ann. Phys. (Paris) **8**, 319 (1973).

27. R. Orbach and L. A. Vredevoe, Phys. **1**, 91 (1964).

28. G. O. Smith, T. Juhasz, W. E. Bron and Y. B. Levinson, Phys. Rev. Lett. **68**, 2366 (1992).

29. W. E. Bron, S. Mehta, J. Kuhl and M. Klingenstein, Phys. Rev. **39**, 12642 (1989).

# CONTACT-FREE CHARACTERIZATION OF ELECTRONIC AND OPTOELECTRONIC DEVICES WITH ULTRASHORT LASER PULSES

Jürgen Kuhl

Max-Planck-Institut für Festkörperforschung
Heisenbergstr. 1, D-7000 Stuttgart 80, FRG

## 1. INTRODUCTION

During the last 10-15 years the steadily increasing requirements in the transfer rate and operation speed of analog and digital data processing have stimulated the development of optical communication techniques, which principally provide bandwidth much greater than 100 GHz [1] (corresponding to a time resolution of a few picoseconds) because of the high carrier frequency of electromagnetic waves in the optical regime. This situation spurred semiconductor physics and technology to develop electronic and optoelectronic devices which can switch on or off within a few picoseconds and can be utilized as ultrafast transmitters or receivers performing functions like signal modulation or demodulation and amplification. Rapid improvements of device structure, functionality and speed were rendered possible by novel semiconductor growth (e.g. molecular beam epitaxy (MBE) or metal-organic chemical vapor deposition (MOCVD)) and processing techniques (like ion implantation) as well as advances in device fabrication. Optical and electron beam lithography permit device structures with micron and submicron dimensions.

Owing to the combined efforts of semiconductor physics, technology and material science the response speed of new ultrafast electronic and optoelectronic circuits is now well beyond the time-resolution capability of conventional electronic test equipment like pulse generators, spectrum analyzers, sampling oscilloscopes, network analyzers or wafer probes [2]. At present, the time resolution attainable with these measurement systems is still limited to 10-20 ps corresponding to a bandwidth of 20-40 GHz. In practice, the performance of electronic measurement instrumentation is further constrained by the need of physical contacts between the instrument and the device under test. Stray capacitances and inductances associated with the connecting leads can severely reduce the bandwidth and disturb the measurement.

Examples of such high speed devices are photoconductive switches [3], metal-semiconductor-metal (MSM) photodiodes [4,5], modulation-doped field effect transistors [6], permeable base transistors and [7] superconducting Josephson junctions [8].

*Ultrashort Processes in Condensed Matter*, Edited by
W.E. Bron, Plenum Press, New York, 1993

The appearance of a continuously increasing gap between the response speed of novel semiconductor devices and the measurement capabilities of electronic instrumentation has stimulated intensive search for new technologies to solve this measurement problem. Optical sampling techniques with picosecond and subpicosecond laser pulses have proven to be very powerful tools to bridge this gap [9-14]. Utilizing these techniques, electronic wave form probing has benefitted from the rapid progress in ultrashort pulse generation obtained in the optical frequency regime by mode-locking of lasers with broad band gain media [15]. For instance, laser pulses as short as 6 fs have been generated [16]. Using 100 fs laser pulses which can now be generated even with commercially available laser systems, time-resolved observation of electrical phenomena on a pico- and subpicosecond time scale becomes feasible. This result represents an improvement of at least one order of magnitude compared to purely electronic methods.

Interaction of optical pulses with electrical signals can be accomplished via different physical mechanisms like photoconductivity [9-11], photoemission [17] or the electro-optic effect [12-14]. All these phenomena have been exploited to develop new characterization techniques for picosecond electrical transients, which are now applied in many semiconductor physics and technological laboratories where they provide substantial contribution to rapid advances in the development of high-speed devices. Besides higher time resolution, spatial resolution of a few micrometers and probing without mechanical contacts (minimal disturbance and loading of the circuit by the measurement process) are the most attractive features of optical sampling techniques.

In the first part of this article the principles, necessary instrumentation and present status of photoconductive (Chap.2) and electro-optic sampling (Chap.3) of electrical waveforms will be briefly reviewed. The limitations in sensitivity and bandwidths for both techniques are discussed in chapter 4. Chapter 5 summarizes the problems associated with the propagation of THz electrical pulses on high-speed transmission lines.

In the second part of the paper we present a comprehensive theoretical and experimental analysis of the carrier transport in Metal-Semiconductor-Metal (MSM) photodetectors (Chap. 6) and briefly review several experiments investigating the switching properties of ultrafast transistors in the picosecond time domain (Chap. 7). The results discussed in Chaps. 6 and 7 demonstrate the enormous potential of photoconductive and electro-optic sampling for characterizing ultrafast electronic and opto-electronic devices.

## 2. PHOTOCONDUCTIVE SWITCHING

Photoconductive switching combines subpicosecond optical technology and high speed photoconducting materials to generate and measure ultrafast electrical pulses [9-11,18]. Photoconductive switches fabricated on semiconductor material with a high concentration of deep defect levels are the crucial component for the conversion of ultrafast optical into ultrafast electrical signals. Free electrons and holes excited in the conduction and valence band of such a material via absorption of a pico- or subpicosecond laser pulse create transient photoconductivity for a very short time interval since the carriers are rapidly trapped and localized at deep defect states after excitation. For an excitation pulse of 100 fs or less, the rise of the photoconductivity internal of the material follows the integrated pulse intensity since the creation process proper of an electron hole pair after absorption is very fast and occurs on a time scale of the order of $10^{-15}$s. The signal decay is limited by the free carrier lifetime.

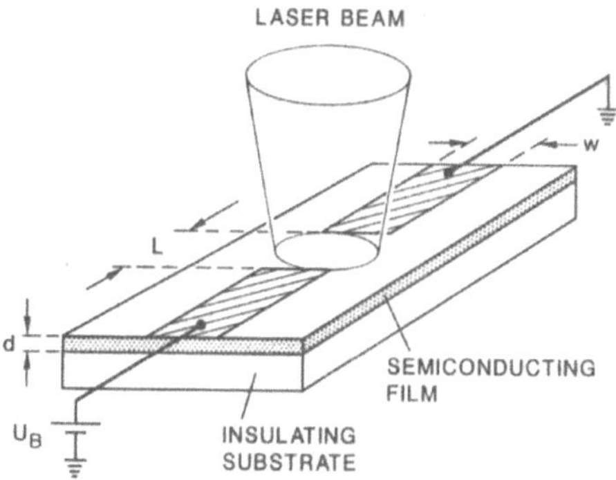


**Fig. 1** Scheme of a thin-film photoconductive switch. The active region is the gap between the electrodes.

Figure 1 depicts the principle structure of a photoconductor. Assuming a homogeneous electric field between the switch contacts, the time dependent conductance G(t) is a function of the geometrical length L and width W of the semi-insulating material and of the concentration and mobility of electrons ($n_e, \mu_e$) and holes ($n_p, \mu_p$)

$$G(t) = \frac{qW}{L} \int_x (\mu_e \cdot n_e(x,t) + \mu_p \cdot n_p(x,t))dx \tag{1}$$

with

$$\int_x n_e(x,t)dx = (1-R) \cdot \frac{q}{h\nu} \cdot \frac{1}{WL} \int_{-\infty}^{t} P_{opt}(t') \cdot e^{-\frac{t'}{\tau_t}} dt' \ . \tag{2}$$

Here h$\nu$ is the photon energy, $P_{opt}$(t) the time-dependent power of the optical pulse, R the reflectivity of the semiconductor surface, $\tau_t$ the carrier trapping time and $q$ the elementary charge. Photoexcitation creates a transient current pulse across the semi-insulating gap of the photoconductor with a current density

$$j(t) = G(t) \cdot E_B \ , \tag{3}$$

where $E_B$ is the electrical field associated with the external bias $U_B = E_B \cdot L$.

## 2.1. Materials for Ultrafast Photoconductors

The intrinsic lifetimes of electrons and holes in a direct semiconductor determined by radiative recombination range between 100 ps and a few nanoseconds. This speed is by far too slow to produce an ultrafast electrical response. However, picosecond and subpicosecond carrier lifetimes are achieved in materials with large densities of naturally occurring defects like polycrystalline and amorphous semiconductors, since these defects act as effective centers for rapid carrier trapping and nonradiative recombination. The large disorder present in these materials results in a drastic reduction of the carrier mobilities which determine the amplitude of the switching signal and of the trapping time $\tau_t$ which is related to the trap density $N_t$

$$\tau_t = \frac{1}{N_t \cdot \sigma \cdot \langle v \rangle} \ , \tag{4}$$

where $\sigma$ is the capture cross-section and $\langle v \rangle$ the average carrier velocity.

Manufacturing of a photoconductive switch usually starts with a crystalline material and the defect level is adjusted by controlled incorporation of impurities or radiation damage.

Fast photocurrent decay times in Si and GaAs, which are the most important materials for the fabrication of electronic devices and circuits, can be achieved by irradiation with $Si^+$ or $O^{++}$ ions [19-22,3] and protons or $He^+$ ions [22-25] in the case of Si and GaAs substrates, respectively.

As the typical capture cross-sections amount to $\sigma = 10^{-13}$ cm$^2$ (GaAs) and $\sigma = 10^{-15}$cm$^2$ (Si), defect densities of $10^{18}$ cm$^{-3}$ (GaAs) and $10^{20}$ cm$^{-3}$ (Si) are necessary for a response time of approximately 1ps. Figure 2 depicts measured carrier lifetimes versus ion implantation dose for silicon-on-sapphire (SOS) and GaAs-switches determined by time-resolved reflectivity and photoconductivity experiments. The data exhibit an almost linear decrease of $\tau_t$ at low ion implantation doses and a subsequent saturation value of 0.5 ps for $Si^+$ doses $> 3 \times 10^{14}$ cm$^{-2}$ and $H^+$ doses $> 10^{15}$ cm$^{-2}$ for the SOS and GaAs wafers, respectively. At these saturation doses the materials reveal a transition to the completely amorphized structure and the effective trap density does not further grow.

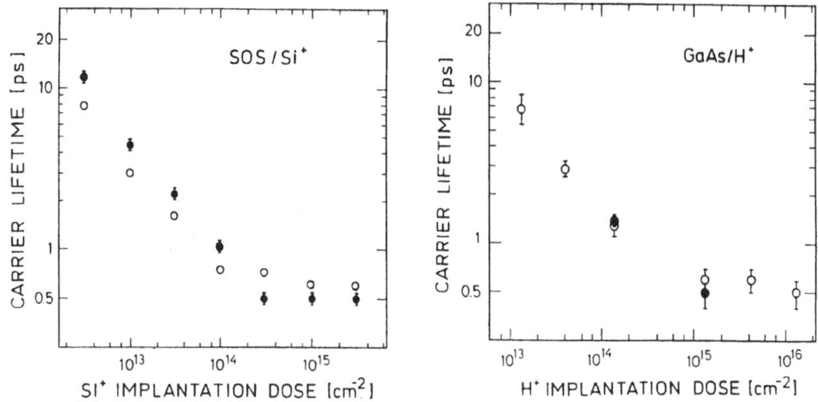


**Fig. 2** Carrier lifetime versus ion implantation dose for SOS (left Part) and GaAs(right part) photoconductive switches obtained by time-resolved reflectivity (o) and photoconductivity experiments (•).

The slight differences of the data points obtained by optical and electrical measurements are explained by the fact that the photoconductivity experiment detects only the highly mobile carriers in the extended band states whereas the optical reflectivity measures changes of the dielectric function caused by both free and localized photoexcited carriers.

The photocurrent amplitude and thus the sensitivity of the switch attainable at a given excitation density increases with carrier mobility (see Eq.(3)). This mobility drops with defect density $N_t$ because of elastic scattering.

$$\mu = \frac{1.4 \times 10^{22}}{N_t} \frac{m^*}{m_o} \cdot \frac{\varepsilon}{\varepsilon_o} \; [cm^2/Vs] \; , \tag{5}$$

where $m^*/m_o$, and $\epsilon/\epsilon_o$ are the effective mass of the carriers and the dielectric constant of the material, respectively.

Consequently, faster response implies a lower amplitude of the photocurrent pulse. Optimization of a switch thus requires a compromise between response time and sensitivity, in particular ion implantation doses above the saturation values for $\tau_t$ should be avoided. Mobilities of 350 $cm^2/Vs$ for GaAs and 5 $cm^2/Vs$ for Si for doses providing 1 ps carrier trapping times are estimated from Eq. (5).

Experimentally, we observe no large difference for the total mobility $\mu = \mu_e + \mu_p$ in the SOS and the GaAs switch. From the maximum of the voltage amplitude which amounts to a few hundred millivolts for a bias of about 10 V in both cases, we estimate the following values for the mobility $\mu = 4.8\times$ $cm^2/Vs$ for SOS ($Si^+$ dose = 3 $\times 10^{14}$ $cm^2$) and $\mu = 8.8$ $cm^2/Vs$ for the GaAs ($H^+$ dose = $4 \times 10^{14}$ $cm^2/Vs$). This discrepancy can be attributed to the generation of defect states with a broad energy distribution by the ion implantation process. Whereas deep traps result in a short free carrier lifetime $\tau_t$, shallow traps have little influence on $\tau_t$ because rapid remission via phonon absorption and multiple trapping events produce a long trailing edge of the current pulse. On the other hand, both shallow and deep traps contribute to carrier scattering and decrease the mobility. Our results seem to indicate that a large fraction of defects created by proton implantation in GaAs are rather shallow. This interpretation is supported by the low dark resistance of GaAs observed after proton bombardment at doses of $10^{15}$ $cm^{-2}$ and above.

Recently a different approach of introducing defects in crystalline GaAs has been reported which seems to create preferentially deep traps and much less shallow defects thus improving he mobility without significant loss of speed [26-29].

This progress has been accomplished by growing GaAs at low (approximately 200°C) substrate temperature in a molecular beam epitaxy process. At such low temperatures nonstoichiometric growth leads to about 1% excess interstitial As in the low temperature (LT)-GaAs layer corresponding to formation of deep defect states with concentrations of $10^{20}$ $cm^{-3}$ and an increase of the lattice constant by 0.1%. Subsequent annealing of the sample at 600° C for 10 min. in the As atmosphere of the growth chamber decreases the defect-concentration to $< 10^{18}cm^{-3}$ since the excess As accumulates to As-clusters with 4nm diameter and an average separation of about 15 nm. This material is distinguished by high resistivity ($> 10^7\Omega m$) greater than that of semi-insulating GaAs and an electron mobility of more than 200 $cm^2/Vs$. These features have been utilized to generate electrical pulses with a duration of less than 1 ps and peak amplitudes of up to 6 V in a coplanar transmission line with 10 $\mu$m line separation [28-29]. The high dielectric breakdown voltage of $3 \times 10^5$ V/cm and pulse biasing of a 100 $\mu$m gap on LT GaAs permits even photoconductive switching of 825 V electrical pulses with 1.4 ps risetime and 4.0 ps duration [30].

Further materials frequently used for the fabrication of photoconductive switches are amorphous Si- or Ge-films and GaAs- or InP-crystals doped with Cr or Fe, respectively. The amorphous materials achieve fast response ($\sim$ 0.5ps) but their sensitivity is quite small because of the high-disorder implying low carrier mobility. In the doped crystals the carrier trapping amounts to several 10ps. 480 fs pulses and high sensitivity have been observed in CdTe grown by metal-organic chemical vapor deposition on sapphire and quartz substrates [31]. This photoconductor may find interesting applications in ultrafast electronics since it can be easily grown on GaAs substrates as well.

## 2.2. Ultrafast Electrical Transmission Lines

The shape of the current pulse observed at the contacts of the switch is obtained by convoluting the transient current pulse excited in the semiconducting film with the RC limited response of the device determined by the charging time $\tau_c$ of the gap capacitance $C_g$. Thus, the switch has to be mounted in a high speed transmission line. These lines which connect the switch to other electronic and optoelectronic devices have to provide low dispersion and loss for electrical pulses comprising frequency components from zero up to several hundred GHz and even to THz. In order to prevent excessive broadening of the pulse, different geometric configurations like microstrip lines, coplanar strip lines and coplanar waveguides have been used [32-34]. Microstrip lines (see Fig. 3) are formed by a narrow (width w) microstrip electrode on the surface of a thin semiconductor crystal and an extended ground plane on the backside. The proper photoconductor consists of a small gap (width 20-30 $\mu$m) in the top microstrip. The high brittleness of semiconductor crystals requires a substrate thickness h of at least 200 $\mu$m and the width w of the microstrip has to be approximately equal to h to achieve transmission line impedance $Z_o = 50\Omega$ [35].

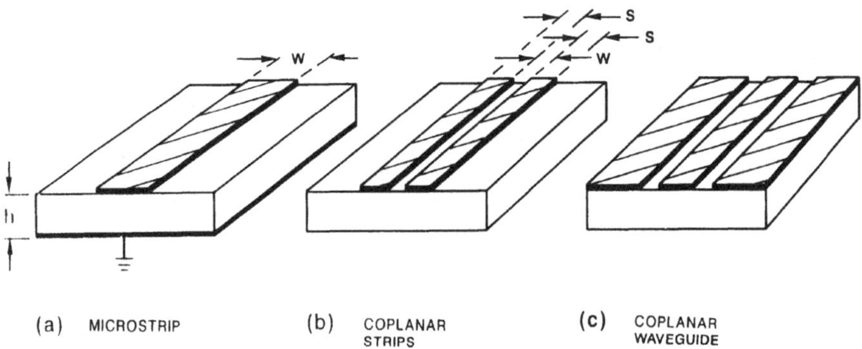


**Fig. 3** High-speed transmission line structures for picosecond optoelectronics.

A 200 $\mu$m wide microstrip leads to a gap capacitance of 60-80 fF which limits the RC response time of the device to 3-4 ps. Moreover, large separation $h$ between the signal conductor and the ground electrode plane involves a low cutoff frequency $\nu_{TE}$ for the lowest TE slab mode [35]

$$\nu_{TE} = c / \left(4h\sqrt{\epsilon - 1}\right) \tag{6}$$

Recently Roskos et al. [36] have demonstrated that the bandwidth of a microstrip line can be increased from about 30 GHz up to 500 GHz if a highly conducting burried silicide layer is used as the ground electrode because of the attainable reduction of h to less than 10$\mu$m.

Still larger bandwidth is provided by coplanar transmission lines (CTL) consisting of two parallel lines on the surface of a semiconducting substrate and coplanar waveguides consisting of 3 parallel lines (see Fig. 3b and c). The center line of the waveguide carries the signal and the two outer lines (often shaped as "semi-infinite" planes) are grounded. The typical dimensions are only a few micrometer. Both the signal and ground lines are fabricated in a single step by means of optical lithography and lift-off processing. The line separation which is equal to the substrate thickness in the microstrip line can be made more than 10 times smaller in the coplanar geometry without influencing the mechanical stability of the structure at all. Furthermore, the small dimensions imply a remarkable reduction of the dispersion and damping of transient signals with a bandwidth of more than several 100 GHz (see Chapter 5).

In a very frequently used configuration providing subpicosecond response, the CTL consists of two 5 $\mu$m wide and 0.5 $\mu$m thick metallic (Al, Au) lines separated by 10 $\mu$m. This geometry results in a specific capacitance of approximately 100 fF/mm (equivalent to $C_g$ = 1 fF for an excitation spot with 10 $\mu$m diameter) and an impedance of $Z_o = 100\Omega$. Local excitation of the gap between the two charged lines connected to a bias of several 10 V switches the conductance by many orders of magnitude. At very high excitation densities the residual resistance of the switch is limited by the contact resistance between the metallic lines and the semi-insulating substrate to the order of 100 $\Omega$.

## 2.3 Photoconductive Sampling

The shape of the electrical pulse excited on a CTL can be analyzed by cross-correlating the photoconductivity response of two switches.

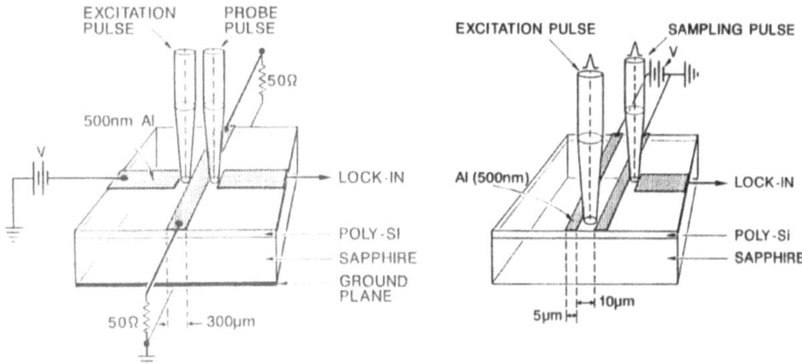


**Fig. 4** Measurement of short elecrical pulses on microstrip (left) and coplanar transmission lines (right) by photoconductivity cross-correlation.

In the sliding contact configuration originally suggested by Ketchen et al. [3] (see Fig. 4), this is accomplished by illuminating the gap between the main transmission line and the perpendicular "sampling" line (width 5 $\mu$m, distance from main line 10 $\mu$m), which is connected to an electrical probe (e.g. a lock-in amplifier with a current sensitive preamplifier), with a synchronized optical pulse from the same laser. In this geometry, the voltage pulse propagating in both directions along the line after the main line is shortened by absorption of an ultrafast laser pulse, acts as a transient bias to the second gap. This second gap operates like an optically controlled electronic gate. If the current pulse on the main line and the photoconductance of the second gap temporarily overlap, part of the charge is directed into the lock-in input. The pulse shape is sampled by recording the average current detected by the lock-in as a function of the delay $\tau$ between the optical excitation of the main line and the gating pulse. The lock-in output represents a convolution of the response of the pulse generator and the sampling gate.

The $\tau$-dependent part of the time-integrated charge detected by the lock-in is described by

$$Q(\tau) \approx \frac{U_B}{C_g} \int_{-\infty}^{+\infty} dt\; G(t-\tau) \int_{-\infty}^{t} dt' G(\tau') \left[ exp(\frac{-A}{Z_o \cdot C_g}(t-t') - exp(-\frac{B}{Z_o \cdot C_g}(t-t')) \right] \tag{7}$$

with A = 0.76, B = 5.24.

This expression has been derived under the assumption that the conductances as well as the capacitances of the two gaps are equal.

$$G_1 = G_2 = G \quad and \quad C_1 = C_2 = C_g \tag{8}$$

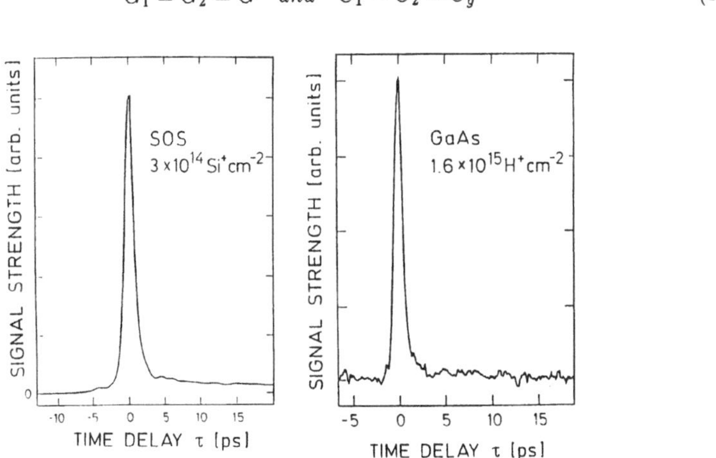


**Fig. 5** Photocurrent cross-correlation curves for silicon-on-sapphire with total irradiation dose of 3 x $10^{14}$ $Si^+$ $cm^{-2}$, FWHM: 1ps (left) and GaAs with total proton irradiation dose of 1.6 x $10^{15}$ $cm^{-2}$, FWHM: 1ps (right).

Figure 5 depicts typical examples of cross-correlation curves measured on a high speed SOS and GaAs switch for a separation between the excitation and the sampling point of only 100 $\mu$m and optical pulses of 100 fs duration from a colliding pulse mode-locked dye laser (photon energy 2 eV, repetition rate 100 MHz). Both curves have a FWHM of only 1 ps. The residual tail on the trailing edge can be attributed to multiple trapping processes. The best fit to these experimental traces calculated with a $sech^2$ profile for the optical pulse (FWHM = 100 fs), and the theoretical values of $C_g$ = 0.62 fF and $Z_o$ = 126$\Omega$ for our CTL yields a free carrier lifetime in both materials as short as $(0.5 \pm 0.1)$ ps.

Photoconductive switches are ideal tools for generating pico- and subpicosecond electrical pulses. Their application as a sampling gate for such signals suffers, however, from the fact that charge has to be removed from the circuit under study. Thus optical probing requires the integration of test points (photoconductive switches) into the circuit design and an electrical link between these test points and an external current or voltage probe. This coupling may become a difficult problem, since the signal carrying lines in an integrated circuit have dimensions of only a few micrometers. Moreover, one has to ensure that the measurement probe does not distort the signal containing frequencies of many 100 GHz up to 1 THz. At such high frequencies, the different wavelengths contributing to the signal are comparable to the dimensions of the connectors. Under this condition, proper design of the connection has to achieve matching of the line geometry and impedance as well as of the dielectric constants of the materials in order to preserve the signal characteristics.

Very recently, J. Kim et al. [37] described a novel free-standing photoconductive sampling tip which can be moved to any position on a current carrying line on a wafer. This probe consists of a interdigital photoconductive switch on LT-GaAs as the sampling gate in a "probe" line connected to the input of a current probe (see Fig. 6).

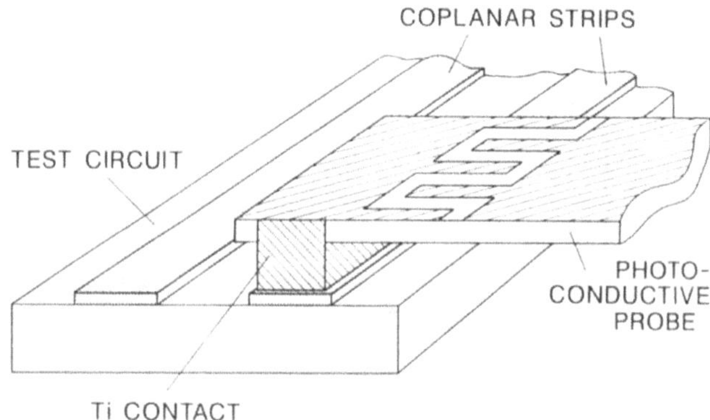


**Fig. 6** Scheme of the free-standing photoconductive sampling tip.

The second end of this probe line has a 3 $\mu$m high and 8 $\mu$m diameter Ti-tip which is pressed to the selected test position on the line and mediates the electrical contact. The signal sampling is achieved in the usual manner by activating the sampling gate with a short optical pulse with variable delay. The first experiments demonstrated a time-resolution of 2.3 ps and 1 $\mu$V sensitivity. The authors elucidate that the invasiveness of their probe tip is very small because of the low capacitance (10 fF) and high resistivity (10 M$\Omega$).

## 3. ELECTRO-OPTIC SAMPLING

Electro-optic sampling [12-14,38], which utilizes the Pockels effect, i.e. the change of the optical birefringence in electro-optic crystals in the presence of an electric field, for the detection and analysis of electrical transients, is a contact-free technique and provides much higher flexibility than photoconductive sampling.

The electric field of a short pulse propagating through a complex circuit induces a rapid variation of the birefringence in electro-optic materials localized in the vicinity of the conducting lines, which can be monitored by the polarization rotation of a light beam passing through the crystal.

If the electro-optic crystal is mounted between crossed polarizers, the field-induced birefringence can be probed by measuring the transmission of a polarized optical beam with and without the electric field being present.

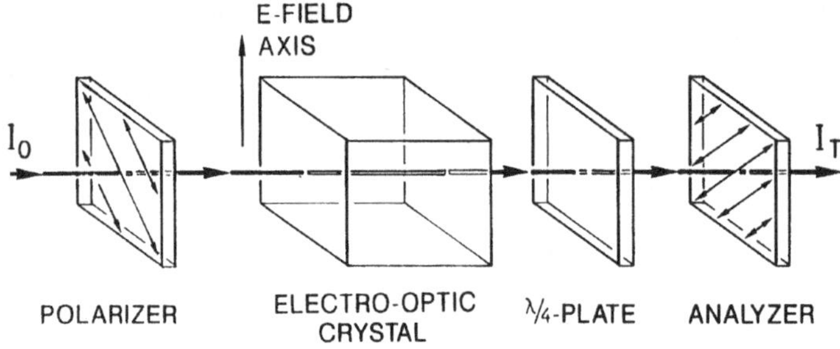


**Fig. 7** The transverse electro-optic Pockels effect. The optical beam polarized at 45° with respect to the optical z-axis propagates normal to the z-axis through the crystal. The electric field is applied along the z-axis.

In $LiTaO_3$ which is a widely used material for electro-optic sampling the highest sensitivity is achieved if the electric field E is parallel to the optic axis (z-axis) and the optical beam propagates perpendicular to the field direction and is polarized at 45° with respect to this direction (see Fig. 7). The field-induced birefringence is given by

$$n_z - n_x = (n_e - n_o) - \frac{1}{2}\,(n_e^3 r_{333} - n_o^3 r_{113}) \cdot E_3\ , \tag{9}$$

where $n_o$ and $n_e$ are the ordinary and extraordinary indices of refraction and $r_{ijk}$ are the coefficients of the electro-optic tensor. In the case of a homogeneous electric field and a cw probe beam, the phase difference between the optical fields polarized parallel to the x and z direction accumulates to

$$\Gamma = \frac{2\pi}{\lambda} \cdot L(\Delta n(0) + \Delta n(E)) = \Gamma_o + \Delta\Gamma\ , \tag{10}$$

where $\Delta n(0)$ and $\Gamma_0$ are the birefringence and phase difference, respectively, in the absence of the electric field, L is the crystal length, $\lambda$ the wavelength of the probe light, $\Delta n(E)$ the field-induced birefringence. The transmitted intensity of a light beam (intensity $I_o$) varies in dependence on the phase difference as

$$I = I_o \cdot sin^2\ (\frac{\Gamma_o + \Delta\Gamma}{2})\ . \tag{11}$$

If $\Gamma_o$ is set to $\frac{\pi}{2}$ by an additional Soleil-Babinet compensator between the crystal and the detector,

$$I = \frac{I_o}{2}\ (1 + sin\ \Delta\Gamma)\ , \tag{12}$$

i.e. the signal is almost proportional to the field for small values of E.

In picosecond optoelectronics, the Pockels effect is applied to resolve the temporal shape of a picosecond electrical pulse. Time resolution is accomplished by applying a femtosecond pulse train from a mode-locked laser as the probe and by systematically varying the temporal overlap between the electrical and the optical pulse in the crystal by means of a variable optical delay line. The time-averaged transmitted intensity is recorded by a comparatively slow photodiode and a lock-in amplifier.

For this case, the phase difference $\Delta\Gamma$ accumulated by the interaction of the probe pulses with the electrical field is given by

$$\Delta\Gamma(\tau) = \frac{\pi}{2\lambda}\left(n_e^3 \cdot r_{333} - n_o^3 r_{113}\right) \frac{\int dt \int d\vec{r} E_k(\vec{r},t) I(\vec{r},t+\tau)}{\int dt \int d\vec{r} I(\vec{r},t)}, \tag{13}$$

where $\tau$ is the delay between the electrical and optical pulse.

$LiTaO_3$ is the most widely used material for electro-optic sampling because it has relatively large electro-optic coefficients, small and almost temperature independent static birefringence and low absorption in the GHz and THz regime. The much larger electro-optic effect, e.g. in $BaTiO_3$ and $Sr_{0.75}Ba_{0.25}Nb_2O_6$ is of little practical value since the large dielectric constant of these crystals implies a strong distortion of the pulse propagation if such a crystal is approached to the signal carrying lines.

### 3.1. Electro-Optic Sampling Configurations

Various types of electro-optic (e-o) sampling schemes have been developed utilizing either the longitudinal or transverse electro-optic effect and being appropriate for different electrode structures like microstrip-, coplanar strip-, and balanced strip transmission lines which are applied in high speed electronic and optoelectronic circuits. The most common configurations are schematically depicted in Fig. 8.

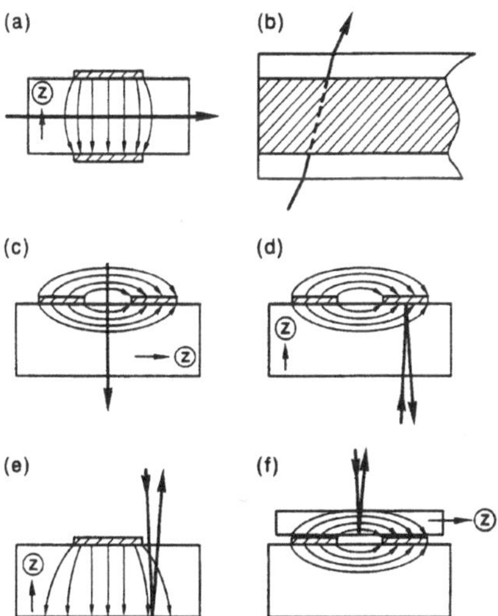


**Fig. 8** Geometries for electro-optic sampling of electrical transients in various transmission line structures. a) Transverse sampling in a balanced line; b) velocity-matched probing with the balanced line modulator; transverse c) and longitudinal sampling; d)in a coplanar transmission line; e) longitudinal sampling in a microstrip line; f) sampling with an electro-optic superstrate on a coplanar transmission line.

The balanced line modulator (Fig. 8a) consists of two equal size electrodes on the top and bottom of a thin ($\sim 0.5\mu m$) electro-optic substrate (e.g. $LiTaO_3$) slightly broader than the metallic stripes [39]. The crystal axis is perpendicular to the electrode surface.

The sampling beam polarized 45° degrees with respect to the z-axis detects the electric field via the transverse e-o effect. The width of the electrodes (200-300 $\mu$m) implies a non-negligible influence of the finite interaction geometry (convolution of the optical with the electrical pulse) on the time resolution. The latter can be remarkably improved if the laser beam transverses the crystal under an angle $\neq$ 90° with respect to the electrode stripes in order to achieve velocity matching (Fig. 8b).

The coplanar-line modulator [40] (Fig. 8c) is easier to fabricate and its dimensions can be reduced to a few microns. Its inherent dispersion is much smaller, and the measurement needs no velocity matching since the extension of the electric field normal to the surface is comparable to the electrode separation, i.e. the interaction length is quite small. The optical axis is in the plane of the substrate surface and perpendicular to the electrodes, i.e. parallel to the field between them. If the optical beam passes the crystal normal to the electrode plane, the measurement implements the same transverse electro-optic coefficients as in the balanced line modulator. Alternatively, the optical beam can enter the substrate from the backside and be reflected from one of the metallic electrodes, thus sampling the electric field via the longitudinal electro-optic coefficient (Fig. 8d).

The electro-optic probing schemes of Figs. 8a-d presume an electro-optic substrate material. As silicon (contrary to GaAs, see below), which is still the dominating material for the fabrication of electronic devices, is electro-optically inactive, application of the sampling methods described above to the characterization of Si-circuits requires broad-band dispersion- and loss-free interconnections between the device on a Si-substrate with an external e-o sampler fabricated e.g. on $LiTaO_3$. This procedure is not easily practicable and can lead to unknown disturbances of the electronic signal under study.

Most of these problems are avoided by the transverse superstrate modulator (Fig. 8f) consisting of an electrode-free slice of e-o material placed on top of the circuit to be evaluated. The orientation of the optical axis is again parallel to the substrate surface and perpendicular to the electrodes. The optical beam entering from the top is reflected at the bottom of the superstrate coated by a dielectric mirror and thus samples the fringent electric field between the electrodes which penetrates into the crystal.

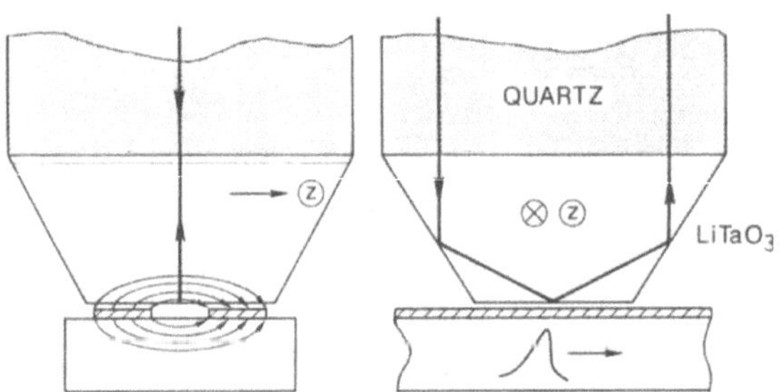


**Fig. 9** View of the external electro-optic wafer prober along (left) and normal to the coplanar transmission line (right). The depicted geometry corresponds to a transverse electro-optic crystal.

A special embodiment of the superstrate modulator is the electro-optic wafer probing tip (Fig. 9) developed by Valdmanis et al. [41]. In this case, the crystal has the shape of an inverted truncated pyramid with typical dimensions of the top plane (positioned in contact with the wafer surface) of 50 $\times$ 50$\mu m^2$. The tip is supported by a quartz rod and can be moved to any location in a complex circuit. Capacitive coupling between the probe and the electronic device is low because of the small size of the crystal. The optical beam should copropagate with the electrical pulse through the crystal to minimize the loss of time resolution caused by the transit time through the interaction volume (Fig. 9b). The direction of the probe beam is usually reversed in the crystal via 3 internal reflections from the pyramid faces. The spatial resolution of the wafer prober depends on the focal spot size of the laser beam at the apex and achieves values of 10 $\mu$m or less.

Besides $LiTaO_3$, GaAs has been applied by several groups to fabricate probe tips with a geometry suitable for sensing the longitudinal component of the fringing field. This configuration should be less sensitive to electric fields from adjacent circuit lines. Calculations reveal that the sensitivity decreases with increasing width of the air gap between the prober and the wafer and increasing dielectric constant of the probe material in accordance with experimental observations [43].

The longitudinal microstrip modulator is depicted in Fig. 8e. This geometry represents the preferred configuration for direct substrate sampling of GaAs circuits [44,45], since the transverse e-o effect vanishes in this material. In this case, the fringing field adjacent to the top electrode strip is sampled via the longitudinal e-o effect.

Whereas subpicosecond temporal resolution has been achieved with most configurations depicted in Figs. 8-9, 2.3ps are the lower limit obtained for the substrate sampling on GaAs wafers. This limit is set by the optical transit through the electric field in a 100 $\mu$m thick substrate.

An attractive novel approach for electro-optic sampling of electronic circuits fabricated on Si or other electro-optically inactive semiconductor materials are poled organic layers deposited on top of the circuit [46]. The temperatures required for processing organic electro-optic materials are compatible with integrated circuit technology and their low permitivity results in low circuit loading. A probe beam propagating normal to the surface of the circuit through the organic film on top acquires phase modulation caused by the longitudinal electro-optic coefficient. These phase modulations are converted to amplitude modulation by comparison with a reference beam.

### 3.2. Electro-Optic Sampling in GaAs

Since GaAs, which is still the most important semiconductor material for fabricating high frequency devices, is electro-optically active, the intrinsic electro-optic effect in the substrate can be utilized for sampling the shape of ultrashort voltage and current pulses in GaAs integrated circuits [44,45]. This technique can be considered as the ideal application of ultra-short light pulses to the characterization of high-speed electrical wave forms, since it represents a perfect noncontact and noninvasive approach and permits *in situ* sampling at any location on the circuit. Since e-o sampling requires photons with an energy below the band gap of the e-o material, mode-locked Nd-YAG or Nd:YLF lasers with an external fiber grating pulse compressor which reduces the pulse width to a few ps are the most frequently used sources. Mode-locked color center, soliton, or Ti : $Al_2O_3$ lasers, and gain-switched or mode-locked semiconductor lasers are new attractive alternatives.

An optical beam entering the substrate from the front or backside can interact only with electrical fields parallel to the propagation direction because GaAs wafers have usually $\langle 100 \rangle$ orientation and the transverse electro-optic effect for a beam propagating along the 100 direction vanishes [47]. The electronic circuit can be either excited by electrical pulses optoelectronically derived from the same laser which is applied for sampling or more generally by a separate synthesizer phase-locked to the RF source that drives the laser. In the latter case, suitable feedback loops are required for dynamical adjustment of the phase of the mode-locked source in order to suppress the timing jitter of actively mode-locked lasers, which achieves values of about 10ps. Such random timing fluctuation would severely degrade the performance of the e-o sampler.

## 4. Temporal Resolution, Sensitivity and Invasiveness of Optical Sampling Techniques

The time-resolution of any photoconductive or electro-optic device depends on the optical pulse duration, the intrinsic frequency bandwidth of the photoconductive or electro-optic crystal and on the geometry and size of the region where the electrical and optical pulses interact [48].

If a femtosecond laser system is used, the most severe limitation to the time resolution is imposed either by the free carrier trapping time of 0.5ps of the photoconductor or by the far-infrared resonance absorption of the electro-optic crystal. As discussed in Chap. 2, the carrier trapping time of 0.5ps in the fastest materials sets a lower limit for the time resolution attainable with photoconductive sampling. The optical phonon in $LiTaO_3$ at 6.3 THz together with a weak resonance at 2.7 THz [48] result in an absorption of 10 $cm^{-1}$ at 0.5 THz equivalent to a penetration depth of 1mm. Thus, the path of the THz-field through the crystal should not exceed a fraction of a mm in order to avoid excessive attenuation and dispersion of the pulse. The experimentally observed maximum speed of a $LiTaO_3$ electro-optic prober of approximately 300fs [42] can be attributed to the damping rate ($\Gamma = 1.1$ THz) of the optical phonon.

The carrier trapping time or the effect of the lattice vibration is distinctly larger than the influence of the finite interaction time if the sampling geometry is optimized. The geometry yields two contributions to the time resolution: 1) the transit time of the optical beam through the field of the electrical pulse and 2) the time it takes the electrical waveform to propagate through the beam waist of the optical pulse. For a 10$\mu$m wide CTL and a beam focused to a 10$\mu$m diameter the electrical and optical transit time are estimated to be $\Delta t_O = n \cdot \frac{2w}{c} \approx 150$ fs and $\Delta t_E = \frac{d}{c}\,(2 \ln 2 \cdot \epsilon_{eff})^{1/2} \approx 100$ fs. $2w$ and $d$ are the length of the path through the electric field and the diameter of the optical beam at the sampling point, respectively. The situation is schematically shown in Fig. 10 for sampling by a probe tip.

Finally, it should be noted that the time resolution severely deteriorates, if the probe position is separated from the pulse generation point by more than a few 100 $\mu$m since the inherent dispersion and damping properties of the connecting electrode configuration can become important. The origin and magnitude of these effects are discussed in detail in chapter 5.

Photoconductive sampling of transient electrical waveforms with mode-locked lasers at repetition rates of 100 MHz achieves an average current detection limit of 0.2 pA corresponding to pulse peak currents of about 1nA for pulse durations around 1ps. The attainable extensive signal averaging results in effective noise suppression making possible micro-volt sensitivity for integration times of 1s [45].

At a first glance, the typical magnitude of electro-optic coefficients of $10^{12}$ m/V

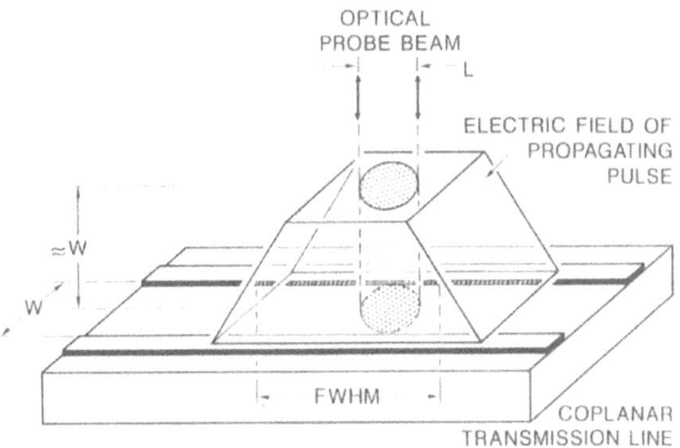


**Fig. 10** Geometric effects on the time resolution of an electro-optic wafer prober described by the optical and electrical transit times.

implicating a half-wave voltage of the order of 1kV for a Pockels cell modulator may suggest a very low sensitivity of the electro-optic effect for the detection of voltage pulses with amplitudes of a few 100 mV, inducing an intensity change of less than $10^{-3}$. However, the high repetition rate of mode-locked ultrafast laser systems of typically 100 MHz allows effective signal averaging and rapid separation of the signal from the noise which is mainly due to intensity fluctuations of the laser. Since the noise spectrum of a mode-locked dye laser exhibits a drop of the noise amplitude by about 30-40 dB for frequencies above a few MHz, the laser noise becomes negligible if high frequency (5-10 MHz) synchronous detection is applied. Under this condition intensity modulations as small as $10^{-8}$ can be detected. A detector shot noise limited sensitivity of 10-100 $\mu V/\sqrt{Hz}$ is attainable [14].

Quantitative experimental studies of the signal reproduction capability and invasiveness of electro-optic sampling with a $LiTaO_3$ probe tip have manifested the important influence of the external probe geometry [48,50]. Substantial inaccuracies can be introduced by reflection of the electrical pulse at dielectric discontinuities which interfere with the signal in the time window of the measurement. The main sources of such reflections are 1) the surfaces through which the electric field enters and leaves the $LiTaO_3$ crystal, 2) the interface between the $LiTaO_3$ and its quartz support and 3) the bottom of the transmission line substrate (see Fig. 1 of Ref. [50]). The experiments of Frankel et al. [50] have disclosed that measurement errors can be drastically reduced, if the crystal dimension and thus the signal integration path are constrained to a few tens of microns. The superior accuracy is achievable with little loss in sensitivity, since the electric field strength drops rapidly with increasing distance from the transmission line surface.

Concerning speed and sensitivity, the optical sampling techniques exceed the performance of present state conventional ultrafast electronic equipment by a factor of roughly 100 (resolution limit $\sim 2 \times 10^{-13}$s compared to $10^{-11}$ s, detection limit 10 $\mu$V compared to 1mV). The dynamic range for the generation and detection of ps electrical transients covers 9 orders of magnitude (10 $\mu$V-10 kV) instead of 4 (1mV-10V). The high repetition rate of $10^8$ Hz (compared to $10^4$ Hz) renders possible effective noise suppression by very high sampling rates. Negligible timing jitter in all applications, where the same laser pulse is used for generating the electrical pulse and for sampling the response of the circuit, and the potential to use these methods over a wide temperature range down to liquid He temperature, since no high-speed electrical connections to external devices are required, are further advantages. Most of the experiments reported in the literature so far have used laboratory scale laser systems such as colliding pulse mode-locked dye lasers or mode-locked solid state lasers. The price and complexity of these lasers represent the major disadvantages of optical sampling compared to purely electronic instrumentation and prevent their widespread applications in the semiconductor industry. The optical power requirements are so low, however, that these noninvasive measurements can be easily performed with low cost, small-sized, gain-switched or mode-locked semiconductor lasers [51-54]. The main impediment for the replacement of the large laser systems by semiconductor laser diodes is due to the typical pulse duration of > 10 ps of the latter. Novel developments in subpicosecond pulse generation from semiconductor lasers [55,56] hold great promise that this situation may change in the near future.

## 5. PULSE PROPAGATION ON COPLANAR TRANSMISSION LINES

Applications of pico- and subpicosecond electrical pulses generated by photoconductive switches to ultrafast electronics necessitate broadband interconnections to other electronic and optoelectronic devices which permit undistorted transport of a signal comprising a spectral bandwidth of many hundreds of GHz up to a few THz over macroscopic distances of several mm without severe dispersion and losses. Here, we shall restrict the discussion to coplanar lines and waveguides, since their fabrication is highly compatible with the processing technology of planar semiconductor microelectronics and their size can be reduced to extremely small dimensions. Experimental and theoretical studies of the dispersion and loss of microstrip lines have been published in references [57-59].

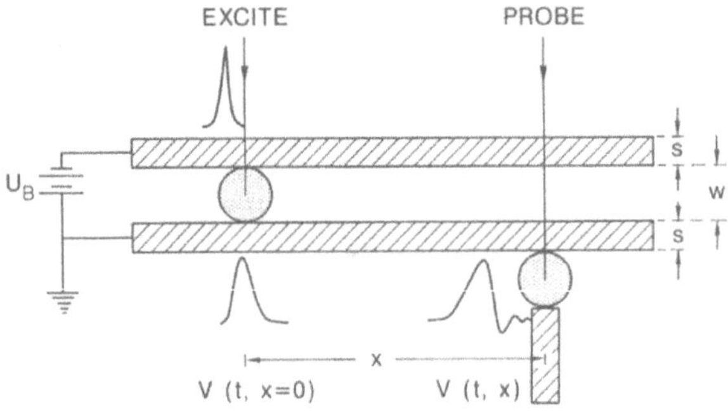


**Fig. 11** Experimental set-up for studying temporal broadening and damping of ultrafast electrical pulses on a coplanar transmission line.

The temporal broadening and damping of an ultrafast electrical pulse on a coplanar waveguide or transmission line can be easily investigated by the sliding contact excitation method (see Fig. 11) allowing simple variation of the distance between the signal sampling gate and the excitation point. The temporal (V(t,x=0)) and spectral profile ($U(\omega$,x=0)) of the electrical transient excited by the focused laser pulse at x = 0 are determined by the optical pulse shape, the material response and the electrical properties (capacitance $C_g$ of the gap and impedance $Z_o$) of the line structure.

$$V(t,x=0) = \frac{G_o \cdot U_B}{2C_g} \left| \frac{exp(-t/\tau_t) - exp(-t/\tau_c)}{\frac{1}{\tau_c} - \frac{1}{\tau_t}} \right| \tag{14}$$

Here, $G_o$ is the peak photoconductance of the gap, $U_B$ the bias, $\tau_t$ the carrier trapping time and $\tau_c = 2Z_oC_g$.

Fourier transformation of V(t,x = 0) yields the frequency distribution

$$U(\omega, x=0) = \frac{1}{\sqrt{2\pi}} \int_{-\infty}^{+\infty} V(t, x=0) e^{i\omega t} dt \tag{15}$$

The amplitude and phase of each individual Fourier component after propagation over a distance x is given by

$$U(\omega, x) = U(\omega, x=0) \cdot e^{i\beta(\omega)x} , \tag{16}$$

where the complex phase factor

$$\beta(\omega) = \beta'(\omega) + i\beta''(\omega) \tag{17}$$

with $\beta'(\omega) = \frac{\omega}{c}\sqrt{\epsilon_{eff}(\omega)}$ comprises the frequency dependence of the phase velocity

$$v_p = \frac{c}{\sqrt{\epsilon_{eff}}} \tag{18}$$

as well as of the attenuation coefficient $\beta''(\omega)$. The pulse shape at x is calculated by the inverse Fourier transformation.

$$V(t,x) = \frac{1}{\sqrt{2\pi}} \int_{-\infty}^{+\infty} U(\omega, x) e^{-i\omega t} d\omega \tag{19}$$

The dispersion of the real part of $\varepsilon_{eff}(\omega)$ of the effective dielectric constant results in temporal spreading of the frequency components with increasing x. Hasnain et al. [60], who calculated the dispersion of electrical pulses on a coplanar transmission line give analytical expressions for $\epsilon_{eff}$ in dependence on the dielectric constant of the substrate $\epsilon_s$ and the line geometry.

Figure 12 depicts the frequency dependence of $\sqrt{\varepsilon_{eff}}$ for Al coplanar strips on an SOS substrate for coplanar lines with w = 10 $\mu$m, s = 5 $\mu$m and s = w = 50 $\mu$m. For low frequencies $\varepsilon_{eff}$ and thus the phase velocity of the wave is almost constant, since the relative spatial distribution of the electric field to the air above the lines and to the substrate is independent of frequency. Above a critical frequency $\nu_{TE} = c/(4w\sqrt{\epsilon_s - 1})$ which depends on the strip electrode geometry, the field is more and more constricted to the substrate. This effect implies an increase of $\varepsilon_{eff}$ with frequency and a smaller phase velocity for the corresponding spectral components and is more pronounced for wider lines. This problem originates from the dielectric asymmetry of the conducting lines. The open electrode geometry prevents formation of a true TEM mode rather it excites extraneous surface modes which travel at different phase velocities. The problem can, in principle, be overcome by covering the coplanar line by a superstrate with the same dielectric constant as the substrate. For this condition the modal dispersion is

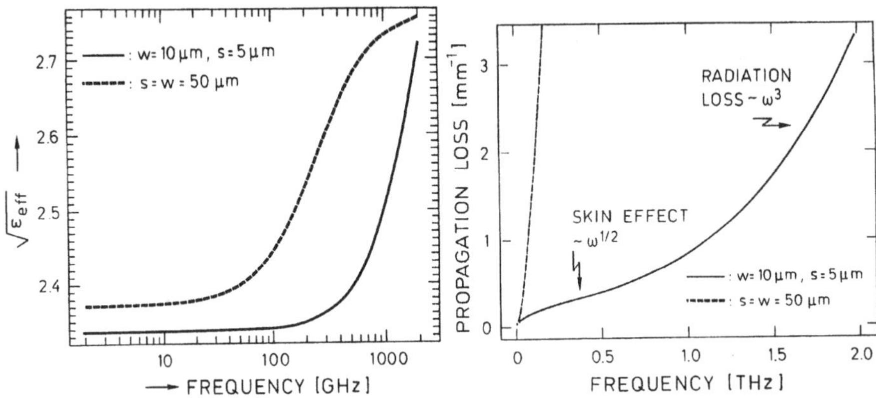


**Fig. 12** Frequency dependence of $\sqrt{\epsilon_{eff}}$ (left part) and damping coefficient $\beta''(\omega)$ (right part) for coplanar transmission lines with s = 5 $\mu$m, w = 10 $\mu$m (solid line) and s = w = 50 $\mu$m (dashed line).

strongly reduced, but the material dispersion remains unchanged and the speed of the pulse decreases since the effective $\varepsilon$ approaches the value of the bulk material.

An alternative approach to remove the air-substrate asymmetry is the coplanar air transmission (CAT) line [61] in which the substrate dielectric is removed around the conductors via etching. As the dominant part of the electric field is now propagating in air, both the modal and material dispersion drop and propagation velocities as high as 86% of the speed $c$ of light in vacuum are attainable, as compared to 0.38 $c$ for the conventional unetched coplanar line. On these CAT lines, signals with rise times as short as 1ps can be propagated over distances of approximately 3mm without significant distortion, thus rendering possible intrachip communication in digital circuits with rise times as short as 1ps.

Problems encountered in propagating high-speed signals on lossy transmission lines are analyzed in Ref. 62. The frequency dependent losses of the CTL are composed of two contributions [33]

$$\beta''(\omega) = \beta''_s(\omega) + \beta''_c(\omega) . \tag{20}$$

The first term of Eq.(19) represents the damping caused by the ohmic resistance of the metallic lines. Its frequency dependence originates from the skin effect, i.e. the displacement of the current from the interior to the surface of the conductor with increasing frequency. The concomittant reduction of the effective cross-section of the conductor implies increasing ohmic losses proportional to the square root of the frequency.

As

$$\beta''_s(\omega) \sim \mathrm{w}^{-1} \cdot \omega^{1/2} , \tag{21}$$

wider CTL's are superior.

These resistive losses can be drastically reduced if superconducting transmission lines operated at cryogenic temperatures below the critical transition temperature are used [63]. As the conduction losses vanish only for those frequencies which are smaller than the energy gap of the superconductor, extremely high-speed (bandwidth of a THz or more) necessitates utilization of the high-$T_c$ ceramic superconductors

($YBa_2Cu_3O_{7-x}$) which have an energy gap of more than 10 THz, if BCS type behavior is assumed [64,65]. Distortion-free propagation of electrical pulses with a few 100 GHz has been observed on such superconducting lines at temperatures of 2K. However, for frequencies approaching 1 THz, pulse broadening due to modal dispersion and radiative losses (see below) become dominating. Additionally, surface roughness and microstructure defects of the superconducting films contribute to the high-frequency losses of a "real" CTL.

The second term in Eq. (20) describes the attenuation of the pulse due to the emission of radiation into the dielectric substrate which becomes very important for frequency components with wavelengths comparable (or smaller) to the distance between the two lines. The radiative losses are due to an electromagnetic shock wave which is excited, if an ultrashort current or voltage pulse is propagating on a CTL. The moving electrical pulse represents an electric dipole which propagates with a group velocity which is larger than the phase velocity of electromagnetic THz radiation in the substrate [66-68]. This situation is analogous to the classical Cerenkov effect observed, if an electric monopol propagates through a dielectric medium (refractive index n) with a group velocity larger than c/n. Therefore, the moving dipole radiates an electromagnetic shock wave in the shape of a Cerenkov-type cone into the substrate. Rutledge et al. [69] have predicted a cubic frequency dependence of the attenuation and a quadratic variation with the total width of the line under quasi-static approximations.

$$\beta_c''(\omega) \sim (w + 2s)^2 \cdot \omega^3 \tag{22}$$

According to the model of Rutledge et al. the radiation losses can be drastically reduced by choice of small line dimensions and a substrate with a low dielectric constant.

Experiments of Grischkowsky et al. [70] and Fattinger and Grischkowsky [71] proved that the radiative losses become dominant for frequencies above 200 GHz for CTL dimensions of a few tens of micrometers.

Figure 12 depicts the total loss versus frequency for two different transmission lines fabricated on SOS. For the CTL with s = 5 $\mu$m, w = 10 $\mu$m, the skin effect dominates below 400-500 GHz and for higher frequencies, we observe the rapid rise of the loss caused by the cubic dependence of $\beta_c''$ on the frequency. The radiative losses are drastically smaller on a CAT, where the major part of the high-$\epsilon$ dielectric between the electrodes is removed and the electric field is predominantly guided in the uniform air gap.

It should be noted that owing to the different dependence of $\beta_s''$ and $\beta_c''$ on $\omega$ and w an optimum w yielding the lowest loss exists for each frequency.

Using the formulas for $\varepsilon_{eff}(\omega)$, $\beta_s''(\omega)$, and $\beta_c''(\omega)$, we have calculated the shape of a pulse with an exponential rise of 100 fs (determined by the width of the optical excitation pulse) and an exponential decay of 500 fs (given by the carrier trapping time in heavily radiation-damaged SOS) propagating on an Al-CTL (s = 5 $\mu$m, w = 10$\mu$m, thickness = 0.5 $\mu$m) on an SOS substrate over a distance of 4 mm.

The results are displayed in Fig. 13. Figure 13b considering solely the influence of dispersion ($\beta'' = 0$), shows a distinct broadening of the rising edge and main peak of the pulse as well as strong oscillations on the trailing edge. All these features reveal the frequency chirp of the pulse caused by the dispersion of $\epsilon_{eff}$. The strong oscillations are almost completely removed and the main peak in further broadened, if the damping is taken into account (Fig. 13c). Finally, Fig. 13d depicts the cross-correlation of the pulses in Figs. 13a and c, i.e., the signal obtained in the experiment, where the pulse shape is measured by photoconductive sampling.

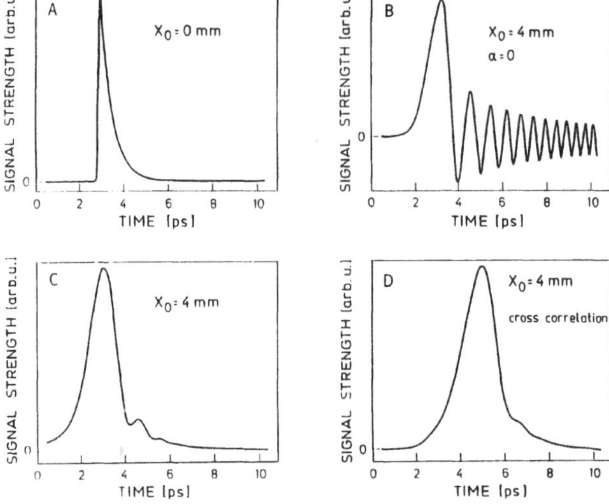


**Fig. 13** Theoretical broadening of an electric pulse with a rise and decay time of 100 fs and 500 fs, respectively. (A) After 4mm of propagation along a CTL (s = 5 $\mu$m, w = 10 $\mu$m) under the influence of dispersion (B) or dispersion and damping (C). (D) depicts the cross-correlation of c with the response function of the sampling gap taking into account the influence of the gap capacitance.

The measured typical waveforms presented in Fig. 14 agree fairly well with the theoretical predictions with respect to the cross-correlation width (2ps) as well as the appearance of a long trailing edge with weak oscillations.

Figure 15 shows a plot of the measured FWHM of the cross-correlation function versus propagation distance for two different CTL's. The data demonstrate the much faster broadening of the pulses with distance for the wider line as expected from the dependence of the modal dispersion and radiation losses on the line dimensions. The broader pulse width observed close to the excitation point for the wide line is due to the higher capacitance.

Very recently, Frankel et al. [72] have demonstrated that the semi-empirical formula for the radiation losses derived by Rutledge requires some frequency dependent corrections, if THz frequency components begin to play a dominant role in order to take into account the influence of non-quasi static effects. The authors have carried out a quantitative comparison between experiment and theory for both dispersion and attenuation. Their work has proved the validity of the modified analytic expressions up to THz frequencies.

Finally, it should be mentioned that several suggestions for the implementation of nonlinear transmission lines into picosecond electronic circuits have been published [73-75]. These nonlinear transmission lines consist of a high-impedance coplanar waveguide on semi-insulating GaAs with a periodic array of metal-semiconductor diodes which have a voltage dependent capacitance. Because the capacitance per unit length of the line determines the propagation velocity of electrical signals on the line, the integration of the nonlinear capacitors results in a voltage dependent propagation velocity. Such devices would permit repetitive compression of electrical pulses and enable soliton-like pulse propagation which meanwhile is easily achieved in the optical fre-

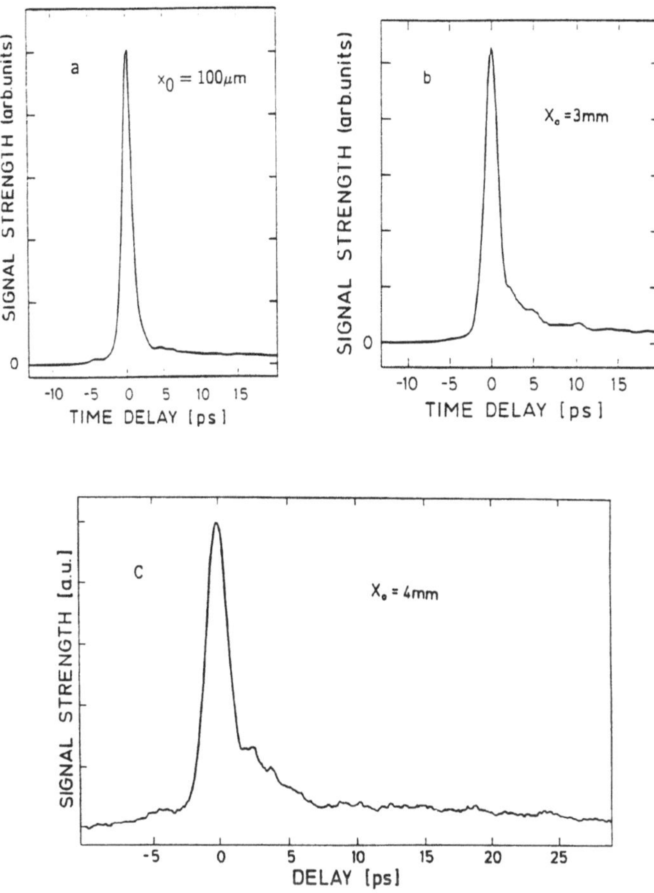


**Fig. 14** Experimentally measured photoconductivity cross-correlation traces for various separations between the excitation and sampling point on a CTL with s = 5 $\mu$m, w = 10 $\mu$m fabricated on an SOS substrate.

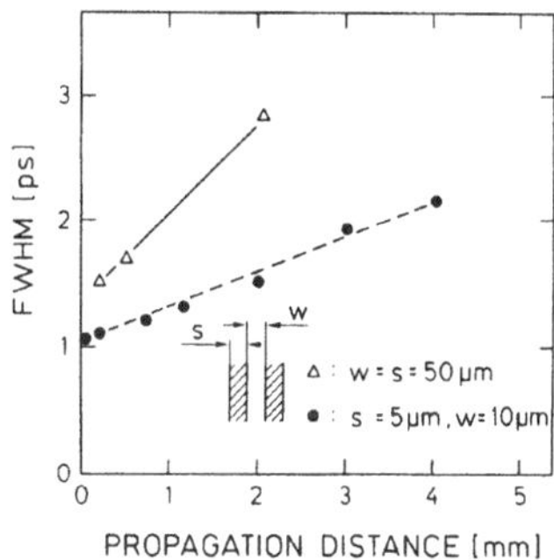


**Fig. 15** FWHM of the cross-correlation trace versus propagation distance for two CTL's with s = 5 $\mu$m, w = 10 $\mu$m ($\bullet$) and s = w = 50 $\mu$m ($\Delta$) fabricated on an SOS substrate.

quency regime. Effective compression of wide (100 ps), low amplitude electrical pulses into larger amplitude, narrow width (5ps) output pulses has been successfully demonstrated on homogeneous as well as inhomogeneous (tapered) nonlinear transmission lines [76]. Madden et al. [77] have generated voltage shock waves with 1.6ps fall time and 6V amplitude on a hyperabrupt doped Schottky diode monolithic GaAs nonlinear transmission line, by injecting a 10 GHz sine wave at one end.

## 6. CHARACTERIZATION OF METAL-SEMICONDUCTOR METAL PHOTODETECTORS

### 6.1. Metal Semiconductor-Metal Photodetectors

Interdigitated metal-semiconductor-metal (MSM) Schottky contact diodes fabricated on GaAs [4,78-80] or InGaAs [81-88] are attractive photodetectors for multigigabit optical communication systems. The compatibility of the fabrication of MSM diodes with the process technology of planar high-speed electronic devices like field-effect transistors permits easy integration of the detector and the signal processing electronic logics on the same chip [89-94]. The most appealing features for the application of MSM photodetectors as the front end of integrated high-frequency optoelectronic receivers are their fast response, high sensitivity, and low dark current.

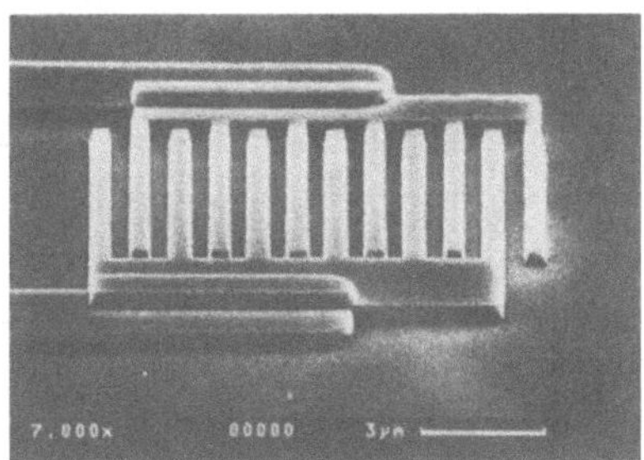

**Fig. 16** Scanning electron microscope photograph of a MSM photodetector with 0.5 $\mu$m finger width and separation.

The diodes are fabricated by means of electron beam lithography and lift-off processing on bulk LEC-GaAs using Ti/Pt/Au as the Schottky contacts with an active area of $10 \times 10 \mu m^2$, and typically 0.5 - 1.5 $\mu$m wide fingers with 0.5 - 1.5 $\mu$m distance between them. For the Monte Carlo simulation of the response the semiconductor material is assumed to be unintentionally n-type doped at a level of $5 \times 10^{13} cm^{-3}$. Figure 16 shows a scanning electron microscope view of the detector surface.

The photodiode is integrated with a photoconductive switch on the same GaAs chip (see Fig. 17) which is locally ion damaged with a total dose of $4.11 \times 10^{14}$ $H^+$ $cm^{-2}$ resulting in a carrier lifetime of 0.8 ps [25]. The transmission line consists of two 5 $\mu$m wide strips with 10 $\mu$m spacing. The 5 $\mu$m wide sampling strip is located at 2.5 mm distance from the diode. Photoconductive or electro-optic sampling (with a small $LiTaO_3$ tip at a distance of 100 $\mu$m from the diode) are applied to analyze the detector response after excitation by 100 fs optical pulses from a dispersion-compensated CPM dye laser

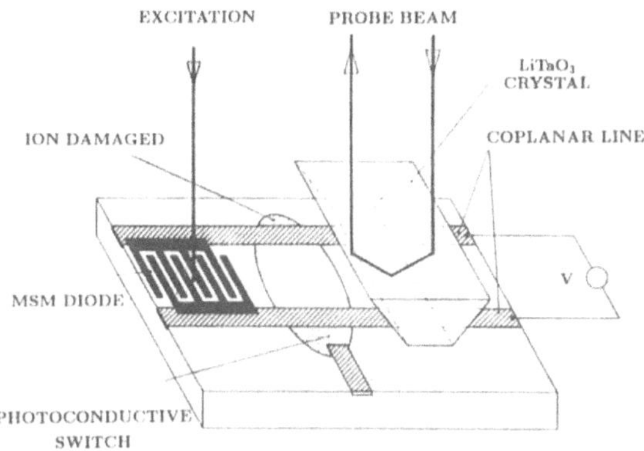


**Fig. 17** Experimental set-up for characterization of MSM photodetectors by photoconductive or electro-optic sampling with ultrashort laser pulses.

(wavelength 620 nm, repetition rate 116 MHz). The time resolution was determined to be 0.8 ps and 0.3 ps for photoconductive and electro-optic sampling, respectively.

## 6.2. MONTE CARLO SIMULATION OF THE ELECTRON AND HOLE TRANSPORT

Several theoretical attempts to determine the fundamental limitations of the response speed set by the scattering of the carriers in the semiconductor material [95] have been published recently. Koscielniak et al. [96,97] have modelled the electron and hole transport in GaAs MSM diodes by a one-dimensional Monte Carlo simulation. This model predicts a clear temporal separation of the contributions of the two species of carriers to the response current and a duration for the electron and hole pulse of a few picoseconds and several tens of picoseconds, respectively. Extension of these calculations to detectors with 100nm electrode patterns have been shown to require a quantum-mechanical analysis using numerical solutions of the time-dependent Schrödinger equation [98,99]. These quantum effects, however, do not play a significant role in the present study which is concerned with interdigital separations above 0.5 $\mu$m. The 2D-simulation of the electron and hole drift in InGaAs MSM diodes reported by Soole and Schumacher [100] did not take into account the modifications of the electric field associated with free carriers. The results of their study, which is particularly focused onto the variation of the transport parameters with finger spacing and thickness of the active layer, thus correspond to the limit of low light levels. Monte Carlo calculations of the carrier motion in GaAs surface space charge fields by Zhou et al. [101] have revealed, however, strong modifications of the transport by field screening originating from spatial separation of the photo-generated electron-hole pairs.

Here, we describe the electron and hole motion by a self-consistent 2D Monte Carlo simulation and discuss the detector response to a 70 fs optical pulse at 2eV photon energy in terms of the distribution of carriers and electric fields [102,103]. Calculations are performed for different detector geometries and excitation conditions in order to illustrate the dependence of the output pulse shape and the attainable frequency bandwidth on the interdigital distance, electric field and the excitation density. This method determines solutions of Boltzmann's transport and Poisson's field equations, both in space and time. Since Poisson's equation is repetitively solved every 40 fs over a rectangular mesh with 7.8 nm×20nm and 23.4 nm×20nm for diodes with 0.5 $\mu$m and 1.5 $\mu$m finger distance, respectively, the local electric field is exactly calculated for each time step in a self-consistent way and therefore screening of the external field by space charge fields originating from the spatial separation of photo-excited electrons and holes are correctly taken into account.

The Schottky contacts are assumed to be perfect barriers which only allow transport of electrons and holes from the semiconductor into the anode and cathode, respectively, but prevent injection of carriers from the electrodes.

The initial vertical electron-hole (e-h) pair distribution $n(y)$ created close to the illuminated detector surface by absorption of the incident laser photons is determined by the optical absorption coefficient $\alpha$

$$n(y) = n_o exp(-\alpha y), \tag{23}$$

where $n_o$ represents the pair density at the surface and $\alpha = 4.279 \times 10^4/\text{cm}$ for 2 eV photons [104]. $y$ is the distance from the detector surface. The lateral distribution of the carrier injection between the electrodes is assumed to be homogeneous ($n(x,z) = const$) and $n(x,z) = 0$ below the opaque metal fingers.

The motion of the photogenerated electrons and holes under the influence of the electric field is computed accounting for stochastic scattering with optical, acoustic, and intervalley phonons and ionized impurities. The procedure permits direct tracing of the motion of individual particles and yields a profound insight into the microscopic physical processes inside the device. Intra- and intervalley transitions for electrons and intra- and interband transitions for holes are taken into account. The model includes the $\Gamma$-, X-, and $L$-minima of the conduction band and the heavy-hole, light-hole, and split-off valence band. The analysis disregards carrier-carrier scattering since the net momentum of the electronic system remains unchanged by electron-electron (hole-hole) collisions and electron-hole scattering has been shown to be negligible for excitation levels [105] below $10^{17}/\text{cm}^3$. The light intensities in future optical data communication systems will surely range below this level.

Owing to limitations in computer power and capacity, the simulation of the transport has to be confined to 10 000 individual particle pairs. Each simulated particle which will be referred to as a "superparticle" [106] in the following is an artificial object. When solving Boltzmann's equation the superparticle is regarded as a single charge carrier. For the recalculation of the electric field by means of Poisson's equation and for estimation of the current, one superparticle is considered to consist of many real charged particles and the number depends on the light pulse intensity. More exactly, in the two-dimensional analysis each superparticle corresponds to an infinitely long rod of charges perpendicular to the cross-section of the detector. Thus, the charge of the superparticles is charactrized by a linear charge density $\varrho$. The value of $\varrho$ varies with excitation density. For the highest value of $n_0 = 5.1 \times 10^{17}/\text{cm}^3$ each superparticle represents 1.803 $\times 10^3$ electrons or holes per cm (corresponding to $\varrho = 2.88\times10^{-16}\text{C/cm}$) for the 1.5-$\mu$m finger spacing structure and one third of this value for the detector with 0.5 $\mu$m finger spacing. The current measured in the external circuit after excitation of a photoconductor is composed of two contributions. First, the particle current caused by carriers recombining at the contacts and secondly, the displacement current originating from the variation of the electric field at the contact surface associated with the displacement of charged carriers inside of the semiconductor material.

$$I(t) = \int_A \left[ \vec{j}(t) + \frac{1}{4\pi} \cdot \frac{\partial \vec{E}}{\partial t} \right] d\vec{f} \tag{24}$$

$\vec{j}$ is the density of the particle current, $d\vec{f}$ the area element, A the contact area.

For the calcuation of the total current, the MC simulation has to add up the contributions of the electrons drifting to the anode and the holes drifting to the cathode. The self-consistent MC-simulation permits simultaneous calculation of both terms in Eq.(24). The electron particle current at the anode, for instance, is determined by counting the total number of electrons hitting this electrode within the unit of time (40fs). The displacement current is obtained by integrating the change of the electric field over the contact surface. It should be mentioned that the theorem of Ramo [107] represents an equivalent alternative technique for the calculation of the photocurrent.

## 6.3. THEORETICAL RESULTS

Figure 18 shows the spatial distribution of electrons, holes, equipotential lines, and the lines of constant magnitude of the electric field between two neighboring fingers at five different times after the illumination. This calculation was performed for a finger separation of 1.5 $\mu$m, bias of 2V and an illumination intensity which produces a surface density of $n_o = 5.1 \times 10^{15}\text{cm}^{-3}$. The initial carrier distribution at $t = 0$ is given by Eq. (23). Under the influence of the electric field which peaks at the edges of the cathode and the anode, the electrons and holes start to separate. The electrons move towards the anode with saturation drift velocity [108], the holes with a noticeably slower speed to the cathode. As the drift velocity of holes increases with the field strength at least up to 60 kV/cm at room temperature [109], the local electric field is in almost every case too weak for the holes to reach saturation drift velocity. After 16ps most of the electrons have entered the anode and have left behind a positive space charge of holes. At later times the holes yield the main contribution to the photocurrent.

Figure 19 depicts the temporal evolution for the pure particle parts of the electron (•) current, the hole (o) current and the total (- - -) current predicted by this model for an MSM diode with a finger separation of 1.5 $\mu$m and a bias voltage of 2V at a relatively modest excitation level of $5.1 \times 10^{15}\text{cm}^{-3}$. The time dependence of the total current exhibits a first rapid decay (1/e-time 8ps) which can be attributed to the electron sweep-out and a second slow decay (1/e-time 30 ps) which results from the hole sweep-out. The signal rise time and the full-width at half-maximum amount to 4ps and 11ps, respectively.

Figure 20 depicts the displacement current contribution obtained for the same diode parameters and excitation condition used for the calculation of the particle current in Fig. 19. The displacement current is mainly due to the carriers located in the vicinity of the contacts and consists of a strong peak (FWHM = 1.3ps) immediately after the absorption of the optical pulse when the created electrons and holes are extracted from these regions. In the experiments the calucated fast rising edge will be broadened to about 1.4ps by the capacitance and propagation effects on the electrode structure.

The main effect of the displacement current on the measured signal is a steepening of the rising edge, the change of the amplitude amounts to less than 10%. The same situation is expected for all detectors fabricated on LEC or high quality MBE GaAs with finger distances between 0.5-1.5 $\mu$m and excitation densities in the range between

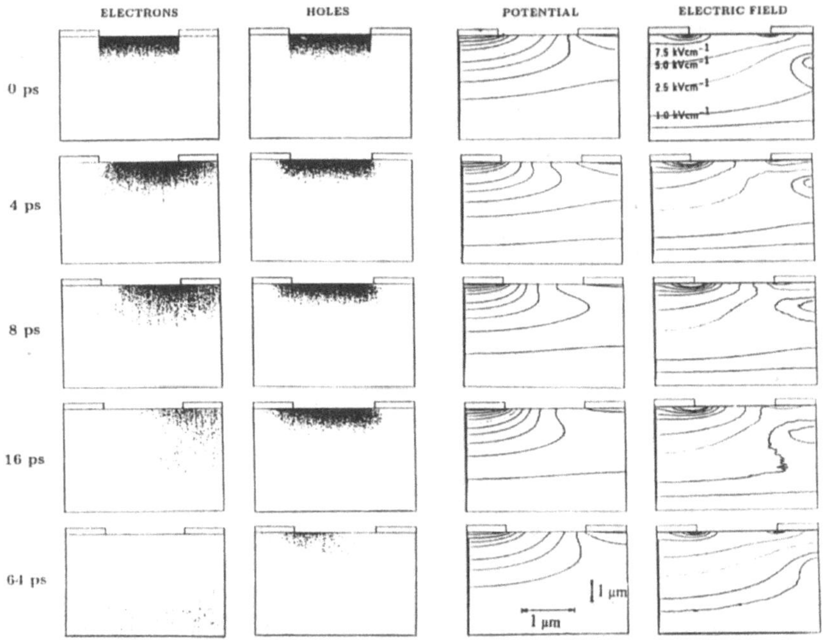


**Fig. 18** Distribution of electrons, holes, electric potential and magnitude of the electric field in a MSM-photodetector (finger spacing = finger distance = 1.5 $\mu$m, bias 2V) after excitation with an electron-hole density of $5.1 \times 10^{15}$ cm$^{-3}$ at 0,4,8,16, and 64 ps after absorption of a 100fs laser pulse.

$5 \times 10^{15}$ - $5 \times 10^{17}$cm$^{-3}$ investigated in the course of our experiments. Therefore, the displacement current has been neglected in the corresponding Monte Carlo simulations and all theoretical curves shown represent solely the particle current.

The curves in the left part of Fig. 21 reveal considerable changes of the carrier transport at higher intensity levels. The model predicts an initially faster decay of the photocurrent compared to the situation at lower excitation and a substantial slowing down of the transport after about 25ps. At these later times the electron and hole population decays coincide. This behavior is explained by Fig. 22 which depicts theoretical results for the temporal evolution of he carrier distribution in the same detector as in Fig. 18 but illuminated by a 10-times higher intensity. The fourth column of this figure reveals the formation of a field-free region near the anode after 4ps caused by the initial spatial separation of electrons and holes which tend to screen the electric field between the anode and the cathode. The carrier distributions displayed in the left two columns of Fig. 22 illustrate the formation of a screened electron-hole plasma in this area. Even after 64ps when the current has dropped to less than 10% of its peak value, a large fraction of photo-generated electrons and holes still exist in the field-screened area near the anode.

In these low-field regions, the coexisting electrons and holes form an electron-hole plasma which coheres by mutual collective attraction between the oppositely charged particles. The electric field is so weak that an electron may need a time of more than a

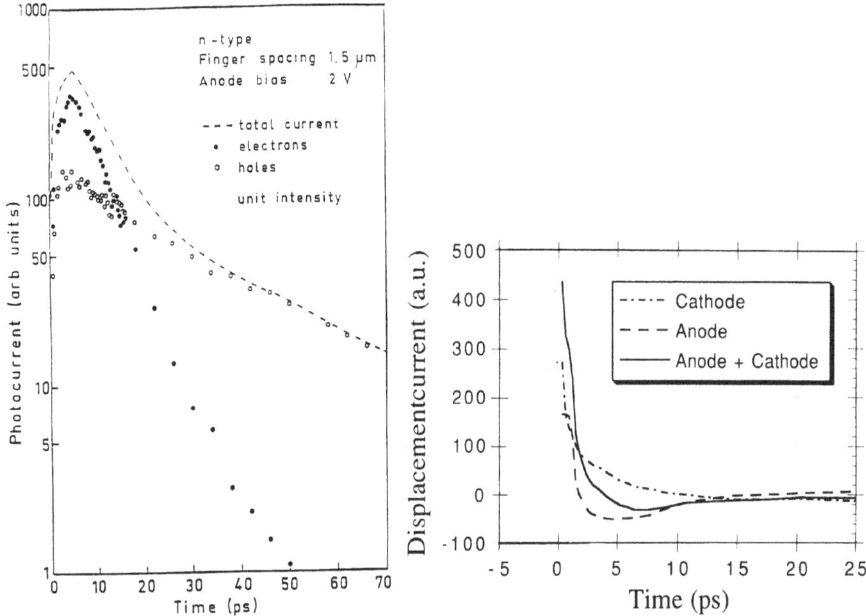


**Fig. 19** Calculated electron (•), hole (o) and total particle current (- - -) for the detector and excitation conditions of Fig. 18.

**Fig. 20** Calculated displacement current at the anode and cathode and the sum of both contributions for the detector and excitation condiction of Fig. 18.

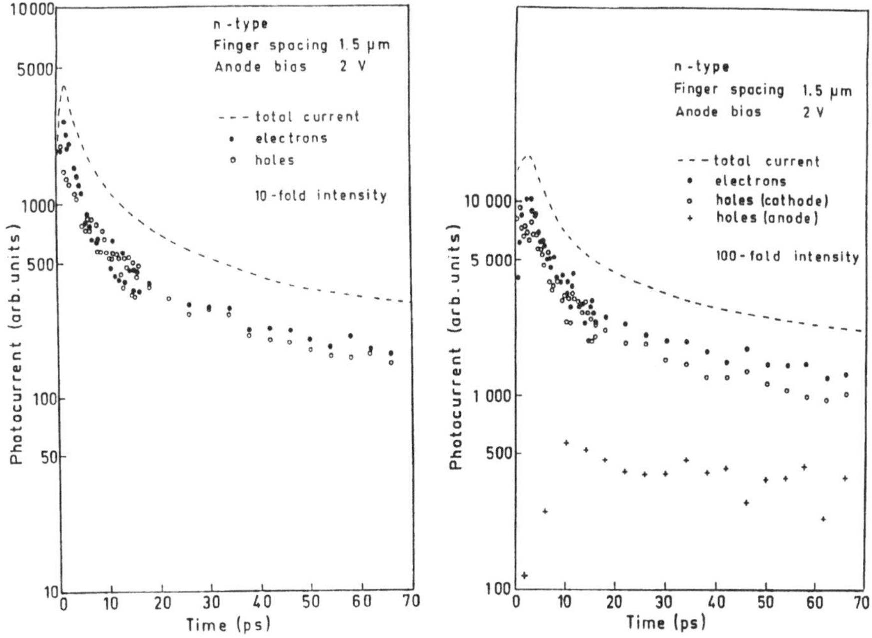


**Fig. 21** Calculated electron (•), hole (to cathode:(o), to anode:(+) and total particle current (- - -) for the 1.5 $\mu$m MSM-detector (bias 2V) at excitation density of $5.1 \times 10^{16}$ e-h pairs/cm$^3$ (left) and $5.1 \times 10^{17}$ e-h pairs/cm$^3$ (right).

nanosecond to reach the anode. Thus, the plasma may preferentially decay via carrier diffusion into the high-field regions or electron-hole recombination. The latter processes have not been investigated in the present work, however, since the carrier motion was simulated for only 72ps after illumination.

At a further ten-fold increase of the excitation density (i.e., at one-hundred times the intensity applied in Fig. 18), the transport properties are totally governed by screening effects which lead to an almost complete compensation of the field across the entire space between anode and cathode after 2ps. In this situation the collection rates of electrons and holes are almost the same. The output current decays to 30% of its peak value after 12ps and then forms a long tail which extends over a few 100ps (right part of Fig. 21).

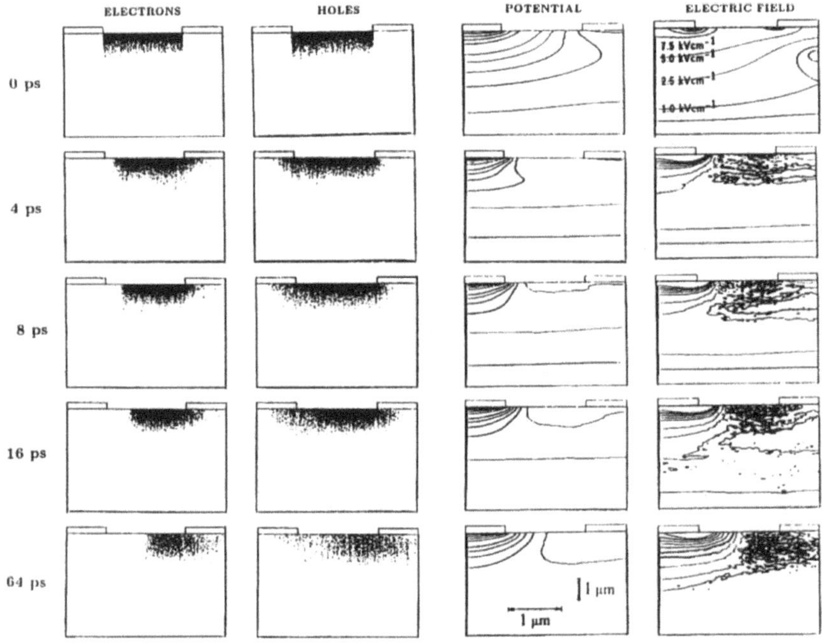


**Fig. 22** Distribution of electrons, holes, electric potential and magnitude of the electric field in the MSM-photodetector (finger spacing = finger width = 1.5 $\mu$m, bias 2V) after excitation with an electron-hole density of $5.1 \times 10^{16}$ cm$^{-3}$ at 0,4,8, 16 and 64 ps after absorption of a 100 fs laser pulse .

Field screening becomes an even more severe problem in the case of low external bias, since the weak internal field now gets immediately screened even by relatively low carrier densities. The current curve (Fig. 23) exhibits only a small peak sitting on top of a long pulse. Again, no difference for the collection rates of electrons and holes is observed. In this case, and in that of the very high excitation (Fig. 21, right part) the screened plasma diffuses in the region under the anode so that a small amount of holes can escape through the anode via scattering.

The current pulse narrows noticeably if the electrode separation of the MSM-structure is reduced to 0.5 $\mu$m. For an excitation density of $n = 5.1 \times 10^{15}$ e-h pairs/cm$^3$

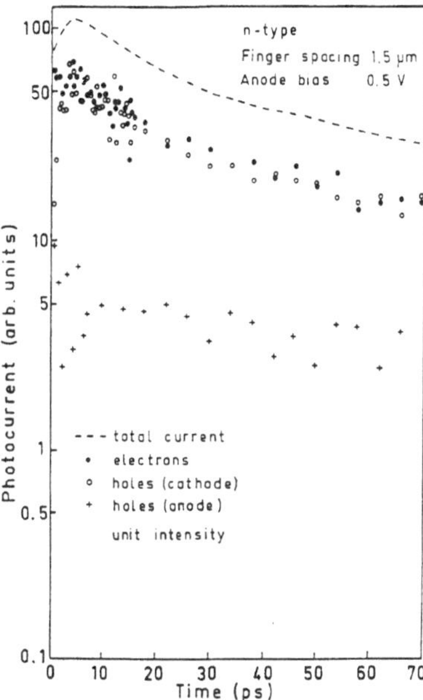


**Fig. 23** Calculated electron (•), hole (to cathode: (o), to anode:(+) and total particle current (- - -) for a 1.5$\mu$m MSM detector at low bias (0.5V) and an excitation density of 5.1 x $10^{15}$ cm$^{-3}$.

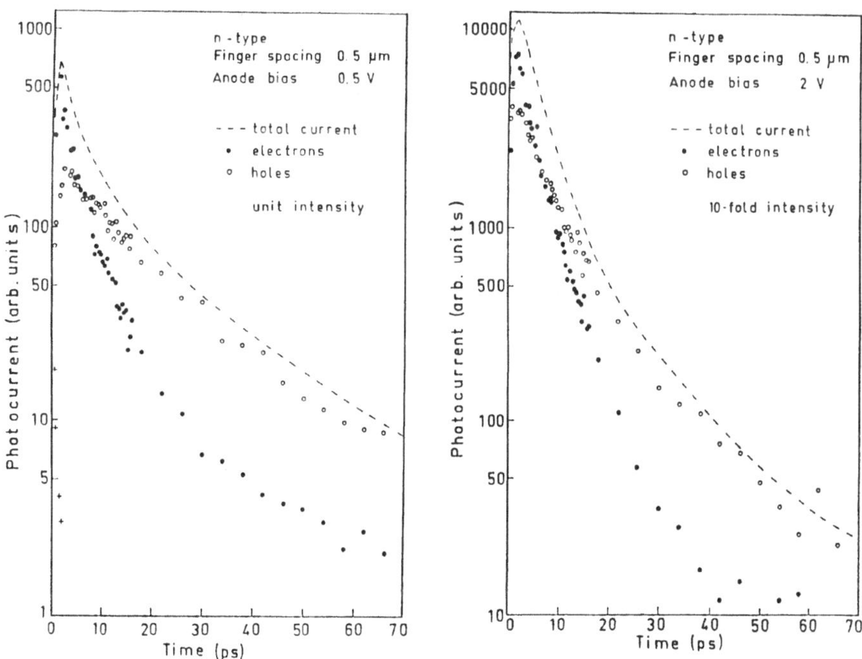


**Fig. 24** Calculated electron (•), hole (o) and total particle current (- - -) for detector with small electrode separation (finger width = finger distance = 0.5$\mu$m) and moderate excitation density (5.1 x $10^{15}$ cm$^{-3}$) at a bias of 0.5V (left) and 2V (right).

and a bias of only 0.5V, the electron and hole parts of the current are temporally separated because of the higher field strength in the device. The pulse of Fig. 24 exhibits a rise time of 1.8ps, a fast and slow decay time of 4.5 and 20 ps, respectively, and a FWHM of 6.3ps. Thus, the computed speed of the detector is even higher than that of the detector in Fig. 19 in spite of the 25% smaller average electric field between the anode and the cathode. The faster carrier collection is explained by the shorter transit times. The response, in particular the hole sweep-out, becomes still faster if the bias is increased to 2V (Fig. 24, right part).

If the excitation density is increased to $5.1 \times 10^{16}$ e-h pairs/cm$^3$ at the surface, the field screening reappears also in the 0.5 $\mu$m diodes (Fig. 25). Both the electron and the hole current are slowed down but the reduction of the response speed is much less than in the case of Fig. 21, owing to the higher field in the device.

### 6.4. Experimental Results and Discussion

The sample design described in chapter 6.1 allows to apply either photoconductive sampling at the position of the 5 $\mu$m wide side gap at a distance of 2.5 mm apart from the MSM-diode or electro-optic sampling with a small $LiTaO_3$ tip which could be approached as close as 100 $\mu$m to the diode. All data presented in this paper were taken with the output from a dispersion-compensated CPM laser providing pulses of less than 100 fs at 620 nm with a repetition rate of 116 MHz and 40 mW of average power.

Figure 26 depicts a direct comparison of experimental data (dots) measured for a 0.5$\mu$m diode (bias 2V) by electro-optic sampling with the particle current calculated by the Monte Carlo method. The excitation density was $5 \times 10^{15}$cm$^{-3}$ in order to eliminate space charge contributions. The measured rise time of the photocurrent amounts to 1.7ps (10-90%), the first rapid decay to 4.5ps, the subsequent slower decay to 10ps, and the FWHM to 6.6ps.

Although the peak amplitude is the only adjustable parameter, excellent agreement between experiment and theory is obtained except for a long tail in the measured trace (note the different zero level for the measured and computed curve in Fig. 26) which is not predicted by the theory. This result proves that the response of the diode mainly depends on the carrier collection time and justifies the neglecting of the displacement current in the Monte Carlo simulation. The slightly steeper rise of the experimental profile reflects the short duration of the displacement current as predicted by the self-consistent MC model (see Fig. 20).

The quantitative reproduction of the pulse shape by the theory demonstrates that RC time constant effects associated with the parasitic capacitance of the diode must be negligible on a ps time scale in accordance with the estimated value of 6fF for our structure [110]. The resulting RC time of 600 fs (transmission line impedance 100 $\Omega$) is still a factor of 3 shorter than the observed signal rise time.

The slow photocurrent tail observed for all our experimental signals and extending over approximately 200ps can be most likely explained as a trap-assisted electron tunneling process from the cathode to the semiconductor material. This interpretation is based on the following observations:

1. The tail amplitude increases with the integrated photocurrent density excited by the main pulse and the ratio of the tail amplitude to the pulse peak amplitude is almost independent of excitation intensity and temperature.

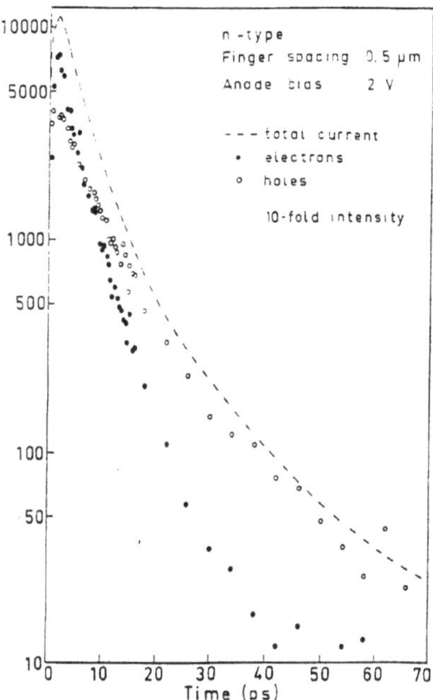


**Fig. 25** Calculated electron (•), hole (o) and total particle current (- - -) for a detector with 0.5$\mu$m-finger separation and width (bias 2V) at an excitation density of $5.1 \times 10^{16}$ $cm^{-3}$.

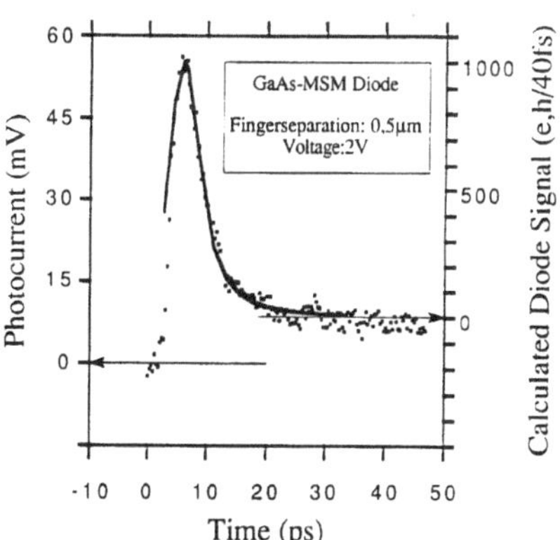


**Fig. 26** Comparison of the experimental photocurrent curve (•) measured by electro-optic sampling and the Monte Carlo calculation of the total current (solid line) in an MSM photodetector (finger separation 0.5 $\mu$m, bias 2 V, excitation: $5.1 \times 10^{15}$ e-h pairs per $cm^3$).

2. In case of double pulse excitation within a time short compared to 200ps, the tail amplitude behind the second pulse is just the sum of the two tails.

3. Diodes on LEC and MBE-substrates (impurity concentration $< 10^{14}$ $cm^{-3}$) exhibit the same tail, indicating that the phenomenon cannot be explained by bulk impurities but rather has to be attributed to defects created in the vicinity of the metallic contacts during the fabrication process.

4. Detectors manufactured on a GaAs substrate with a 20nm top $Al_{0.6}Ga_{0.4}As$ cladding layer show the same tail. This cladding layer is expected to have no influence on the electron transport, since electrons are excited to the L- and X-valley in the strong electric fields near the contacts. Holes, however, should be unable to surpass the (Al,Ga)As layer.

All these experimental observations are consistently and quantitively explained if we assume that holes are trapped at defect states (created during manufacturing of the contacts) close (10nm) to the cathode. The conduction band bending near the cathode induced by the trapped positive charge renders possible a two-step tunneling process of electrons from the metal contact to an excited state of the hole trap from where it may either tunnel into the bulk of the semiconductor or recombine with the trapped hole.

### 6.4.1. Influence of Excitation Density

Immediately after photoexcitation of the detector by the 100 fs laser puls, electrons and holes coexist at the same spatial position, so that the net electric field associated with the presence of charged particles vanishes. The subsequent motion of the electron cloud to the anode and of the hole cloud towards the cathode separates the centers of masses for the positive and negative charges, thus creating an internal electric field opposed to the external field, which consequently is effectively screened by the space charge fields.

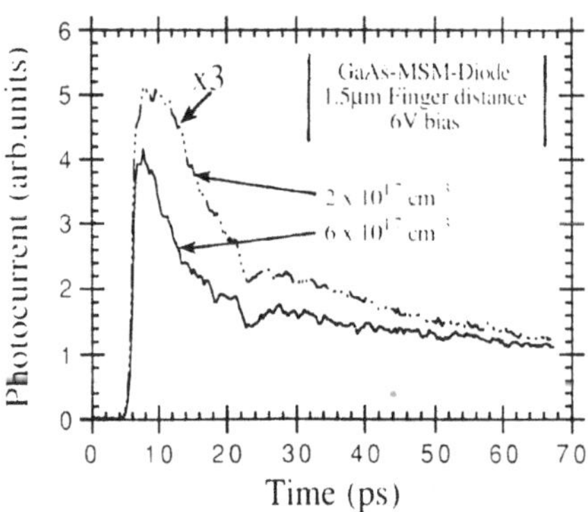


**Fig. 27** Photocurrent response of a MSM-photodetector (1.5 $\mu$m finger width and distance, bias 6V) for excitation densitites of $2 \times 10^{17}$ $cm^{-3}$ (dashed line) and $6 \times 10^{17}$ $cm^{-3}$ (solid line). For easy comparison the photocurrent measured at the lower excitation is multiplied by 3.

Figure 27 compares the measured photocurrent for a 1.5$\mu$m diode (biased to 6V) for two excitation densities differing by a factor of 3. For comparison, the pulse measured at the lower intensity has been multiplied by 3 to determine the pulse shape expected in the linear response regime. The curve measured at the higher intensity exhibits a considerably faster decay of the signal after 1.5 - 2ps. The MC simulation predicts an electric field > 5 kV/cm for the 500 nm surface layer where 90% of the electron/hole pairs are created. Evidently the space charge at the higher excitation screens this field to values below 2 kV/cm so that the electron drift velocity drops below the saturation value of $10^7$cm/s. This "breakdown" of the electric field rather than carrier depletion is responsible for the faster decay of the signal.

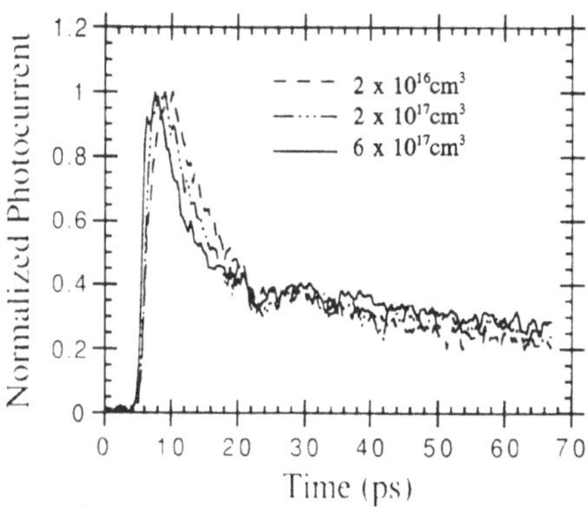


**Fig. 28** Normalized photocurrent pulse shapes of a MSM-photodetector (1.5$\mu$m finger distance, bias 2V) for three different excitation densities.

Figure 28 shows photocurrent curves for the 1.5 $\mu$m diode (bias 2V) at three excitation densities. The increase of the excitation to $2\times10^{17}$cm$^{-3}$ causes a field screening sufficient to slow down the electron motion, resulting in a faster decay. This effect is even more pronounced at the still higher excitation of $6\times10^{17}$cm$^{-3}$. The sublinear increase of the pulse peak amplitude with excitation density (see Tab. 1) is much more pronounced than the corresponding nonlinearity of the time-averaged current which has been utilized by Carruther et al. [111] to characterize the response time of such MSM-detectors by a double pulse correlation technique. The slightly slower (1-2ps) decay of the peak and the higher tail observed at $2\times10^{16}$cm$^{-3}$ compared to $2\times10^{15}$cm$^{-3}$ (not shown in Fig. 28) is mainly due to the reduction of the hole drift velocity since the space charge fields associated with these densities are too small to affect the motion of the electrons. The drop of the peak current is mainly compensated by a longer tail, whereas alternative carrier relaxation channels like recombination in the bulk and at the surface play only a minor role.

The steeper rise of the amplitude between $2\times10^{17}$ and $6\times10^{17}$cm$^{-3}$ compared to that at lower densities can be attributed to the contribution of the displacement current which gains growing importance at higher densities because of stronger and faster variations of the electric field caused by the initial separation of the photo-created electrons and holes.

The experimental dependence of the peak amplitude on the excitation level agrees fairly well with the theoretical prediction. Contrary to the experimental data, the MC model predicts, however, an increase of the FWHM at the highest densities after the initial decrease. This discrepancy is not yet understood.

Table 1. Experimental and Theoretical Dependence of the Pulse Peak Amplitude on the Excitation Density

| Experiment | Carrier Density [$cm^{-3}$] | Amplitude [$\mu$V] | | |
|---|---|---|---|---|
| $N_0$ | $2.5 \times 10^{15}$ | 75 | $A_0$ | |
| $8 \times N_0$ | $2 \times 10^{16}$ | 500 | $6.6 \times A_0$ | Experiment |
| $80 \times N_0$ | $2 \times 10^{17}$ | 1050 | $13.2 \times A_0$ | |
| $240 \times N_0$ | $6 \times 10^{17}$ | 1800 | $24 \times A_0$ | |
| $N_1$ | $5.1 \times 10^{15}$ | – | $A_1$ | |
| $10 \times N_1$ | $5.1 \times 10^{16}$ | – | $8 \times A_1$ | Theory |
| $100 \times N_1$ | $5.1 \times 10^{17}$ | – | $24 \times A_1$ | |

### 6.4.2. Influence of Finger Separation

Variation of the finger distance $d$ is expected to change the pulse duration for a transit time limited detector. Figure 29 compares the pulse shape of three detectors with $d = 0.5\mu$m, 1.0 $\mu$m and 1.5 $\mu$m biased at 2V, 4V and 6V, so that the average electric field strength remains constant. The excitation density of $1 \times 10^{16}cm^{-3}$ keeps the influence of space charge effects small.

Whereas the risetime of all pulses is the same, the decay time varies and the FWHM increases from 7.2ps (0.5 $\mu$m) to 10ps (1$\mu$m) and 13ps (1.5$\mu$m diode). The sublinear increase of the FWHM with finger separation is explained by the light penetration depth of 250 nm which is approximately equal to the average distance of photo-excited electrons and holes from the anode and cathode, respectively, for the 0.5 $\mu$m diode and involves a longer propagation path for the carriers excited in the depth. The extension of the electric field perpendicular to the crystal surface gets smaller with decreasing electrode separation. Therefore, an increased percentage of the carriers is photo-excited in low electric field regions in the 0.5 $\mu$m device and will move with lower velocity towards the contacts. The experimental current pulse shapes of the 0.5 $\mu$m and 1.5 $\mu$m diode are represented together with the theoretical curves in the right part of Fig. 29. The only adjustable parameter is the amplitude of the theoretical curve for 0.5 $\mu$m which has been fitted to match the experimental. Nevertheless, the experimental and theoretical data agree fairly well for both curves with respect to signal height and width. The slight deviation of the experimental amplitude from the theoretical predictions for the 1.5 $\mu$m diode can be attributed to the stronger influence of surface recombination.

### 6.4.3. Influence of the Bias

The onset of detectable field screening by space charges critically depends on the field strength. At moderate excitation densities and bias voltages well above the value, creating an electric field where the electrons propagate with their saturation drift velocity, the photocurrent pulse shape is almost independent of the bias. For a fixed excitation density, the field can be much more easily screened if the bias is lowered. Figure 30 shows the variation of the photocurrent pulse for a 1.5 $\mu$m diode excited at a level of $2 \times 10^{17}$ cm$^{-3}$, if the bias is reduced from 6V tc 2V. The given excitation must create screening fields of 5-7 kV/cm because the average field at 2V bias amounts to roughly 7.5 kV/cm. At 6V the same screening reduces the average field from 22.5 kV/cm to about 15kV/cm where electrons still propagate with their saturation velocity towards the anode. Consequently the pulse duration observed at 6V remains constant, if the excitation level is reduced. At still lower voltages (1V for the 1.5 $\mu$m detector), the intensity of the long tail is significantly smaller.

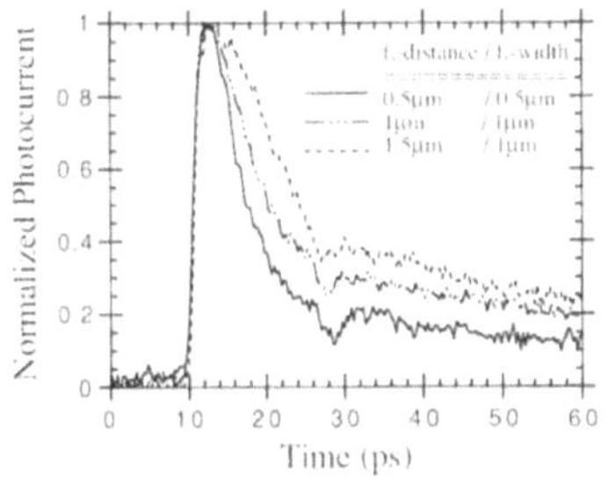


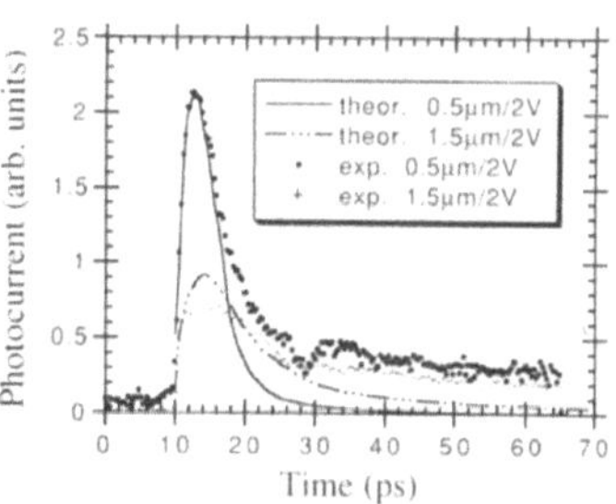


**Fig. 29** Left: normalized photocurrent pulse shape for MSM-photodetectors with 0.5 $\mu$m (bias 2V), 1 $\mu$m (bias 4V) and 1.5 $\mu$m (bias 6V) finger distance. Excitation density $10^{16}$ cm$^{-3}$. Right: comparison between experimental and theoretical photocurrent pulse shapes for MSM detectors with 0.5 $\mu$m and 1.5 $\mu$m finger distance (bias 2V).

### 6.4.4. Double Pulse Excitation

Data transfer and processing at rates well beyond 20 GHz necessitates efficient, fast collection of all photogenerated electrons and holes and complete depletion of the detector within a time shorter than the pulse separation, i.e. $\tau < 50$ ps. The theoretical and experimental investigations of the pulse shape in dependence on excitation density revealed, however, that the pulse shape can be significantly affected by field screening originating from space charge effects after separation of the electrons and holes. The data shown in chapter 6.4.1 demonstrate that decay of the photocurrent signal does not necessarily imply sweep-out of the carriers, but may alternatively be caused by a collapsing electric field within larger parts of the detector volume.

These results raise the question whether the resolvable pulse repetition rate can be limited by occurrence of a "deadtime" after the pulse and a disturbing cross-talk between succeeding pulses.

For investigations of this problem, the output pulse of the CPM laser is split into two equal parts which are sent through the two arms of a variable optical delay and then superimposed on the detector surface. This arrangement permits a systematic study of the amplitude and duration of the second pulse in dependence on the delay $\Delta\tau$ with respect to the first pulse and on the excitation density.

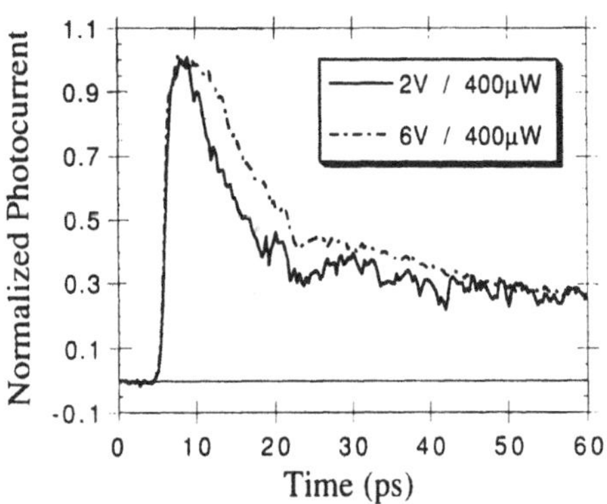


**Fig. 30** Normalized photocurrent response of a 1.5 $\mu$m detector for a bias voltage of 2V (solid line) and 6V (dashed line). Excitation density $2 \times 10^{17}$ cm$^{-3}$.

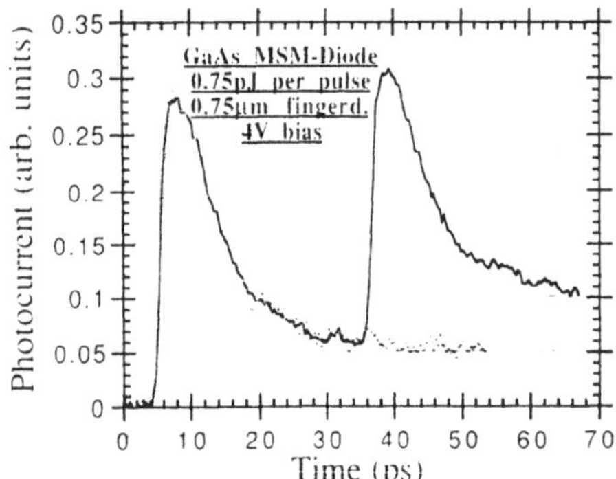


**Fig. 31** Photocurrent output of a MSM-detector (finger distance 0.75 $\mu$m, bias 4V) after illumination by two subsequent pulses (excitation density 0.75 pJ/pulse) separated by 33ps. The dotted curve shows the response to excitation by a single pulse.

Figure 31 depicts the response of a 0.75 $\mu$m/0.75 $\mu$m MSM detector (bias 4V) to excitation by two 100 fs pulses with a pulse energy of 0.75 pJ separated by 33 ps, corresponding to a transfer rate of 30 Gbit. The combined signal corresponds to the exact additive superposition (including the long tail) of two independent pulses. The amplitude and shape of the second pulse completely resembles the amplitude and shape of the first pulse proving the absence of severe cross-talk at these experimental conditions. This behavior agrees fairly well with the predictions of the Monte Carlo simulation for the density dependence of the photocurrent response. The only detectable disturbance of the second by the first pulse is caused by the long tail due to low frequency gain (see Chap.6.4) and results in a shift of the zero level. Thus, the zero level varies in dependence on the bit sequence within a pulse train. (The background for a series of $n$ times bit "1" will be much higher than for the same number of "0"). This effect may severely limit the useful bandwidth of the detector.

At higher excitation densities or lower bias, space charge fields left behind by incomplete extraction of carriers from the first pulse strongly modify both the peak amplitude and the duration of the second pulse for small delays. Figure 32 displays the ratio of the probe pulse amplitudes measured with and without the presence of the first pulse as a function of the delay for 3 different experimental conditions. For a pulse energy of 0.6 pJ corresponding to an excitation density of $1.5 \times 10^{16}\text{cm}^{-3}$ and a bias of 4V, we observe a small decrease of the amplitude of pulse 2 if $\Delta\tau$ increases from 0-10 ps which can be attributed to field screening caused by the increasing spatial separation of electrons and holes from the first pulse. For $\Delta\tau > 10$ps the major fraction of carriers created by pulse 1 has been collected before arrival of pulse 2.

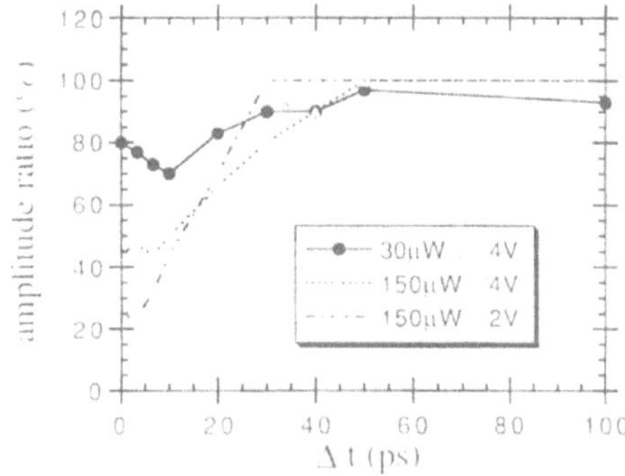


**Fig. 32** Ratio of probe pulse to pump pulse amplitude (equivalent to the probe pulse amplitude in the absence of the pump) response versus time delay between the two pulses for different excitation densities and bias voltages.

For the five times higher excitation density ($8 \times 10^{16}\text{cm}^{-3}$), the drop of the peak amplitude of pulse 2 in the presence of pulse 1 is much stronger even at very small delay. The apparently complete recovery of the detector within 50 ps seems to contradict the Monte Carlo calculation which predicts the formation of a long-living electron hole plasma in the vicinity of the anode for this excitation level. This discrepancy is explained by the fact that the initial pulse reflects just the depletion of carriers from the high field regions, whereas the long-living plasma in the screened regions of the electric field form only a weak, barely detectable long tail. Therefore, space charge effects in the first place involve a reduction of the detector sensitivity defined as the ratio of the peak voltage to the optical pulse energy, whereas the pulse duration reveals an unexpectedly weak dependence on the intensity.

Further enhancement of the screening corresponding to a further reduction of the "active" detector volume achieved by decreasing the bias from 4V-2V supports this interpretation. Field screening appears still faster and covers a larger volume of the detector. Carrier depletion from the remaining unscreened regions which are confined to the vicinity of the contacts appears even faster and explains the complete recovery of the pulse for delays as short as 30ps.

### 6.4.5. Temperature Dependence of the Photocurrent Pulse Shape

In order to identify the response limiting mechanisms, we investigated the temperature dependence of the output pulse shape [112]. For these experiments the diode (0.75 $\mu$m finger spacing, 4V bias, e-h pair density $3.5 \times 10^{16}cm^{-3}$) fabricated on LEC GaAs was mounted in a He cryostat and the photocurrent pulse shape was analyzed by photoconductive sampling (proton implantation dose of the area of the photoconductive switch $4.1 \times 10^{14}cm^{-3}$).

Figure 33 depicts a plot of the measured FWHM of the current pulses (triangles) versus temperature T. The pulse duration decreases from 10.8 ps at 300K to 5.6 ps at 100K, remains almost constant between 100K-50K and increases again to 11.1 ps if the temperature is decreased to 10K. The distinct decrease of the pulse duration between 300-50K has to be ascribed to a remarkable increase of the carrier transport velocity at lower temperatures, since the time resolution of the photoconductive sampling and the parasitic capacitance of the diode structure are not expected to vary strongly with temperature. For the high $H^+$ implantation dose, multiple trapping of carriers in shallow traps will be of minor importance similar to the situation in Si switches [113] where the carrier lifetime has been shown to be independent of $T$ for irradiation doses exceeding $10^{13}$ $Si^+$ $cm^{-2}$.

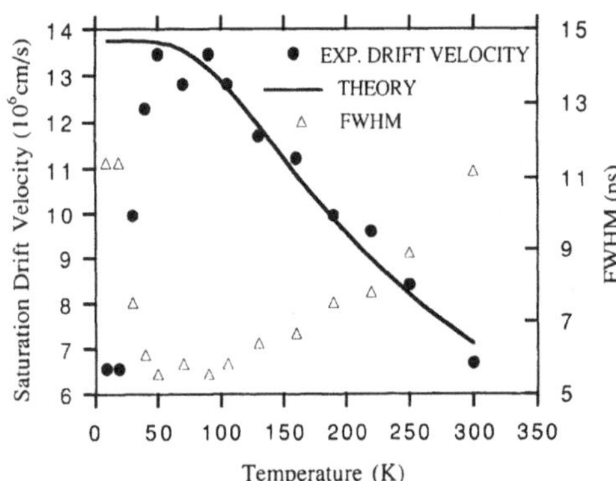


**Fig. 33** FWHM of the photocurrent pulses (triangles, right scale), measured (black dots) and calculated (solid line) saturation drift velocity of electrons in the MSM-detector (left scale) vs temperature.

At a bias of 4V and a finger distance of 0.75 $\mu$m, the electric field in the 250 nm thick semiconductor surface layer corresponding to the penetration depth of 2eV photons is larger than 10kV/cm. Thus, almost all electrons propagate with saturation velocity towards the anode and the transport will be strongly influenced by intervalley phonon scattering ($\Gamma \rightarrow L/L \rightarrow \Gamma/L \rightarrow L$). As our analysis is confined to the temperature dependence of the FWHM, the hole contribution to the photocurrent can be neglected because of its small influence onto this quantity. An approximate experimental measure for the electron saturation drift velocity is obtained, if the finger distance of the diode is divided by the FWHM of the photocurrent pulse response. This quantity is depicted by the black dots in Fig. 33.

The temperature dependence of the electron drift velocity can be described by a simple model [114] based on the change of the intravalley and intervalley phonon scattering (LO phonons at the Γ point, respectively, LO, LA, and TA phonons at L and X points [115] with temperature. The scattering comprises phonon absorption [rate proportional to phonon occupation number N(T,$\hbar\omega_{ph}$)] and phonon emission processes [rate proportional to N(T,$\hbar\omega_{ph}$)+1] with a weighted average phonon energy $\hbar\omega_{ph} = 30$ meV [116]. Therefore, the total intervalley scattering rate $r_{sc}$ is proportional to

$$r_{sc} \propto 2N(T, \hbar\omega_{ph}) + 1, \tag{25}$$

and the temperature dependence of the saturation drift velocity $v_s$ becomes

$$v_s(T) = v_s(0)/[2N(T, \hbar\omega_{ph}) + 1]. \tag{26}$$

Here $N(T, \hbar\omega_{ph})$ is the Bose distribution of phonons:

$$N(T, \hbar\omega_{ph}) = \frac{1}{exp(\hbar\omega_{ph}/kT) - 1}. \tag{27}$$

The function $v_s$(T) is drawn as the solid line in Fig. 35 with the saturation drift velocity at T=0 K, $v_s(0)$, being a fit parameter. The value $v_s(0) = 1.38 \times 10^7$cm/s obtained in our fit is in good agreement with Ref. 117. The experimental data for the electron drift velocity (black dots in Fig. 33) are described very well by the theoretical curve (solid line) for T > 50K.

The theory for intravalley and intervalley phonon scattering predicts an almost constant electron drift velocity for T < 70K. Contrarily, the experimental data exhibit a rapid increase of the response time if T decreases below 50K. The rise of the FWHM of the diode output can be attributed to the growing importance of ionized impurity scattering at low temperatures. This scattering mechanism starts to dominate the electron transport in LEC-GaAs below 50K [118]. Our interpretation is supported by intensity-dependent measurements of the response time in the two temperature regimes. Figure 34a shows the diode response for two different $e-h$ densities at T = 100K. The response time for t > 10ps is longer for the higher $e-h$ density. This effect is explained by screening of the externally applied electric field by the $e-h$ plasma and can be observed at all T > 50K. The opposite effect for increasing excitation density is observed at 10K (Fig. 34b) providing strong indication for the presence of a different scattering mechanism. The ionized impurity scattering rate drops with increasing carrier density because of more effective screening of the ionized impurities by the plasma [118]. Extrapolation of electron mobility data of LEC-GaAs [119] from 80K to lower temperatures reveals that below 50K the scattering rate due to ionized impurities exceeds the optical phonon intervalley scattering rate. The extrapolated mobility at 50K $\mu$(50K)≈1000 cm$^2$/Vs is of the same order of magnitude as the mobility $\mu$(50K)≈700 cm$^2$/Vs estimated from our experiments (assuming an average electric field of 20 kV/cm).

These temperature dependent experiments definitely prove that the measured pulse duration is completely explainable by the motion of the carriers from their excitation point to the metallic contacts and the limitation of this transport is due to electron-phonon interaction. The capacitance of the structure or parasitic resistances seem to play no role on a time scale of a few ps.

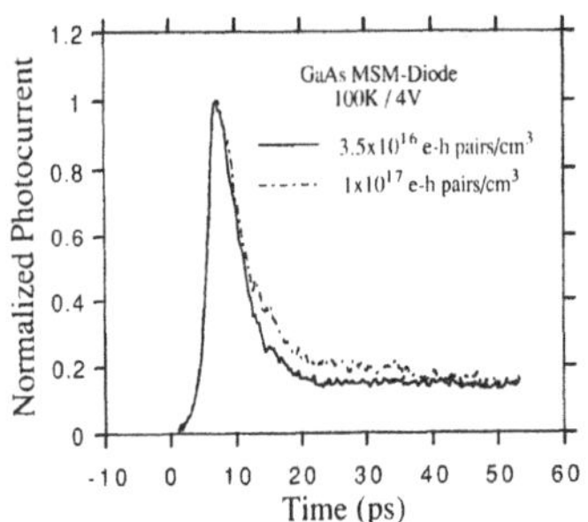


**Fig. 34a** Normalized photocurrent vs time delay for a GaAs MSM diode (finger distance 0.75 $\mu$m) at T = 10K and two different carrier densities [$3 \times 10^{16}$ e-h pairs/cm$^3$ (solid line) and $1.2 \times 10^{17}$ e-h pairs/cm$^3$ (dotted line)].

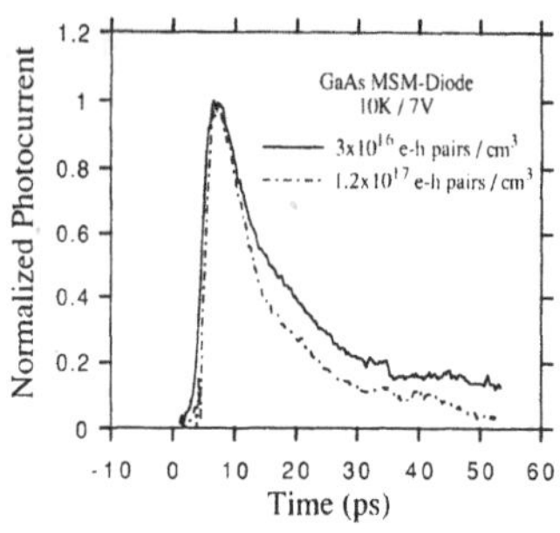


**Fig. 34b** Normalized photocurrent vs time delay for a GaAs MSM diode (finger distance 0.75 $\mu$m) at T = 100K and two different carrier densities [$3.5 \times 10^{16}$ e-h pairs/cm$^3$ (solid line) and $1 \times 10^{17}$ e-h pairs/cm$^3$ (dotted line)].

### 6.4.6. MSM Detectors on LT-GaAs

Further reduction of the detector response time may be accomplished either by decreasing the finger distance (shorter carrier transit time) or by reduction of the free-carrier lifetime (switching-off the photocurrent via carrier recombination or trapping in a time shorter than the carrier collection time). Smaller finger distances imply an increasing finger number and a corresponding increase of the device capacitance if the sensitivity of the detector should be kept constant. Reduction of the finger distance/finger width from 500 nm/500 nm to 160 nm/40 nm results in an increase of the RC time constant from 0.7 ps to 3ps for a 100 $\Omega$ transmission line. Thus, the decrease of the carrier transit time is mostly compensated by the higher RC time.

Of course, shorter carrier lifetimes usually involve a lower detectivity, too. On the other hand, fast carrier trapping provides the additional advantage of effective suppression of the long signal tail, since all carriers injected from the contacts will be trapped almost instantaneously. Recently, we have demonstrated that MSM photoconductors fabricated on LT GaAs which combines carrier recombination times of approximately 1ps and high carrier mobility permit a substantial shortening of the pulse response compared to photoconductors on normal GaAs [120].

The diodes have been fabricated by means of electron-beam lithography on LT-GaAs using Ti/Pt/Au as contacts. The 2.5 $\mu$m-thick LT-GaAs layer was grown at 200°C substrate temperature and annealed for 10 min. at 600°C. The diodes under investigation had an active area of $10 \times 15$ $\mu m^2$ and ten 0.75-$\mu$m-wide fingers with 0.75-$\mu$m distance between them. Electro-optic sampling with a small $LiTaO_3$ tip was used to measure the diode response. Figure 35 depicting the response of such a diode at 4V bias to a 2.4 pJ pulse of 100 fs duration at 620 nm (focused to 100 $\mu m^2$, gives a risetime of 1.3 ps, a 1/e decay time of 2.5 ps for the first rapid decay and a FWHM of 3.3 ps. Neglecting the residual tail Fourier transformation of the pulse shape yields a bandwidth of 70 GHz.

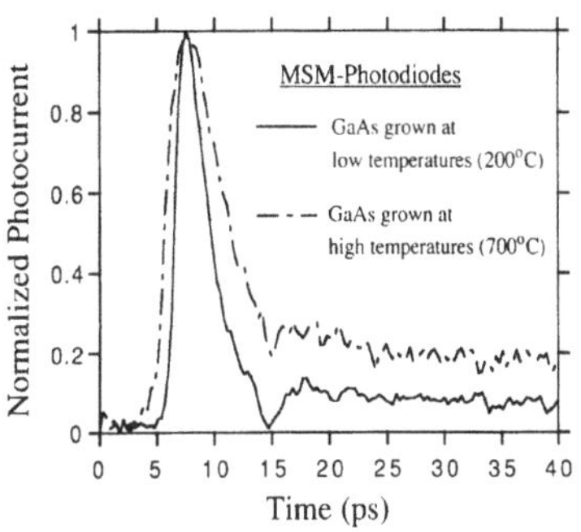


**Fig. 35** Normalized photocurrent of a MSM photodiode fabricated on LT-GaAs (solid line) (0.75$\mu$m finger distance/4V bias/2.4 pJ pulse energy) and on HT-GaAs (dotted line) (0.5 $\mu$m finger distance/3V bias/0.3 pJ pulse energy) after excitation by a 100 fs pulse at 620 nm vs time.

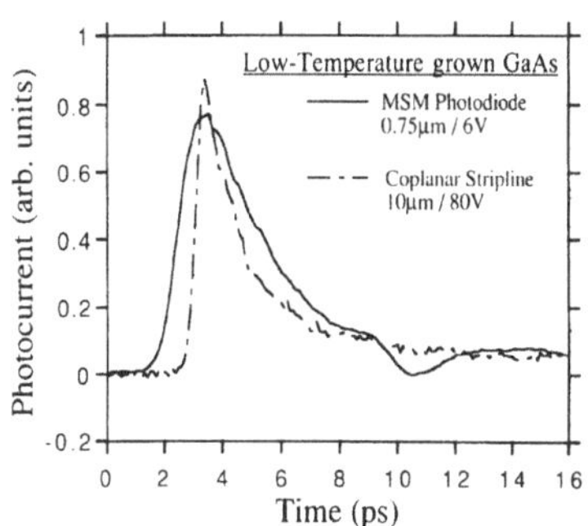


**Fig. 36** Photocurrent signal of a MSM diode (solid line) (0.75 $\mu$m finger distance/6V bias) and a coplanar stripline (dotted line) (10$\mu$m line distance/80V bias) vs time. The elements were fabricated on LT-GaAs and illuminated by 24 pJ pulse energy.

These characteristic times of the diode remain unchanged for a variation of the pulse energy from 0.6 to 24 pJ and of the voltage from 1 to 6V and the signal amplitude is approximately proportional to the voltage and the intensity in these ranges.

For comparison, Fig. 35 shows the fastest response from an MSM diode fabricated on HT-GaAs as the dash-dotted line. In spite of the smaller finger separation (0.5$\mu$m), all the characteristic time constants (risetime 2.2 ps, 1/e decay 4.8ps and FWHM 6ps) are distinctly longer than the data obtained from the LT-GaAs detector. The faster response of the LT-GaAs based device has to be paid off by a smaller quantum efficiency since only part of the photo-generated electrons and holes recombine at the contacts and most of them are trapped at the metallic As clusters. This trapping results in a drop of the time-integrated photocurrent (quantum efficiency) by more than one order of magnitude. The drop of the pulse amplitude which represents the important quantity for digital applications is much less (only a factor of 2-3) because trapping mainly suppresses the trailing edge of the pulse. For optimum operation conditions (3V bias/pulse energy $< 0.6$pJ), the diode on HT-GaAs produces a signal amplitude of 200 mV/pJ pulse energy (for optical pulses short compared to the FWHM of the electrical response) which already decreases to 115 mV/pJ at 2.4 pJ energy because of field screening effects (see Chap. 6.4.1). The LT-GaAs diode has a sensitivity of 85 mV/pJ at 0.6 pJ energy, which goes down slightly to 65 mV/pJ at 2.4 pJ and 50 mV/pJ at 24 pJ.

Using 0.2-$\mu$m-spaced interdigitated electrodes on LT GaAs, Chen et al. [5] observed a response time as fast as 1.2ps (FWHM) corresponding to a 3dB-bandwidth of 375 GHz. This detector generated signal amplitudes up to 6V and achieved a responsivity of 600 mV/pJ at a bias voltage of 8V. This sensitivity agrees fairly well with the value obtained from our detectors if one takes into account the 7.5 times larger average electric field applied in the experiments of Ref. 5.

Comparison of the response of the LT-GaAs diode (solid line in Fig. 36) with that measured for excitation of the 10 $\mu$m wide gap between the strips of the coplanar transmission line on the same substrate (dashed line in Fig. 36) allows to explore, whether the particle current or the displacement current provides the dominant contribution to the signal. The coplanar line was biased to 80V so that the average electric field was the same as in the diode (0.75 $\mu$m, 6V). The signal excited in the gap of the coplanar line has a risetime of 0.6 ps close to the resolution limit of the electro-optic sampler and a FWHM of 1.4 ps.

The signal from the diode should have half the amplitude of the signal received from the striplines, if only the displacement current contributes to the response, since half of the diode surface is covered by the metallic fingers. If the particle current yields the dominant contribution to the diode signal, the amplitude should be seven times larger (and the pulse duration 14 times shorter) than the signal from the striplines, because the contact area of the diodes being seven times larger (and the contact separation 14 times smaller). The fact that both signal amplitudes in Fig. 36 are almost identical supports the assumption that mainly the displacement current contributes to the photo-current response of LT-GaAs diodes. This assumption is consistent with Monte Carlo calculations of the displacement current. Our result implies that the differences between the shapes of the LT-GaAs diode and stripline pulses are mostly due to electrical parasitics and the propagation time of the electrical pulse on the electrodes of the interdigitated diode structure. The deconvolution of both signals gives an electrical diode response function having a FWHM of 1.5 ps. This value agrees fairly well with estimated capacitances (6-10fF) and signal propagation times on the diodes. Compared to a simple stripline photodetector, the interdigitated diode structure offers the advantage of a significantly lower bias voltage. 4-6V are sufficient to achieve a sensitivity of 85 mV/pJ instead of 80V needed in a stripline. The successful growth of device-quality GaAs on top of a LT-GaAs layer [121] and the low bias permit an easy integration of LT-GaAs diodes with high-speed electronic circuits.

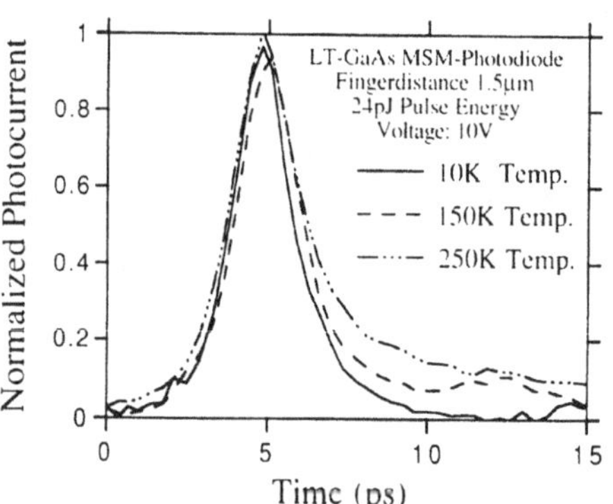


**Fig. 37** Normalized photocurrent of MSM photodiode fabricated on LT-GaAs (1.5 $\mu$m finger distance/10 V bias) for 10K (solid line), 150K (dashed line), and 250K (dotted line).

Temperature-dependent measurements of the response of an LT-GaAs diode (see Fig. 37) reveal that the long tail of the signal for this detector is due to hopping processes. The change of the hopping conductivity $\sigma_h$, which is proportional to the current with temperature is given by

$$\sigma_h = \sigma_0 \, e^{-(\Delta E/kT)}, \tag{28}$$

with $\Delta E$ being the energy difference between adjacent hopping states. $\sigma_0$ depends on the density of states at the Fermi energy, the phonon spectrum and the spatial overlap of the wave functions of the hopping states. The analysis of the photocurrent shows an activation energy of $\Delta E \approx 7$ meV over the whole temperature range down to 50K. This value is much smaller than the activation energies obtained from dark current measurements [122], which we could reproduce with our samples. This discrepancy can be explained by the strong and highly transient nonequilibrium occupation of traps during the first picoseconds after the excitation by ultrashort optical pulses. Therefore, hopping processes of photo-excited carriers from occupied trap states which are energetically closer to the empty states are very likely in this transient regime. The hopping conductivity continuously decreases if the carriers relax to deeper traps.

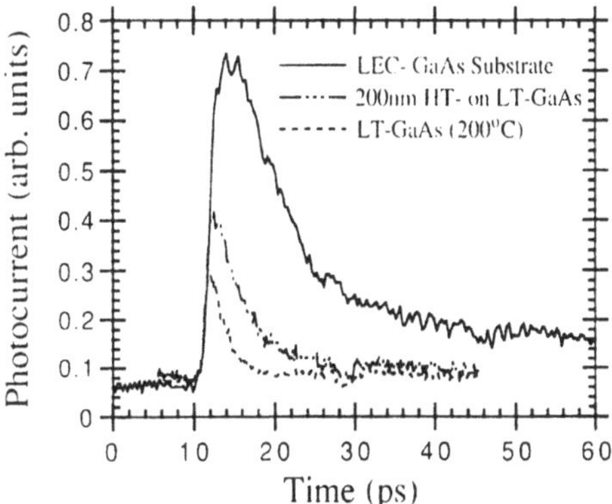


**Fig. 38** Comparison of the photocurrent impulse response of 3 different MSM-photodetectors (finger distance 1$\mu$m, bias voltage 6V) fabricated on a LEC GaAs substrate (solid line), on LT-GaAs (dashed line) and on a 200 nm thick HT-GaAs grown epitaxially on a 1.5 $\mu$m thick LT-GaAs layer. Excitation density $2 \times 10^{16}$ cm$^{-3}$.

In order to increase the sensitivity we have finally fabricated a detector on a "hybrid" substrate with a 200 nm high quality expitaxial GaAs film grown on top of a 2 $\mu$m LT-GaAs layer. The signal amplitude of this detector exceeds that of the LT GaAs based device by 50%. The FWHM, however, increases by 35% (Fig. 38). The curves reveal the simultaneous reduction of the pulse response duration and sensitivity of the detector if the particle current contribution which dominates in the "normal" GaAs is reduced with respect to the displacement current prevailing in the LT-GaAs layer. Tailoring of the HT-GaAs layer thickness grown on the LT-GaAs substrate, thus offers a selectable compromise between time-resolution and sensitivity within broad ranges.

## 7. Characterization of High-Speed Transistors

The combination of a photoconductive switch operating as a generator for ultrafast electrical pulses with an electro-optic or photoconductive sampling gate represents a versatile ultrahigh-bandwidth measurement system for small signal characterization of high-speed transistors or integrated circuits, which permits direct analysis of the response to ultrashort pulse excitation in the time domain with picosecond resolution.

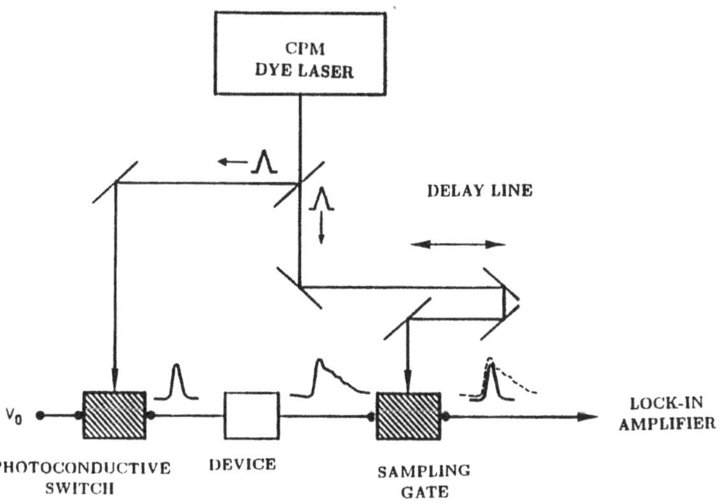


**Fig. 39** Ultrafast optoelectronic measurement system for measuring the response of high-speed electronic devices.

The principal scheme of the experimental configuration is illustrated in Fig. 39. The device under test is driven by the output pulses of the switch and its impulse response is sampled by an electro-optic wafer prober. Because the optical pulses driving the pulse generator and interrogating the sampling gate are derived from the output of the same laser via beam splitting, the technique is free of jitter and allows very accurate measurements of the shape and delay of the output signal with respect to the incident pulse.

In one of the first pioneering experiments Smith et al. [123] applied this method to study the impulse response of a GaAs FET. Their experimental set-up consisted of radiation-damaged SOS photoconductive switches mounted in a microstrip transmission line structure and providing electrical transients or a respective gate resolution of 15ps. The gate and drain of the FET were connected to the main transmission lines of the switches. These lines simultaneously applied the gate-source and drain source bias voltages (see Fig. 40).

Gap 1 represents the proper pulse generator injecting the signal into the gate via the main transmission line after illumination by a ps optical pulse. Gap 2 is used to

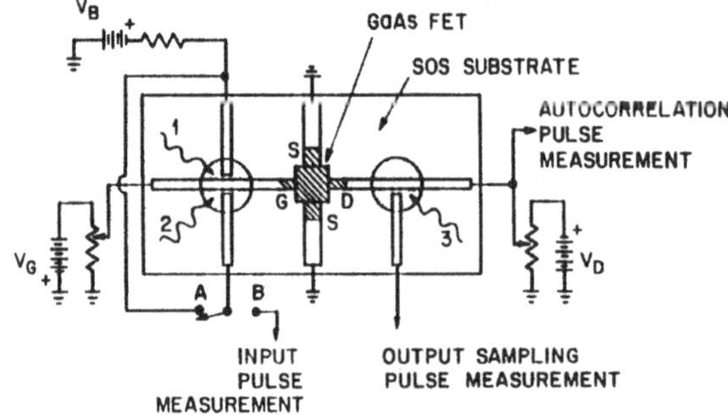


**Fig. 40** Scheme of the optoelectronic circuit used to characterize the picosecond electronic response of a GaAs FET (adapted from Ref. 123).

investigate the input signal waveform as well as the signal reflected by the gate and gap 3 serves for analyzing the output waveform. The gain, temporal broadening and delay of the signal have been evaluated by comparing the response measured at gap 3 in the presence of the FET to that observed when the FET is replaced by a 50 $\Omega$ transmission line. The analysis of the two cross-correlation traces yields a broadening of the 15 ps-pulse injected into the gate to 40 ps at the drain and the formation of a long tail extending to 75 ps, a gain of 3.7dB and a delay of 19ps. The nonlinearity of the response has been investigated by using both, photoconductor 1 and 2 as pulse generators and recording the total time averaged current as a function of the delay between the two pulses. This nonlinear electronic correlation experiment revealed a response time of 20 ps.

Meyer et al. [6,41] sandwiched a GaAs MESFET between an optically triggered GaAs:Cr photoconductive switch in a microstripline and a coplanar waveguide fabricated on an $LiTaO_3$ electro-optic crystal serving as the sampling gate. After injection of a 50 mV signal with a rise time of 5.4 ps an inverted and slightly amplified pulse with a risetime of approximately 16 ps was observed at the drain output. This risetime corresponds to a current gain-cutoff frequency of 23 GHz in accordance with values obtained from normal $S$ parameter measurements.

With a very similar experiment Dykaar et al. [7] observed a risetime as short as 5.3 ps at the output of a permeable base transistor after injection of a step function pulse. These results prove that the PTB is one of the fastest transistor structures and supports theoretical models which predict oscillation frequencies in excess of 200 GHz for this device.

Frankel et al. [124] have applied electro-optic sampling with 100 fs optical pulses for the development of a vector network analyzer providing a bandwidth as large as 100 GHz for the small-signal characterization of high-speed transistors. They embedded photoconductive switches for the generation of ps electrical pulses into a high-frequency CTL (50 $\mu$m wide conductors separated by 50 $\mu$m) fabricated on a radiation-damaged SOS-substrate. An AlGaAs/InGaAs/GaAs heterojunction field-effect transistor (HFET) was wire-bonded into this test fixture, as shown in Fig. 41.

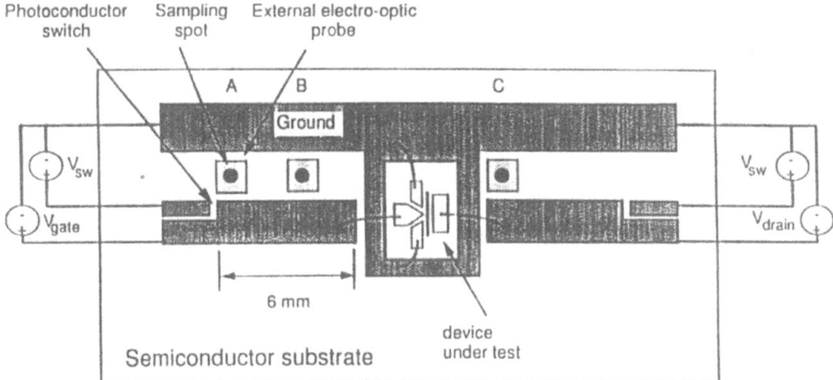


**Fig. 41** Experimental scheme for ultrahigh-bandwidth S-parameter analysis of fast transistors (adapted from Ref. 124).

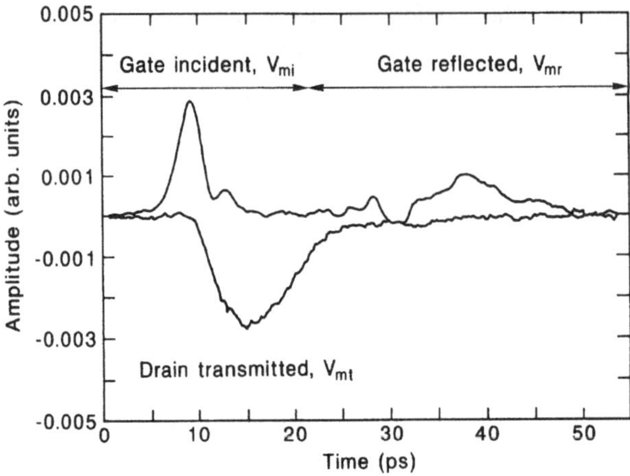


**Fig. 42** Response of the HFET measured at the gate and drain transmission lines using the experimental set-up of Fig. 41 (adapted from Ref. 124).

The dc bias voltages and the broadband electrical pulses are connected to the transistor under test by the same strips. The shape of the input and output signal propagating on the gate and drain transmission line, respectively, after optical excitation of the photoconductive switch on the left side are measured by an electro-optic wafer prober located at either position B or C. Typical experimental curves are depicted in Fig. 42.

The highly symmetric configuration of the set up permits a full S-parameter analysis of the device under test. The reflection scattering parameter (S parameter $S_{11}$ is calculated from the ratio of the Fourier transform of the reflected and the incident time domain signal. The ratio of the Fourier transform of transmitted and incident time domain signals yields the forward transmission scattering parameter $S_{21}$. The reverse scattering parameters can be obtained if the photoconductive switch in the output (right) part of the structure is illuminated. The amplitude and phase of the S-parameters measured by the electro-optic probing technique agree fairly well with data taken with a conventional network analyzer up to 40 GHz.

The methods described in this chapter mark a remarkable progress for testing semiconductor electronic devices in the picosecond time regime, since the parameters of a single device can be evaluated. Because conventional electronic measurement techniques do not have the required time resolution, transistor switching speeds are usually determined by ring oscillators which average the response of many individual components.

Here, we have described a few selected experiments concerned with the characterization of discrete transistor devices. For more information about the basic principles as well as the present status of optical probing techniques for testing ultrafast electronic devices, in particular high-speed integrated circuits, the reader is referred to the comprehensive review article of Wiesenfeld and Jain [125].

## 8. Summary

This article has summarized the present status of photoconductive switching and electro-optic sampling with pico- and subpicosecond laser pulses. These techniques have proven to be powerful tools for the generation and characterization of fast electrical transients with unprecedented bandwidths in excess of 1 THz. The enormous potential and excellent versatility for solving a variety of measurement problems in ultrafast electronics are expected to stimulate the implementation of optical probing techniques into the test equipment of the semiconductor industry.

**Acknowledgements**

The results of our own work presented in this article have been obtained in close cooperation with several colleagues at the *Max-Planck-Institut für Festkörperforschung* (MPI-FKF) in Stuttgart, and the *Fraunhofer Institut für Angewandte Festkörperphysik* (IAF) in Freiburg. I am particulary indebted to M. Klingenstein (MPI), M. Lambsdorff (MPI), K. Ploog (MPI), J. Rosenzweig (IAF), C. Moglestue (IAF), K. Köhler (IAF), A. Hülsmann (IAF), Jo. Schneider (IAF), and A. Axmann (IAF) for the stimulating collaboration. The important contributions and expert technical assistance of M. Lang, M.L. Andres, K. Glorer, A. Hadad, P. Hoffmann, M. Krieg, J. Rüdiger, A. Schulz, H. Stützler and R. Osorio are greatfully acknowledged. Finally, I would like to express my thanks to Professors H.S. Rupprecht and H.J. Queisser for their continuous encouragement during the course of this project. Parts of our research work have been financially supported by *The Bundesministerium für Forschung und Technologie.*

# References

1. E. Desurvire, *Scientific American* 96, January 1992.
2. Measurement of High-Speed Signals in Solid State Devices, *in:* "Semiconductors and Semimetals", Vol. 28, Ed. R.B.Marcus, Academic Press, New York (1990).
3. M.B. Ketchen, D. Grischkowsky, T.C. Chen, C.C. Chi, I.N. Duling, N.J. Hales, J.M. Halbout, J.A. Kash, and G.P. Li, *Appl. Phys. Lett.* 48:751 (1986).
4. C. Moglestue, J. Rosenzweig, J. Kuhl, M. Klingenstein, M. Lambdsdorff, A. Axmann, Jo. Schneider, and J. Hülsmann, *J. Appl. Phys.* 70:2435 (1991).
5. Y. Chen, S. Williamson, T. Brock, F.W. Smith, and A.R. Calawa, *Appl. Phys. Lett.* 59:1984 (1991).
6. K.E. Meyer, D.R. Dykaar, and G.A. Mourou, *in:* Picosecond Electronics and Optoelectronics, Eds. G.A. Mourou, D.A. Bloom, and C.H. Lee, "Springer Series in Electrophysics", Vol. 21, Springer Verlag, Berlin-Heidelberg-New York (1985).
7. D.R. Dykaar, R. Sobolewski, J.F. Whitaker, T.Y. Hsiang, G.A. Mourou, M.A. Hollis, B.J. Clifton, K.B. Nichols, C.O. Bozler, and R.A. Murphy, *in:* "Ultrafast Phenomena V", Eds. G.R. Fleming and A.E. Siegman, Springer Verlag, Berlin-Heidelberg-New York, p. 103 (1986).
8. P. Wolfe *in:* "Picosecond Electronics and Optoelectronics", Eds. G.A. Mourou, D.M. Bloom, and C.H. Lee, Springer Verlag, New York, p. 236 (1985); and D.G. McDonald, R.L. Peterson, C.A. Hamilton, R.E. Harris, and R.L. Kantz, IEEE Trans. Electr. Dev., ED-27:1945 (1980).
9. D.H. Auston, Picosecond Photoconductors, *in:* "Picosecond Optoelectronic Devices", Ed. C.H. Lee, Academic Press, New York (1984).

10. D.H. Auston, Ultrafast Optoelectronics, *in*: "Ultrashort Laser Pulses and Applications", Ed. W. Kaiser, Topics in Appl. Phys., 60, Springer Verlag, Berlin-Heidelberg-New York (1988).

11. D.H. Auston, Picosecond Photoconductivity: High-Speed Measurements of Devices and Materials, *in:* "Semiconductors and Semimetals", Vol. 28, Ed. R.B. Marcus, Academic Press, New York (1990).

12. J.A. Valdmanis and G.A. Mourou, Subpicosecond Electrical Sampling and Applications *in:* "Picosecond Optoelectronic Devices", Ed. C.H. Lee, Academic Press, New York (1984).

13. J.A. Valdmanis and G.A. Mourou, "Electro-Optic Sampling: Testing Picosecond Electronics", Laser Focus/Electro-Optics, Feb. 1986, p. 84 and March 1986, p. 96.

14. J.A. Valdmanis, Electro-Optic Measurement Techniques for Picosecond Materials, Devices and Integrated Circuits, *in:* "Semiconductors and Semimetals", Vol. 28, Ed. R.B. Marcus, Academic Press, New York (1990).

15. C.V. Shank, Generation of Ultrafast Optical Pulses *in:* "Ultrafast Laser Pulses and Applications", Ed. W. Kaiser, Springer Verlag, Berlin-Heidelberg-New York (1988), p. 5.

16. R.L. Fork, C.H. Brito Cruz, P.C. Becker, and C.V. Shank, *Opt.Lett.* 12:483 (1987).

17. A.M. Weiner and R.B. Marcus, Photoemissive Probing *in:* "Semiconductors and Semimetals", Vol. 28, Ed. R.B. Marcus, Academic Press, New York (1990).

18. D.H. Auston, *Appl. Phys. Lett.* 26:101 (1975).

19. P.R. Smith, D.H. Auston, A.M. Johnson, and W.M. Augustyniak, *Appl. Phys. Lett.* 38:47 (1981).

20. D.H. Auston, *IEEE J. Quant. Electron.* QE-10:639 (1983).

21. F.E. Doany, D. Grischkowsky, and C.-C. Chi, *Appl. Phys. Lett.* 50:460 (1987).

22. M. Lambsdorff, Thesis, University of Kaiserslautern, Germany (1990).

23. H.B. Hammond, N.G. Paulter, and H.S. Wagner, *Appl. Phys. Lett.* 45:289 (1984).

24. M.B. Johnson, T.C. McGill, and N.G. Paulter, *Appl. Phys. Lett.* 54:2424 (1989).

25. M. Lambsdorff, J. Kuhl, J. Rosenzweig, A. Axmann, and Jo. Schneider, *Appl. Phys. Lett.* 58:1881 (1991).

26. F.W. Smith, H.Q. Le, V. Diadiuk, M.A. Hollis, A.R. Calawa, S. Gupta, M. Frankel, D.R. Dykaar, G.A. Mourou, and T.Y. Hsiang, *Appl. Phys.* 54:890 (1989).

27. A.C. Warren, N. Katzenellenbogen, D. Grischkowsky, J.M. Woodall, M.R. Melloch, and N. Otsuka, *Appl. Phys. Lett.* 58:1512 (1991).

28. A.C. Warren, J.M. Woodall, J.L. Freeouf, D. Grischkowsky, D.T. McInturff, M.R. Melloch, and N. Otsuka, *Appl. Phys. Lett.* 57:1331 (1990).

29. S. Gupta, M.Y. Frankel, J.A. Valdmanis, J.F. Whitaker, G.A. Mourou, F.W. Smith, and A.R. Calawa, *Appl. Phys. Lett.* 59:3276 (1991).

30. T. Motet, J. Nees, S. Williamson, and G.A. Mourou, *Appl. Phys. Lett.* 59:1455 (1991).

31. M.C. Nuss, D.K. Kisker, P.R. Smith, and T.E. Harvey, *Appl. Phys. Lett.* 54:57 (1989).

32. S. Ramo, J.R. Whinnery, T. van Duzer, Fields and Waves in Communication Electronics, Wiley, New York (1984).

33. K.C. Gupta, R. Garg, and I.J. Bahl, Microstrip Lines and Slotlines, Artech House, Deadheam, USA (1979).

34. T.C. Edwards, Foundations for Microstrip Circuit Design, Wiley, New York (1981).

35. E. Yamashita, K. Atsuki, and T. Ueda, IEEE Trans. Micr. Th. Techn. MTT-27:1036 (1970).

36. H. Roskos, M.C. Nuss, K.W. Goosen, D.W. Kisker, A.E. White, K.T. Short, D.C. Jacobson, and J.M. Poate, *Appl. Phys. Lett.* 58:2604 (1991).

37. J. Kim, S. Williamson, J. Nees, S. Wakane, to be published.

38. J.A. Valdmanis and G.A. Mouron, *IEEE J. Quant. Electr.* QE-22:69 (1986).

39. J.A. Valdmanis, G.A. Mourou, and C.W. Gabel, *Appl. Phys. Lett.* 41:211 (1982).

40. G.A. Mourou and K.E. Meyer, *Appl. Phys. Lett.* 45:492 (1984).

41. K.E. Meyer and G.A. Mourou, *in:* "Picosecond Electronics and Optoelectronics", Eds. G.A. Mourou, D.A. Bloom, and C.H. Lee, Springer Verlag, Berlin-Heidelberg-New York (1985).

42. J.A. Valdmanis, *Elect. Lett.* 23:1308 (1987).

43. M.H. Heutmaker, G.T. Harvey, and P.F. Bechtold, *Appl. Phys. Lett.* 59:146 (1991).

44. B.H. Kolner and D.M. Bloom, *Electr. Lett.* 20:818 (1984).

45. B.H. Kolner and D.M. Bloom, *IEEE J. Quant. Electr.* QE-22:79 (1988).

46. P.M. Fermi, C.W. Knapp, C. Wu, J.T. Yardley, B.-B. Hu, X.-C. Zhang, and D.H. Auston, *Appl. Phys. Lett.* 59:2651 (1991) and J.I. Thackara, D.M. Bloom, and B.A. Auld, *Appl. Phys. Lett.* 59:1159 (1991).

47. S. Namba, *J. Opt. Soc. Am.* 51:76 (1961).

48. T. Nagatsuma, T. Shibata, E. Sano, and A. Iwata, *J. Appl. Phys.* 66:4001 (1989).

49. D.H. Auston and M.C. Nuss, *IEEE J. Quant. Electr.* QE-24:184 (1988).

50. M.Y. Frankel, J.F. Whitaker, G.A. Mourou, and J.A. Valdmanis, *IEEE Microwave and Guided Wave Lett.* 1:60 (1991).

51. J. Nees and G.A. Mourou, *Electron. Lett.* 22:918 (1986).

52. I.H. White, D.F.G. Gallagher, M. Osinski, and D. Bowley, *Electron. Lett.* 21:197 (1985).

53. A.J. Taylor, J.M. Wiesenfeld, G. Einstein, R.S. Tucker, J.R. Talman, and U. Koren, *Electr. Lett.* 22:61 (1986).

54. A.J. Taylor, J.M. Wiesenfled, G. Eisenstein, and R.S. Tucker, *Appl. Phys. Lett.* 49:681 (1986).

55. Y.K. Chen, M.C. Wu, T. Tanbun-Ek, R.A. Logan, and M.A. Chin, *Appl. Phys. Lett.* 58:1253 (1991).

56. P.J. Delfyett, C.H. Lee, L.T. Florez, N.G. Stoffel, T.J. Gmitter, N.C. Andreadakis, G.A. Alphonse, and J.C. Conolly, *Opt. Lett.* 15:1373 (1990).

57. D.E. Cooper, *Appl. Phys. Lett.* 47:33 (1985).

58. K.W. Goosen and R.B. Hammond, *IEEE Trans. Micr. Th. Techn.* MTT-37:469 (1989).

59. G.A. Mourou *in:* "High Speed Electronics", Eds. Kallbach and Benching, Springer Verlag Series in Electronics and Photonics 22, Springer Verlag, Berlin-Heidelberg-New York (1986), p. 191.

60. G. Hasnain, A. Dienes, and J.R. Whinnery, *IEEE Trans. Micr. Th. Techn.* MTT-34:738 (1986).

61. D.R. Dykaar, A.F.J. Levi, and M. Anzlower, *Appl. Phys. Lett.* 57:1123 (1990).

62. A. Deutsch, G.V. Kopcsay, V.A. Ravieri, J.K. Cataldo, E.A. Galligan, W.S. Graham, R.P. MacGouey, S.L. Nunes, J.R. Paraszczak, J.J. Ritsko, R.J. Serino, D.Y. Shin, and J.S. Wilczynski, *IBM J. Res. Develop.* 34:601 (1990).

63. W.J. Gallagher, C.-C. Chi, I.N. Duling III, D. Grischkowsky, N.J. Halas, M.B. Ketchen, and A.W. Kleinsasse, *Appl. Phys. Let.* 50:350 (1987).

64. D. R. Dykaar, R. Sobolowski, J.M. Chwalek, J.F. Whitaker, T.Y. Hsiang, and G.A. Mourou, *Appl. Phys. Lett.* 52:1444 (1988).

65. M.C. Nuss, P.M. Mankiewich, R.E. Howard, B.L. Straughn, T.E. Harvey, C.D. Brandle, G.W. Berkstresser, K.W. Goosen, and P.R. Smith, *Appl. Phys. Lett.* 54:2265 (1989).

66. D.H. Auston, K.P. Cheung, J.A. Valdmanis, and D.A. Kleinman, *Phys. Rev. Lett.* 53:1555 (1984).

67. D.A. Kleinman and D.H. Auston, *IEEE J. Quant. Electr.* QE-20:964 (1984).

68. D.H. Auston, *Appl. Phys. Lett.* 43:713 (1983).

69. D.B. Rutledge, D.P. Neikirk, and D.P. Kasilingham, *in:* "Infrared and Millimeter Waves", Vol. 10, Ed. K.J. Button, Academic Press, New York (1983).

70. D. Grischkowsky, I.N. Duling III, J.C. Chen, and C.-C. Chi, *Phys. Rev. Lett.* 59:1663 (1987).

71. C. Fattinger and D. Grischkowsky, *Phys. Rev. Lett.* 62:2961 (1989).

72. M.Y. Frankel, S. Gupta, J.A. Valdmanis, and G.A. Mourou, *IEEE Trans. Micr. Th. Tech.* MTT-39:910 (1991).

73. D. Paulus, B. Wedding, A. Gasch, and D. Jäger, *Phys. Lett.* 102A:89 (1984).

74. D. Jäger, *Int. J. Electron.* 58:649 (1985).

75. M.J.W. Rodwell, M. Kamegawa, R. Yu, M. Case, E. Carman, and K.S. Giboney, *IEEE Trans. Micr. Th. Tech.* MTT 39:1194 (1991).

76. M. Tan, C.-Y. Su, and W.J. Anklam, *Electron. Lett.* 24:213 (1988); M. Case, M. Kamegawa, R. Yu, M.J.W. Rodwell, and J. Franklin, *Appl. Phys. Lett.* 58:173 (1991); M. Case, E. Carman, R. Yu, M.J.W. Rodwell, and M. Kamegawa, *Appl. Phys. Lett.* 60:3019 (1992).

77. C.J. Madden, R.A. Marsland, M.J.W. Rodwell, D.M. Bloom, and Y.C. Pao, *Appl. Phys. Lett.* 54:1019 (1989).

78. W. Roth, H. Schuhmacher, J. Kluge, H.J. Geelen, and H. Beneking, *IEEE Trans. Electron Devices* 32:1034 (1985).

79. B.J. van Zeghbroeck, W. Patrick, J.-M. Halbout, and P. Vettinger, *IEEE Electron Device Lett.* 9:527 (1988).

80. S.M. Sze, D.J. Coleman, and A. Loya, *Sol. Stat. Electron.* 14:1209 (1971).

81. D.L. Rogers, J.M. Woodall, G.D. Pettit, and D. McInturff, *IEEE Trans. Electron Devices* 34:2382 (1987).

82. H. Schuhmacher, H.P. Leblanc, J. Soole, and R. Bhat, *IEEE Electron Device Lett.* 9:607 (1988).

83. O. Wada, H. Nobuhara, H. Hamaguchi, T. Mikawa, A. Takeuchi, and T. Fujii, *Appl. Phys. Lett.* 54:16 (1989).

84. L. Yang, A.S. Sudbo, and W.T. Tsang, *Electron. Lett.* 25:1479 (1989).

85. C.J. Wei, D. Kuhl, E.H. Böttcher, D. Bimberg, and E. Kuphal, *IEEE Electron. Device Lett.* 11:334 (1990).

86. D. Kuhl, F. Hieronymi, E.H. Böttcher, T. Wolf, A. Krost and D. Bimberg, *Electron. Lett.* 26:2107 (1990).

87. M. Zirngibl, J.C. Bischoff, D. Theron, and M. Ilegems, *IEEE Electron Device Lett.* 10:336 (1989).

88. D. Kuhl, F. Hieronymi, E.H. Böttcher, T. Wolf, D. Bimberg, J. Kuhl, and M. Klingenstein, *J. Lightwave Techn.* 10:753 (1992).

89. M. Ito and O. Wada, *IEEE J. Quant. Electron.* QE-22:1073 (1986).

90. M. Ito, O. Wada, K. Nakai, and T. Sakurai, *IEEE Electron. Device Lett.* 5:531 (1984).

91. W.S. Lee, G.R. Adams, J. Mun, and J. Smith, *Electron. Lett.* 22:147 (1986).

92. D. Rogers, *IEEE Electron. Device Lett.* 7:600 (1986).

93. H. Hamaguchi, M. Makiuchi, T. Kumai, and O. Wada, *IEEE Electron. Device Lett.* 8:39 (1987).

94. C.S. Harder, B. van Zeghbroeck, H. Meier, W. Patrick, and P. Vettinger, *IEEE Electron. Device Lett.* 9:171 (1988).

95. X. Zhou and T.Y. Hsiang, *J. Appl. Phys.* 67:7399 (1990).

96. W.C. Koscielniak, J.L. Pelouard, and M.A. Littlejohn, *IEE Photonics Technol. Lett.* 37:1632 (1990).

97. W.C. Koscielniak, J.L. Pelouard, R.M. Kolbas, and M.J. Littlejohn, *IEEE Trans. Electron. Devices* 37:1632 (1990).

98. W.C. Koscielniak, M.A. Littlejohn, and J.L. Pelouard, *IEEE Electron. Device Lett.* 10:209 (1989).

99. W.C. Koscielniak, J.L. Pelouard, and M.A. Littlejohn, *Appl. Phys. Lett.* 54:567 (1989).

100. J.B.D. Soole and H. Schumacher, *IEEE Trans. Electron. Devices* 37:2285 (1990).

101. X. Zhou, T.Y. Hsiang, and. R.J.D. Miller, *J. Appl. Phys.* 66:3066 (1989).

102. C. Moglestue, *IEE Proc.* Part I: Solid State and Electronic Devices, 131:103 (1984).

103. C. Moglestue, Intl. Conf. on Simulation of Semiconductor Devices and Processes, Swansea 1984, Eds. K. Board and D.R.J. Owen, p. 153.

104. D.E. Aspnes, S.M. Kelso, R.A. Logan, and R. Bhat, *J. Appl. Phys.* 60:754 (1986).

105. M.A. Osman and D.K. Ferry, *J. Appl. Phys.* 61:5330 (1987).

106. C. Moglestue, *IEEE Trans. CAD Integrat. Circuits Systems* 5:326 (1986).

107. S. Ramo, Proc. I.R.E., 27:584 (1939).

108. P.A. Houston and A.G.R. Evans, *Sol.Stat.Electron.* 20:197 (1977).

109. V.L. Dalal, A.B. Dreeben, and A. Triano, *J. Appl. Phys.* 42:2864 (1971); V.L. Dalal, *Appl. Phys. Lett.* 16:489 (1970).

110. G.D. Alley, *IEEE Trans. Micr. Th. Tech.* 18:1028 (1970).

111. T.F. Carruther and J.F. Weller, *Appl. Phys. Lett.* 48:460 (1986).

112. M. Klingenstein, J. Kuhl, J. Rosenzweig, C. Moglestue, Jo. Schneider, and A. Hülsmann, *Appl. Phys. Lett.* 58:2503 (1991).

113. T. Pfeiffer, J. Kuhl, E.O. Göbel, and L. Palmetshofer, *J. Appl. Phys.* 62:1850 (1987).

114. K. Hirakawa and K. Sakaki, *J. Appl. Phys.* 63:803 (1988).

115. S. Zollner, S. Gopalan, and M. Cardona, *Sol. Stat. Commun.* 76:877 (1990).

116. Landolt-Börnstein, *New Serie III*, 17a:234, Ed. K.-H. Hellwege, Springer Verlag, Berlin-Heidelberg-New York (1982).

117. P.A. Houston and A.G.R. Evans, *Sol. Stat. Electron.* 20:197 (1977).

118. C.M. Wolfe, G.E. Stillman, and W.T. Lindle, *J. Appl. Phys.* 41:3088 (1970).

119. T.W. Hickmott, *IEEE Trans. Electron. Dev.* ED-31:54 (1984).

120. M. Klingenstein, J. Kuhl, R. Nötzel, K. Ploog, J. Rosenzweig, C. Moglestue, A. Hülsmann, Jo. Schneider, and K. Köhler, *Appl. Phys. Lett.* 60:627 (1992).

121. F.W. Smith, A.R. Calawa, C.L. Chen, M.J. Manfra, and L.J. Mahoney, *IEEE Electron. Dev. Lett.* 9:77 (1988).

122. M. Kaminska and E.R. Weber, *in:* "Proc. 20th Int. Conf. on Physics of Semiconductors", Eds. E.M. Anastassakis and J.D. Joannopoulos, World Scientific, Singapore (1990), p. 473.

123. P.R. Smith, D.H. Auston, and W.M. Augustyniak, *Appl. Phys. Lett.* 39:739 (1981).

124. M.Y. Frankel, J.F. Whitaker, G.A. Mourou, and J.A. Valdmanis, *Sol. Stat. Electr.* 35:325 (1992).

125. J.M. Wiesenfeld and R.K. Jain, Direct Optical Probing of Integrated Circuits and High-Speed Devices, in: "Semiconductors and Semimetals", Vol. 28, Ed. R.B. Marcus, Academic Press, New York (1990).

# VIBRATIONAL RELAXATION STUDIED WITH LIGHT

Ad Lagendijk[1,2]

[1]FOM-Institute for Atomic and Molecular Physics
Kruislaan 407, 1098 SJ Amsterdam
The Netherlands

[2]Van der Waals-Zeeman Laboratorium
Valckenierstraat 65-67
1018 XE Amsterdam
The Netherlands

## 1 ABSTRACT

In this lecture I will deal with optical studies of vibrational dynamics. I will address the question of what (minimal number of) parameters characterize vibrational motion and under what conditions the relaxation-time approximation is valid. Within this assumption the vibrational dynamics are represented by, besides the eigenfrequency, the decay constants $T_2$ and $T_1$. More of these constants will be needed when different oscillators are coupled.

Following we will consider the extraction of these quantities from optical experiments. Relevant optical techniques are linear and nonlinear absorption spectroscopy, spontaneous Raman scattering and forced or "coherent" Raman scattering. The complications of strong coupling of the oscillators to the light field, in absorption spectroscopy referred to as polariton formation, will be dealt with.

*Ultrashort Processes in Condensed Matter*, Edited by
W.E. Bron, Plenum Press, New York, 1993

It will be demonstrated that the conventional response theory that is employed to calculate line shapes has serious shortcomings. I refer to these "Kubo-type" response principles as to "one-way response theory" in contrast to, what I have coined, "two-ways response theory".

## 2 INTRODUCTION

When one considers nowadays the many scientific papers in natural science one might get confused and a little desperate. The fact that a major fraction of these writings deals with optics and lasers, that is to say our field, adds to this discomfort. One should not be surprised that there are numerous publications on scientific applications of lasers, and in particular on *non*linear optics, because many very clever scientific workers have bought many lasers and all the necessary auxiliary equipment. Optics is in fashion. Almost any physicist, and chemist for that matter, that respects himself is in some way associated with research in which lasers and nonlinear optics are involved. Whatever he will find or not find, he will write a publication on it. No more microwave spectroscopy, no more magnetic resonance spectroscopy, no more acoustics, no, just lasers. The problem with lasers is that they work on optical principles. Suddenly everybody realizes that before one can embark in this trendy field on nonlinear optics one has to learn something about this dull field of linear optics. Linear optics has long been abandoned as a basic subject, not only in (Dutch) highschools, but also in (Dutch) curricula for university degrees in physics. Many scientists working in the field of nonlinear optics feel the need to reinvent the wheel, not realizing that a great deal of the formalism for linear and nonlinear optical response was developed much earlier for magnetic resonance in the 1960's. Famous names associated with major contributions during that period are for instance P.W. Anderson, N. Bloembergen, R. Kubo, and A.G. Redfield. When somebody wants to learn the subject my advice is: just buy and read a good book on magnetic resonance like C.P. Slichter's "Principles of Magnetic Resonance".[1] The problem with the field of nonlinear optics is that the subject is of enormous extension. Is it a technique, or is it a phenomenon, or a class of phenomena? Should one discuss the possible applications of the technique, or should one look at purely optical phenomenon like nonlinear light propagation? Books on nonlinear optics necessarily only describe a tiny part of the field. Nonlinear optics is also often approached from the viewpoint of an engineer. For an engineer relaxation times like the well-known, and for us crucial, $T_2$ and $T_1$, are just some parameters determining the quality factor of a lumped circuit, in the same way as for instance the reflectivity of a mirror does, but for a solid state physicist these relaxation times differ like day and night from the imperfections of a mirror.

Let us first find out what type of physics you are interested in and let us see if nonlinear optics has a role to play in satisfying your curiosity about nature. The emphasis throughout this lecture will be: how can we *use* optical principles, and if necessary nonlinear optical experiments, to learn more about our material.

Optical studies of vibrational dynamics are numerous. This is not a review paper, so I cannot give proper credit to the vast and excellent work carried out by many colleagues.

Sometimes, for educational purposes, I will need and discuss some specific examples. In the majority of cases these demonstrations will be drawn from our own work. This implies nothing whatsoever about (my opinion on) the quality of other work, but is purely done on the basis of convenience (for me).

## 3 TYPE OF PHYSICS

Before we switch on our laser, let us first discuss what type of information we want to obtain from our experiment. This is a very important contemplation as otherwise we get lost in the tremendous complications of laser systems and nonlinear optical phenomena. A laser is a very bright source of light and as such able to introduce all kinds of nonlinear couplings and effects, many of which are a nuisance and are unwanted. Nonlinearity is found in many branches of physics besides optics, and its presence invariably means complications. If you can avoid nonlinearities when gathering your information, please do.

I assume you are interested in some "system", consisting of many particles or degrees of freedom. This corpus is a large collection of atoms or molecules if you are a solid-state physicist and a smaller number of atoms if you are a molecular physicist. What quantities are of interest? When do you "understand" the behavior of your crowd of particles? For any system one can define many, often even an infinite number of, observables and one cannot be interested in all of them, leave alone to know everything about them. Understanding implies among other things characterization and prediction of the behavior of your object using a minimal number of observables. Consequently we will have to put some limitation on our ambition for knowledge of everything.

This lecture is about vibrational dynamics, so apparently we want to understand more about the displacement of atoms around their equilibrium positions. Quantities of concern could be for instance simple expectation values like $\ll u^2 \gg$, the mean-squared displacement of an atom at a particular site, but also the more informative functions describing spatial correlations between displacements. Examples of these (static) correlation functions are: $\ll \mathbf{u}(\mathbf{r}_1)\mathbf{u}(\mathbf{r}_2) \gg$, $\ll \mathbf{u}(\mathbf{r}_1)\mathbf{u}(\mathbf{r}_2)\mathbf{u}(\mathbf{r}_3) \gg$, $\ll \mathbf{u}(\mathbf{r}_1)\mathbf{u}(\mathbf{r}_2)\mathbf{u}(\mathbf{r}_3)\mathbf{u}(\mathbf{r}_4) \gg$ and higher-order functions. The expectation value $\ll \cdots \gg$ denotes an averaging over some relevant distribution function, like the quantum-mechanical expectation value over the ground state, or a thermal average (quantum or classical). Sometimes we would like to Fourier transform our spatial correlation functions, to take maximal advantage of translational symmetry, and wavevectors will become the relevant variables like for instance: $\ll \mathbf{u}(\mathbf{p}_1)\mathbf{u}(\mathbf{p}_2) \gg$. In case of translational symmetry this function is proportional to $\delta(\mathbf{p}_1 + \mathbf{p}_2)$. We have used the label $\mathbf{p}$ rather than $\mathbf{k}$ for the momentum variable as later $\mathbf{k}$ will be used for light propagation.

All expectation values are equilibrium expectation values, and they are only determined by the, time- independent, Hamiltonian of your system. We call the correlation functions discussed so far *static*, because they do not depend on time. One can, and for the study of vibrational *dynamics* one should, also be interested in *dynamic* correlation functions, that depend on time (or frequency). An important dynamic correlation function is

$$\ll \mathbf{u}(\mathbf{r}_1, t_1)\mathbf{u}(\mathbf{r}_2, t_2) \gg, \tag{1}$$

which describes correlations in both spatial and temporal variables. This dynamic correlation function can also be generalized to situations where more displacement operators are involved in the correlation function. Note that these correlation functions are again defined in terms of equilibrium expectation values, and time is only a parameter. The distribution function over which the averaging is performed, does not depend on time. Fourier transforming over time is of course very frequently performed as nature looks commonly much simpler in the frequency domain than in the time domain. Fourier transforming the lowest-order dynamic displacement correlation function (1) with respect to space and time defines $\sigma(\mathbf{p}, \Omega)$,

$$\sigma(\mathbf{p}, \Omega)\delta(\mathbf{p}_1 + \mathbf{p}_2) = \int e^{i\Omega t} \ll \mathbf{u}(\mathbf{p}_1, t_1)\mathbf{u}(\mathbf{p}_2, t_1 + t) \gg dt\,, \tag{2}$$

where the delta function comes from the assumption of spatial translational symmetry. Integrals without explicit indication of the domain of integration are to be taken from $[-\infty, +\infty]$. I have chosen to use $\Omega$ rather than $\omega$ for frequency as $\omega$ is often used to describe the deviation from the central frequency (bandwidth) of incoming light.

Up to now I have mainly considered correlation functions containing displacement operators. This is a professional deformation of workers in optics. The coupling between light and our type of matter is primarily through the displacement operator. People studying vibrational dynamics with neutron scattering always consider correlation function of *density* operators. The obvious reason is here that the interaction between neutrons and atoms is through a very local potential (often represented as a delta-function pseudo-potential). Using the first Born approximation to describe the scattering of neutrons off nuclei is an excellent approach and immediately leads to the study of correlation functions of the type

$$\ll \rho(\mathbf{r}_1, t_1)\rho(\mathbf{r}_2, t_2) \gg, \tag{3}$$

in which $\rho$ is the density operator: $\rho(\mathbf{r}) \equiv \sum_{i=1}^{N} \delta(\mathbf{r} - \mathbf{r}_i)$, where $N$ is the number of particles. A rather famous function is the double (space and time) Fourier transform of correlation function (3), which after performing the spatial Fourier transform explicitly looks like

$$S(\mathbf{p}, \Omega)\delta(\mathbf{p}_1 + \mathbf{p}_2) = \int e^{i\Omega t} \ll \rho(\mathbf{p}_1, t_1)\rho(\mathbf{p}_2, t_2) \gg dt\,, \tag{4}$$

This function $S(\mathbf{p}, \Omega)$ is the most important function describing neutron scattering (where it is better known as $S(\mathbf{k}, \omega)$, and sometimes referred to as the "scattering law"). Under mild conditions, to be discussed later, the correlation function $S(\mathbf{p}, \Omega)$ has a Lorentzian shape[2] (as a function of $\Omega$) and we call its center, $\Omega_0(\mathbf{p})$, the frequency, and the width of the line, $\tau(\mathbf{p})$, we refer to as the damping of the oscillator. The displacement correlation function $\sigma(\mathbf{p}, \Omega)$ will have approximately the same shape as $S(\mathbf{p}, \Omega)$.[3]

In many cases higher-order correlation functions can be decoupled into lower-order correlation functions. This alleviates the task of characterizing your system as only some lower-order correlation functions need to be known under these circumstances. One cannot in all seriousness claim that a correlation function consisting of a product of thirty-six displacement operators will contain much different information from one holding thirty-eight or thirty-four displacement operators. Many of these operators will fluctuate (almost) independently and can, besides some exceptional pathological cases, safely be decoupled into a product of lower-order correlation functions. The correlation functions discussed so far are intrinsic to your system, that is to say these correlation functions are all defined in the absence of the probes. I have indicated that information can be obtained about these intrinsic correlation functions by coupling your system to a probe and perform an experiment. This is only true for investigations that can be described as linear-response experiments. We will discuss this question in more detail in Section 8. In the philosophy of linear response the influence of the probe is as weak as possible. When using bright light sources linear-response concept is often not valid any more. The good thing is that a stronger coupling between probe and system will potentially lead to a situation where more knowledge can be accumulated. The bad news is that the correlation functions being measured are much more complicated and not any longer intrinsic to your system as the separation of probe and system is not longer possible.

I want to stress that if you are a condensed matter physicist who is interested in vibrational dynamics you do not want to know everything about chirps, bandwidth-limited pulses, regenerative gain, amplified stimulated emission, unstable cavities, mode locking etc. These effects and phenomena are indeed very important, but for you only to the extent that they enable you to study your vibrational system. If you like to study all these nonlinear optical effects in great detail and just for the sake of themselves, fine, but do not call yourself a condensed-matter physicist. A butcher should sell meat, and not bread. Many specialists in vibrational dynamics do not know anything about lasers but it is of crucial importance to keep on communicating with them. By using too much "optical jargon" it is quite simple to alienate oneself from this community. A condensed-matter physicist should have an understanding of the general concepts and methods of calculation of correlation functions for many-body systems. He should know about Green's functions for phonons and electrons. Only if you speak that language, and present your new results using that terminology your discoveries will get the full attention and appreciation of the condensed-matter community at large.

## 4 HOW COMPLICATED IS YOUR SYSTEM?

If your system is not complicated it is not interesting because many other people will already have studied it in great detail, even before you were born. If a system has only a "discrete" spectrum, containing a finite or at most a countable infinity of spectral points, the dynamics are often solved trivially and uninteresting. In the systems you will be interested in this is never the case. You will always have to deal with a continuous spectrum and in many circumstances you also will have to handle a many-body system (which is the physicist understatement for an infinite-body system). When I explain to colleagues that they work with an infinite-body system they often look very surprised. In their view they only study a few oscillators having some irreversible behavior (for instance expressed by a dephasing time). But they do not realize that in order to obtain irreversible behavior from a microscopic point of view one has to couple the oscillator to degrees of freedom having a continuous spectrum, frequently a many-body system. Otherwise the system could be described with a finite number of modes and no irreversible behavior could be established.

In all situations a large number, and in many cases an infinite number of operators and degrees of freedom is involved. We want to understand our system, and we will come a long way in comprehension when we determine the behavior of some simple correlation functions. One always speaks of the *relevant* or *slow* variables in contrast to the fast or irrelevant variables. The influence of the fast variables can usually be taken into account as a (generalized) damping force on the slow variables. One looks for the slow variables and treats them in great detail and the fast variables are taken into account only in an approximate way as one expects the details of their dynamics not to be very important. Conserved variables always belong to the collection of slow variables but in many cases there is quite some ambiguity about which variables are allowed to be classified as slow and which not. Sometimes it is a matter of taste and often one really does not know whether in a theory all slow variables have been taken into account properly. When an important slow variable has been overlooked we are frequently warned because when this has happened the damping forces show unphysical behavior (like a very strange resonance structure) signalling an unjustified neglect of a slow variable. There are several systematic approaches to incorporate more and more, less and less slow, variables in a consistent (and hopefully convergent) procedure and one such a method is the Mori method.[4,5]

## 4.1 Introducing $T_2$ and $T_1$

We will outline the general conditions under which the temporal behavior of a system can be described by a single (or just a few) relaxation time(s) and the frequency response can be described by a single (or just a few) Lorentzian line(s). This is very important as the description of a complicated system in these simple terms is a tremendous simplification and a big step towards understanding the system. The Hamiltonian of a system, $H_{sys}$ can very often be partitioned in two commuting parts and a third coupling or interaction term. We will call them subsystem, bath and interaction term:

$$H_{sys} = H_{subsystem} + H_{interaction} + H_{bath} \equiv H_{sub} + H_{int} + H_{bath}. \tag{5}$$

We suppose that $H_{sub}$ is a very simple system, like consisting of only a few bound states or just an undamped harmonic oscillator, and we assume $H_{bath}$ to fluctuate on very short time scales. If this is not true we will switch their roles and call the bath the system and vice versa. If this does not work your system is either trivial or very complicated. Skipping the trivial case the failure of a partitioning like expression (5) can only be due if both $H_{sub}$ and $H_{bath}$ are nontrivial many-body systems. But in that case you can probably forget all about their coupling because they have enough problems to take care of their own dynamics and will have no time to influence each other. Under these circumstances you can disregard the "bath" and can continue with $H_{sub}$ which you should try to partition in its own subsystem and bath as indicated in expression (5).

I hope by now that I have convinced you that partitioning (5) is indeed very general and poses hardly any limitation on the systems you can describe. I expect the reader also to recognize that partitioning (5) complies with the principle of separation of the variables into slow and fast variables. The subsystem takes care of the slow variables and the rest of the fast variables. In some approaches $H_{int} + H_{bath}$ are taken together into one term in which the bath variables do not appear as operators, but as classical fields having an explicit time dependence. This more limited scheme is often the basis of stochastic methods. The line of attack that we propose to deal with $H_{sys}$ can easily be adjusted so as to apply to these more limited cases also. In the following I will give one explicit example of a stochastic model and indicate the parallel with the approach based on Hamiltonian dynamics.

We assume that our system can be depicted as a few levels in interaction with a continuum. In principle one could define dephasing for a single level, but the situation is much clearer for a two-level system. Let us suppose we have two levels, the excited state $| e >$ and the ground state $| g >$ both interacting with a continuum. The total wavefunction of the compound system can then be described as $| i > \otimes | \{bath\} >$, where a symbolic notation for the bath wavefunctions has been introduced and $| i >$ stands for either the excited or the ground state. Due to the presence of a high density of continuum states the state $| i > \otimes | \{bath\} >$ will be (almost) degenerate with many other states which differ only in their bath states. After the system is excited in one of these compound states it will go through many of the other compound states almost degenerate with it, because the interaction term will couple them and will take care of the energy balance. After some characteristic time the phase of this initial state will not be related anymore to the phase at a later time. The typical time associated with this dephasing mechanism is called the pure dephasing time $T_{deph}$ and is best described to be associated with quasi-elastic processes. Elastic because the subsystem component of the wavefunction is unaltered and remains $| i >$, and quasi because the coupling energy takes care of small, but not vanishing, energy mismatches.

A second dynamic process can be visualized which could be classified as inelastic. Let us again discuss the case of two levels for the time being: $| e >$ and the subsystem ground state $| g >$. Both these levels will be coupled to the continuum and would be dephased by it. However the state $| e > \otimes | \{bath\} >$ can now also be degenerate with a state of the type $| g > \otimes | \{bath\} >$ if the right continuum states are excited. The state $| e > \otimes | \{bath\} >$ will decay into these states but after this kinetic process the excited compound state has ended, in contrast to a pure dephasing process, in a *different* subsystem state. The bath will take care of extreme fast de-excitation of the excited continuum states and we will wind up with the subsystem in the ground state. This mechanism represents an inelastic process and the characteristic time will be called $T_{1,eg}$. The reader might complain that this inelastic process is also, in a way, described as a dephasing process, albeit one in which the subsystem component has changed. This is a very fundamental problem with present day (quantum) physics. Building one's models using hermitian Hamilton operators will always lead to energy conservation and the inability to formally describe inelastic processes. The only technique to represent inelastic behavior is by relating it to dephasing into a larger space.

The method to obtain the simplest possible theoretical expressions for the relaxation times from Hamiltonian (5) goes under many names, my favorite is *the weak-coupling approach.* All these methods can be viewed as variations on the theme of "expansion in the interaction Hamiltonian $H_{int}$". I will estimate higher-order corrections later but will now indicate the lowest-order result. The dephasing time is measured when a spectroscopic measurement which induces transitions and for this reason the dephasing time is a measure for the *difference* in dephasing of the two levels. The answer, correct to lowest order in $H_{int}$, for $T_{deph}$ can also be found by applying Fermi's Golden Rule, and the result is given by

$$\frac{1}{T_{deph}} = E_{int}^2 \rho(E = 0), \tag{6}$$

in which $E_{int}$ is a typical energy scale for $H_{int}$ (or difference in diagonal matrix elements of $H_{int}$ between excited and ground states), and where $\rho(E)$ is the density of states of the continuum or bath at energy $E$. I cannot be more specific about the precise magnitude of $E_{int}$ unless I would treat in much more detail the exact form of the interaction Hamiltonian. It is important to realize that $E_{int}$ is not at all influenced or determined by $H_{bath}$. People familiar with motional-narrowing theory in magnetic resonance know that $E_{int}^2$ is also equal to the second moment of a relevant operator of the subsystem (the relevant operator is the simplest operator of the subsystem that has matrix elements between the ground state and the excited state).[1] This second moment is not determined by the bath Hamiltonian because this Hamiltonian commutes with any operator of the Hilbert space of $H_{sub}$. The density of bath states has as dimension *per energy* so it denotes a time scale. For this reason many people refer to it as $\tau_c(E = 0)$, the correlation time of the bath. In the latter notation the dependence on the frequency $E$ is often dropped. With the inverse of $E_{int}$ one can also associate a time scale and this could be called the time scale of the interaction term.

Knowing the expression for the dephasing times makes it rather simple to anticipate the expression for the inelastic time $T_{1,eg}$:

$$\frac{1}{T_{1,eg}} = E_{int}^2 \rho(E = E_e - E_g) \equiv E_{int}^2 \tau_c(E = E_e - E_g). \tag{7}$$

As expected in this case the density of states has to be evaluated at the energy difference between the two levels as the energy has to be supplied or received somewhere. Of course $T_{1,eg}$ and $T_{1,ge}$ are connected through detailed balance in case of thermal equilibrium:

$$\frac{T_{1,eg}}{T_{1,ge}} = e^{(E_g - E_e)/k_B T}. \tag{8}$$

This relation is essentially a condition on the density of states of the continuum. The magnitude of $E_{int}$ appearing in equation (7) for the inelastic rate constants can differ by factors of order one from the factor appearing in equation (6) for the pure dephasing kinetics.

Up to now we have discussed the relaxation times in terms of levels. In a many-level system each two levels have their own dephasing times and all levels are potentially connected through inelastic channels. The relaxation times then really become matrices where the indices denote the levels.[6] Experiments often measure lumped combinations of these times and for these reason some linear combinations of these times are well-known. When one measures the decay of the population *difference* between two levels $i$ and $j$ the decay time, defined as $T_1$, would be given by the parallel sum , $1/T_1 \equiv 1/T_{1,ij} + 1/T_{1,ji}$.

For a multilevel system it makes more sense not to combine them in lumped relaxation times. In a spectroscopic measurement, like optical absorption, the linewidth is defined to be $1/T_2$. The pure dephasing process gives rise to an energy uncertainty of $1/T_{deph}$, and the inelastic process gives rise to an energy uncertainty of $1/(2T_1)$. These processes are independent and add, and the linewidth is given by

$$\frac{1}{T_2} = \frac{1}{T_{deph}} + \frac{1}{2T_1}. \tag{9}$$

The factor of two appearing in this equation can be explained in the following way. Dephasing is a property of the probability amplitude (or wavefunction or vibrational amplitude) and population dynamics is connected with the amplitude *squared.*

I have presented enough information to make it possible to indicate how higher-order corrections to the weak-coupling result would look like. Higher-order corrections to the relaxation times can be visualized by writing expression (6) for $T_{deph}$ in a different way,

$$\frac{1}{T_{deph}} = E_{int}\left\{E_{int}\rho(E=0)\right\} = E_{int}\left(E_{int}\tau_c\right). \tag{10}$$

The dimensionless parameter $E_{int}\tau_c$ is the expansion parameter. In a theory incorporating higher-order corrections to the weak-coupling theory higher-order powers of $E_{int}\tau_c$ would be added to the term between brackets in equation (10). The expansion parameter must be small for our approach to be valid and we see that the request is that the bath should be much faster than the time connected with interaction Hamiltonian $E_{int}^{-1}$. For many types of baths this will be an excellent approximation and we have done a superb job. We are also able to understand the term motional narrowing. If the bath would be extremely slow the dynamics of the subsystem would be totally determined by $E_{int}$, and the linewidths and inverse decay times would be of the order of $E_{int}$, so the width is narrowed by the motion of the bath by a factor $E_{int}\tau_c$ .

In the foregoing it was assumed that the continuum could supply or absorb the necessary energy for the inelastic process to go. Here a practical difference between optics and magnetic resonance shows up. In the standard magnetic resonance techniques, like NMR and ESR, the spin energies are usually so small that $\tau_c(E=0) \approx \tau_c(E_e - E_g)$. In that case the quasi-elastic dephasing times and the inelastic times are approximately equal. In optics the situation is very often the other way around. The energies are usually much larger than $k_BT$ and for that reason the $T_1$'s are usually orders of magnitude larger than the dephasing times. The situation could even be more dramatic since it is possible that the density of states is zero at the required energy. This is because the degrees of freedom of the bath have a certain energy scale and for much larger energies than this energy scale there are less and less states and beyond the edge of the density states there will be no states at all. Under these conditions the energy gap is too large and the continuum cannot absorb or supply the necessary energy. In the language of quantum mechanics one would say that one quantum of the bath is not enough to bridge the energy gap and more quanta are necessary, until we have enough of them to span the energy gap. It is clear that we have to go to higher-order perturbation theory with the interaction as perturbation in order to obtain a finite transition probability. There are in principle two ways of doing this. We could use the same interaction energy and go to higher-order in weak-coupling theory but this is very cumbersome. A simpler way to go about is to realize that in the Hamiltonian (5) the interaction term is also usually a first term in an expansion in bath coordinates. If we would allow higher-order term there we could still do simple weak-coupling theory to obtain expression for the linewidths.[7-9] Let me be more specific. Suppose we consider some subsystem interacting with its environment. We could symbolically denote the environment with some lumped variable $\{R\}$. The coordinates of our simple subsystem will be represented by the coordinate $q$. We can formally define the "displacement" $\delta R$ of the bath operators by

$$R \equiv R_0 + \delta R, \tag{11}$$

where $R_0$ is some equilibrium or reference set of coordinates for the bath. Let us imagine some adiabatic approach (for instance by assuming for the time being that the masses of the degrees of freedom of the bath are infinite). We can then calculate the energy of the total system. This energy is a function of $\{R\}$ and $q$. Assuming now that we can expand in small bath displacements we will have the following terms (in symbolic notation !!)

$$\begin{aligned} U(\delta\{R\},q) \quad - \quad U(\delta\{R\}=0,q) = \left(\frac{\partial U(\delta\{R\},q)}{\partial\delta\{R\}}\right)_{\delta\{R\}=0} \delta\{R\} + \\ \frac{1}{2!}\left(\frac{\partial^2 U(\delta\{R\},q)}{\partial\delta^2\{R\}}\right)_{\delta\{R\}=0} \delta^2\{R\} + \cdots \ . \end{aligned} \tag{12}$$

All the terms on the rhs. of this expansion will give rise to coupling between bath and subsystem as they depend on both the dynamic coordinates $q$ and $\delta\{R\}$. In the higher-order terms successively more bath operators (or bath quanta) are involved. If we use weak-coupling theory for each separate coupling term we would find that the successive densities of states we would need in the expression for the linewidths (See for instance equation (6)) consists of a compound density of states which can usually simply be described as a (convolution) product of the single-excitation density of bath states. For the first term in the rhs. of the equation above we need the single-excitation density of bath states and for the second term we would need the convolution of two single-excitation density of states, approximately twice as wide in the frequency domain as the single-excitation density of states. If we would go to higher order we would be able to bridge any energy gap. If this is after $N$ terms we would describe the vibrational relaxation in terms of a $N$-excitation (or $N$-phonon) process. One calls this the energy-gap law as one expects an exponential decrease in magnitude of the successive terms in expansion (12).[10] The sharp drop for each next order implies that if the process goes for a certain $N$, one does not need to take into account additional higher-order processes as their contributions will be exponentially smaller.

In multilevel systems some new and interesting phenomena can occur. We will use a three-level system as a generic example: $\mid i>, \mid j>$, and $\mid g>$ presented in order of decreasing subsystem energy. Excitation to a compound $\mid i > \otimes \mid \{bath\} >$ can be followed by direct decay to the ground state but it can also decay through the intermediate excited compound state $\mid j > \otimes \mid \{bath\} >$. Important factors determining the branching ratio's of the two channels are the respective energy gaps and the magnitudes of the matrix elements of the interaction Hamiltonians. This new situation can again simply be handled in the weak-coupling situation. A system having two excited levels can be described with the two coordinates $q_1$ and $q_2$. The potential energy of this system will be of the form $U(q_1, q_2, \{R\})$. The actual dependence of this energy on $\{\delta R\}$ will determine what type of bath excitations will drive the transition. An expansion of this energy in small deviations $\{\delta R\}$ will give rise to terms like $q_1 q_2 \delta\{R\}$, a bath-induced harmonic coupling between normal modes, but also higher-order terms like bath-induced anharmonicity would show up in this expansion and would provide the right matrix elements to allow for a path in which a low-lying excited subsystem level gets populated before the transport to the ground state takes place.

In addition to introducing new inelastic channels a third level can also have interesting effects on linewidths and line positions. These two quantities are closely related as lineshifts and linewidths are given by respectively the sine and cosine transform of the same dynamic correlation function. From motional narrowing theory in magnetic resonance it is known that if an absorption spectrum consist of two narrow separate lines and a kinetic process exist connecting the two transitions that is faster than the inverse of the separation in frequency between the two lines one "motionally averaged" line will appear at some weighted average position.[11] The weight depends on the occupation probability of the initial states of the two separate transitions. When these initial states differ in energy, a Boltzmann factor will determine their occupation probability and so also the position of the averaged line. Changing the temperature will then change the line position. This is a situation often encountered in optics where a high-energy mode is anharmonically coupled to a low-energy mode that has a fast $T_1$. The fast occupation dynamics of the low-energy mode (usually referred to as the "exchanging mode") will pull on the frequency of the high-energy mode. The existence of an fast exchanging mode will also have effects on the linewidth. The structure of the contribution to the linewidth must again be of the form $E_{int}^2\tau_c$. $E_{int}$ will be of the order of the anharmonic coupling between the two modes and $\tau_c$ will be the correlation time of the kinetic process weighted by a Boltzmann factor containing the energy of the exchanging mode.[12] Under these circumstances measurement of the lineposition and width of a certain oscillator as a function of temperature could give information on the energy of a different mode. In Figure (1) I depict the the linewidth of the CN stretch vibration 2095 $cm^{-1}$ in a ferroelectric crystal as a function of temperature, data obtained with spontaneous and forced Raman scattering. From the slope one infers that this stretch vibration is anharmonically coupled to a mode at 346 $cm^{-1}$.[13]

## 4.2 Relaxation of an Oscillator

We will discuss a model that is more phenomenological than our previous example that existed of a few low-lying quantum-mechanical levels. The new model we introduce for the description of a system coupled to a continuum is the very popular harmonic oscillator. An oscillator is a generic model used in physics from degrees of freedom having energies in the range of milli-electron-volts (nuclear magnetism) up to Giga-electron-volts (high-energy physics). In our case we specifically mean the displacement of an atom around its equilibrium position, but much of what we will discuss has a much wider applicability. A major advantage of the oscillator model is that the classical interpretation hardly differs from a quantum-mechanical one. We will assume that our degree of freedom has some generalized coordinate $q$ (we will use $q$ rather than $u$ to comply with a convention popular in the optical literature). In an experiment one measures a collection of these oscillators and we define the "coherent" amplitude

$$Q \equiv \sum_{i=1}^{N} q_i. \tag{13}$$

A careful reader would notice that in the beginning we addressed wavevector-dependent displacement operators of the type $\mathbf{u}(\mathbf{p})$ and now we are only considering displacement operator (13), which is essentially the $p = 0$ component of $\mathbf{u}(\mathbf{p})$. The reason is that due to the discrepancy between atomic length scales and the wavelength of light only the $p = 0$ displacement operator will be probed with light. This is a tremendous disadvantage compared to other techniques like inelastic neutron scattering when investigating vibrational dynamics in the solid state, but less in liquids where (almost) no dispersion (that is $\mathbf{p}$-dependence) exists.

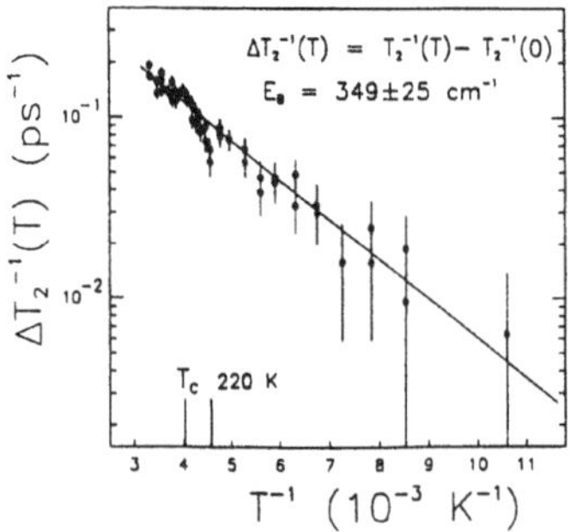


Figure 1. *Logarithmic plot of the temperature dependent contribution to the Raman linewidth of the CN stretching mode in the ferroelectric crystal* $K_4Fe(CN)_6 \cdot 3H_2O$ . *The solid line is a fit to exponential behavior which yields an activation energy of* $349 \pm 25\text{cm}^{-1}$. *The interpretation is that the CN mode at* 2095 $cm^{-1}$ *is anharmonically couple to a water mode having this low energy. After Reference (13).*

I will treat only a single coordinate $Q$ . It is trivial to extend this into regime of more coupled systems, but I do not want to write out matrices all of the time. The equation of motion of the simple damped harmonic oscillators is

$$\frac{d^2Q(t)}{dt^2} + \frac{2}{T_2}\frac{dQ(t)}{dt} + \Omega_0^2 Q(t) = 0. \tag{14}$$

We have neglected a possible distribution ("inhomogeneous linewidth") in the frequencies of the oscillator. The coordinate $Q$ is a generalized coordinate. To emphasize the general character of the present approach I will point out that $Q$ could be the displacement of a lattice vibration (which we have called $u$ earlier), but it could also be the excitation of a bound electron, or it could be the excitation of a magnon, etc. We could even generalize the meaning of $Q$ more. $Q$ could denote some non-equilibrium expectation value introduced by some probe at earlier times and equation (14) would describe the return to equilibrium.[14] $Q$ could also denote some equilibrium displacement-displacement correlation function $\ll Q(0)Q(t) \gg$ .

The damping has been introduced in a phenomenological way. In the previous section we have outlined how to justify the concept of damping times in a microscopic way using one of the possible interpretations of $Q$ when we are dealing with a weakly-coupled system. The coordinate $Q$ describes the phase of the oscillators and that is why the damping time contains the total dephasing time $T_2$. Equation (14) cannot represent population dynamics of the oscillator and so $T_1$ is not present in it. The extension to include the population dynamics will be discussed later. The solution of equation (14) is simple and requires two initial (or final) boundary conditions i.e. $Q(t = 0)$ and $\left(\frac{dQ(t)}{dt}\right)_{t=0}$. In the case the interpretation is in terms of correlation functions these initial conditions are determined by the equilibrium ensemble.

I have outlined that equation (14) is a very general equation. Now I want to stress the limitations of this equation. This equation can never be an operator equation. That is to say equation of motion (14) describes the motion of $Q$, but one cannot infer from the behavior of, for instance, $Q^2$ from it. The reason is that equation of motion (14) can only be obtained from a microscopic equation after some averaging has been performed. This averaging could have been sneaked in our calculation either because quantum-mechanical matrix-elements have been taken, or because a statistical-mechanical averaging has been carried out, or because an averaging over some stochastic process has been invoked. Equation (14) can never be a microscopic equation, as no microscopic origin of irreversible behavior exists for a finite system. A better practice would have been to use $\langle q \rangle$ and $\langle Q \rangle$ in equation of motion (14) rather than $q$ and $Q$, to emphasize that some averaging procedure has been introduced.

A simple example will demonstrate the phenomenological character. The dynamical coordinate of our oscillator will be denoted by $\tilde{q}$ (and their sum by $\tilde{Q}$). This coordinate does not represent an averaged coordinate and would for instance appear in the Hamiltonian. The oscillator is coupled to other degrees of freedom. This makes the whole situation into an interacting many-body system. A simple, and ad hoc way, to introduce the rest of the system is to propose a random force, that modulates the frequency,

$$\frac{d^2\tilde{q}(t)}{dt^2} + +\Omega_0 \left(\Omega_0 + E_{int}\xi(t)\right) \tilde{q}(t) = 0, \tag{15}$$

where $\xi$ is a zero-mean random function with correlation time $\tau_c$. The strength of the random-force has been parametrized with the energy parameter $E_{int}$. It will turn out that this parameter maps exactly on the $E_{int}$ we have been using extensively in the weak-coupling approach when describing relaxation with a microscopic approach using Hamiltonians. The solution of the stochastic equation is straightforward. Essentially the same expansion in $E_{int}\tau_c$ as in the case of weak-coupling can be inserted. (See van Kampen[15] and his excellent book[16]). The result for the coordinate $\langle \tilde{q} \rangle$, where the brackets mean averaging over the stochastic process, to lowest order in an expansion of $E_{int}\tau_c$ is that one obtains exactly the phenomenological equation of motion (15) when we make the identification: $\langle \tilde{q} \rangle \equiv q$ and $\langle \tilde{q} \rangle \equiv Q$, and the most important result $1/T_2 \equiv (\pi/8)E_{int}^2\tau_c$. I hope the reader sees that this stochastic approach is very analogous to the Hamiltonian approach within the weak-coupling scheme. It is clear that the solution of the Langevin equation for $\langle \tilde{q}^2 \rangle$ not simply is given by $\langle \tilde{q} \rangle^2$. A stochastic fluctuation of the frequency is related to pure dephasing. It is also possible to include inelastic channels in this stochastic approach. For instance an anharmonic force proportional to $\tilde{q}^2\xi(t)$ would lead, after the assumption of $\tilde{q}^2 \approx \langle \tilde{q} \rangle \tilde{q}$ to a $T_1 - type$ contribution to the linewidth $T_2$.

I will now explain why I have emphasized so much the fact that the phenomenological equation of motion can only be obtained aver averaging. Later we will couple this oscillator to an electromagnetic field, and if we would interpret the phenomenological equation as a real dynamical equation, the oscillator cannot absorb light but only scatter light elastically. In the latter case the dephasing time $T_2$ must correspond to the radiative lifetime of the level as the radiation field would be only reservoir that could cause dephasing of the oscillator. We must allow for the possibility that our oscillators absorb light (that is convert radiation into non-radiative degrees of freedom), and consequently we have to accept that equation of motion (14) has been obtained after averaging over some bath and it describes the motion of the *averaged* amplitude.

Higher-order correlation function are in general notoriously difficult to calculate. Under many circumstances they can be decoupled into a product of lower-order correlation functions, and the only independent correlation functions contains two operators. In the weak-coupling situation we have found that $\ll Q(0)Q(t) \gg$ decays exponentialy with a decay time of $T_2$. The decoupling of correlation functions is allowed in more general circumstances and certainly in the weak-coupling situation. A decoupling procedure would look like

$$\ll Q(0)Q(t')Q(t'')Q(t''') \gg \approx \ll Q(0)Q(t') \gg \ll Q(t'')Q(t''') \gg + \dots \quad , \tag{16}$$

and we see that in the case of exponential decay of the $\ll Q(t')Q(t'') \gg$ the higher-order correlations also fall off exponentially.

## 4.3 Beyond the Relaxation-Time Approximation

Very often $E_{int}\tau_c$ is very small and this is why in the frequency domain simple Lorentz lines and in the time domain simple decay are so often observed. In a way one should also be a little disappointed when confronted with simple exponential decay in a time-resolved experiment as one does not get to know much of the bath as the correlation time $\tau_c$ only reflects global behavior of the bath. I call this type of findings about the bath second-hand information (or hear-say witness) If the weak-coupling expansion parameter is not small your line shapes will certainly not be like Lorentzians and you have stumbled on an extremely interesting case. The price you will have to pay when the system allows you to obtain first-hand information is that you have to do a much better job than first-order weak-coupling theory to be able to compare your experimental results with theory.

If the weak-coupling approach does not converge very fast (that is when $E_{int}\tau_c$ is not a small number) one can use the so-called memory function formalism. The central object in this approach is an operator which has many names in different branches of physics. To mention a few: mass operator, or self-energy, or memory function, and it is all the same stuff. We will indicate it in our approach for the oscillator. Equation of motion (14) is of the Markov type. The dynamics of $Q(t)$ only depends on $Q$, and some of its derivatives, at the same value of $t$. There is no memory in the system. This means that a much better and more general equation of motion would look like

$$\frac{\mathrm{d}^2Q(t)}{\mathrm{d}t^2} + \int_0^t \Sigma(t-\tau)\left(\frac{\mathrm{d}Q(t')}{\mathrm{d}t'}\right)_{t'=\tau} d\tau + \Omega_0^2 Q(t) = 0, \tag{17}$$

where the memory function $\Sigma(t)$ has been introduced. The simple weak-coupling result can be recovered when the memory function is approximated by a delta function. A finite response time of the bath can be taken into account in this way. Very often it can be proved that the *form* of equation (17) is exact. This means of course that all our problems have been sublimated in this memory function. The advantage of this approach is that the memory function is a very convenient vehicle to make approximations for. One of the simplest simplifications that recently got quite popular in optics, but is already known much longer,[17] is to perform a continued-fraction expansion.

An expansion in the coupling does not make sense at all in case the bath is extremely slow. A very important and popular theory developed to deal with this situation for magnetic resonance has never become popular in optics, for reasons beyond my understanding. I am referring to the Kubo and Tomita theory.[18] This theory is essentially a theory based on cumulants. An important aspect of this idea is that it is very easy to obtain either a Gaussian or a Lorentzian lineshape. To obtain the same results in weak-coupling-type theories or memory-function approaches is very involved and only emerges as the solution of a set of difficult self-consistent equations, often referred to as the mode-coupling approach.[4,19] A disadvantage of the Kubo-and-Tomita theory is that an ultra-slow bath always gives a Gaussian lineshape, whereas the correct result should reflect the distribution of the slow bath fields, not necessarily characterized by a Gaussian stochastic process. Another possible method, and alternative for the weak-coupling expansion, would be to use an asymptotic series expansion much like the WKB method by expanding in the kinetic energy of the bath.[20] In a weak-coupling approach an expansion in the interaction Hamiltonian is performed and this is essentially a Born expansion. The asymptotic approach is only expected to be a sensible strategy for a slow bath.

## 5 PROBE

Once we have decided what kind of information we want to accumulate we will have to figure out how to go about it? Physics is an empirical science, consequently we perform a measurement. That is to say we introduce a probe. Since we know that the conservation of energy is one of the main laws of physics we always write down Hamiltonians or Hamiltonian operators. For our situation the relevant energy operator seems to be

$$H_{world} = H_{system} + H_{probe} \equiv H_{sys} + H_{probe}. \tag{18}$$

This Hamiltonian is not very satisfactory, however, because there is no coupling between system and probe as these terms commute and could treated independently. There must be an additional, *coupling*, term. A coupling term is a part of the Hamiltonian that depends on both the variables of the probe field and the system degrees of freedom. We finally arrive at the Hamiltonian:

$$H_{world} = H_{sys} + H_{probe} + H_{coupling} \equiv H_{sys} + H_{probe} + H_{cou}. \tag{19}$$

You already decided that you were mainly interested in $H_{sys}$, your hobbyhorse with your favorite particles, and your only motivation to introduce a probe is to get to know more about them. This immediately puts some serious restraints on your probe:

- you must know the probe dynamics in great detail.
- the coupling mechanism must be as simple as possible.

Otherwise you do not know how to interpret your signals. With this in mind it seems that it does not make much sense at all to go to nonlinear optical response. Linear response seems already difficult enough. The whole classic book of Born and Wolf[21] is devoted to *part* of linear optics. To treat nonlinear optics to the same depth one would at least need a multiple of these books. When you start to do nonlinear optics you really open a can of worms. Superficially the form of Hamiltonian (19) is very similar to partitioning (5). Nevertheless there is quite a distinction between the two cases, and for that reason I have called the coupling term here $H_{cou}$ rather than $H_{int}$. Here $H_{sys}$ stands for the Hamiltonian of the total system which is an interacting many-body system that very likely can be partitioned into form (5). $H_{probe}$ should be as simple as possible, preferably consisting of one, or only a few, well-defined modes. An experimentalist excites this mode and registers at which rate energy is absorbed by $H_{sys}$ from the excited mode of the probe. In this case one looks for the response of a many-body system to the excitation of a simple mode to which it is coupled. In the narrowing Hamiltonian (5) one looks for the reaction of a simple oscillator or simple two-level system on a coupling with an interacting many-body system, the bath. The similarity between the two situations is that, often against the will or even the awareness of the experimentalist, $H_{probe}$ has many more modes than the one being excited by the experimentalist. An example of this situation is the existence of radiative lifetimes that find their origin in the same coupling to the electromagnetic field that takes care of the absorption of light in a spectroscopic measurement, albeit that the coupling is to different modes of that field. I will come back to this difficulty in some detail later, when problems with respect to response theory will be dealt with.

An efficient way of getting information on a particular phenomenon is to get your probe in resonance with a relevant degree of freedom of your system. Usually a probe has a certain characteristic length (for instance the wavelength of the probe) and a characteristic time (or inverse frequency of the probe). Your degree of freedom probably can also be characterized by similar parameters. Optimal coupling is achieved when you match them: probe and system are in resonance. It is very clumsy in general to try to get information on nuclear magnetism from x-ray absorption. So if for some reason you are committed to do optical experiments your primary length scale is the wavelength of light and your primary frequency scale is the optical frequency. Other important factors are pulse width and resolution of your probe, which quantities are constrained through a Fourier-transform uncertainty product.

## 6 CONTROL OF EXPERIMENT

Any physical system contains at least as many as $10^{23}$ degrees of freedom and usually an infinite number of them. Can you control them all? Can you change them all? No, of course not! How then can you do a reproducible experiment? Well we use some assumptions about the realization of the microsystems. In the first place the presence of thermal equilibrium takes care of the fact that nature goes through all the possible microstates and you are just measuring the average. This deals with a lot of the uncontrolled parameters. However, almost always there is still some uncontrollable quenched disorder. In solids we talk about random fields, random strains etc, in liquids we talk about slow density fluctuations. It always means that there are non-thermal degrees of freedom. To deal with this situation we assume that we know the kind of stochastic law that the field is described by. We either average in our theory over all these realizations or we pick a typical realization depending on what is better in agreement with the way the experiment is being done (not of course which fits the experimental results better). In practice we assume a stochastic law that is much simpler than the real situation. Many effects, like interaction between impurities and dislocations, are neglected. Usually so much averaging goes on in a disordered system, that the details of the structure of the disorder do not matter much. The exception to this rule is nowadays the subject of intensive study of electron dynamics in condensed matter physics, but does not need to bother us here.

## 7 INTERACTION LIGHT AND MATTER

We have decided to do an optical experiment. This introduces some problems with respect to getting the probe in resonance with the scatterer. If you are interested in atoms they have dimensions much less than the wavelength of visible light. So the probe cannot simply be put in resonance as far as the length scales is involved. The only way to getting them into resonance is that if there is an internal degree of freedom that couples to the light. Happily atoms and molecules have internal degrees of freedom and in this way resonance coupling can be achieved by matching the energy of the probe with the energy of the internal degrees of freedom.

The, linear and nonlinear, interaction of light with vibrational degrees of freedom of matter, is always treated within, what I would call, a Kubo-response formalism.[22] Not widely recognized is that this framework is very limited, not at all exact, and not at all as general as (formal) scattering theory.[23] The general shortcomings and break-down of Kubo's response theory is probably only fully appreciated in mesoscopic condensed mater physics, mainly due to the pioneering work of Landauer.[24] A major advantage of response theory is its simplicity, clearly demonstrated by the fact that no precise description of the probe is required in Kubo's response theory. This should already warn us about the limitations of this procedure as we would expect that the precise form of the probe Hamiltonian should matter. To emphasize the limitations of Kubo's response theory I will refer to it as "one-way response theory" in contrast to what I will call, and explain later, "two-ways response theory". In scattering theory the whole Hamiltonian is necessary and the precise forms of all the coupling terms are essential. This demand immediately points out that scattering theory is much more general than Kubo's one-way response theory.

One-way response doctrine is explicitly, or implicitly, used in almost all books on nonlinear optics. To emphasize the limitations of one-way response theory, I will describe in some detailed the Hamiltonian of a radiation field and how formal scattering theory deals with this problem. This presentation is not meant to be an introduction to scattering theory and the equations presented are not meant to be understood in detail, and their consequences will not be dealt with in this paper. I will only give you a flavor of the limitations of one-way response theory.

Let us look at the Hamiltonian describing the probe dynamics, $H_{probe}$, which in this case is given by the radiation Hamiltonian $H_{rad}$,[25,26]

$$H_{rad} = \frac{\epsilon_0}{2}\left(\frac{\partial \mathbf{A}}{\partial t}\right)^2 + \left(\frac{\nabla \times \mathbf{A}}{2\mu_0}\right)^2 - \frac{\epsilon_0}{2}(\nabla\phi)^2, \tag{20}$$

in which $\mathbf{A}$ and $\phi$ are the vector potential and the scalar potential of the electromagnetic field respectively. In equation (20) the coulomb gauge $\nabla \cdot \mathbf{A} = 0$ has been used. All the equations of motion of the free field could be obtained from this Hamiltonian using a Lagrangian formalism.[27]

The interaction with an atom at the origin of nuclear charge $Z$ is through the charge and current densities,[26]

$$\sigma(\mathbf{r}) = -e\sum_j \delta(\mathbf{r} - \mathbf{r}_j) + Ze\delta(\mathbf{r}), \tag{21}$$

$$\mathbf{J}(\mathbf{r}) = -e\sum_j e\dot{\mathbf{r}}_j\delta(\mathbf{r} - \mathbf{r}_j). \tag{22}$$

In optical studies the polarization (density) $\mathbf{P}(\mathbf{r})$ and magnetization (density) $\mathbf{M}(\mathbf{r})$ are useful variables, and they are defined by

$$\sigma(\mathbf{r}) = -\nabla \cdot \mathbf{P}(\mathbf{r}), \tag{23}$$

$$\mathbf{J}(\mathbf{r}) = \dot{\mathbf{P}}(\mathbf{r}) + \nabla \times \mathbf{M}(\mathbf{r}). \tag{24}$$

The above set of equations are not simple and I will discuss later why you will hardly ever find for instance Hamiltonian (20) featuring in a text on optical studies of nature. I have treated the field classically which is a very good starting point. Only for experiments performed with low light intensities and in which higher-order electric-field correlation functions are measured, the need arises to introduce the complication of a quantum-mechanical description of light.[28] As the matter part still has to described quantum-mechanically for vibrational degrees of freedom, one calls this treatment "the semi-classical approach".

A tremendous simplification can be obtained in the coupling Hamiltonian when one realizes that the wavelength of light is much larger than the dimensions of the atom. This calls for an expansion in the ratio of these two length scales. To this end one calculates the potential energy of an atom in an electromagnetic field.[26] We only need the transverse part of the electric field as the longitudinal part is simply be described by electrostatics. This interaction energy is

$$H_{cou} = -\int \mathbf{E}_T(\mathbf{r}) \cdot \mathbf{P}(\mathbf{r})d\mathbf{r}, \tag{25}$$

in which the subscript $T$ refers to the transverse part.

An expansion in atomic length scale over wavelength amounts to an expansion of the polarization in multipoles and only the first term, containing the dipole moment $\mu$ is retained in the "dipole approximation",

$$\mathbf{P}(\mathbf{r}) = -e\sum_j \mathbf{r}_j\delta(\mathbf{r}) \equiv \mu\delta(\mathbf{r}), \tag{26}$$

and $\mathbf{E}_T(\mathbf{r})$ will be expanded around $\mathbf{r} = 0$, the center of the atom. Substituting these expansions into expression (25) gives as first term,

$$H_{dip} = -\mu \cdot \mathbf{E}_T(0). \tag{27}$$

So our simplified Hamiltonian describing the probing of our system with radiation looks like

$$H_{world} = H_{sys} + H_{dip} + H_{rad} \ , \tag{28}$$

where we have summed over all atoms. This result should bring about a big disappointment. We have been able to simplify the coupling to the electromagnetic field considerably without loosing any accuracy, but we are still left with the complicated radiation Hamiltonian featuring in $H_{world}$ and so the whole system is still very complicated. In the one-way response theory, to be discussed in the next section, the radiation Hamiltonian will be absent.

The only way I know to handle Hamiltonian (28) properly is scattering theory. In a scattering formalism one describes the scattering of a wave from a potential. To use this approach for our case we have to identify $H_{cou}$ with this potential. In scattering theory the wavefunction is partitioned in a scattered and an unscattered wave,

$$\mathbf{\Psi} = \mathbf{\Psi}_{incoming} + \mathbf{\Psi}_{scattered} \equiv \mathbf{\Psi}_{inc} + \mathbf{\Psi}_{sca}, \tag{29}$$

In scattering theory the solution is obtained by introducing the transition matrix T,

$$\mathbf{\Psi} = (1 + \mathbf{G}_0\mathbf{T})\mathbf{\Psi}_{inc}, \tag{30}$$

where $\mathbf{G}_0$ is the Green's function of the system without scatterers (that is for $H = H_{sys} + H_{rad}$). So $\mathbf{G}_0$ describes the propagation of the system and the free electromagnetic field. The $T$-matrix takes care of the coupling between them.[23] We will deal in more detail with these concepts in Section (9.1). Obtaining the full $T$-matrix amounts to obtaining an exact solution, so very often approximations are used, and one such an approximation is an expansion in the potential (Born series), another one is the (asymptotic) expansion in the kinetic energy (WKB method). Retaining only the first term in the, not necessarily converging, Born expansion of the $T$-matrix is labeled the first Born approximation.

A simplification very often introduced in scattering theory is to mimic the vector field by a scalar field. The justification for this is beyond the scope of the present lecture, but it is quite clear that all polarization phenomena are thrown away when adopting this simplification. The wave equation for the free field for scalar waves looks like

$$\left(\varepsilon(\mathbf{r})\frac{\partial^2}{\partial t^2} - \nabla\right)\Psi(\mathbf{r},t) = 0\,, \tag{31}$$

in which $\varepsilon = 1/c^2$ is the dielectric constant.

I emphasize again that this is not an article on scattering theory and the scattering equations are only presented here to point out the shortcomings of conventional response theory.

## 8 (NON)LINEAR RESPONSE THEORY

Using the first term of the Born expansion of the $T$-matrix, for both the loss and the gain contribution to a a transition probability provides identical results as obtained by the familiar Fermi's Golden Rule. Fermi's Golden Rule is equivalent to Kubo's one way response theory. In a way one-way response theory is the reformulation of Fermi's Golden Rule in the time domain. If the exact answer is contained in the $T$-matrix, why does, presumably rigorous, one-way linear response theory only gives the first term of the Born expansion of the $T$-matrix? The answer is that standard one-way linear-response and one-way nonlinear response theory are very incomplete. It comes about when Hamiltonians (19) and (28) are treated as

$$H_{world} = H_{sys} + H'_{cou}. \tag{32}$$

$H'_{cou}$ differs in a principal way from $H_{cou}$. In the original Hamiltonian the coupling term contained the field degrees of freedom (for instance the electric field) as dynamic variables, the dynamics of which was controlled by the probe Hamiltonian. In $H'_{cou}$ these dynamic degrees of freedom are replaced by massive, adiabatic fields, the magnitude of which is described and controlled by the experimentalist. For our optical example the coupling Hamiltonian given by the dipole approximation $H_{dip}$ is replaced by $H'_{dip}$,

$$H'_{dip} = -\mu \cdot \mathbf{E}_{external}(0), \tag{33}$$

according to the one-way response dogma. The dynamic electric field $\mathbf{E}_T$ has now turned into an external field that does not seem to care about the response of the system. The probe dynamics has been turned off. This is a serious flaw of this approach.

In one-way optical response approach one seeks next an expansion in the response of the system in powers of the applied field. This scheme is possible because the field has been converted from a dynamic variable into a parameter. Collecting powers of the electric field in the response defines linear response and nonlinear response, expressed in terms of the linear and, notorious, nonlinear susceptibilities $\chi_n$. When adhering to this strategy one encounters the following symbolic expansion for the polarization density,

$$\mathbf{P} = \chi_1 \mathbf{E} + \chi_2 \mathbf{EE} + \chi_2 \mathbf{EEE} + \cdots, \tag{34}$$

to be found in any book on nonlinear optics. The expressions given there for these susceptibilities are invariably the result of one-way response theory and thus incomplete because the response formalism does not allow, amongst other things, for a back reaction of the system to the field.

Which effects are not included in one-way response methodology? I will give some examples. Any system being probed is coupled to many more modes of the probe field than the one being excited by the experimentalist. In case of optical spectroscopy, the propagation from one mode of the probe field to another field mode is exactly what we call light scattering. It follows that all forms of probe field scattering are neglected in one-way response theory. If one would be a purist one could even maintain that the influence of absorption on the propagation of the probe field is not contained in one-way response theory. Changes in the index of refraction and polariton behavior are not described by one-way response theory. Cavity pulling is the effect, well-known to occur in high-$Q$ ($Q$=quality factor) microwave and optical cavities, that the eigenfrequencies of system and (probe) cavity perturb each other. Cavity pulling cannot be described by one-way response theory. The depletion of an incoming field is also not included in the one-way response doctrine. Some of these shortcomings that I mentioned can be treated, and have been treated, in a phenomenological way. One should realize that these phenomenological approaches are essentially patches to one-way response concepts.

One-way linear response ideology does not describe all the linear response to the field, but it only describes the effects that are linear in the coupling constant. A simple additional example should convince the reader. If low-intensity light is multiply elastically scattered in an inhomogeneous medium, like white paper, and then absorbed in the medium, this effect is linear in the incoming field. This manifestly linear absorption is not described by one-way linear response principles as the scattering is not contained in that approach. The failure of the one-way response concept is even worse in the quantum case which I am not going to examine.

Under what conditions can one use the one-way response philosophy? The material should be transparent for all the optical frequencies being used or generated. The quality factor of the probe cavity should be very small and the mass, or energy capacity, of the probe should be very large. In addition one should introduce some ad hoc patches like wave-vector and frequency dependent "local-field" corrections, that will account for effects like the change in index of refraction due to the presence of matter. This approach certainly does not deserve a price on a beauty contest but that is what is being done all the time. Given these conditions it is simple to contemplate situations where the one-way response hypothesis fails dramatically. A correct theory, which includes back reaction to the field, I would refer to as a "two-ways response theory". It seems obvious to me that such a theory will involve the full $T$-matrix. For non-propagating fields, like a DC electric field, two-ways response theory amounts to correctly accounting for the boundary conditions under which the experiment is being performed. Application of one-way response theory to these cases assumes implicitly one particular boundary condition out of the many possible under which the experiment can be performed. If one deals with nonlinear optics and one considers the generation of more than one field the reaction of all these fields to each other should be included, and I would refer to this as "all-ways response theory". An example where implicitly all-ways response theory has been used is the case of four-wave mixing in a disordered medium.[29]

## 9 DIRECTLY DRIVEN OSCILLATOR

Let us make a connection between the coupling Hamiltonian and our vibrational degree of freedom. This degree of freedom has some (generalized) coordinate $q_j$. The $j^{th}$ oscillator is coupled to the electromagnetic field through its dipole moment. We assume the dipole moment to be a function of the coordinate $q_j$, and perform a Taylor expansion,

$$\mu_j = \mu(q_j = 0) + \left(\frac{\partial \mu_j}{\partial q_j}\right)_{q_j=0} q_j + \cdots. \tag{35}$$

When we have $N$ identical oscillators the total dipole moment is represented by

$$\mathbf{p} = \sum_j^N \mu_j. \tag{36}$$

The interaction of the oscillators with the optical field is given by $E_{cou} = -\mathbf{p}\cdot\mathbf{E}$, and by using expansion (35) we find that there is a contribution proportional to $q_j$. As minus the gradient of an interaction represents a force, this term gives rise to a total force on the oscillators,

$$\mathbf{F} = N\mathbf{E} \cdot \left(\frac{\partial \mu}{\partial q}\right)_{q=0} \tag{37}$$

Apparently we can drive an oscillator directly by an optical field only if its dipole moment changes during vibration. Our original equation of motion for the coherent amplitude $Q$ of the vibrator did not contain an external force. Extending the equation of motion with an external driving force in the equation of motion results in

$$\frac{\mathrm{d}^2 Q(t)}{\mathrm{d}t^2} + \frac{2}{T_2}\frac{\mathrm{d}Q(t)}{\mathrm{d}t} + \Omega_0^2 Q(t) = \left(\frac{N}{m}\right)\mathbf{E} \cdot \left(\frac{\partial \mu}{\partial q}\right)_{q=0} \tag{38}$$

The solution of this equation of motion contains two classes of responses. The first type is the transient response and the second is the force-induced reaction. The transient response will have died away after the elapsed time exceeds several damping times and the driven reaction is the only lasting effect. This driven motion has two components, one in phase with the driving force and one 90° out of phase with the driving field. The simplest and most convenient way to represent these results is to introduce a Fourier-Laplace transform of a time-dependent function $g(t)$,

$$\begin{aligned} g(z) &\equiv \int_0^\infty g(t)\mathrm{e}^{izt}, \\ z &= \omega + i\epsilon, \epsilon > 0 \ . \end{aligned} \tag{39}$$

The solution of equation (38) formulated with the help of this transform is

$$Q(z) = \left(\frac{1}{\Omega_0^2 - z^2 - iz2/T_2}\right)\left(\frac{N}{m}\right)(\partial\mu/\partial q)_{q=0} \cdot \mathbf{E}(z) \ , \tag{40}$$

$$= (1/2) \quad (\Omega_0^2 - 1/T_2^2)^{-1/2}\left(\frac{1}{z - z_+} - \frac{1}{z - z_-}\right)\left(\frac{N}{m}\right)(\partial\mu/\partial q)_{q=0} \cdot \mathbf{E}(z), \tag{41}$$

in which

$$\begin{aligned} z_\pm &= [\pm(\Omega_0^2 - 1/T_2^2)^{1/2} - i/T_2], \\ &\approx (\pm\Omega_0 - i/T_2). \end{aligned} \tag{42}$$

We will follow closely Loudon[26] and try to make our results look as much as possible equivalent to case of absorption of visible light by atoms. Using Eqs. (36) and (40) shows that we will have an induced polarization ($\equiv$ dipole moment per volume $V$),

$$\mathbf{P}(z) = (\partial\mu/\partial q)_{q=0} Q(z)/V \equiv \epsilon_0 \chi_1(z)\mathbf{E}(z). \tag{43}$$

The definition for the linear, second-rank tensor, susceptibility implies

$$\chi_1(z) = \left(\frac{N}{mV\epsilon_0}\right)(\partial\mu/\partial q)_{q=0}^2 \left(\frac{1}{\Omega_0^2 - z^2 - iz2/T_2}\right). \tag{44}$$

To compare the magnitude of this susceptibility with other optical transition probabilities one could introduce the oscillator strength

$$\mathrm{f} \equiv f_0(\partial\mu/\partial q)_{q=0}^2/e^2, \tag{45}$$

where $f_0$ is introduced to allow for the fact that not all oscillators have frequency $\Omega_0$ and $f_0$ represents the fraction that does. The oscillator strength can, and should, be given a full quantum-mechanical interpretation. In principle $\mathbf{f}$ is a second-rank tensor. From now on we will drop the vector and tensor notation in this section as different combinations of cartesian components can easily be traced back. We rewrite the expression for the linear susceptibility by using the definition of the oscillator strength and arrive at

$$\chi_1(z) = \left(\frac{fNe^2}{m\epsilon_0 V\Omega_0^2}\right)\left(\frac{\Omega_0^2}{\Omega_0^2 - z^2 - iz2/T_2}\right). \tag{46}$$

$$\equiv S\frac{\Omega_0^2}{\Omega_0^2 - z^2 - iz2/T_2}, \tag{47}$$

where I have defined, following Loudon's treatment[26] the dimensionless parameter $S$, $S \equiv (fNe^2)/(mV\epsilon_0\Omega_0^2)$. A good estimate of the size of $S$ will be derived later, here it suffizes to say that the answer is that $S$ is approximately equal to the difference between the dielectric constant at a frequency much lower and the dielectric constant at a frequency much higher than the eigenfrequency of the oscillators. The, somewhat artificial, introduction of the oscilator strength has the virtue that a number of our following equations apply too many other situations, besides the case of *vibrational* degrees of freedom coupled to an electromagnetic field, as well. Only when actual numbers have to be computed differences will appear and the oscillator strength has to be evaluated for each instance separately.

It is obvious that the susceptibility and the induced displacement have two resonances, situated at $z \approx \pm\Omega_0 + i\epsilon$. Concentrating on the resonance at positive frequency (that is neglecting the counter-rotating term) yields

$$\chi_1(z \approx \Omega_0 + i\epsilon) = \left(\frac{S\pi\Omega_0}{2}\right)\left(\frac{1}{\pi}\frac{1}{\Omega_0 - z - i/T_2}\right), \tag{48}$$

$$\equiv S\frac{\pi\Omega_0}{2}L(z;\Omega_0,T_2), \tag{49}$$

where I have defined the normalized "complex" Lorentzian $L(z;\Omega_0,T_2)$. The linear susceptibility $\chi_1$ can be partitioned in a real and an imaginary part according to

$$\chi_1(z = \Omega + i\epsilon) \equiv \chi_1'(\Omega) + i\chi_1''(\Omega). \tag{50}$$

In the following we will indicate $\Omega + i\epsilon$ by $\Omega^+$. (The use of the infinitesimal complex number $\epsilon$ is just a mathematical trick to write the sine and cosine transform of a time-dependent function in a convenient and compact way in one formula.) If we would use as external drive for our system of harmonic oscillators a harmonic driving force, $F(t) \propto \mathbf{E}(t) = Re(\mathbf{E}_0 e^{i\Omega t}) \rightarrow F(z) \propto i/(z+\Omega)$, the real and imaginary part would give the in-phase and out-phase component, respectively. We notice that the out-phase component $\chi_1''$ is proportional to a normalized Lorentzian lineshape,

$$L''(\Omega;\Omega_0,T_2) \equiv \frac{1}{\pi T_2}\frac{1}{(\Omega_0 - \Omega)^2 + (1/T_2)^2}, \tag{51}$$

and the in-phase component $\chi_1'$ is proportional to, what is sometimes called, a "dispersive Lorentzian"

$$L'(\Omega;\Omega_0,T_2) \equiv \frac{1}{\pi}\frac{(\Omega_0 - \Omega)}{(\Omega_0 - \Omega)^2 + (1/T_2)^2}. \tag{52}$$

Standard arguments based on energy flow, show that the rate at which energy (per volume and cycle-averaged) is absorbed from a time-varying electric field with frequency $\Omega$ is[30]

$$\frac{\partial U}{\partial t} = \ll \mathbf{E}(t) \cdot \frac{\mathrm{d}\mathbf{P}}{\mathrm{d}t} \gg \propto \chi''(\Omega). \tag{53}$$

This outcome is typically a fruit of one-way response theory, and in the next section I will point out that it only holds under limiting conditions.

According to equation (53) the observation of absorption out of a harmonic field as a function of the frequency of this field supplies us, experimentalists, with both $\Omega_0$ and $T_2$. In addition to these parameters we also would like to infer $T_1$ from our investigations. $T_1$ relates to population dynamics which are expected to become significant only at higher light intensities. Up to now there is no nonlinear response to the external field. A damped harmonic oscillator does not have any nonlinear response. Consequently there is also no saturation, which seems rather unphysical. The only way to encompass nonlinear response in a model for a classical oscillator is to introduce anharmonicities, but anharmonicities are notoriously difficult to deal with. Happily if we use a quantum-mechanical picture the situation becomes simpler. In practice energy scales and temperatures are such that in the study of vibrational dynamics it is hardly ever allowed to envisage the dynamics from the standpoint of classical (statistical) mechanics anyway. The simplest quantum-mechanical picture of an anharmonic oscillator is to limit oneself to the two lowest levels of the oscillator. This means we are discussing as a model for our oscillator a two-level system. A two-level system can be considered the embody the "ultimate anharmonic oscillator". The saturation behavior can now be incorporated by explicitly discussing the population dynamics between the two levels. This can be done in the following way. The proportionality factor $N$ in the force on the oscillator will be interpreted as the (dynamic) difference in population between excited and ground state, $\Delta N(t)$. An additional equation for the population dynamics is then easily established. We arrive at[30]

$$\frac{\mathrm{d}^2 Q(t)}{\mathrm{d}t^2} + \frac{2}{T_2}\frac{\mathrm{d}Q(t)}{\mathrm{d}t} + \Omega_0^2 Q(t) = \frac{\Delta N(t)}{m}\mathbf{E} \cdot \left(\frac{\partial \mu}{\partial q}\right)_{q=0}, \tag{54}$$

$$\frac{\mathrm{d}\Delta N(t)}{\mathrm{d}t} + \frac{\Delta N(t) - \Delta N_0}{T_1} = \frac{-2}{\hbar\Omega_0}\mathbf{E}(t) \cdot \frac{\mathrm{d}\mathbf{P}(t)}{\mathrm{d}t}, \tag{55}$$

in which $\Delta N_0$ is the Boltzmann equilibrium population difference. In the rhs. of equation (55) the rate is presented at which energy (in units $\hbar\Omega_0$) is absorbed from the external electric field. Per transition the population difference changes by a factor of two and the negative sign comes from the fact that an increasing field intensity will lead to a decreasing population difference. The steady-state solution to both equations can best be obtained by introducing as ansatz, $Q(t) = Q_0(t)e^{i\Omega_0 t}$, and keeping only the lowest-order time derivatives. In this way the effect of the counter-rotating field has been eliminated. Requesting that $\frac{\mathrm{d}\Delta N(t)}{\mathrm{d}t} = \frac{\mathrm{d}Q_0(t)}{\mathrm{d}t} = 0$ gives as solution of Eqs. (54) and (55)

$$\mathbf{P}(z) \equiv \epsilon_0 \chi(z)\mathbf{E}(z), \tag{56}$$

in which the full nonlinear susceptibility

$$\chi(z \approx \Omega_0^+) = \left(\frac{S\Omega_0}{2}\right)\left(\frac{1}{\Omega_0 - z - i/T_2 + \frac{\Delta^2 T_1/(2T_2)}{\Omega_0 - z + i/T_2}}\right), \tag{57}$$

in which $\Delta$ is the Rabi frequency, $\Delta \equiv \frac{(\partial\mu/\partial q)_{q=0}E_0}{2\hbar}$. It is evident that our coupled system shows nonlinear behavior and saturation for large Rabi frequencies. We notice that from the saturation behavior we can determine both $T_2$ and $T_1$, and these were the relaxation parameters we were after.

Eqs. (54) and (55) with solution (57) are well-known in magnetic resonance, where they are called the Bloch equations. To make contact with the formalism of nonlinear susceptibilities used in optics, we will apply a simple mathematical trick. The Rabi frequency is located in the denominator of solution (57). The Bloch solution can be expanded in a power series of the Rabi frequency. We will rewrite the expression for the susceptibility as

$$\chi(z \approx \Omega_0^+) = \left(\frac{S\Omega_0}{2}\right)\left(\frac{1}{\Omega_0 - z - i/T_2}\right)\left(\frac{1}{1+\frac{\Delta^2 T_1/(2T_2)}{(\Omega_0-z)^2+1/T_2^2)}}\right). \tag{58}$$

The last, dimensionless, factor on the rhs. of equation (58) is of the form $1/(1+\alpha)$, and can be expanded in powers of $\alpha$ when $\alpha$ is small. I hope that the reader recognizes that this indeed generates an odd power series in $E_0$ with the successive terms represent $\chi_1$, $\chi_3$, and higher-order susceptibilities. This procedure makes immediately transparent that for saturation we need *all* the nonlinear susceptibilities, up to infinite order. In addition we come to the somewhat surprising conclusion from our expansion that if the Rabi frequency is very large the series expansion of the polarization density in nonlinear susceptibilities does not even converge.

I have stated that higher-order correlation functions of the operator $Q$ could be decoupled in the weak-coupling scheme. The existence of decoupling within the weak-coupling situation could already be concluded from the solution to the Bloch equations. We have solved the Bloch equations for all powers of the electric field. When we expanded this solution in powers of the electric field we found *explicit* answers for all the nonlinear susceptibilities $\chi_n$. A formalism calculating these susceptibilities from one-way linear response theory demonstrates that the higher-order nonlinear susceptibilities are represented by higher-order correlation functions of the dipole-moment operator, and through Taylor expansion (35), by higher-order correlation functions of the displacement operator $Q$. Apparently we have calculated all these higher-order correlation functions, and this is only possible if somewhere the decoupling approximation has crept inside. The answer is that the Bloch equations (54) already imply a decoupling of the dynamic correlation functions, as first the averaging over the bath has been introduced and then the influence of the electric field. A proper, but much more dificult, way would have been to reverse the order of these two actions.

From the experimental observation of saturation behavior of the susceptibility of the vibrational degrees of freedom the eigenfrequency and the relaxation parameters can be obtained for infrared active oscillators. Such experiments could be realized in the frequency as well as the time domain. Given the fact that high light intensities are easier to come by with pulsed light sources, the experiments are commonly carried out in the time domain. The simplest implementation is a one-color two-pulse infrared experiment, in which the first, intense, infrared *pump* pulse saturates the transition and the increased transmission of a delayed, possibly weak, *probe* pulse monitors the recovery of the population difference. For this purpose one needs a tunable, and powerful, pulsed infrared source. One possibility, and one of the most convenient realizations, is to use parametric down conversion in nonlinear optical crystals of the light beam at the fundamental frequency of a Nd-Yag laser.[31] Handicaps of this technique are the rather small maximum wavelengths of about 4.5 $\mu$ and the rather long pulse widths of about 20 picoseconds.

As explained earlier vibrational dephasing times are under normal conditions (temperatures much larger than $4K$) much shorter than inelastic times. When monitoring a population difference on its characteristic $T_1$ time scale all coherence has been lost and the dynamics of the coherent amplitude $Q(t)$ can be neglected. In such experiments the rate processes to deal with are only the population and depopulation kinetics of the various levels. Many types of experiments have been conceived to study the kinetics of level occupation and it is absolutely impossible to mention even a fraction of them here, leave alone review or explain them. I will make one exception and, as explained in the introduction, work of our own group is involved. Bakker has performed an extensive study of the dynamics C-H stretch vibrations in various liquids using infrared saturation spectroscopy.[32] An interesting feature has been discovered in these studies. In many situations the return of a transferred population to the ground state was found to evolve via an intermediate excited vibrational state of the same molecule. This mechanism is referred to as IVR (Intra-molecular Vibrational Relaxation). Subsequently this excited low-energy mode (or modes) redistributes its energy among the many surrounding liquid molecules. This decay mechanism is referred to as IET (Intermolecular Energy Transfer). The time constants associated with IVR and IET could both often easily be obtained when monitoring the transmission of the probe beam as a function of the time delay because this intermediate vibrational state of the molecule has a transition probability for the probe pulse differing from the transition probability of the molecule in the ground state. This can most easily be interpreted in terms of a shift of the absorption band induced by anharmonicities. Depending on the sign of the change in spectral line position and on the frequency of the infrared light with respect to the center of the absorption band, the probe beam can be pushed more into or more out of resonance. If the mode is driven more into resonance one observes the amusing feature that during the decay of the non-equilibrium population transmission of the probe can temporarily be *less* than the long-time equilibrium situation. See Figure (2).

Many more sophisticated nonlinear optical techniques, besides simple saturation spectroscopy, have been devised in order to extract the same relaxation parameters. I have not treated the complication of inhomogeneous broadening. Some of the nonlinear techniques, like photon-echo spectroscopy, allow one to get rid of the complications of inhomogeneous broadening, and admit a determination of the pure relaxation parameters. The coupling of light to an infrared active oscillator is usually strong. This explains why many nonlinear techniques can be and have been applied with great success. Sometimes the coupling to the light field is so strong that the propagation of the light is seriously influenced, an effect not taken into account yet.

## 9.1 Polaritons

The coupling of light to a collection of oscillators can have a profound effect on the propagation of light. We call this polariton behavior. Polariton action is connected to the fact that infrared active oscillators radiate light back into the system. Such a phenomenon will never be included in a one-way response theory and indeed this back reaction to the field was neglected when deriving the one-way response formula (53). The complication of the presence of polaritons has to be fully appreciated by the experimentalist as an ill-understood propagation behavior would lead to erroneous conclusions about the susceptibility of the oscillators, and would lead to wrong deductions about the relaxation time $T_2$. There is quite some confusion about what a polariton essentially is. Any resonance structure in the index of refraction should be called polariton behavior, in my opinion. Any coupling to an infrared active oscillator will lead to polariton behavior. Sometimes this coupling is described as resulting in a hybridized mode in which a photon is coupled to another quantized field, for instance a phonon, and the new mixed modes are called polaritons. This representation gives the impression that the concept of a polariton finds its origin in quantum mechanics. However, the classical

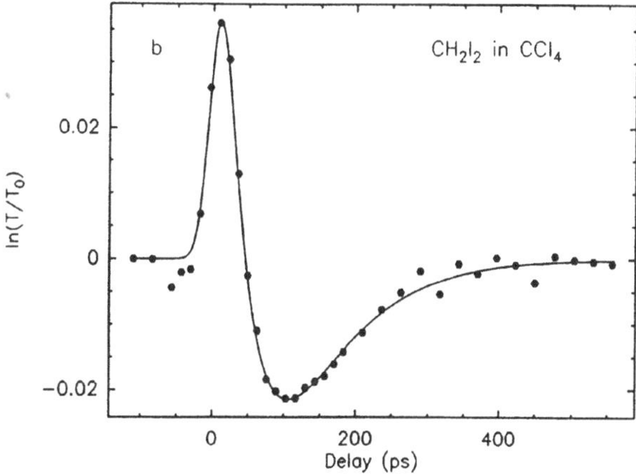


Figure 2. *Relative transmission,* $\ln(T/T_0)$*, of an infrared probe pulse as a function of the delay between probe and pump pulse for the* 3065 *cm*$^{-1}$ *asymmetric* $CH_2$ *stretch vibration of* $CH_2I_2$ *dissolved in* $CCl_4$*. Pump and probe have the same frequency spectrum and are tuned below the maximum of the absorption band. Clearly two kinetic recovery proceses can be identified. After Reference (32).*

Maxwell equations condone polariton behavior. If one is only interested in the transmission of light and not in the propagation of the material excitation (which is often the case when the dipole approximation is valid) one can integrate out the dynamic material degrees of freedom, and their behavior is lumped into parameters like the index of refraction and the mean free path for light. At this level the microscopic origin of these parameters, either of classical or of quantum-mechanical nature, is of no concern any longer and maybe the term polarization wave would be more appropriate. A very intimate connection exists between polariton behavior and scattering theory, that is very important but hardly ever emphasized, so I will do it here. The close relationship with scattering theory is usually overlooked, as one apparently does not realize that, from the view point of the amplitude of light, there is no distinction between elastic scatterers and real absorbers. Only the sum of these two effects, elastic scattering and real absorption, counts, and this combined effect is called extinction.

The key feature of polariton theory is the fact that the presence of infrared active degrees of freedom influence the dielectric constant of the medium. The best method to incorporate this effect is along the lines of scattering theory, but it is not necessary. We could deduce the change in index of refraction directly from our equation-of-motion approach. A major disadvantage is that in this scheme all the implicit assumptions remain hidden, whereas they have to be made manifest in the scattering approach in a very elegant manner. In addition the partitioning of extinction in elastic scattering and real absorption follows most naturally and harmoniously from a full scattering approach. We will review some multiple-scattering theory as all polariton effects can be classified as multiple-scattering phenomena. Nowadays, there is a tremendous revival of in interest in multiple-scattering events and the realization of these analogies will be beneficial to all of us.[33] For simplicity we will treat the electromagnetic waves in the following as scalar waves with amplitude $\Psi$.

An important aspect of wave dynamics and propagation is Huygens' principle. The mathematical expression for this principle is couched in a function called a Green's function, which in free space is essentially a spherical wave. The free (unperturbed by scatterers) Green's function in real space only depends, besides on the frequency of the wave, on the *difference* of two spatial variables,

$$G_0(\mathbf{r}_1, \mathbf{r}_2; \Omega^+) = -\frac{e^{ik_0^+|\mathbf{r}_1-\mathbf{r}_2|}}{4\pi \mid \mathbf{r}_1 - \mathbf{r}_2 \mid}, \tag{59}$$

where $k_0 = \Omega/c$. Free space is homogeneous and momentum is conserved, which can be fully exploited by going to wavevector space,

$$G_0(\mathbf{p}; \Omega^+) = -\frac{1}{p^2 - (\Omega^+/c)^2}. \tag{60}$$

Let us first look at what one scatterer does. As always we partition the wavefunction in a scattering geometry in an incoming and a scattered wave

$$\Psi = \Psi_{inc} + \Psi_{sca}. \tag{61}$$

The simplest possible scatterer is an isotropic point scatterer. The concept of an isotropic point scatterer in a wave equation is not without mathematical difficulties,[34] but we will not be sidetracked here by requesting too much mathematical rigor. Part of an incoming plane wave, with direction given by the unit vector $\hat{\mathbf{k}}$, will be scattered by the $j^{th}$ point scatterer located at $\mathbf{r}_j$ into a spherical wave and the result is that

$$\Psi(\mathbf{r}; \Omega^+) = e^{ik_0^+\hat{\mathbf{k}}\cdot\mathbf{r}} + \frac{f_j(\Omega^+)e^{ik_0|\mathbf{r}-\mathbf{r}_j|}}{\mid \mathbf{r} - \mathbf{r}_j \mid}, \tag{62}$$

in which the, complex-valued, single-particle scattering length or scattering amplitude $f$ has been specified. The cross-section for scattering is given by

$$\sigma_{sca} = 4\pi \mid f(\Omega^+) \mid^2 \tag{63}$$

The calculation of the flux or the intensity of a wave involves the product of two wave amplitudes. We will only do this in a symbolic way

$$\mid \Psi \mid^2 - \mid \Psi_{inc} \mid^2 = \mid \Psi_{sca} \mid^2 + (\Psi_{sca}\Psi^*_{inc} + c.c.). \tag{64}$$

The first term on the rhs. of this equation containing a product of two scattering amplitudes describes the scattered intensity. The left side contains the total outcoming intensity minus all the incoming intensity, and a nonvanishing negative number would indicate that the scatterer has absorbed part of the radiation. Apparently (minus) the second term on the rhs. of equation (64), characterizing the interference between scattered wave and incoming wave, accounts for all the intensity being taken away from the incoming beam. This describes the extinction. If there is no real absorption and all the extinguished intensity reappears as scattering intensity the lhs. of equation (64) vanishes. This puts a constraint on the scattering amplitudes

$$\mathrm{Im} f(\Omega^+) = k_0 \mid f(\Omega^+) \mid^2, \tag{65}$$

which is known as the optical theorem,[35] taking this simple form because of the assumed isotropic character of the scattering.

Formally we would like to describe the scattering process as scattering of the incoming wave from $\mathbf{r}'$ to $\mathbf{r}''$ and then propagation from $\mathbf{r}''$ to $\mathbf{r}$. Integration over all the primed coordinates would generate the scattered light at $\mathbf{r}$. We arrive at this by defining a single-particle transition matrix $t(\mathbf{r}, \mathbf{r}')$ through

$$\Psi(\mathbf{r}; \Omega^+) = \Psi_{inc}(\mathbf{r}; \Omega^+) + \int \cdots \int G_0(\mathbf{r}, \mathbf{r}''; \Omega^+) t_j(\mathbf{r}'', \mathbf{r}'; \Omega^+) \Psi_{inc}(\mathbf{r}') \mathrm{d}\mathbf{r}' \mathrm{d}\mathbf{r}''. \tag{66}$$

It is evident that our original result (62) can be recovered by using

$$t_j(\mathbf{r}', \mathbf{r}; \Omega^+) = -4\pi f_j(\Omega^+) \delta(\mathbf{r} - \mathbf{r}_j) \delta(\mathbf{r}' - \mathbf{r}_j), \tag{67}$$

and by using as the incoming wave the plane wave defined in equation (62). The two delta functions in equation (67) illustrate convincingly the zero-range of the scattering potential. From now on we will use only a matrix notation in real space as the number of convolution integrals would be to large. This matrix notation implies integration over all dummy spatial variables. The rewriting in symbolic, or matrix, notation of equation (66)

$$\mathbf{\Psi} = (1 + \mathbf{G}_0 \mathbf{t}_j) \mathbf{\Psi}_{inc}, \tag{68}$$

has a much simpler look indeed. A similar equation was alread presented in Section (7), compare equation (30), where we were dealing with the interaction between light and matter.

What happens when we have many scattering centers? All waves can be scattered once, twice up to an infinite number of times. All these events will be contained in the many-particle $T$-matrix. In the same symbolic notation we define

$$\mathbf{\Psi} = (1 + \mathbf{G}_0 \mathbf{T}) \mathbf{\Psi}_{inc}. \tag{69}$$

All we have to do is to find the full $T$-matrix, which is unfortunately a horrendous undertaking. In order to make progress some dramatic assumption has to be advanced. All single-scattering events can symbolically be represented by $\sum_j \mathbf{t}_j$, and the double events by $\sum_j \sum_k \mathbf{t}_j \mathbf{G}_0 \mathbf{t}_k$ etc. The total $T$-matrix is given by

$$\mathbf{T} = \sum_j \mathbf{t}_j + \sum_j \sum_k \mathbf{t}_j \mathbf{G}_0 \mathbf{t}_k + \sum_j \sum_k \sum_l \mathbf{t}_j \mathbf{G}_0 \mathbf{t}_k \mathbf{G}_0 \mathbf{t}_l + \cdots \quad . \tag{70}$$

The scattering centers are located at random positions and we have to average over all positions (indicated by $\langle \cdots \rangle$). This was discussed in Section (6). To take the mean over all particle positions becomes very cumbersome because of the existence of strong correlations when in the products a particular particle appears more than once. The simplest possible hypothesis is to request that we retain only those contributions in the averaging of the terms in summation (70) of the total $T$-matrix where *no specific particle occurs more than once.* With this sledgehammer approach we find

$$\langle \mathbf{T} \rangle \approx \langle \sum_j \mathbf{t}_j \rangle + \langle \sum_j t_j \rangle \mathbf{G}_0 \langle \sum_k t_k \rangle + \langle \sum_j t_j \rangle \mathbf{G}_0 \langle \sum_k t_k \rangle \mathbf{G}_0 \langle \sum_l t_l \rangle + \cdots \quad . \tag{71}$$

All recurrent scattering events are ignored, and we have called this assumption the "independent scattering approximation". It seems that with this supposition this series can easily be summed as it has the looks of a geometric series with $\langle \sum_j \mathbf{t}_j \rangle$ as the first term and $\langle \sum_j \mathbf{t}_j \rangle \mathbf{G}_0$ as the multiplicative factor. Now our symbolic notation has tricked us because, if we would restore the spatial coordinates we would still find a number of convolution integrals. Happily, this cluttering can easily be entangled by proceeding to wavevector space, where due to translational symmetry, all convolution integrals will become simple products. The averaged Green's function will then be

$$G(\mathbf{p}; \Omega^+) = \frac{\frac{-1}{p^2 - (\Omega^+/c)^2}}{1 + \frac{(N/V) t(\Omega^+)}{p^2 - (\Omega^+/c)^2}}, \tag{72}$$

$$= \frac{-1}{p^2 - (\Omega/c)^2 + (N/V) t(\Omega^+)}. \tag{73}$$

where the first, awkwardly looking, equation has been presented for educational reasons, because it demonstrates lucidly that we are really dealing with summing a geometric series. The $t$-matrix element in this equation is factually $t(\mathbf{p}, \mathbf{p}; \Omega^+)$ but the momentum variables have been been dropped as for a point scatterer there is no momentum dependence. The optical theorem can now be rephrased in terms of the $t$-matrix,

$$\operatorname{Im} t(\Omega^+) = -\frac{k_0}{4\pi} \left| t(\Omega^+) \right|^2 . \tag{74}$$

If not all the scattering is elastic this relation does not hold. The ratio of these two quantities (rhs. over lhs.) is called the albedo $a$ , and is a number between one and zero. When the albedo equals one the scattering cross-section equals the total cross-section. A vanishing albedo implies a total cross-section equal to the cross-section for absorption and a vanishing cross-section for scattering. To give an example, for white paint the albedo is about 0.9999 for visible light, whereas I will show that for our oscillators it is much smaller for infrared light. Equation (73) shows immediately that one can define an effective, complex valued, dielectric constant $\varepsilon(\Omega^+)$ ,

$$G(\mathbf{p}; \Omega^+) = \frac{-1}{p^2 - \varepsilon(\Omega^+)(\Omega/c)^2}, \tag{75}$$

where

$$\varepsilon(\Omega^+) \equiv 1 - (N/V)(c/\Omega)^2 t(\Omega^+). \tag{76}$$

This equation should be compared with empty-space equation (59), that can also be obtained from equation (75) by setting in equation (76) the density of scatterers, $N/V$, equal to zero. The implications of a complex valued dielectric constant become clear when we go back to real space. To facilitate the process of spatial Fourier transforming we define the effective, complex valued, index of refraction $n$,

$$n^2(\Omega^+) \equiv \varepsilon(\Omega^+) \equiv \varepsilon'(\Omega) + i\varepsilon''(\Omega), \tag{77}$$

$$n(\Omega^+) \equiv \eta(\Omega) + i\kappa(\Omega). \tag{78}$$

To avoid too many formulas that look clumsy and complex the frequency arguments will often be dropped from the real and imaginary parts of the dielectric constant $\epsilon$ and from the index of refraction $n$. Definitions (77) and (78) imply

$$\varepsilon' = \eta^2 - \kappa^2, \tag{79}$$

$$\varepsilon'' = 2\eta\kappa. \tag{80}$$

The propagation of light in real space is represented by the real-space Fourier transform of equation (75)

$$G_0(\mathbf{r}_1, \mathbf{r}_2; \Omega^+) = -\frac{e^{i\eta k_0^+ |\mathbf{r}_1 - \mathbf{r}_2|} e^{-\kappa k_0^+ |\mathbf{r}_1 - \mathbf{r}_2|}}{4\pi \mid \mathbf{r}_1 - \mathbf{r}_2 \mid}. \tag{81}$$

This very sensible outcome represents a damped wave with a renormalized wavevector

$$k \equiv k' + ik'' \equiv k_0(\eta + i\kappa). \tag{82}$$

The imaginary part $\kappa$ of the index of refraction reflects the exponential decay in length and its effect is usually described in terms of a mean free path $\ell_{ext} \equiv 2/(\kappa k_0)$, where we have added the subscript to emphasize that we are monitoring the extinction rather than the scattering.

Let us explicitly find the expressions for $\eta$ and $\kappa$ in terms of $\varepsilon'$ and $\varepsilon''$. Using

$$\eta = \frac{1}{\sqrt{2}}\sqrt{\eta^2 - \kappa^2 + \sqrt{(\eta^2 - \kappa^2)^2 + 4\eta\kappa)}} \ , \tag{83}$$

we find that

$$\eta = \frac{1}{\sqrt{2}}\sqrt{\varepsilon' + \sqrt{\varepsilon'^2 + \varepsilon''^2}} \ , \tag{84}$$

$$\kappa = \varepsilon''/(2\eta). \tag{85}$$

Given the complex index of refraction allows us to define the phase velovity $v_p$ as

$$ck'/k_0 \equiv v_p = c/\eta. \tag{86}$$

For pulse propagation the group velocity is relevant,

$$v_g = \frac{\partial \Omega}{\partial k'} = 1/(\partial k'/\partial \Omega), \tag{87}$$

$$v_g = \frac{1}{\eta + \Omega(\partial \eta/\partial \Omega)}. \tag{88}$$

Infrared active oscillators scatter and possibly absorb light, so each of them can be described by a single-particle $t$-matrix. We have not treated our oscillators within the framework of $t$-matrices, but we have used an approach based on an explicit equation of motion of the vibrators. We should by now have become very curious about what $t$-matrix applies actually to our oscillator. This is a crucial step in making the connection between standard polariton theory and scattering theory. The relationship can be found easily though the use of the well-known link between the dielectric constant and the linear susceptibility,

$$\varepsilon(z) = 1 + \chi_1(z), \tag{89}$$

$$= 1 + S\left(\frac{\Omega_0^2}{\Omega_0^2 - z^2 - iz2/T_2}\right), \tag{90}$$

$$= \quad 1 + S\left(\frac{(\Omega_0^2)(\Omega_0^2 - z^2)}{(\Omega_0^2 - z^2)^2 + z^2 4/T_2^2}\right) + iS\left(\frac{(\Omega_0^2) z 2/T_2}{(\Omega_0^2 - z^2)^2 + z^2 4/T_2^2}\right), \quad (91)$$

$$\approx \quad \varepsilon'_\infty + (\varepsilon'_0 - \varepsilon'_\infty)\left(\frac{\Omega_0^2}{\Omega_0^2 - z^2 - iz2/T_2}\right), \quad (92)$$

where the last equation has been presented to make contact with a convention that is popular in the solid state physics community, and corroborates our earlier identification of $S$ with dielectric constants. From relations (76) and (89) we conclude that the $t$-matrix of an infrared active oscillator

$$t(\Omega^+) = -(V/N)\chi_1(\Omega^+)(\Omega/c)^2, \quad (93)$$

$$= -\left(\frac{fe^2}{m\epsilon_0 c^2}\right)\left(\frac{\Omega^2}{\Omega_0^2 - \Omega^2 - i2\Omega/T_2}\right), \quad (94)$$

$$\equiv -4\pi r_o\left(\frac{\Omega^2}{\Omega_0^2 - \Omega^2 - i2\Omega/T_2}\right), \quad (95)$$

$$t(\Omega^+ \approx \Omega_0^+) = -2\pi r_o\left(\frac{\Omega}{\Omega_0 - \Omega - i/T_2}\right), \quad (96)$$

in which the length $r_o$ has been introduced. Our search path for an expression for the $t$-matrix really has been a round about. I have preferred this approach over the one in which I would have solved the scattering of light of one oscilator directly. Of course our "derivation" of the $t$-matrix does not depend on whether or not relation (94) holds for any density since the identification may be made for arbitrarily small density.

The cross-section for scattering of the oscillator is given by

$$\sigma_{sca} = (1/4\pi)\,|t(\Omega^+)|^2 = 4\pi r_o^2\left(\frac{\Omega^4}{(\Omega_0^2 - \Omega^2)^2 + 4\Omega^2/T_2^2}\right) \quad (97)$$

With these results we can find out what the albedo is of our oscillator.

$$a \equiv \frac{-\frac{k_0}{4\pi}\,|t(\Omega^+)|^2}{\mathrm{Im}t(\Omega^+)} \approx \frac{r_o k_0 \Omega T_2}{2}, \quad (98)$$

in the neighborhood of the resonance. Before we will estimate this number we will ask ourselves when is the albedo equal to one and are we talking about elastic scattering. The answer is of course that in that case the only dephasing comes from the radiation field and $T_2$ will be the radiation time $T_{rad}$,

$$T_{rad} = 2/(r_o k_0 \Omega). \quad (99)$$

Knowing that radiation times in the visible are varying roughly from nano to microseconds, and realizing that the scaling is with $\Omega^3$ (from the Einstein coefficient), we estimate these times in the infrared to vay between micro and milliseconds. Given dephasing times of the order of picoseconds we find that the albedo roughly varying between $10^{-6}$ and $10^{-9}$ . Consequently all scattering by our oscillators is absorption and the elastic scattering can safely be neglected.

The relation between effective dielectric constant and effective dielectric index of refraction has some important complications. This can best be seen when there is no dephasing (infinite $T_2$). Under this (unphysical) condition $\varepsilon'' = 0$ and

$$\eta = \left[1 + S\left(\frac{\Omega_0^2}{\Omega_0^2 - \Omega^2}\right)\right]^{1/2}, \quad \text{if } \varepsilon'' = 0. \quad (100)$$

When we look at a frequency slightly larger than the resonance frequency we will find that we cannot take the square root as the argument will be negative. There is a large counteracting polarization as all the oscillators lag the optical force and they interfere destructively with the incoming field. There will be total reflection described by an evanescent wave. The decay length of this evanescent wave is related to fully imaginary root of equation (101). For very large frequency the argument of the square root will be positive again. There will be a "stopband" that is located at $\Omega_0 < \Omega < \Omega_0(\varepsilon_0'/\varepsilon_\infty')$.

If the damping is not zero, the situation becomes more intricate. We now define the dimensionless "polariton parameter", $\mathcal{P}$,

$$\mathcal{P} \equiv 8S\Omega_0 T_2 \tag{101}$$

The factor of eight is for cosmetic reasons. The smaller $\mathcal{P}$ the lesser the polariton behavior. An expansion in small $\mathcal{P}$ gives.

$$\eta \approx 1 + \frac{S}{4}\frac{\Omega_0(\Omega_0 - \Omega)}{(\Omega_0 - \Omega)^2 + 1/T_2^2}, \tag{102}$$

$$\kappa \approx \frac{S}{4}\frac{\Omega_0/T_2}{(\Omega_0 - \Omega)^2 + 1/T_2^2}, \tag{103}$$

$$1/\ell_{ext} = 2\kappa k_0 \approx \frac{S}{2}\frac{\Omega_0^2/(cT_2)}{(\Omega_0 - \Omega)^2 + 1/T_2^2} \tag{104}$$

The smaller the damping or the higher the density of scatterers the more pronounced the polariton behavior is. Our previous example of the stopband for a dampingless system corresponds to $\mathcal{P} = \infty$. In contrast to the case of $\mathcal{P} = \infty$ we will always have a solution for both $\eta$ and $\kappa$ when $\mathcal{P}$ is finite and no real stopband seems to occur, as there is always some propagation going on, characterized by a real $\eta$. However we should still define the stopband to occur when $\varepsilon' < 0$, implying $\eta$ to become smaller that $\kappa$. In this situation the wave is overdamped (the damping length is shorter than the wavelength divided by $2\pi$), and indeed no substantial propagation is going on. We have defined $\mathcal{P}$ in such a way that $\varepsilon' < 0$ corresponds to $\mathcal{P} > 2$.

When $\mathcal{P}$ is not small the dispersion relation of light has a funny look as it displays an S-curve. Figure (3). Here the frequency has been plotted as a function $\eta$.

I will now show that the expansion parameter $\mathcal{P}$ is equivalent too the much more famous expansion parameter $(k'\ell_{ext})^{-1}$ in multiple scattering. A stopband exists for $\mathcal{P} > 2$, but in terms of $\eta$ and $\kappa$ it occurs when $\eta < \kappa$ or when $\eta/(2\kappa) \equiv k'\ell_{ext} < 0.5$. Indeed $\mathcal{P}$ and $(k'\ell_{ext})^{-1}$ are equivalent and this identification makes an interesting connection with multiple-scattering theory. In localization of waves one says that it occurs when $k'\ell_{sca} < 1$, where $\ell_{sca} \equiv (1/a)\ell_{ext}$ There are several points to be made here. Localization is concerned with transport properties rather than amplitude properties like dielectric constants. Localization only occurs when real absorption is very small, and consequently it is not relevant for the present case of polaritons where all interaction is absorption rather than scattering, and the scattering mean free path is much longer than the extinction mean free path .

How good is our bulldozer approach (71), the independent scattering approximation, to the total $T$-matrix. It can be shown that here again $(k'\ell_{ext})^{-1}$ is the expansion parameter. So for strong polariton behavior we should improve our basic assumption and

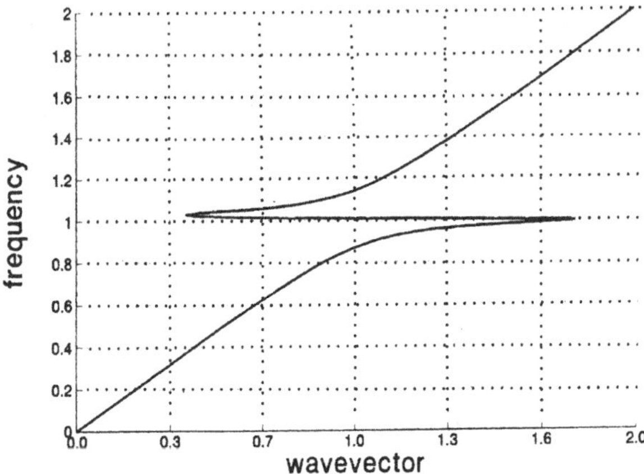


Figure 3. *Dispersion relation for polariton. The following parameters have been used:* $\Omega_0 = 1$, $S = 0.075$, $(1/T_2) = 0.01$. *The wavevectors are given in units of* $\Omega_0/c$.

come up with a much better treatment of the $T$-matrix than our bulldozer ansatz. We call these effects dependent-scattering effects: the scattering properties of one center depend on the presence of other scatterers, an effect which is difficult to deal with. In principle it is not even possible anymore to *define* an effective dielectric constant as the $T$-matrix is going to be momentum dependent. We have performed some dependent-scattering corrections[36] for elastic scattering and large corrections at $k'\ell_{sca} \sim 1$ have been found. We also see the large advantage in relating polariton behavior to multiple scattering. In our equation of motion approach we never realized that there was a serious assumption being made. Where did we make our equivalent sledgehammer assumption in the equation of motion? This happened of course when we assumed the force to be proportional to the number of particles. Here the independent-scattering approximation has sneaked in. It is obvious that improving on our equation of motion is very difficult, whereas the $t$-matrix formalism is ideally suited for this.

If a transition is infrared active either through scattering or absorption one must be careful as the coupling is always strong and the propagation is always perturbed. Much of the misery will occur at the surface as the light propagation is seriously hampered. One way to circumvent this suffering is to try to get $\mathcal{P}$ as small as possible by using a low concentration of infrared active material in an otherwise inactive solvent or host. In that case equation (105) can be used. This still amounts to an exponentially attenuated beam (Lambert-Beer law) $I \propto e^{L/\ell_{ext}}$, $L$ being the length of the sample. The only way to recover the one-way response result (53) is to request in addition that $L << \ell_{ext}$. I hope that the reader by now has appreciated the limited value of one-way response theory.

It is quite clear that when the polariton parameter is large it is quite involved to obtain the relaxation parameters from experiment as the full polariton behavior has to be accounted for.

## 10 RAMAN-ACTIVE OSCILLATORS

Up to now we have assumed that we could indeed excite our oscillators directly with a single time-varying electric field. Our oscillator was infrared active. This has advantages and disadvantages. The drawback is that the transition must be allowed and secondly you must have light at the right frequency. For the majority of oscillators this means light in the infrared region. Fields in the visible are much easier to come by than in the infrared.

Many oscillators are not infrared active, but they are polarizable. That is they can scatter light (Raman and Rayleigh). The theory of spontaneous Raman scattering belongs to quantum scattering theory proper. If the scattering is dealt with in the, for this case of a weak scattering potential excellent, first Born approximation it is found that the scattered intensity $S(\mathbf{p}, E)$ is described by the polarizability-polarizability correlation function

$$S(\mathbf{p}, \Omega)\delta(\mathbf{p}_1 + \mathbf{p}_2) = \int \mathrm{e}^{i\Omega t} \ll \alpha(\mathbf{p}_1, t_1)\alpha(\mathbf{p}_2, t_1 + t) \gg dt\,, \tag{105}$$

The polarizability, which is a second-rank tensor, is assumed to be a function of the coordinate $q_j$ of the $j^{th}$ oscillator,

$$\alpha_j = \alpha(q_j = 0) + \left(\frac{\partial \alpha_j}{\partial q_j}\right)_{q_j=0} q_j + \cdots \quad . \tag{106}$$

From this Taylor expansion the contribution of the oscillators to the polarizability correlation function can be evaluated. Apparently the polarizability-polarizability correlation function can be related to $\langle Q(0)Q(t)\rangle$. If for the oscillators the relaxation-time approximation is valid, equation of motion (14), that is the equation of motion *without* external force, can be used to evaluate this correlation function. One finds two Lorentzian lines, one at the Stokes and one at the Anti-Stokes frequency, their frequency shift given by the eigenfrequency $\Omega_0$ of the oscillator and their width by $1/T_2$. The temperature can be deduced from the Boltzmann factor determining the ratio of the two intensities. This property can be deduced from time-symmetry properties of the polarizability-polarizability correlation function.[5] The relaxation time and frequency can be obtained. I will discuss later the difference between the information obtained from this technique of spontaneous Raman scattering and the socalled "coherent Raman techniques".

In nonlinear Raman experiments the oscillators are *driven* by the Raman effect and that effect can be described classically. In this case the oscillator can be excited by two frequencies. Let us look how this works.

The interaction of this oscillator with an external field $\mathbf{E}$ is given by

$$U = -\frac{1}{2}\mathbf{E} \cdot \alpha \cdot \mathbf{E} \quad . \tag{107}$$

Using expansion (106) in the above equation we find a term proportional to $q_j$. Completely analogous to the case of the infrared active oscillator this gives rise to the following equation of motion for a Raman-active oscillator

$$\frac{\mathrm{d}^2 Q(t)}{\mathrm{d}t^2} + \frac{2}{T_2}\frac{\mathrm{d}Q(t)}{\mathrm{d}t} + \Omega_0^2 Q(t) = \left(\frac{N}{m}\right)\mathbf{E} \cdot \left(\frac{\partial \alpha}{\partial q}\right)_{q=0} \cdot \mathbf{E} \quad . \tag{108}$$

So you now see this oscillator is not driven by one field, but by two fields. This makes many more effects possible. Suppose now that our incoming field has two frequency components: $\Omega_1$ and $\Omega_2$, $\Omega_1 > \Omega_2$. The field looks like

$$\mathbf{E} = Re(\mathbf{A}e^{i\Omega_1 t} + \mathbf{B}e^{i\Omega_2 t}). \tag{109}$$

The bilinear driving field in the equation of motion (109) is now a combination of two optical fields. This field has four possible frequencies, $\pm\Omega_1 \pm \Omega_2$. This indicates that the induced coordinate $q$ has also these four contributions,

$$q(t) \propto Re\left(e^{(\pm\Omega_1 \pm\Omega_2)t}\right). \tag{110}$$

From equation (107) we perceive that this implies that the polarizability has the same time components

$$\alpha(t) \propto Re\left(e^{(\pm\Omega_1 \pm\Omega_2)t}\right). \tag{111}$$

The polarization is the polarizability times the electric field and has therefore the following components

$$\mathbf{P}(t) \propto Re\left(e^{(\pm\Omega_1 \pm\Omega_2 \pm\Omega_{1,2})t}\right), \tag{112}$$

which should be considered a $\chi_3$ effect as the product of three electric fields, albeit not all at the same frequencies, is involved. In experiments with the two frequencies in the visible one always picks the sign combination that causes the polarizability to be forced to oscillate at the frequency $\pm \mid \Omega_1 - \Omega_2 \mid$. This combination allows for a resonant driving of the oscillator when this frequency difference fits the transition frequency. Two final possible sign combinations are very popular for the study of oscillators. One option for the signs implies that the frequency $2\Omega_1 - \Omega_2$ arises in the polarization. This is a polarization at a new frequency, and consequently gives rise to radiation at a new frequency. One could say that the experimentalist supplies the laser field and the Stokes field and the system returns the Anti-Stokes field. For this reason this techniques is called CARS, Coherent Anti Stokes Raman Scattering. This method requires phase matching as the wavevector and the frequency of the polarization are determined by the two incoming fields. This polarization field can generate propagating light only if wavevector and frequency are matched. This can be fulfilled for instance by choosing a particular angle between the two incoming beams. The other useful sign combination, besides the combination relevant for CARS, for the induced polarization density, returns an induced polarization at either $\Omega_1$ or $\Omega_2$. At those frequencies we already have our incoming lasers and consequently they are then amplified or attenuated due to the driven nonlinear polarization. This is called stimulated Raman scattering and in the form when using two laser beams it is called Raman gain (and loss). Raman gain is self-phasematched. All these experiments can both be performed in the time domain and the frequency domain. In the frequency domain one of the two incoming laser beams will have its frequency scanned such that the difference fits the eigenfrequency of the oscillator. In the time-resolved version pump and probe pulses have to be applied. The probe pulse can have two pulsed incoming laser beams and the probe pulse one (CARS) or two (Raman Gain). If the bandwidth of one pulse is large enough to find the necessary combination $\Omega_1 - \Omega_2$ within the frequency content of one pulse, a single pump and a single probe pulse will be enough (impulsive Raman scattering[37]). The time-resolved CARS experiments will be discussed in detail in the lectures by Professor Bron.

What information can be obtained from these forced Raman spectroscopies? In as much as our equation of motion of the oscillator is valid, irrespective of the type of Raman technique or force being used force (like CARS, Raman Gain, Impulsive Scattering or else) we can measure both the frequency $\Omega_0$ and the dephasing time $T_2$. Different techniques measure the *same* parameters in different ways. Analogously to the infrared active oscillators we could define a susceptibility associated with our driving polarization. This will be a $\chi_3$ susceptibility. The fact that this is a fourth-rank tensor implies that its definition and tracing will request a lot of bookkeeping. The various techniques will measure different parts (real part, imaginary part or absolute value squared) of this nonlinear suceptibility, but their result pertains basically to the same information.

Analogously to the situation of infrared active oscillators there is no saturation and $T_1$ cannot be obtained. Of course I could again introduce the equation for the population dynamics and find the higher-order Raman susceptibilities. It is clear that the observation of forced Raman polarization beyond the lowest-order $\chi_3$ would in principle give information on $T_1$. I will not discuss the extension with the population dynamics because the "Raman force" is quite weak and in practice in condensed matter it is very difficult to saturate transitions in this way. One practical reason is that there will always be some other degree of freedom present (possibly due to impurities) that will be excited directly by the light. This usually means a damage threshold that low that no sizeable Rabi frequencies can be induced through the Raman effect.

## 11 NEW INFORMATION FROM NONLINEAR OPTICS

Is there any advantage by going to a nonlinear technique rather than using a linear technique? To answer this problem we will have to make a distinction whether or not the expansion of the polarizability in terms of the normal coordinate is allowed (See for instance Eq. (107). Implicitly an adiabatic approximation is introduced when using this expansion. I will not discuss the situation when this adiabatic approach fails, as this condition is much more relevant for electronic degrees of freedom than for vibrational degrees of freedom. So we assume the expansion in the normal coordinate of the polarizability to be valid. The situation is then quite clear with respect to the difference between spontaneous and driven Raman techniques. CARS and other coherent Raman are techniques where an oscillator is *driven* by a combination of electric fields. In such a technique one measures the balance between *gain* and *loss* and that is why the susceptibility describing this effect is a *commutator* of correlation functions. In spontaneous Raman experiments, which is a quantum-mechanical scattering technique, one measures gain and loss separately: in Raman language they are called Anti-Stokes and Stokes. So in the latter case one measures the two correlation functions corresponding to loss and gain separately. This makes it for instance possible to measure the temperature of one mode by comparing the intensity of the Stokes and Anti-Stokes signal. Consequently there is more information in the spontaneous-Raman experiments. As these experiments are often more easier to perform than the nonlinear Raman experiments it is quite clear what experiments have to be performed first, even if they are not so fancy as the nonlinear Raman experiments.

When linewidths become extremely narrow in the frequency domain it can become quite cumbersome to measure them accurately due to the limitations imposed by the instrumental resolving power. In that case experiments in the time domain become much simpler. Only the driven coherent Raman techniques can be performed in the time domain. This can be an important reason to opt for a nonlinear Raman technique.

As I indicated in the previous section the situation becomes different when the driving Raman fields are so strong that the response can be saturated. In that case the susceptibilities beyond third order have to be taken into account, and much more information can be obtained. There is no anologue of saturation in spontaneous Raman scattering.

## 12 ACKNOWLEDGEMENTS

I am very grateful to Huib Bakker, Marco Brugmans, Ron Kroon, Paul Planken, Rudolf Sprik, Mark Switser, and Bart van Tiggelen for discussions.

The work described in this paper is part of the research program of the "Stichting voor Fundamenteel Onderzoek van de Materie" (Foundation for Fundamental Research on Matter) and was made possible by financial supoport from the "Nederlandse Organisatie voor Wetenschappelijk Onderzoek" (Netherlands Organization for the Advancement of Research)

# References

[1]C.P. Slichter."Principles ofMagnetic Resonance," Harper, New York (1965) 1st edition.

[2]the far wings of a spectral line will never have a lorentzian shape as otherwise the moments would diverge.

[3]S.W. Lovesey in "Dynamics of Solids and Liquids by Neutron Scattering,", edited by S.W. Lovesey and T. Springer, Springer, Berlin (1977).

[4]H. Mori, Progr. Theor. Phys. **33**, 423 (1965).

[5]D. Forster. "Hydrodynamic Fluctuations, Broken Symmetry, and Correlation Functions," Benjamin, Reading (1975).

[6]P.A. Madden and R.M. Lynden-Bell, Chem. Phys. lett. **38** 163 (1976).

[7]A. Nitzan, S. Mukamel, and J. Jortner, J. Chem. Phys. **60**, 3929 (1974).

[8]A. Nitzan and J. Jortner, Mol. Phys. **25**, 713 (1973).

[9]A. Nitzan and R.J. Silbey, J. Chem. Phys. **60**, 4070 (1974).

[10]A. Nitzan, S. Mukamel, and J. Jortner, J. Chem. Phys. **63**, 200 (1975).

[11]A. Carrington and A.D. McLachlan "Introduction to Magnetic Resonance, " Hper, New York (1967).

[12]C.B. Harris, R.M. Shelby, and P.A. Cornelius, Phys. Rev. Lett. **38**, 1415 (1977).

[13]R. Kroon, R. Sprik, and A. Lagendijk, Phys. Rev. B **42**, 2785 (1990).

[14]A. Laubereau and W. Kaiser, Rev. Mod. Phys. **50**, 607 (1978).

[15]N.G. van Kampen, Phys. Rep. 24C, 171 (1976).

[16]N.G. van Kampen, "Stochastic Processes in Physics and Chemistry," North-Holland, Amsterdam (1981).

[17]H. Mori, Progr. Theor. Phys. **34**, 399 (1965); M. Dupuis, Progr. Theor. Phys. **37**, 502 (1967).

[18]R. Kubo and K. Tomita, J. Phys. Soc. Jpn. **9**, 888 (1954).

[19]W. Götze and K.H. Michel, in "Dynamical Properties of Solids," edited by G.H.Horton and A.A. Maradudin, North-Holland, Amsterdam (1974).

[20]an outline of such an approach has been given recently in H. Bakker, submitted to J. Chem. Phys.

[21]M. Born and E. Wolf. "Principles of Optics," Pergamon, Oxford 6th edition, 1987).

[22]R. Kubo, J. Phys. Soc. Jpn., **12**, 570 (1957).

[23]C.J. Joachain,"Quantum Collision Theory," North-Holland (1975).

[24]R. Landauer in "Analogies in Optics and Micro Electronics", edited by W. van Haeringen and D. Lenstra, Kluwer, Dordrecht (1990).p. 243.

[25]S.R. de Groot and L.G. Suttorp, "Foundations of Electrodynamics," North-Holland, Amsterdam (1972).

[26]R. Loudon, *The Quantum Theory of Light* (Clarendon, Oxford, 1973) first edition. (The second edition differs considerably from the first edition.)

[27]W.K.H. Panofsky and M. Phillips, "Classical Electricity and Magnetism," Addison-Wesley, Reading (1962); P.M. Morse and H. Feshbach,"Methods of Theoretical Physics," McGraw-Hill, New York (1953).

[28]"Nonclassical Effects in Quantum Optics,", edited by P. Meystre and D.F. Walls, American Institute of Physics, New York (1991).

[29]V.E. Kravtsov, V.I. Yudson, and V.M. Agranovich, Phys. Rev. **B 41**, 2794 (1990).

[30]A.E. Siegman, "Lasers," Oxford, Oxford (1986) .

[31]A. Laubereau, L. Greiter, and W. Kaiser, Appl. Phys. Lett. **25**, 87 (1974);H. Graener and A. Laubereau, Appl. Phys B **29**, 213 (1982).

[32]H.J. Bakker, P.C.M. Planken, and A. Lagendijk, Nature **347**, 745 (1990); H.J. Bakker, P.C.M. Planken, L. Kuipers, and A. Lagendijk, J. Chem Phys. **94**, 1730 (1991); H.J. Bakker, P.C.M. Planken and A. Lagendijk, J. Chem. Phys. **94**, 6007 (1991).

[33]"Scattering and Localization of Classical Waves in Random Media," edited by Ping Sheng, World Scientific, Singapore, (1990).

[34]Th.M. Nieuwenhuizen, A. Lagendijk and B.A. van Tiggelen. Phys. Lett. A. **169**, 191 (1992).

[35]A. Messiah, "Quantum Mechanics," Vols. I and II, North-Holland, Amsterdam (1961); H.C. van de Hulst, "Light Scattering by Small Particles,", Dover, New York (1981).

[36]B.A. van Tiggelen, A. Lagendijk, A. Tip, and G.F. Reiter, Eur. Phys. Lett. **15**, 535 (1991).

[37]Y. Yan and K.A. Nelson, J. Chem. Phys. **87** 6240 (1987).

# RELAXATION OF FRENKEL-TYPE ROTATIONAL AND VIBRATIONAL EXCITONS IN DIATOMIC MOLECULAR CRYSTALS

Etienne Goovaerts

Department of Physics
University of Antwerp (U.I.A.)
Universiteitsplein 1,
B - 2610 Wilrijk, Belgium

## 1 INTRODUCTION

The study of vibrational and rotational relaxation in condensed phases has acquired a new dimension with the development of ultrashort laser pulses and the corresponding techniques for time-resolved optical measurements. Nowadays, infrared- as well as Raman-active transitions can be interrogated with shorter than picosecond (1 ps = $10^{-12}$ s) time-resolution using, e.g., infrared absorption saturation [1-4] and transient hole-burning [5] spectroscopy and a variety of time-resolved techniques of stimulated Raman scattering [6-10]. They are complementary to methods in which the linewidths and -shapes of the transitions are measured with high spectral resolution, such as in infrared absorption and persistent [11] hole-burning spectroscopy, and in spontaneous and stimulated Raman scattering [8] [12, Chap. 4]. In most optical investigations only the $\boldsymbol{k} \simeq \mathbf{o}$ excitations can be investigated because of the small wavevector of the light and the requirement of wavevector conservation. The goal of these experiments is to acquire a detailed understanding of the relaxation mechanisms of the molecular excitations in solids, relating them to different microscopic effects which are specific for the system under study: anharmonic interactions, impurity scattering, compositional or structural disorder, etc.... Excellent reviews have been published during the past several years, covering the study of the relaxation of internal and external vibrational modes in molecular crystals, and its experimental methods [7, 13-16].

In this paper we will focus on the pure dephasing of a specific class of rotational and vibrational excitations in diatomic molecular crystals, which are model systems

*Ultrashort Processes in Condensed Matter*, Edited by
W.E. Bron, Plenum Press, New York, 1993

for the dynamics of Frenkel-type excitons. Special attention will be given in Sec. 2 to the techniques of time-resolved (TR) Stimulated Raman Gain (SRG) spectroscopy and TR Coherent Anti-Stokes Raman Scattering (CARS), which were employed in our experiments. The interpretation of the results from time-resolved and spectral experiments, and the relation between them, will be briefly recapitulated. Specifically we will discuss two types of excitations about which many new results have been obtained in our group and elsewhere. The *rotons* —coupled quantum rotations of $H_2$ molecules— in parahydrogen (para-$H_2$) crystals (Sec. 6) and the *vibrons* —stretching vibrations coupled by intermolecular interactions— in solid para-$H_2$ and in nitrogen in the $\alpha$ crystalline phase (Sec. 7), have been studied both by picosecond time-resolved and by high-resolution spectral measurements. Comparison will be made with calculated results for both the rotons in para-$H_2$ [17-19] and the vibrons in pure $\alpha$-$N_2$ [20]. Finally, the influence of atomic Ar impurities on the vibron dephasing will be illustrated in Sec. 7 by our recent TR CARS investigations in $Ar_x(N_2)_{1-x}$ crystals.

## Rotational and Vibrational Relaxation in Solids

In the conventional description of dephasing relaxation [12, Chap. 2] of molecular excitations, one considers an ensemble of two-level systems, only the ground state and one excited state per molecule, coupled to a bath of fast variables. One differentiates between the effects of inhomogeneous distribution of the local fields by which the transition frequencies are statistically spread over a frequency width $\Delta$, and the homogeneous dephasing effects described by an exponential dephasing time $T_2$. The latter effects contain both the decay of the excited state population (or diagonal relaxation), with population decay time $T_1$, and the so-called pure dephasing (non-diagonal relaxation), characterized by $T_2'$, leading to $1/T_2 = 1/(2T_1) + 1/T_2'$. In the experiments discussed below, only the total dephasing time is determined. Specific techniques have been proposed to directly measure the the population relaxation time $T_1$, using infrared absorption or spontaneous Raman scattering in the probing process [6].

Starting from a microscopic Hamiltonian, including the harmonic part and a perturbation term representing anharmonic interactions, it is possible to analyse the dephasing relaxation of a quasi-harmonic excitation by interaction with other excitations in the solid [7, 13, 16]. Diagrammatic representations of the relaxation processes are frequently employed. Different terms in the expansion can be related to either population or pure dephasing relaxation, and each of them possesses a characteristic temperature dependence. In order to identify the dephasing mechanism the analysis of this temperature dependence has been applied to a number of internal vibrations in different molecular crystals, including the vibrons and librons in $\alpha$-$N_2$ [21, 22].

Crystals with relatively large molecules possess a large number of vibrational modes of comparable frequency, providing a large number of population decay channels. In this case the decay of the investigated mode is often found to proceed by up- or down-conversion processes involving only a few vibrational quanta, and pure dephasing relaxation cannot be observed. Examples are benzene [23-25], naphthalene [26-28], anthracene [29, 30] and amino acids and peptides [31]. Even relatively small molecule crystals, such as $NH_3$, with a modest number of internal and external modes, fall in this category [32]. Also, in a series of relatively simple crystals, such as $NaNO_3$, $CaCO_3$, $K_2SO_4$, $KClO_4$ and $RbClO_4$ [33-38], the intramolecular modes are strongly coupled with the lattice modes, a mechanism which yields fast and $T_1$ dominated dephasing relaxation. An additional complication occurs in such ionic crystals as a result of the long range electrostatic interactions. An interesting effect is the coupling of a high-frequency vibrational mode with a continuum, the Fermi resonance, such as for the $\nu_1$ vibration in $CO_2$ [39-43] and $CS_2$ [44], and the phonon-polaritons in $NH_4Cl$ [45]. The Fermi resonance, and the resulting relaxation processes, could be turned on and of by variation of hydrostatic pressure [46], and of polariton $\boldsymbol{k}$-vector [45], for $CO_2$ and $NH_4Cl$, respectively.

The influence of crystal defects and impurities on vibrational dephasing was investigated in several molecular crystals by the Hochstrasser group, and reviewed in [14]. Different possible dephasing effects were experimentally demonstrated: Scattering by static defects in $\alpha$-$N_2$ [47], and by ortho-$H_2$ impurities in para-$H_2$ [48] lead to a non-exponential coherence decay of the $A_g$ vibrons in the nanosecond time range (see Sec. 6). The investigated $\boldsymbol{k} = \mathbf{o}$ vibrons are located at a singular point for which the density of states vanishes, which would be at the origin of the observed non-exponential decay. More recent picosecond experiments on the vibrons in $\alpha$-$N_2$ will be discussed in Sec. 6. Trapping of the vibrons at isotopic impurities ($C_6D_6$ or $^{13}C^{12}C_5H_5$) —a population decay process— has been observed for the $\nu_1$ $A_g$ (991 $cm^{-1}$) vibrons in benzene [23] (for the other modes the mechanisms remain to be exactly identified). Both for scattering and trapping at impurities, the efficiency of the process strongly depends on the energy difference between the vibron band and the impurity level. For example, the vibration of ortho-$H_2$ is only 3-6 $cm^{-1}$ below the para-$H_2$ vibron band, which possibly explains the dominance of scattering processes —enhanced by the near-resonance condition— over trapping effects [14].

## Pure Dephasing Relaxation in Molecular Crystals

The nature of the rotons in para-$H_2$ and vibrons in para-$H_2$ and $\alpha$-$N_2$ (and $H_2$) permit the detailed investigation of pure dephasing processes in solids. Indeed, population decay is very inefficient because of the large energy difference with other excitations in the crystal —it requires the emission of a large number of quanta of the lattice modes: about 5, 50 and 30 for the rotons and the vibrons in para-$H_2$ and the vibrons in $\alpha$-$N_2$, respectively— and because of the large degree of decoupling from the other modes. This is experimentally supported by the extremely long decay times measured for the stretching vibration in liquid $N_2$ ( $T_1$ = 56 s [49]) and estimated for the vibrons in $\alpha$-$N_2$ (110 s [50]). For the vibrons in para-$H_2$ a faster relaxation (10 $\mu$s [51]) was measured, but still far beyond the picosecond time range. To our knowledge, no data are available for the rotons. Similar circumstances of relaxation dominated by pure dephasing can possibly be found in simple molecular crystals such as $O_2$, $CH_4$, CO, ..., with high-frequency internal modes, provided that the intermolecular forces, Van der Waals, dipole-dipole and/or repulsive interactions, are sufficiently weak.

Solid hydrogen and nitrogen are built of the simplest possible molecules, held together by Van der Waals forces, with relatively well-known intermolecular potentials [52, 53]. In both cases Raman excitations can be investigated which are well-separated in frequency from other modes and hence only weakly couple to them. The theoretical description is relatively easy in such circumstances, making the excitations under investigation into ideal model systems for the study of rotational and vibrational dephasing in solids. As a result detailed calculations of the relaxation properties are feasible, such as the scattering efficiency of the $J = 2$ rotons in para-$H_2$ by impurities [18, 19], the intrinsic homogeneous roton linewidth in pure para-$H_2$ [17], and the homogeneous linewidths of the librons [22] and very recently of the vibrons [20] in pure $\alpha$-$N_2$.

Impurities can be introduced in both crystals as additional probes for the relaxation mechanisms: ortho-$H_2$ or HD in para-$H_2$, and $^{15}N_2$ or Ar atoms in $N_2$. At low concentrations they do not change the crystal structure and only weakly perturb the intermolecular interactions. Their influence on the relaxation behavior was found to be very different in the following three cases:
(i) For the rotons in para-$H_2$ the wave-like excitations are scattered by the ortho-$H_2$ and HD impurities [19, 54-56], which yields an exponential decay because the Raman-active rotons are located in the middle of the roton band. The residual relaxation rate in pure crystals, although first ascribed to anharmonicity effects within the roton system [17], is probably due to either scattering by phonons into other roton states within the band, or to population decay into several phonons.
(ii) The long-time tail of the dephasing decay of the vibrons in $\alpha$-$N_2$ is dominated by scattering from static disorder in the crystals [14, 47], yielding non-exponential decay curves. From picosecond TR and high-resolution Raman measurements, the low temperature dephasing in pure $\alpha$-$N_2$ is attributed to dephasing by coupling of the vibron to lattice modes, especially the librons [57, 58]. No effect of scattering from or trapping at isotopic impurities is observed. Increasing doping levels of Ar in $\alpha$-$N_2$ drastically change the vibron dynamics in a complex manner, and finally suppresses the collective excitations [59, 60]. This is also the case for increasing temperature, near the phase transition, where vibron-libron interactions are dominant.
(iii) A distinct case is that of the vibrons in para-$H_2$: Scattering by ortho-$H_2$ impurities, with vibrational frequency very close to the vibron band, seems to be dominant in the slow non-exponential dephasing [48], even at the lowest ortho-$H_2$ concentrations [61]. In para-$H_2$, even more than in $\alpha$-$N_2$, the vibrons are decoupled from rotational and translational excitations (rotons and phonons).

There is a close analogy between the roton and vibron states considered here, and the Frenkel-type [62] excitons which are build from localized electronic excitations in molecular crystals. In the latter case, the exciton band is formed as a result of the exchange interactions between different molecular sites, which are much weaker than the exciton energy. The coherent motion and relaxation of excitons has been extensively investigated, including the effects of static disorder and of exciton-phonon coupling at increasing temperature [63, 64]. Specific theoretical methods have been developed to describe the dynamics of these systems [65-68], and it would be fruitful to apply them to the present cases. Indeed, the roton and vibron bands are formed in a similar way from single molecule excitations, rotational and vibrational, respectively, by the action of intermolecular interactions between the corresponding excited states. These resulting energy bands are well separated from the other excitations in the solid, to a better extent than for the electronic excitons. The intermolecular interactions, which give rise to the roton and vibron bands and to coupling with other excitations in the crystals (phonons or librons) can be accurately derived from microscopic models for the hydrogen and nitrogen intermolecular potentials [52, 53, 69-76]. As a result they form appropriate systems to test the results of the theoretical treatments of exciton motion.

The conventional picture of dephasing relaxation [12, Chap. 2] separates inhomogeneous from homogeneous effects, which correspond to non-exponential (often gaussian) decay and exponential decay functionals, respectively, or in the spectral domain, to non-lorentzian (gaussian) and lorentzian lineshapes. This has been very useful for the qualitative and quantitative evaluation of many experimental results. This picture was also applied to the analysis of the roton and vibron decay as described in Secs. 5 to 7. However, it would be worthwhile to reconsider the results in the framework of exciton dynamics. Several characteristic exciton features are nicely illustrated in this work, e.g., spectral line narrowing, asymmetric lineshapes and transition from coherent to incoherent motion under the influence of thermal motion and disorder.

# 2 RAMAN TECHNIQUES FOR FAST RELAXATION MEASUREMENTS

## In the Time Domain

The development of high-intensity lasers with very short pulses has permitted the discovery and investigation of many non-linear optical effects. Since the early seventies, stimulated Raman scattering has been applied to measurements of the time response of vibrational excitations in liquids and solids [6]. The time-resolution of these experiments was closely following the available laser pulsewidths, going from nanoseconds in the early measurements down to about 50 femtoseconds in more recent impulsive excitation experiments. Two picosecond techniques, which were employed in our studies of para-$H_2$ and $N_2$ crystals, will be briefly described and compared: the time-resolved (TR) versions of coherent anti-Stokes Raman scattering (CARS) —probably the most popular stimulated Raman technique— and stimulated Raman gain (SRG), which are based on homodyne and heterodyne signal detection, respectively.

Stimulated Raman scattering is described [6] by the third-order optical susceptibility $\chi^{(3)}$ of the material, which relates the induced polarization $P^{(3)}$ to the applied total electric field $\boldsymbol{E}$,

$$P_i^{(3)}(\omega) = \sum_{jkl} \chi_{ijkl}^{(3)}(-\omega, \omega_1, \omega_2, \omega_3) E_j(\omega_1) E_k(\omega_2) E_l(\omega_3), \tag{1}$$

which is given in terms the Fourier transform quantities. This corresponds to processes with annihilation or emission of four photons. Light is emitted at frequency $\omega$ by the third-order electric polarization. In order for the light emitted by the polarization $P_i^{(3)}(\omega)$ at different points in the interaction volume to interfere constructively, the propagation directions have to fulfill the phase-matching condition $\boldsymbol{k}_1 + \boldsymbol{k}_2 + \boldsymbol{k}_3 - \boldsymbol{k}_4 = \mathbf{o}$. In condensed matter this is a serious restriction because of the dispersion of the light, $\omega/k = n(\omega)c$, in which $n(\omega)$ is the refractive index and $k$ is the size of the wavevector $\boldsymbol{k}$. The spatial coordinates were omitted in (1) as a matter of simplification, which is valid when all light beams are considered to propagate along the same direction. However, in order to derive the phase-matching condition and calculate the absolute signal intensity, their inclusion is indispensable.

Raman resonances occur in the imaginary part of $\chi^{(3)}$ whenever the difference between two frequencies equals that of the Raman transition, $\Omega$. This is exploited in different stimulated Raman techniques through the application of several laser fields with suitable differences in frequency, and the selection of the frequency at which the detection is performed. In CARS as well as in SRG measurements two laser pulses with frequencies $\omega_L$ and $\omega_S$ (often called laser- and Stokes-beam), with $\omega_L - \omega_S = \Omega$, are applied to generate the coherent excitation. The CARS signal induced by a delayed probe pulse of the laser beam at $\omega_L$, is detected at the anti-Stokes frequency $\omega_{AS} = \omega_L + \Omega$. This corresponds to $\chi^{(3)}(-\omega_{AS}, \omega_L, \omega_L, -\omega_S)$, with two photons absorbed at $\omega_L$, one emitted at $\omega_S$ and one at the anti-Stokes frequency, $\omega_{AS}$. For SRG, the decay is probed by a pair of pulses, identical to but delayed with respect to the pump pair. The SRG signal is superimposed on the Stokes beam at $\omega_S$, and it derives from the susceptibility $\chi^{(3)}(-\omega_S, \omega_S, \omega_L, -\omega_L)$, with one photon absorbed and one emitted, both at $\omega_L$ and $\omega_S$. The resonances are schematically depicted in a level scheme in Fig. 1.

Considering a Raman-active transition at frequency $\Omega$, the susceptibility can be split into a resonant and a non-resonant part, $\chi^{(3)} = \chi^{(3)R} + \chi^{(3)NR}$. The latter one describes the instantaneous response of the electrons to the electric fields and contributes in time-resolved measurements only at zero delay. In frequency-domain

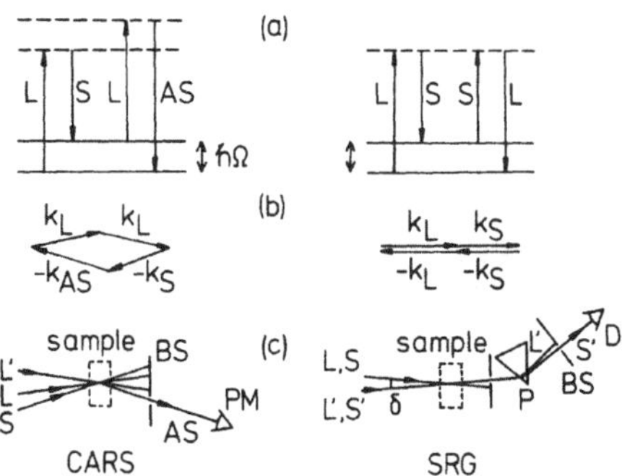


**Figure 1.** (a) Level scheme representations and (b) phase-matching configurations for the CARS and SRG techniques. (c) Excitation and detection geometries in the time-resolved experiments. Laser, Stokes and anti-Stokes beams are labeled L, S, and AS, respectively, and primes indicate the probe beams. BS: beam splitter, P: prism, D: fast photodiode, PM: photomultiplier and $\delta$: small angle in quasi-collinear configuration.

experiments both terms interfere, giving rise to a non-resonant background and complicated lineshapes [12]. For a Raman transition in a transparent material (i.e., $\chi^{(3)}$ exhibits only rotational and vibrational resonances) $\chi^{(3)R}$ is related to the matrix elements of the electronic polarizability, $\boldsymbol{\alpha}$, between the nuclear states, i.e., to the Raman tensor of the transition,

$$\alpha_{ij}^{gf}(\omega) = \frac{2\pi}{h} \sum_e \left( \frac{\langle f|d_i|e\rangle\langle e|d_j|g\rangle}{\omega_{ef} - \omega} + \frac{\langle f|d_j|e\rangle\langle e|d_i|g\rangle}{\omega_{eg} - \omega} \right), \tag{2}$$

in which $g$ and $f$ label the ground and final state of the transition and $e$ the intermediate electronic states, $\boldsymbol{d}$ is the electric dipole operator and $\omega_{ef}$ and $\omega_{eg}$ are virtual electronic transition frequencies. For CARS,

$$\chi_{ijkl}^{(3)R}(-\omega_{AS}, \omega_L, \omega_L, -\omega_S) \propto [\alpha_{ik}(\omega_{AS})\alpha_{jl}^*(\omega_S) + \alpha_{ji}(\omega_{AS})\alpha_{kl}^*(\omega_S)], \tag{3}$$

while for SRG,

$$\chi^{(3)R}_{ijkl}(-\omega_S, \omega_S, \omega_L, -\omega_L) \propto \alpha_{jl}(\omega_L)\alpha^*_{ki}(\omega_S). \tag{4}$$

In this way, the selection rules for stimulated Raman scattering can be derived from those valid for spontaneous Raman scattering.

In the Born-Oppenheimer approximation, which is suitable for transparent materials, the non-linear susceptibility can be strongly simplified by the application of time-dependent perturbation theory, which yields the time-dependent third-order polarization in terms of the total applied electric field [77, p. 16],

$$P_i^{(3)}(t) = \sigma_{ijkl}E_j(t)E_k(t)E_l(t) + E_j(t)\int_{-\infty}^{t} dt' d_{ijkl}(t-t')E_k(t')E_l(t')$$

in which the coefficient $\sigma_{ijkl}$ of the instantaneous term corresponds to $\chi^{(3)NR}$, and $d(t)$ describes the response of the nuclear contribution to $P^{(3)}$. This nuclear response function can be expressed as an equilibrium correlation function, $\langle[\chi_{ij}(t), \chi_{kl}(0)]\rangle$ (evaluated in the absence of radiation) of the linear susceptibility, $\chi_{ij}$. Taking into account a local field correction factor $[(n^2+2)/3]$, the latter is proportional to the electronic polarizability density $\alpha_{ij}$ [12, Chap. 1], and is dependent on the nuclear operators. The results (3) and (4) are easily recovered in this treatment, if only a single normal mode is considered. This is also an interesting application of the fluctuation-dissipation theorem, taking products of electric fields $E_iE_j$ as effective driving fields between the radiation and the matter system, coupled via the susceptibility $\chi_{ij}$. The nuclear response function, $d$, is a good starting point to compare the results obtained from TR CARS, TR SRG and spontaneous (frequency-domain) Raman scattering. For simplicity of the expressions the carthesian indices will be dropped from now on, which does not essentially influence the following discussion.

We will take the nuclear response to be oscillating at a frequency $\Omega$, with a slowly varying envelope function $A$ and phase shift $\phi$,

$$d(t) = A(t)e^{i(\Omega t + \Phi(t))}\theta(t) + c.c.,$$

in which $\theta = 0$ and 1, for $t < 0$ and $t > 0$, respectively. The instantaneous term, which is known to interfere with the measurement of the Raman response function at very short times [9, 10], will be neglected. Assuming (i) that the duration of the laser pulses in the TR techniques is short compared to the time-scale of the relaxation, i.e., $A$ and $\phi$ are quasi-constant during the pulse duration, and (ii) that their frequency difference is centered around that of the material response, $\omega_L - \omega_S = \Omega$, simple expressions are derived for the relevant part of the third order polarization induced by the pump pair (at $t = 0$) and by the probe pulse (at $t = t_D$).

For TR CARS,

$$P^{(3)}(t) \propto E_{L'}E_LE_SA(t)e^{i\Phi(t)}e^{i(\omega_{AS}t+\phi)} + c.c., \tag{5}$$

in which $\phi$ depends on the detailed shapes and phases of the laser pulses, and $E_B$ stands for the modulus of the peak electric field amplitude of the corresponding beam, $B$. Taking advantage of the phase-matching condition, the light emitted by this term in the polarization is detected separately from the laser beams, and one measures the intensity:

$$S_{CARS}(t_D) \propto I_{L'}I_LI_S|A(t_D)|^2, \tag{6}$$

in terms of the intensities $I \propto |E|^2$. The TR CARS measurement only yields the coherent amplitude of the excitation as a function of time!

In TR SRG experiments, the Stokes field emitted by $P^{(3)}$

$$P^{(3)}(t) \propto E_{L'} E_L E_S A(t) e^{i\Phi(t)} e^{i(\omega_L t + \phi')} + c.c., \tag{7}$$

is mixed with the probe Stokes field. For a fully collinear geometry of the four laser beams, the phase-matching conditions are always fulfilled. The changes in intensity of the Stokes probe beam are given by

$$S_{SRG}(t_D) \propto E_{L'} E_{S'} E_L E_S A(t_D) e^{i(\Omega t_D + \Phi(t_D) + \phi')} + c.c. = E_{L'} E_{L'} E_L E_S d(t_D) e^{i\phi'} + c.c., \tag{8}$$

which in principle contains the complete information on the nuclear response function. As previously demonstrated [9, 10, 78], the time-resolution is given by the width of a fourth-order electric-field correlation function and is shorter compared to TR CARS, where one is limited by the width of an intensity correlation function between the three laser pulses. Finally, a point of interest is the linear functionality on $d$ (or $\chi^{(3)}$) of the signal in (8) compared to the quadratic one in the TR CARS expression (6). In the case of exponential dephasing, $d(t_D) = A\exp(-t/T_2)$, the SRG and CARS signals decay with a time constant $T_2/2$ and $T_2$, respectively. Since the susceptibility depends linearly on density, the signals would also decrease linearly and quadratically on concentration, respectively.

For the detection of the relatively small SRG signal in (8), superimposed on the probe beam $S'$, a combination of high-frequency modulation of one of the pump beams ($\simeq 10$ MHz), and low-frequency modulation of another beam, was necessary to eliminate the intensity noise of the mode-locked dye lasers. The TR SRG signal contains the electric fields, not the intensities as in TR CARS, and therefore, the signal is extremely sensitive to the relative phases, $S_{SRG}(t_D) \propto \exp[(\phi_{L'} - \phi_L) - (\phi_{S'} + \phi_S)]$, which means that the whole set-up acts as a large interferometer. In our set-up [79, 55] the noise arising from instabilities of the relative phases of different laser beams was eliminated by a specific phase-modulation method , which results in very stable operation, but completely eliminates the phase information $\phi(t)$ in $d(t)$. This is not the case in an alternative implementation of the technique [10, 78, 80], in which interferometric stability of the components is assured and use is made of double amplitude modulation. In experiments on diamond and $CS_2$, M. van Exter *et al.* [80, 81] clearly demonstrated phase measurements of coherent excitations by TR SRG: The phase is imposed by the driving laser fields during the $(L,S)$ excitation, and then linearly shifts away from it due to the mistuning of $\omega_L - \omega_S$ compared to $\Omega$. Due to experimental difficulties, it is very hard to extract the phase information with sufficient accuracy, and up to now this has not been employed to obtain new results on the relaxation properties of a Raman transition.

## In the Frequency Domain

Measurements of the linewidth and lineshape of infrared- and Raman-active transitions have for many years formed the technique of choice for the spectroscopic investigation of vibrational and rotational relaxation. With grating spectrometers, the resolution is of the order of one wavenumber, and vibrational and rotational relaxation can be measured in the corresponding time range below about 10 ps. High-resolution Raman measurements [8] have frequently been performed in a tandem configuration of a grating spectrometer and a Fabry-Perot interferometer. The latter can achieve a very high resolution, with transmission peaks as narrow as 0.001 cm$^{-1}$. However, these peaks occur in many orders, at the frequencies $\omega_n = n(\pi c/d)$, for which the interference conditions hold when the optical distance between the mirrors is $d$. The latter can be either pressure-scanned (the gas pressure between the mirrors is varied, and thus the refractive index) or piezo-electrically scanned (the actual distance is varied by moving one mirror) in order to record the spectrum. The grating monochromator is employed as a relatively broad-band filter, a few $cm^{-1}$ wide, in order to select only the region of interest in the Raman spectrum. Its contrast must be sufficient to eliminate most of the laser stray light. Recent applications of this technique to rotons in para-$H_2$ (Sec. 5) and vibrons in $\alpha$-$N_2$ (Sec. 6) will be discussed below.

The spontaneous Raman spectrum is connected to the time-response by simple Fourier transformation [77, p. 20]

$$\begin{aligned} I(\omega-\omega_L) &\propto \int_{-\infty}^{\infty} dt e^{i(\omega-\omega_L)t}\langle\delta\chi_{ij}(t)\delta\chi_{kl}(0)\rangle \\ &\propto \frac{1}{1-\exp[\hbar(\omega-\omega_L)/k_BT]}\int_0^{\infty} dt\cos[(\omega-\omega_L)t]d(t), \end{aligned} \tag{8}$$

in which the $\delta\chi$'s are the fluctuations on the linear susceptibility. Given sufficient resolution and accuracy, the Raman spectrum yields the complete information concerning the relaxation of the Raman-active excitation. This technique is technically straightforward, and should be considered first.

Nevertheless, the application of frequency and time-domain techniques are complementary and must both be envisaged. In particular, when the different response times of a given system are spread over many orders it will be hard to investigate this using a single technique, be it in the time- or the frequency domain. In each case limitations exist on the range in which measurements can be performed, and on the available resolution, accuracy and dynamic range. An example of interest here is the TR investigation of the vibrons in $\alpha$-$N_2$, to be discussed in Sec. 6: The nanosecond stimulated Raman measurements yield different results from our recent CARS experiments with picosecond time-resolution. As noted before, the TR techniques often yield only the envelope function of the nuclear response, and this is sufficient information only if the Raman line is symmetric, or, equivalently, if the response function can be recast into $d(t) = A(t)\cos(\Omega t)$ $(t \geq 0)$. For excitons in solids it was previously shown that asymmetric lineshapes were often observed at low temperatures [65], carrying information about the exciton dynamics. It would be of considerable interest to measure the evolution of the phase $\phi(t)$ of the nuclear response function in this type of systems. For high-frequency transitions like the vibrons in $N_2$ this is very difficult, because of the very short period of oscillations of the nuclear response function $d(t_D)$, which implies a very high precision of the delay line which is used to vary the delay time $t_D$.

Non-linear techniques can also take advantage of the available very narrow-band laser systems. In particular, very high resolutions, down to 10 MHz ($3\times10^{-4}$ $cm^{-1}$), have been achieved using continuous wave (cw) stimulated Raman scattering (SRS), in the CARS, stimulated Raman gain or loss spectroscopy, and other techniques [12]. The results of cw SRS measurements on the vibrons in $\alpha$-$N_2$ [82] and para-$H_2$ [61] will be discussed in Secs. 5 and 6. The cw CARS and SRG techniques are mostly performed in a 2-beam arrangement, and correspond to the same components of the non-linear susceptibility for the corresponding TR technique, as given in (3) and (4), and the same level diagrams as shown in Fig. 1. However, the non-resonant contribution, which in TR CARS is only present when the probe pulse overlaps in time with the excitation pair, interferes with the resonant part, and this eventually complicates the analysis of the lineshapes. Several techniques of cw SRS were devised to avoid this interference effect [12], and 2-beam SRG is one of them. While this technique is insensitive to depolarization occurring in optical elements or in the sample, and automatically provides phase matching, it requires high stability of the lasers, and more sophisticated detection techniques [12].

# 3 PROPERTIES OF SOLID HYDROGEN AND NITROGEN

## From Free Molecules to Molecular Crystals

The rovibrational states of free diatomic molecules are known in great detail [83]. For our purpose, we can initially neglect the small coupling between the rotation and internal (stretching) vibration, and treat them separately. The optical spectrum of the free molecule then consists of a series of rotational transitions at the frequencies $\Omega_r^J = J(J+1)B/\hbar$, in which $J$ is the rotational quantum number, and $B = 1/(2\mathcal{I})$ is the rotational constant, with the inertial moment $\mathcal{I} = \frac{1}{2}MR^2$ expressed as a function of the interatomic distance $R$ and and the nuclear mass M. In the harmonic approximation only one vibrational transition is found at $\omega_{stretch}$ in infrared absorption (heteronuclear molecules) or Raman scattering. The vibrational frequencies and the rotational constants for hydrogen and for nitrogen are given in Table 1. Both are larger for the former molecule, as a result mainly of the much lower mass of the proton compared to the nitrogen atom. In particular, the rotational constants differ by a factor about 30, which has important consequences for the different excitation spectra in the solid, as discussed below. For both hydrogen and nitrogen the vibrational frequency $\Omega_v$ is much higher than that of the lowest rotational transitions as can be derived from Table 1.

Homonuclear diatomic molecules are symmetric under permutation of the two nuclei, and depending on the nature of these nuclei, bosons or fermions, the total nuclear wavefunction of the molecule has to be symmetric or anti-symmetric, respectively, under this operation. As a result, hydrogen exists in two species, para-$H_2$ and ortho-$H_2$ [52]. Para-$H_2$ is a species with $I = 0$, where $I$ is the total spin angular momentum of the two protons (fermions), and can only exist in the states with even quantum number, $J$, of the molecular rotation, because of the constraint of

**Table 1**. Selected properties of the para-$H_2$ and $^{14}N_2$ free molecules. $R$, the internuclear distance; $l$, the main axis and $\xi$, the excentricity of the ellipsoid given by the $|\psi|^2 = 0.002$ surface of the electronic wavefunction; $\Omega_v$, the fundamental vibrational frequency; $B$, the rotational constant; $Q$, the electric quadrupole moment.

| | $R$ (nm) | $l$ (nm) | $\xi$ | $\Omega_v$ (cm$^{-1}$) | $B$ (cm$^{-1}$) | $Q$ (a.u.) |
|---|---|---|---|---|---|---|
| para-$H_2$ | 0.0741 | 0.42 | 0.91 | 4159.4 | 60.8 | 1.05 |
| $^{14}N_2$ | 0.1098 | 0.43 | 0.78 | 2330.5 | 2.010 | 0.486 |

anti-symmetry under permutation of the protons (see Fig. 2). For orthohydrogen the spin state is $I = 1$, and this species can exist only in odd numbered rotational states. Spin-flips of the protons are forbidden to a very high degree if no paramagnetic impurities are present, such that para-$H_2$ and ortho-$H_2$ behave as separate species with a very long lifetime, up to weeks, even in the condensed state. As a result the rotational states of para- and orthohydrogen are totally different, with important consequences for the optical spectra, and also for their thermodynamical properties. Similar species exist for $D_2$, where the nuclei are bosons and the total nuclear wavefunction under permutation must be symmetric. For the heteronuclear case, in the absence of permutation symmetry, all the rotational states are permitted, as is shown for the HD molecule in Fig. 2. For the free nitrogen molecule the same considerations of permutation symmetry are valid. However, the nuclear spin-states have a much shorter lifetime, and do not give rise to two quasi-separate species. The states of the free rotator, which are nearly unperturbed for $H_2$ in the solid state, are completely altered by interactions between the $N_2$ molecules in solid nitrogen. Because of the small rotational constant (compared to other interactions in the solid) the nitrogen molecular orientation is restricted to anharmonic oscillations around its equilibrium orientation.

The relative simplicity of the molecules involved, and the precision by which the intermolecular potentials are known from experimental investigations and from ab-initio and empirical calculations, would suggest that the properties of the molecular solids must be simple and easy to explain. In reality, the hydrogen and nitrogen crystals have complex structural and dynamical properties [52, 53, 69, 74-76], which in a number of cases remain to be fully understood. The solid hydrogens, together with solid $^3$He and $^4$He, are called quantum crystals, because the translational lattice modes deviate strongly from the classical harmonic oscillator description: The small mass and small curvature of the isotropic intermolecular potential result in very large zero-point motions, with an r.m.s. displacement of 18% of the nearest neighbor (n.n.) distance for $H_2$. This should be kept in mind whenever the term phonon is employed for the lattice modes in solid $H_2$. The parameters of the intermolecular interactions have to be renormalized for the zero-point motion by proper quantum-crystal techniques. Also in solid $\alpha$-$N_2$, quantum effects have to be taken into account, especially for the librational modes [71, 75] which have a zero-point motion with an r.m.s. angular displacement of about 12°. Solid hydrogen is one of the very few crystals in which the rotation of the molecules is nearly free, and is accurately

**Table 2.** Selected properties of para-$H_2$ and $\alpha$-$N_2$ crystals including the point and space groups the number of sublattices ($N_{sub}$), the nearest-neighbor intermolecular distance ($a$, in nm), the maximum frequency of the lattice modes ($\Omega_{max}$), the estimated widths of the roton and vibron bands ($\Delta_r$ and $\Delta_v$), the $J = 0$ to $J = 2$ transition frequency in para-$H_2$ ($\Omega_r = 6B$), the frequency of fundamental transition of the intramolecular vibration ($\Omega_v$). Frequencies and linewidths are given in cm$^{-1}$.

| | Structure | $N_{sub}$ | $a$ | $\Omega_{max}$ | $\Delta_r$ | $\Delta_v$ | $\Omega_r$ | $\Omega_v$ |
|---|---|---|---|---|---|---|---|---|
| para-$H_2$ | hcp | 2 | 0.379 | 82 | 23 | 6 | 355 | 4150 |
| $\alpha$-$N_2$ | Pa3–fcc | 4 | 0.404 | 70 | — | 1.5 | — | 2328 |

described by the quantum states of the free rotator (others are $CH_4$ and compounds with nearly free rotating $CH_3$ groups).

The zero-pressure phases of the pure ortho and para species of $H_2$ (see Fig. 3a) and $D_2$ differ very strongly [69]: The even-$J$ species, para-$H_2$ and ortho-$D_2$, are near-to-spherical in the ground state ($J = 0$) and, at low pressure and up to the melting point ($T_m = 13.8$ K for para-$H_2$), they order in an hexagonal close-packed (hcp) crystalline structure. Under increasing hydrostatic pressure the higher rotational states, mainly $J = 2$, are admixed in the ground state, until the symmetry is broken and a phase transition occurs to the so-called broken-symmetry phase, a $Pa_3$ structure with oriented molecules on an fcc lattice. The odd-$J$ hydrogens, ortho-$H_2$ and para-$D_2$, crystallize in a orientationally disordered hcp phase with time-averaged spherical

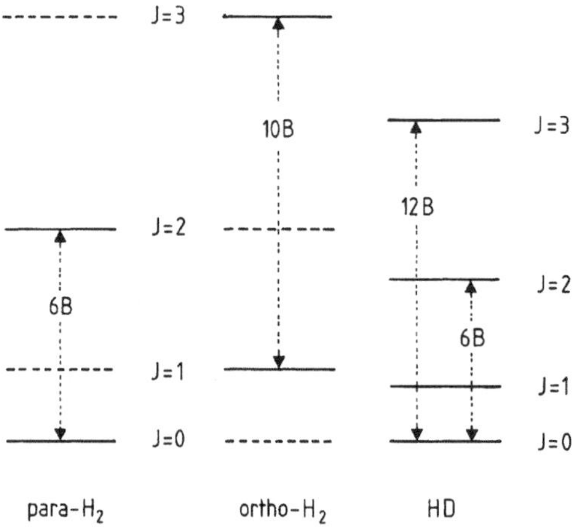


**Figure 2.** Energy level diagram of the rotational states of the hydrogen molecules, $H_2$ and HD.

symmetry of the molecules, but freeze into a orientationally ordered $Pa_3$ structure around 2.8 and 3.8 K, respectively. Furthermore, a quadrupolar glass phase has been reported [69] in ortho-para $H_2$ mixtures at intermediate concentrations (Fig. 3a). High-pressure experiments conducted in the pursuit of a metallic hydrogen phase, possibly metastable and superconducting, have lead to the discovery of a series of further phases with interesting properties [84].

The zero-pressure phases of solid nitrogen [53] bear qualitatively much resemblance to those of the anisotropic $J = 1$ hydrogens. Liquid nitrogen crystallizes at 63 K in an orientationally disordered hexagonal (hcp) phase, called the $\beta$-phase. The details of the molecular reorientation around the hexagonal axis are not fully clarified: either a nearly free classical rotation at energies above the potential barriers, or a jumping motion between three equilibrium orientations. The results of recent *ab initio* calculations [75] provide support for the second alternative. Cooling below 35.6 K induces a structural and orientational phase transition to the cubic $\alpha$-phase which possesses $Pa_3$ symmetry [53]. The molecules are located on an fcc lattice and oriented along the different $\langle 111 \rangle$ directions on four sublattices. It is interesting to note that the orientation of the $J = 1$ hydrogens in the ordered phase is related to the symmetry axis of the rotational ground state (the preferential axis of quantum rotation), in contrast to nitrogen for which it corresponds to the molecular axis, joining the two nuclei. Hydrostatic pressure forces the crystal in an ordered hexagonal phase, $\gamma$-$N_2$ above 3.5 kbars [53].

Solid argon and nitrogen have similar n.n. distances ($a_{N_2} = 0.404$ nm and $a_{Ar} = 0.377$ nm), and mixed crystals can be grown in the entire concentration range, giving rise to different crystalline phases. The low-pressure phase diagram [85] is shown in Fig. 3b. At increasing Ar concentration, up to about 20 %, the $\alpha - \beta$ transition temperature is slowly decreasing. As discussed in detail in Sec. 7, we have introduced

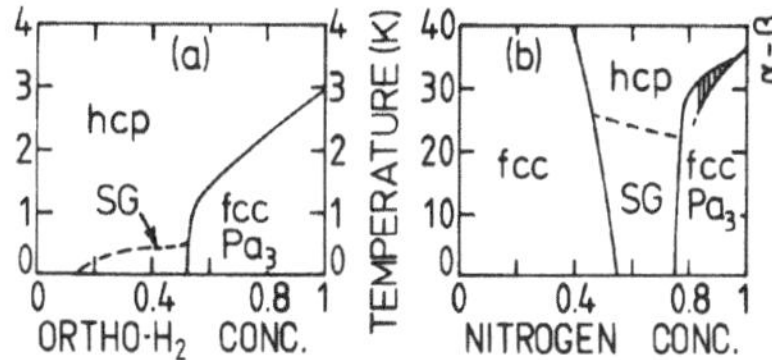


**Figure 3.** The phase diagrams of (a) para-$H_2$:ortho-$H_2$, and (b) $N_2$:Ar mixed crystals at low pressure, for varying compositions and temperatures. Shading indicates the recently discovered orientationally disordered fcc phase in $N_2$:Ar crystals. Also, spin-glass (SG) regions are pointed out.

as an additional perturbation of the vibron dynamics, low concentrations of Ar atoms in solid nitrogen, which do not alter the crystalline structure. At higher Ar concentrations, the $\alpha$-phase disappears in favor of an interesting quadrupolar glass phase [85, 86] in the 25-40 % range, similar to the case of solid para-$H_2$:ortho-$H_2$ mixtures [69].

It is a quite general feature that the orientational ordering transition of diatomic molecular crystals occurs simultaneously with a structural transition. However, a new phase has recently been discovered [87] in $Ar_x(N_2)_{1-x}$ mixtures at low Ar concentrations $x < 20$ % (see Fig. 3b). It is intermediate between the $\alpha$ and $\beta$-phase, and combines dynamical orientational disorder with a cubic (fcc) crystalline structure. Spectroscopic evidence of this phase has recently been provided by temperature dependent measurements of the librational and vibrational Raman spectra [88]. This offers the opportunity to investigate separately the effects of the order-disorder and of the structural transition on the existence and properties of the vibrons in these crystals.

## Comparison of the Excitation Spectra in the Solids

The lowest rotational transition of para-$H_2$ is well above the cut-off frequency of the lattice modes in the crystal (see Table 2), which derive essentially from the isotropic intermolecular potential [52, 69]. As a result, the translational and rotational modes in the crystal are only weakly coupled. The dominant anisotropic interactions which couple the rotations of neighboring $H_2$ molecules are the electrostatic interactions between their quadrupole moments, the electric quadrupole-quadrupole (EQQ) interactions [52]. The translational zero-point motion yields a decrease of the strength of the EQQ interactions by 10 to 15 %, compared to a pair of free molecules. With respect of the rotational energies (of the order $\Omega_r = 6B \simeq 355\text{cm}^{-1}$) the EQQ interactions are a small perturbation, as can be judged from the bandwidth $\Delta_r \simeq 23\text{cm}^{-1}$ of the $J = 2$ excitations induced in the solid (Table 2). The EQQ interactions couple excitations on different sites into rotons, which are Frenkel-type excitons with the $J = 2$ states as localized excitations [52, Ch. 4]. Let us consider only states $|2M, j\gamma\rangle$ with a single molecule excited at site $(j, \gamma)$, in which $M$ is the magnetic quantum number and $\gamma = \alpha$ or $\beta$ indicates the sublattices of the hcp crystal possessing $N$ unit cells positioned at $\boldsymbol{R}_j$. Diagonalization for each $\boldsymbol{k}$ value of the EQQ interactions evaluated in the basis of Bloch states

$$\phi_{M\gamma}(\boldsymbol{k}) = \frac{1}{\sqrt{N}} \sum_j |2M, j\gamma\rangle e^{i\boldsymbol{k}\cdot\boldsymbol{R}_j}, \tag{9}$$

yields the eigenstates

$$\Phi_a(\boldsymbol{k}) = \sum_{M=-2}^{2} \sum_{\gamma=\alpha,\beta} A_{M\gamma}^{\boldsymbol{k}a} \phi_{M\gamma}(\boldsymbol{k}), \tag{10}$$

labeled by $a = 1, \ldots, 10$ (two molecules per unit cell with each five $J = 2$ states). The roton dispersion relation and corresponding density of states [89, 19] derived from this model are shown in Fig. 4. Because of the relatively long range of the EQQ interactions, a large number of neighbors (e.g. 92 molecules in the 10 neighboring shells) must be included to obtain an appropriate accuracy. At the center of the BZ the eigenstates in (10) reduce to three even, $\Phi_{M+}(\mathbf{o})$, and three odd, $\Phi_{M-}(\mathbf{o})$, combinations of the Bloch states in (9), $\phi_{M\gamma}(\mathbf{o})$. All of the even rotons are Raman active and give rise to three equidistant transitions (Fig. 5a), split by 2 cm$^{-1}$, and only one of the odd rotons is IR active, which was experimental verified [90-92]. The $J = 2$ roton bandwidth is $\Delta_r \simeq 25$ cm$^{-1}$ in para-$H_2$.

The validity of this description has been further verified by measurements of the width and shape of the Raman spectra of the $J = 2$ two-roton excitations in para-$H_2$ and ortho-$H_2$ [93]. Also, the Raman transitions from the thermally populated $J = 2$ rotons to the narrow $J = 4$ level were measured in ortho-$D_2$ under hydrostatic pressure [94]. This spectrum was identical with the calculated density of states, within the spectral resolution of the experiment. Support for the above model also comes from the study of the pressure-induced hcp-$Pa_3$ phase transition [95, 96].

In solid nitrogen, the rotational modes of the molecules possess very different properties. Due to the much smaller rotational constant (see Tables 1 and 2), compared to the phonon energies and the anisotropic interactions, the rotations are strongly hindered [53, 71]. In $\alpha$-$N_2$ at low temperature, the molecules are ordered along alternating $\langle 111 \rangle$ directions. They perform wide angle librations around these average orientations in a strongly anharmonic potential, and are strongly coupled with the translational modes. The librations are coupled from site to site forming the so-called *librons*. The Raman-active $E_g$ (32.1 cm$^{-1}$) and $T_g$ (36.4 cm$^{-1}$) librons

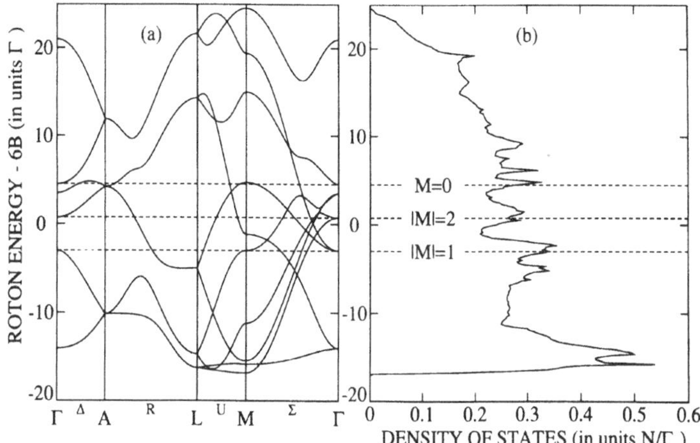


**Figure 4.** The dispersion relations of the $J = 2$ rotons in hcp para-$H_2$ (and ortho-$D_2$) crystals are shown along special directions in the Brillouin zone (BZ), together with the total density of roton states. The energy difference with the free rotator energy, $6B$, is given in units of the EQQ interaction parameter, which in solid para-$H_2$ has an effective value of $\Gamma = 0.575$ $cm^{-1}$ (see Sec. 5). Dashed lines indicate the energies of the three Raman-active rotons, labeled by $|M|$. (From Ref. [19].)

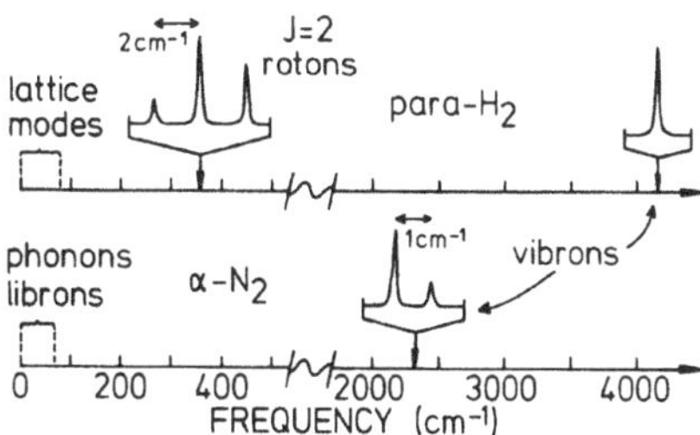


**Figure 5.** Schematic representation of the Raman spectra (a) of solid para-$H_2$ and (b) of solid $\alpha$-$N_2$. Raman linewidths (FWHM) of the $J = 2$ rotons and the vibron in para-$H_2$, and of the vibrons in $\alpha$-$N_2$ are 0.1 $cm^{-1}$, $2.3 \times 10^{-4}$ $cm^{-1}$ and 0.01 $cm^{-1}$, respectively, in the purest crystals at low temperature.

at $\boldsymbol{k} = \mathbf{o}$ [71] are strongly broadened with increasing temperature, and the spectrum abruptly transforms in that of an overdamped oscillator at the phase transition to $\beta$-$N_2$ [60, 97, 98]. Librons decay relatively fast through down- and upconversion processes involving creation and annihilation of other librons and phonons [21, 22, 75]. They are also contributing to the gradual suppression of the vibron states with increasing temperature or argon concentration in $\alpha$-$N_2$ [21, 57-59], as will be further discussed in detail (Secs. 6 and 7).

At very high frequency compared to the translational and rotational modes, we find the stretching vibrations of the molecules, $\omega_v = 4160$ cm$^{-1}$ and 2330 cm$^{-1}$ for free para-$H_2$ and $N_2$, respectively [83]. For the $J = 0$ hydrogens the Raman line is slightly frequency shifted in the solid phase [52, 61, 99] to 4149.8 cm$^{-1}$ in para-$H_2$ (3559.3 cm$^{-1}$ in ortho-$D_2$). In $\alpha$-$N_2$ it is shifted [53] to about 2328 cm$^{-1}$ and split in

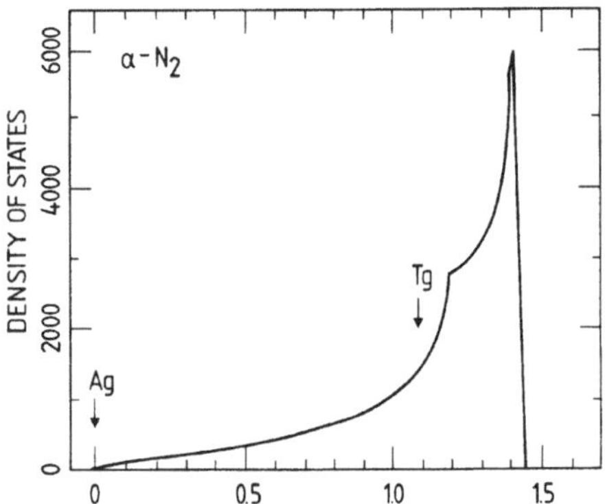


**Figure 6.** Density of states of the vibrons in $\alpha$-$N_2$. The energy scale is relative to the energy of the $A_g$ vibron, $\Omega_A = 2327$ cm$^{-1}$ (from Ref. [100]).

a pair separated by about 1 cm$^{-1}$. The vibron states can be build from vibrational Bloch states, as for the rotons in (9), considering only one state per molecule (the first excited one of the stretching vibration, $\nu = 1$), and two or four sublattices, for the hcp and fcc ($Pa_3$) structures, respectively [53].

The coupling between the excited vibrational states derives from the terms in the pair potential which depend on both intramolecular coordinates. They consist of contributions of dispersive (induced dipole — induced dipole), electrostatic (quadrupole — quadrupole) and repulsive forces [52, 72, 73]. The interaction parameters should not be evaluated simply at the equilibrium position of the centers of mass, or for a specific orientation of the molecules. An averaging procedure of the potential has to be performed over the translational, as well as over the orientational motion of the molecules. At low temperature this reduces to the zero-point motion. The rotational averaging is very different for the two solids: For para-$H_2$ the quantum ground state possesses spherical symmetry and an even integration over the orientation of the molecules is performed [52, Chap. 2]. In $\alpha$-$N_2$, the molecular axis librates around its $\langle 111 \rangle$ equilibrium orientation with a limited amplitude which, however, is increasing with temperature. This is reflected in the decrease of the order parameter [53, Fig. 19][74] $\langle \cos \Theta \rangle$ ($\Theta$ is the angle between the molecular axis and its $< 111 >$ equilibrium orientation), and has its influence on the interaction parameters.

The intermolecular interactions between the stretching vibrations are relatively short-ranged, and it is therefore an acceptable approximation to consider only n.n. interactions in the calculation of the dispersion relations. The density of vibron states calculated in such a simple model is shown for $\alpha$-$N_2$ in Fig. 6. In the $\alpha$-$N_2$ case, both the $\boldsymbol{k} = \mathbf{o}$ vibrons, $A_g$ and $T_g$, are Raman-active, as experimentally observed. The $A_g$ vibron is situated at the lower edge of the vibron band (Fig. 6), where the density of states vanishes. This is expected to inhibit elastic scattering within the vibron band. The vibron bandwidth $\Delta_v \simeq 1.5$ cm$^{-1}$ is smaller than for para-$H_2$, $\Delta_v \simeq 6$ cm$^{-1}$, and much smaller than for the rotons in this crystal. The crystal structure of para-$H_2$ results in only one $\boldsymbol{k} = \mathbf{o}$ vibron, which is Raman active [52, Ch. 3].

The Raman-active vibrons in $\alpha$-$N_2$ are broadening with increasing temperature up to the phase transition, and this is ascribed to vibron-libron coupling. The interactions of the vibrons with the rotational modes are much weaker in para-$H_2$ because of the very different character of the rotons in para-$H_2$ compared to the librons in $\alpha$-$N_2$: nearly free rotations compared to anharmonic librational modes. Moreover, measurements are only possible up to the melting point ($T = 13.8$ K for para-$H_2$ at zero pressure), and in these conditions the rotons in para-$H_2$ are not thermally populated because of their high excitation energies ($\simeq 500$ K) and cannot contribute to the vibron relaxation.

# 4 SAMPLE PREPARATION

The growth of hydrogen and nitrogen single crystals is a non-trivial part of the experiments discussed here. Several techniques have been employed by different investigators, which include slow Bridgman-type growth from the liquid, fast cooling of the liquid to $T \simeq 4$ K, sometimes under mechanical pressure, direct growth from the gas phase by slow sublimation, and thin layer deposition on a sapphire window. The growth methods and the precise thermal and mechanical treatments yield crystals with very different quality.

Fast solidification of hydrogen results in seriously strained crystals, as witnessed by the thread-like cracks which have developed in 3 to 4 hours after the growth [48]. Also, the $\alpha$-$N_2$ crystals produced by a similar procedure but under mechanical pressure ($\simeq 5$ atm.) consisted of relatively large single crystalline areas (about 5 mm diameter) [47]. Similar results are reported for relatively slow growth from the liquid [47] and for the sublimation [21] method. In the case of the thin layer deposition from a gas [101], the layer is probably polycrystalline, and as will be shown, the linewidth is determined by inhomogeneous effects.

In our experiments [54-60, 88], the crystals have been grown[1], by a Bridgman-type method in a quartz tube of 8 mm internal diameter (Fig. 7). The tube end rests on a cold finger, is surrounded by He contact gas at low pressure, and carries a heating element at $\simeq 5$ cm from the tube end. The cold finger consists of electrolytically

[1]Method and apparatus developed by A. Bouwen

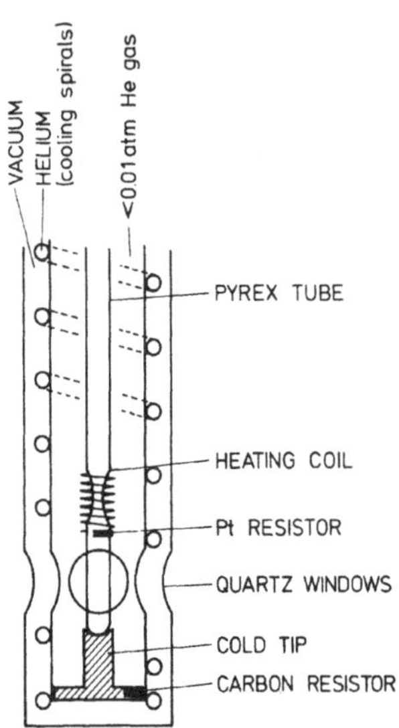


**Figure 7.** The end of the pyrex tube is shown in the optical cryostat, with cold tip, heater and sensing resistors (platinum and carbon) for controlling the temperature gradient.

pure Cu for better thermal conductivity. In this way a controllable and reliable temperature gradient can be obtained, which permits a slow Bridgman-type growth of a single crystal from the melt. Very slow cooling of the cold tip through the melting point is performed in order to avoid multiple seeding. The evolution of the crystal growth can be visually inspected. Even when the growth is completed, cooling to the measuring temperatures must be slow, in order to allow for relaxation of the strains resulting from thermal contraction and sticking of the crystal on the pyrex tube. Visually very clear para-$H_2$ crystals were obtained, which were shown by the Raman measurements to be single crystals in the whole probing region. Problems with internal strain could not be avoided in the nitrogen crystals: Good quality crystals result from the growth procedure, but cooling down from $T$ = 63 K to 7 K, unavoidably lead to varying amounts of thread-like defects parallel to the tube axis. In general, the largest part of the crystals was still available for the optical experiments. Crossing the $\beta$-$\alpha$ phase transition was performed very slowly, to limit the effects of the sudden contraction ($\approx$ 1 vol %). Crystal growth and cooling to $T \simeq 10$ K took 4 to 5 hours for parahydrogen, and 7 to 8 hours for nitrogen.

Prior to crystal growth, the hydrogen gas was converted from normal (about 1:4 para-to-ortho ratio) to para-$H_2$ by diffusing it through a catalyst, Apachi nickel-silica gel [69]. At normal pressures, the Boltzmann factor $\exp(-6B/k_BT_b)$ at the boiling point, $T_b$ = 20.4 K, determines the minimum concentration of ortho-$H_2$: $c_{ortho} = 0.2$ %. We have reached a concentration as low as 0.05 %, by performing the catalysis at lower vapor pressure, which also lowers the attainable temperature for the catalysis to about 17 K. During all subsequent procedures, contact with paramagnetic materials and impurities was carefully avoided. The gas was kept in a glass bulb for subsequent experiments with one day interval which yielded an increase of $c_{ortho}$ of the order of 1 % per day.

Impurity concentrations of hydrogen and nitrogen crystals, para-$H_2$:ortho-$H_2$, para-$H_2$:HD and $(^{15}N_2)_x(^{14}N_2)_{1-x}$ , could be determined from the relative spontaneous Raman intensities, and were found to be in good agreement with gas phase determinations. For the $Ar_x(N_2)_{1-x}$ crystals, the concentration in the gas phase was determined by mixing known gas quantities, using calibrated glass bulbs and accurate pressure gauges. The possible inhomogeneity of the mixtures could not be observed by position-dependent measurements of the concentrations, if available, and of the frequency- and/or time-domain Raman spectra, and were therefore considered negligible.

# 5 ROTON RELAXATION IN PARAHYDROGEN

## Experimental Results

Detailed measurements of the dephasing relaxation of the $J = 2$ rotons in para-$H_2$ were first performed by high-resolution spontaneous Raman spectroscopy [54], using a interferometer-monochromator tandem configuration. No changes of the linewidth could be observed as a function of temperature, as concluded from a series of experiments between $T = 6$ K and the melting point (13.8 K). In contrast, the linewidths were found to be strongly dependent on the ortho-$H_2$ concentration, $c_{ortho}$, as is illustrated in Fig. 8. At low $c_{ortho}$ the linewidth was found to be much narrower than previously reported [91], and to approach the resolution width of the instrument. The latter was limited to about 0.1 $cm^{-1}$, when a sufficiently wide free spectral range (FSR = 6.3 $cm^{-1}$ and finesse $\mathcal{F} \approx 50$) was chosen to measure the roton triplet. Deconvolution with the simultaneously measured instrumental response function was necessary to determine the smallest linewidths. All of the Raman lines could be accurately fit by lorentzian profiles (see Fig. 8), indicating that the roton relaxation is dominated by homogeneous broadening processes. However, the lineshape fitting is not an extremely dependable test in our experiments at the highest and lowest concentration. Indeed, at low $c_{ortho}$ the measured lineshape is influenced by the instrumental response, and at high $c_{ortho}$ by the overlap between the roton lines (Fig. 8).

From TR SRG experiments [55], clear evidence was obtained for the single exponential dephasing decay of the $J = 2$ rotons (see Fig. 9), corresponding to the previously observed lorentzian lineshape. As pointed out before (Sec. 2), the decay over two orders of magnitude in TR SRG ($S_{SRG} \propto A(t)$) is equivalent to four orders in TR CARS ($S_{CARS} \propto [A(t)]^2$), which underlines the quality of the exponential fits. These experiments also yielded more reliable dephasing times, especially for the long decay times, $T_2 \approx 100$ ps, in the purest crystals, with correspondingly narrow linewidths. It was possible to discern small differences between the dephasing times of the three $J = 2$ roton transitions. These improved results were obtained only by a careful selection of the frequency and polarization of the laser beams in order to select only one Raman transition, which simplifies the fitting procedure and makes it more reliable. The polarization geometries for the selection of only one of the modes can be derived from the form of the Raman tensors (Table 3) of the transitions, and from relation (4) between the resonant part of the $\chi^{(3)}$ and the Raman tensors. The selectivity can be enhanced for the outer rotons, $M = 0$ and $|M| = 1$, by tuning the difference frequency $\omega_S - \omega_L$ to the corresponding edge of the spectrum. Indeed, the combined linewidth ($\approx 8$ $cm^{-1}$) of the picosecond laser pulses encompasses the

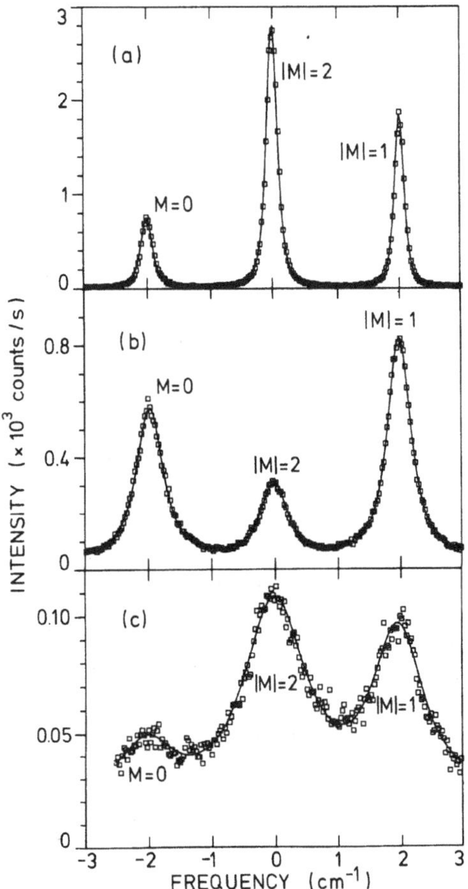


**Figure 8.** Spontaneous Raman spectra (open symbols) and fitting result (solid line) of the $J = 2$ rotons in para-$H_2$ measured at $T = 9$ K in single crystals with ortho-$H_2$ content: (a) $c_{ortho} = (0.05 \pm 0.03)$ %, (b) $c_{ortho} = (2.6 \pm 0.3)$ % and (c) $c_{ortho} = (6.3 \pm 0.6)$ %. The frequency scale is relative to the position of the central peak at 355 cm$^{-1}$. The relative intensities vary as a result of the different orientations of the crystals with respect to the laser polarization. (From Ref. [54].)

whole roton spectrum, and several modes can be simultaneously excited. In that case, a beating pattern [6, 24] is measured in the decay curve as a result of summation of the induced non-linear polarizations of the different excited modes, oscillating at different frequencies [55, 56]. It is also interesting to note that the non-resonant contribution $\chi^{(3)NR}$ discussed in Sec. 2, is negligible in para-$H_2$ compared to the Raman resonant part at the roton transition: This is demonstrated by the absence of any instantaneous peak in the decay curves of Fig. 9.

In Fig. 10 linewidths of the three roton transitions $|M| = 0, 1$ and 2, derived from the measured relaxation times $T_2$, are shown as a function of ortho-$H_2$ concentration [55, 56]. The linewidths, proportional to the relaxation rates $1/T_2$, increase linearly with $c_{ortho}$, starting from a value of about $\Delta\Omega \approx 0.1$ cm$^{-1}$ at zero concentration.

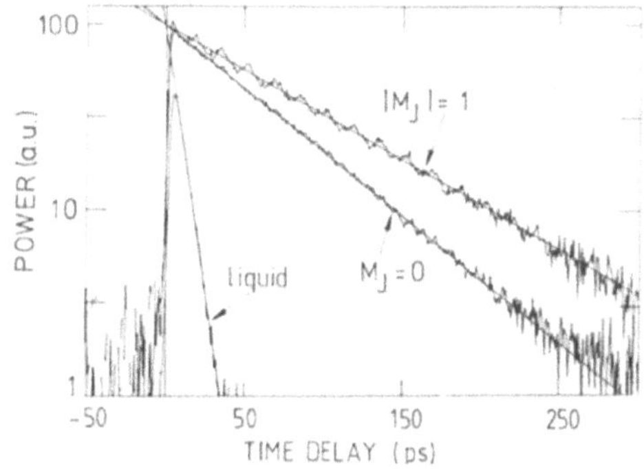


**Figure 9.** Exponential fits of the $J = 2$ roton decays ($T_2 = 63 \pm 2$ ps for $M = 0$ and $T_2 = 89 \pm 3$ ps for $|M| = 1$) in solid para-$H_2$, and of the rotational dephasing decay in the liquid ($T_2 = 7.5 \pm 1.0$ ps) obtained with $zzzz$, $zxzx$ and $zzzz$ polarizations (in the crystals: $z$ along the hexagonal axis), respectively. The measurements were performed at $T = 8$ K in the solid, and $T = 15$ K in the liquid, with $c_{ortho} = 0.14$ %. (From Ref. [55].)

Apart from ortho-$H_2$, the impurity with the highest concentration in these crystals is the isotopic impurity HD, with a natural abundance $c_{HD} = 0.03$ %. The question arose whether this could be responsible for the residual relaxation rate in the purest crystals. TR SRG measurements were performed in HD doped para-$H_2$ crystals [56], demonstrating that the impact of this impurity on the roton dephasing relaxation is exactly equal to that of ortho-$H_2$: The data from the HD containing crystals ly on the same straight lines as for para-$H_2$:ortho-$H_2$, as shown in Fig. 10. In Table 4 the slope, $\delta(\Delta\Omega)$, and the extrapolation to zero of the linewidth, $\Delta\Omega_0$, as a function of concentration are listed for the three roton transitions. From this results it follows that the HD impurities in natural abundance cannot be responsible for the residual linewidth. Finally, the $T_2$ measurements for a series of nearly pure

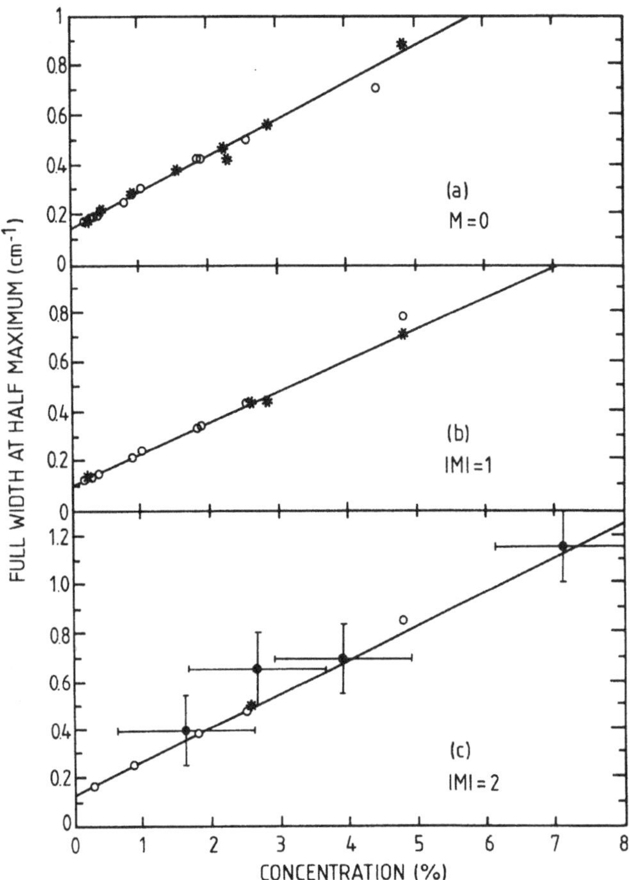


**Figure 10.** Full-width-at-half-maximum (FWHM) of the $J = 2$ roton Raman transitions in para-$H_2$ as a function of total impurity concentration $c_{HD} + c_{ortho}$. The results in HD-doped crystals ($*$) (Ref. [56]) are compared to those in crystals containing only ortho-$H_2$ impurities ($\bigcirc$) (Ref. [55]). The linewidth measurements of the IR-active $|M| = 2$ transition (Ref. [102]) are also given for comparison, including error bars.

para-$H_2$ crystals from different growth procedures were found to give reproducible results, which can be judged from the small scatter in the obtained linewidths (Fig. 10). Therefore the influence of random strain and growth defects can safely be eliminated, since it would be expected to vary from one crystal to another, or from one position to another in the same crystal. We therefore concluded that the extrapolated linewidths result from relaxation effects intrinsic to the perfect para-$H_2$ crystal.

One more experimental result is available about relaxation of the $J = 2$ rotons in para-$H_2$. Linewidth measurements [102] of the $|M| = 2$ odd roton have been performed as a function of $c_{ortho}$, using IR Fourier transform spectroscopy. They are compared in Fig. 10 to our results for the Raman-active $|M| = 2$ even roton, and are consistent with our results within the experimental errors. This is also the case for the upper limit of the linewidth of 0.2 $cm^{-1}$ (FWHM) which was reported in the IR work for the pure crystal (Table 4).

## Scattering from ortho-$H_2$ and HD Impurities

Our experiments in para-$H_2$ crystals have demonstrated that the picosecond dephasing of the $J = 2$ rotons is dominated by homogeneous processes, both in pure crystals and in those purposely doped with ortho-$H_2$ and HD impurities. The linear increase of the relaxation rate suggests scattering of the rotons by non-interacting impurities. The effect is strong compared to the dephasing in pure crystals yielding a tenfold increase in linewidth for $c_{ortho} \approx 5$ %. We have attributed the impurity-induced relaxation to elastic scattering of the $\boldsymbol{k} = \mathbf{o}$ rotons into other rotons in the BZ with equal energy, as is further discussed below.

Two other possibilities have to be considered: First, inelastical scattering from the impurities, with simultaneous creation or annihilation of a phonon, can be excluded. The corresponding relaxation rates would be expected to be temperature dependent, because mainly low frequency phonons ($< 10$ $cm^{-1}$) would contribute to such intraband scattering processes, and these are thermally populated in the measured temperature region ($T = 6 - 13.8$ K). Second, it could be envisaged that the ortho-$H_2$ and HD impurities are mediating in the population decay of the $\boldsymbol{k} = \mathbf{o}$ roton into several (at least five) phonons. This process could be independent of temperature up to the melting point, if mainly optical phonons are involved. However, as a many-phonon process it would be expected to possess a low probability, and the mechanism for phonon-roton coupling at the impurity is unclear. Also, there is no simple argument why ortho-$H_2$ and HD would yield exactly the same induced relax-

**Table 3**. Rotational matrix elements $\langle 2M|\alpha_{ij}|00\rangle$ of the polarizability tensor of an $H_2$ molecule, in terms of $\gamma$, the anisotropic part of the polarizability. The hexagonal crystal axis is labeled $z$. One of the polarization combinations for the selection of one mode in TR SRG is given in the last row.

| $M = 0$ | $M = \pm 1$ | $M = \pm 2$ |
|---|---|---|
| $\frac{\gamma}{\sqrt{45}}\begin{pmatrix} -1 & 0 & 0 \\ 0 & -1 & 0 \\ 0 & 0 & 2 \end{pmatrix}$ | $\frac{\gamma}{\sqrt{30}}\begin{pmatrix} 0 & 0 & \mp 1 \\ 0 & 0 & i \\ \mp 1 & i & 0 \end{pmatrix}$ | $\frac{\gamma}{\sqrt{30}}\begin{pmatrix} 1 & \mp i & 0 \\ \mp i & -1 & 0 \\ 0 & 0 & 0 \end{pmatrix}$ |
| $zzzz$ | $zxzx$ | $yxyx$ |

**Table 4.** The experimental values of the roton linewidth, $\Delta\Omega_0$, for the pure para-$H_2$ crystal and the increase per percent impurity concentration, $\delta(\Delta\Omega)$, are given (in $cm^{-1}$) for the $|M| = 0$, 1 and 2 Raman-active rotons, as derived from the complete set of experiments in crystals with ortho-$H_2$ and HD impurities. Comparison is made with theoretical results. Also, the slopes of impurity-induced linebroadening calculated for the odd roton modes are included.

| Parity | | | $M = 0$ | $\|M\| = 1$ | $\|M\| = 2$ |
|---|---|---|---|---|---|
| Even | $\Delta\Omega_0$ | exp.[a] | $0.14 \pm 0.01$ | $0.101 \pm 0.009$ | $0.125 \pm 0.005$ |
| | | calc.[b] | 0.090 | 0.085 | 0.087 |
| | | exp[a] | $0.148 \pm 0.005$ | $0.126 \pm 0.005$ | $0.140 \pm 0.005$ |
| Even | $\delta(\Delta\Omega)$ | calc[c] | 0.117 | 0.094 | 0.004 |
| | | calc[d] | 0.171 | 0.143 | 0.160 |
| Odd | $\delta(\Delta\Omega)$ | calc[c] | 0.76 | 3.35 | 0.08 |

[a] Ref. [56]. [b] Data from Ref. [17] reduced by 24 % for phonon renormalisation. [c] Ref. [19] reduced by 13 % for phonon renormalisation.
[d] Ref. [18].

ation rates for this mechanism, while this observation can be easily explained on the basis of elastic roton scattering within the band. By elimination it was concluded that pure dephasing effects are dominating the relaxation of the $J = 2$ rotons in para-$H_2$.

Two calculations of the impurity-induced rate of elastic scattering have been performed, one in a simple golden-rule type approximation [19], and one using Green's function techniques [18]. The roton states described by equations (9) and (10) were calculated from the unperturbed Hamiltonian

$$\mathcal{H} = 6B + \sum_{p>q} V_{EQQ}(p,q), \tag{11}$$

in which the first term derives from the free rotor Hamiltonian and is constant for all of the states considered here. The second term is the EQQ interaction between pairs of molecules connected by $\mathbf{R}_{pq}$, with size $R_{pq}$ and direction $\mathbf{\Omega}_{pq}$. If the direction of the molecular axes is given by $\mathbf{\Omega}_p$ and $\mathbf{\Omega}_q$ the contribution of one pair of molecules reads,

$$V_{EQQ}(p,q) = \frac{20\pi}{9}\sqrt{70\pi}\Gamma\left(\frac{R_0}{R_{pq}}\right)^5 \sum_{m,n=-2}^{2} C(224;mn)Y_m^2(\mathbf{\Omega}_p)Y_n^2(\mathbf{\Omega}_q)Y_{m+n}^4(\mathbf{\Omega}_{pq}) \tag{12}$$

in which $R_0$ is the n.n. distance, $C(224;mn)$ are Clebsch-Gordon coefficients, $m$ and $n$ are magnetic quantum numbers, and $Y_m^l(\mathbf{\Omega})$ are the spherical harmonics of order $l$. For the free molecules the EQQ interaction parameter is $\Gamma = 0.660$ $cm^{-1}$ and $0.817$ $cm^{-1}$ for para-$H_2$ and ortho-$D_2$, respectively. The renormalization resulting from the lattice zero-point motion, results in a decrease of the effective EQQ constants which is about 13 % [69]. The site occupied by an ortho-$H_2$ or HD impurity does not carry any excitation at an energy close to the $J = 2$ roton band of para-$H_2$, and it is excluded in the motion of the rotational excitation from site to site. This situation is very different from the one discussed in Sec. 1 concerning the dynamics of

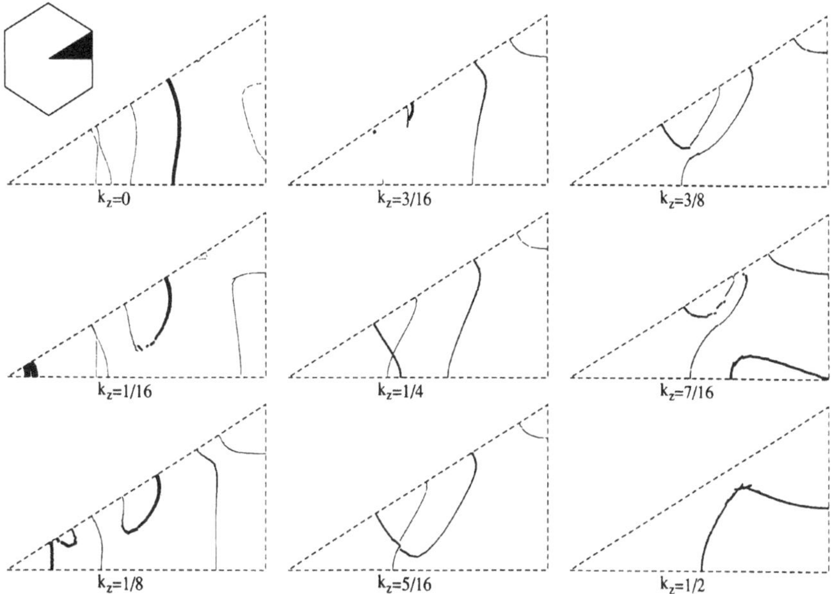


**Figure 11.** The surfaces containing $J = 2$ roton states with energy equal to that of the $|M| = 1$ Raman-active roton are shown by means of horizontal sections through the BZ. The line thickness illustrates their relative contributions to the scattering rate. Only an irreducible part (1/24) of the BZ is shown, as identified in the inset. The sections are labeled by their vertical position, $k_z$, in units of the height of the first BZ. (From Ref. [19].)

a Frenkel-exciton in a system in which the site energies and/or interaction parameters are only weakly perturbed (e.g., [65, 66]). Indeed, in the present case the disorder follows from substitution with impurities possessing an excitation spectrum which is completely different from that of the host molecules. It would be worth adapting the treatment of the separate band limit for mixed crystals by Velsko and Hochstrasser [103] to handle the case of roton scattering.

One way to describe this case is to consider the perfect para-$H_2$ crystal and to decouple the molecule in the impurity site from all the others, by substraction of all the pair interactions with this site [19]. The perturbation Hamiltonian

$$\mathcal{H}' = -\sum_{p \neq 0} V_{EQQ}(p, 0\alpha), \tag{13}$$

then represents a single impurity at the site $q = 0\alpha$, on the $\alpha$ sublattice. The differences between ortho-$H_2$ and HD, such as their eigenstates and their interactions with the surrounding para-$H_2$ molecules, are considered negligible for the present purpose. This is supported by our experimental results which are identical for both impurities. The scattering rate for a given initial state $\Phi_{|M|\pm}(\mathbf{o})$, with energy $E_{|M|\pm}$, was numerically integrated over all states with equal energy. The rotons to which the initial state can be elastically scattered ly on equal energy surfaces in the BZ as illustrated for the $|M| = 1$ roton level in Fig 11. The scattering rate $w$ was calculated from the golden-rule type formula,

$$w = \frac{2\pi}{\hbar} \sum_{\boldsymbol{k},a} |\langle \Phi_{M\pm}(\mathbf{o})|\mathcal{H}'|\Phi_a(\boldsymbol{k})\rangle|^2 \delta[E_{|M|\pm} - E_a(\boldsymbol{k})], \tag{14}$$

which, for the choice of Hamiltonians (11) and (13), strongly simplifies to

$$w = \frac{1}{N}\frac{2\pi}{\hbar}4E^2_{|M|\pm}\sum_{\boldsymbol{k},a}|A^{\boldsymbol{k}_a}_{M\alpha}|^2\delta[E_{|M|\pm} - E_a(\boldsymbol{k})], \tag{15}$$

in which $2N$ is the number of molecules in the crystal, and the $A^{\boldsymbol{k}_a}_{M\alpha}$ are the coefficients of the roton states defined in (10).

Using appropriate BZ integration methods, the impurity-induced relaxation rate of the even and odd $\boldsymbol{k} = \mathbf{o}$ rotons in para-$H_2$ were evaluated (Table 4). The results for the outer roton Raman levels, $|M| = 0$ and 1, were found to be very close to the experimental results, closer than might be expected, but the slope of the $|M| = 2$ relaxation rate with $c_{ortho}$ was drastically underestimated. As shown below, the golden-rule type approximation [Eq. (14)] which is the lowest-order approximation in scattering theory, is not appropriate, because the scattering center does not form a small perturbation compared to the intermolecular interactions: The EQQ interactions of the impurity site with the other molecules are not just altered but completely substracted!

The slope observed for the linewidth of the IR-active roton (see Fig. 10) is very close to that measured for the three Raman-active ones, and is in very good agreement with the calculated result (Table 4), but the latter could be somewhat fortuitous. On the basis of our approximate calculation, we would expect a even stronger impurity-induced broadening of the other odd roton transitions with $M = 0$ and $|M| = 1$ (Table 4). However, no experimental results are available for these two silent modes.

A more sophisticated approach was followed by Igarashi [18], who describes the rotons by boson operators, and explicitly introduces ortho-$H_2$ molecule in the impurity site. Using Green's function techniques, the probability rates are calculated for scattering a Raman-active roton off an isolated ortho-$H_2$ impurity. The scattering probabilities to all the rotons on the corresponding equal energy surface (see Fig. 11) were summed including all higher order processes with virtual states in the $J = 2$ roton band. The treatment of the ortho-$H_2$ impurity states is also more explicit, taking the ground state multiplet into account. However, this should have a minor influence on the relaxation rates if one considers the equivalence of ortho-$H_2$ and HD impurities which was demonstrated by our experiments.

In Igarashi's calculation, the slope obtained for the line broadening as a function of $c_{ortho}$ for each of the three even rotons, $\Phi_{|M|+}(\mathbf{o})$, is in very good agreement with the experiment (Table 4). Although a renormalized EQQ parameter was employed, $\Gamma = 0.55$ cm$^{-1}$, which permits to fit the measured splitting between the even roton transitions, the slopes are slightly overestimated. Nevertheless, the calculation accurately reproduces the small differences in the slopes for the three transitions. Unfortunately, the results of this calculation for the odd $J = 2$ rotons were not given in [18].

An additional test of the proposed model, incorporating only elastic scattering by the "passive" impurity molecule, will come from experiments on the other $J = 0$ hydrogen species, ortho-$D_2$, with para-$D_2$ (or HD) impurities. According to the scattering model the scattering efficiencies, i.e. the slope of the linewidths with impurity concentration, should scale linearly with the EQQ interaction parameter, $\Gamma$. Indeed, the quadratic dependence of the scattering probabilities [Eqs. (12) and (14)] is compensated by the fact that the density of states is inversely proportional to $\Gamma$. The best values for the effective interaction parameters in the solid are $\Gamma = 0.575$ cm$^{-1}$ and 0.713 cm$^{-1}$ [69] for para-$H_2$ and ortho-$D_2$, respectively, and as a result an increase of the scattering rates of about 24 % would be expected.

## Intrinsic Dephasing Relaxation in para-$H_2$

We were able, from the experiments on ortho-$H_2$ and HD doped crystals, to attribute the residual linewidth in pure para-$H_2$ crystals to intrinsic dephasing processes, practically excluding effects from random strain and from other impurities.

The first mechanism which was investigated involves only the $J = 2$ rotational excitations of the para-$H_2$ molecules. Until now, we have only considered states [Eq. (10)] in which a single molecule is excited, but eventually multiple excitations have to be considered. Prior to our experiments, a calculation was performed by Vanhimbeeck *et al.* [17] of the Raman linewidths of the $J = 2$ rotons in para-$H_2$ and ortho-$D_2$, including only the rotational degrees of freedom. The quantum rotations were limited to the ground state and single/multiple $J = 2$ rotational excitations, coupled by the EQQ interactions of Eq. (12). The linewidth was obtained from the second and fourth moments of the relaxation function. The linewidth of the corresponding spectral function was determined using a continued-fraction representation up to the fourth moment of the relaxation function. The moments were obtained in second-order perturbation of the EQQ interactions, using a memory-function formalism. The calculated results for para-$H_2$, reduced according to the renormalization by the translational zero-point motion (24 % because of the quadratic dependence on $\Gamma$), are compared to the experimental results in Table 4. They are in surprisingly good agreement with experiment, taking into account the complexity of the problem. The difference in relaxation rate for the different rotons is somewhat less pronounced in the calculated than in the experimental results.

In spite of this very good numerical agreement between theory and experiment, the discussion on the intrinsic relaxation processes in para-$H_2$ is not closed. The theoretical treatment does not clearly pinpoint the type of relaxation process involved. The calculation can be considered to be performed at $T = 0$ K, and therefore thermal excitations are not involved. The system is limited to only the $J = 2$ excitations, and does not consider interactions with other variables, which could form a thermal bath. As a result, the evolution of the system should conserve energy, within the rotational degrees of freedom. This seems to exclude population decay processes because there are no low-energy excitations available in the model system to which the energy of a $J = 2$ roton could be transferred. Since the perfect crystal is considered here, which possesses translation symmetry, there is no mixing of the $\boldsymbol{k} = \mathbf{o}$ states with other $\boldsymbol{k}$ states. Only multiple $J = 2$ excitations with total vanishing $\boldsymbol{k}$ have to be considered, but they possess a much higher energy. For example, the so-called *two-roton* band, of double $J = 2$ excitations is located at about $6B \approx 355$ cm$^{-1}$ above the $J = 2$ roton levels, and has an estimated width of less than 50 cm$^{-1}$ as a result of the EQQ interactions. This was confirmed by Raman measurements of the two-roton spectrum in very pure ($c_{ortho} \approx 0.7$ %) para-$H_2$ crystals, which shows the density of states of excitations containing two rotons with opposite wavevector ([93] and [104, p.61 to 64]). In a similar way, the width of the three-roton band can be estimated to be much smaller than the distance ($12B$) to the single roton band, and only states with up to three excited molecules are considered in the model, due to the approximations involved.

The interpretation of the calculated moments in Ref. [17] rests on the assumption that the Raman spectrum consists of a single line, and does not take into account the weak two-roton band which is also part of the calculated spectral function. This additional weak two-roton spectrum explains the functional dependence of the calculated $2^{nd}$ and $4^{th}$ spectral moments on $(\Gamma/B)^2$. Starting from the theoretical integrated intensities [104, p.61] of this band, the second and fourth moments obtained for the spectra are close to those of the theoretical treatment by [17]. If this is the correct interpretation of their results, the latter are completely unrelated to the Raman linewidth of the zone-center $J = 2$ rotons.

In the real crystal, a bath of low energy excitations is provided by the translational modes. Roton scattering within the band, by emission of low-energy phonons, is a possible relaxation mechanism which would be temperature-independent in the range of the measurements. The population decay would then be a subsequent, probably slower process. If the roton-phonon scattering is inoperative, the alternative is direct population decay in which a roton is annihilated, and several phonons are created. In both crystals the phonon cut-off frequency is around 82 $cm^{-1}$, and therefore it takes at least 5 and 3 phonons to emit the roton energies of 355 $cm^{-1}$and 179 $cm^{-1}$ in para-$H_2$ and ortho-$D_2$, respectively. The relaxation linewidth of the $J = 2$ rotons in pure ortho-$D_2$ would possibly permit the identification of the relaxation process. The best available Raman measurement [92] in relatively pure ortho-$D_2$, $c_{para} \approx 2.7$ %, was severely limited by a spectral resolution of around 1.5 $cm^{-1}$. Following Ref. [17], the relaxation rate would scale as $(\Gamma/B)^2$, which means a factor 6 from para-$H_2$ to ortho-$D_2$. The large relative changes of the roton energy with respect to that of the phonon spectrum, are also expected to drastically influence the phonon-mediated intraband as well as the population relaxation rates in ortho-$D_2$ compared to para-$H_2$, but this would be expected to yield a different factor.

# 6 VIBRON DECAY IN $\alpha$-$N_2$ AND para-$H_2$ CRYSTALS

## Non-Exponential Decay in the Nanosecond Time-Range

The vibrons in $\alpha$-$N_2$ and para-$H_2$ crystals have been investigated in the late seventies by means of TR stimulated Raman scattering (SRS) [47, 48] in the time range up to hundreds of nanoseconds, with nanosecond time-resolution ($\approx$ 10 ns). This was performed with pulsed nitrogen lasers, and dye lasers pumped by them, using electronic triggering of the intense excitation and probe pulses.

In Fig. 12 the coherence decay of the $A_g$ vibron in $\alpha$-$N_2$ at low temperature ($T$ = 2 K) is shown, as measured by the TR SRS technique [47]. Because of the lower Raman gain, the $T_g$ transition could not be detected. The measurement of the $N_2$ stretching vibration in the liquid, dephasing very fast on this time-scale, illustrates the instrumental response function (Figs. 12a and b, broken line). The $A_g$ SRS signal initially follows the instrumental response, especially in curve b, revealing the presence of fast components in the dephasing relaxation. Subsequently, a non-exponential decay sets-in with an effective rate, $1/\tau$, which is decreasing for later delay times. The decay curve depends on the details of the crystal growth procedure, but the effective rates at delay times longer than about 10 ns are the same for different measurements. The effective exponential decay time is increasing from $\tau \approx 5.5$ ns at early times, to $\tau \approx 14.5$ ns for delay times between 60 and 100 ns.

The vibron decay in $\alpha$-$N_2$ was found to be independent of temperature between 1.33 K and 4.2 K [47]. From this observation Abram *et al.* concluded that the dephasing was not caused by interactions between the vibrons and low-frequency phonons. The influence of the natural abundancy (0.74 %) of $^{14}N^{15}N$ isotopic impurities, which

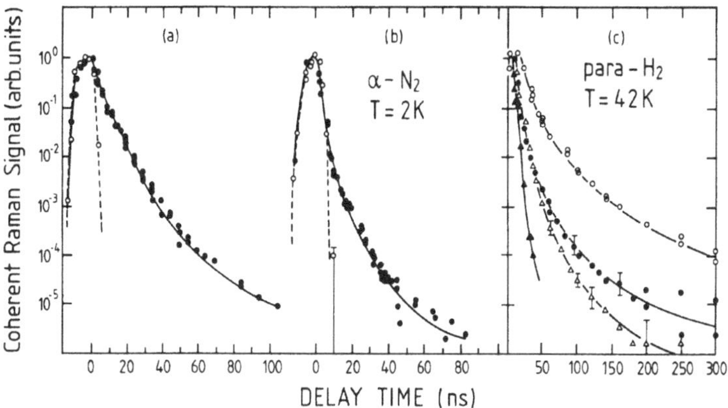


**Figure 12.** Coherence decay of the $A_g$ vibron in $\alpha$-$N_2$ ($\Omega_v$ = 2328 $cm^{-1}$, $T$ = 2 K) measured in crystals grown (a) in 5 min under a pressure of 5 atm, and (b) in 100 min at normal pressure. (c) In para-$H_2$ the dephasing at 4.2 K of the $A_g$ vibron ($\Omega_v$ = 4150 $cm^{-1}$) is shown for several concentrations of ortho-$H_2$ impurities (○ : 0.22 %, • : 0.32 %, △ : 0.37 %, ▲ : 2.7 %). Solid curves serve as a guide to the eye. (From Refs. [47] and [48], with permission.)

have a vibrational frequency 39 $cm^{-1}$ below that of $^{14}N_2$, was estimated to be negligible. This was confirmed by experiments in crystals purposely doped with $^{15}N_2$, with a 79 $cm^{-1}$ lower frequency. The impurity states are lying far outside of the vibron band in both cases. The absence of any influence on the vibron dephasing curves clearly excluded both the possible trapping of the vibron by isotopic impurities, and scattering from them. Eliminating all other possibilities, the non-exponential decay of the vibrons has been attributed to scattering by static structural disorder. The different growth procedures induce a variable amount of random microscopic strain in the crystal, which explains the observation of decay curves in different crystals which differ in the relative importance of the non-exponentially decaying tail.

Very comparable decay curves were measured for the $A_g$ vibron in para-$H_2$ single crystals. They exhibit a non-exponential decay with decreasing effective relaxation rates which were found to be independent of temperature in the 2-4 K temperature range. However, the concentration of the major impurity, ortho-$H_2$, very strongly influences the vibron relaxation, which becomes significantly faster in the concentration range 0.22 – 2.7 %. The free ortho-$H_2$ impurities have a slightly shifted (−5.9 $cm^{-1}$) vibrational frequency compared to para-$H_2$, due to rotation-vibration coupling. In the solid the trap level would be located about 3.5 $cm^{-1}$ below the $A_g$ vibron [52, Chap. 3]. This should be compared to the width of the vibron band, which is about 6 $cm^{-1}$ [52, p.78]. As mentioned before, no temperature dependence of the coherence decay could be observed. Considering the precision of the measurements, Abram *et al.* were able to put a lower limit on the ratio of the probabilities for elastic scattering from and (inelastic) trapping at the ortho-$H_2$ impurities, showing that the trapping processes are completely dominated by near-resonant elastic scattering at these temperatures.

The nanosecond non-exponential relaxation of the vibrons in nitrogen and hydrogen has been extensively treated [14, 47, 48, 66], and we will only briefly recapitulate this discussion. If a vibron state is situated in the middle of the band (e.g., the $J = 0$ rotons in para-$H_2$, see Fig. 4), and for sufficiently weak perturbation of the perfect crystal Hamiltonian, the lowest order treatment of elastic scattering yields a finite scattering rate $w$, given by a Golden-rule expression such as (14). An exponential relaxation at this rate is found, or equivalently a lorentzian lineshape with the corresponding linewidth. The $A_g$ modes, both in $\alpha$-$N_2$ and para-$H_2$, are located at the lower edge of the band, where the density of vibron states is vanishing. Klafter and Jortner [65] have derived, within the so-called *average T-matrix approximation*, the (by now classical) formula for the lineshape of excitons in a crystal with weak static disorder:

$$I(\omega) \propto \frac{\pi D^2 \rho(\omega)}{(\omega_0 - \omega)^2 + [\pi D^2 \rho(\omega)]^2}, \quad D \ll \Delta_b \tag{16}$$

in which $\omega_0$ is the frequency of the optical transition, $\Delta_b$ is the width of the exciton band, $D^2$ is the variance of the gaussian distribution of the site energies (diagonal disorder), and $\rho(\omega)$ is the density of states. This was derived for triplet excitons in molecular crystals, but can be transposed to the vibrons. For sufficiently small $D$, and provided the density of states is only smoothly varying around the exciton frequency, this reduces to a lorentzian shape with linewidth $\pi D \sqrt{\rho(\omega)}$. In this case, the measured linewidth is related to the width of the site energy distribution, but the latter cannot be directly determined from measurements on the exciton transitions. For band edge states with corresponding $\rho = 0$, such as the $A_g$ vibrons, non-lorentzian lineshapes are obtained. Fourier transformation of these results yields decay curves in qualitative agreement with the nanosecond vibron decay in $\alpha$-$N_2$ and para-$H_2$ [47, Fig. 5]

Abram and Hochstrasser [66] have tackled the same problem directly in the time-domain, in order to properly describe the long time limit of the coherence decay, and to get a better understanding of the physical processes involved. This approach is not equivalent to Fourier transformation of (16) because of the different approximations involved in each of the treatments. The initial state prepared by the laser light in the crystal is described as a minimum uncertainty wave-packet of vibron states (Dicke-like collective states), the time-evolution of which is determined by the tight-binding exciton Hamiltonian perturbed by random diagonal disorder. Because of the perturbation the translation symmetry is broken, and the eigenstates consist of a superposition of states with a distribution of $\boldsymbol{k}$ values. Assuming weak disorder without site-to-site correlation, the long time asymptotic form of the $A_g$ vibron decay is found to be

$$I(t) \propto e^{-D^2 \sqrt{t}/\alpha}, \tag{17}$$

in which $\alpha$ is essentially proportional to the intermolecular interactions, or the width of the exciton band. On the basis of this calculated decay function, convoluted with the experimental response function, a very good fit was obtained [66] of the experimentally measured relaxation decay of the $A_g$ vibron in solid $\alpha$-$N_2$. For the heavily strained sample (Fig. 12a), $D = 0.15$ cm$^{-1}$ was derived for the width of the site energy distribution. In heavily $^{15}N_2$ doped crystals an upper limit of $D = 0.25$ cm$^{-1}$ was put on this parameter from linewidth measurements of the impurity stretching vibration. This does not definitely prove the validity of the simple model including only diagonal, uncorrelated random disorder. For the $T_g$ vibron, which is an intraband excitation corresponding to a non-zero density of states, the leading term in $I(t)$ is an exponential decay with the same rate as previously derived [65] in the frequency-domain calculation.

The disorder in para-$H_2$ is caused by the substitution of host molecules by ortho-$H_2$, as was clearly demonstrated by the concentration dependence of the relaxation rates. For such a substitutional disorder, the equivalence with the above model of weak structural disorder was demonstrated for two limiting cases: The amalgamation limit, when the impurity excitation lies within the host exciton band, and the separated band limit for low concentrations of impurities with an excitation energy which is very different from that of the host molecules [103]. Although qualitatively the decay curves in para-$H_2$ are similar to those in $\alpha$-$N_2$, they cannot be satisfactorily described by the above theory. In fact, para-$H_2$:ortho-$H_2$ does not fit in any of these limiting cases, since the ortho-$H_2$ vibration is located at only about 3 $cm^{-1}$ below the 6 $cm^{-1}$ wide vibron band. This near resonance effect could be at the origin of the large scattering efficiency, and it would be interesting to compare this to the cases of HD and $D_2$ impurities in para-$H_2$, which correspond to the separate band limit.

Very recently, high-resolution frequency-domain SRG measurements have been performed on $A_g$ vibron in para-$H_2$ containing only 0.06 % ortho-$H_2$ and 0.04 % HD at liquid helium temperature [61]. The very narrow linewidth of 7 MHz (or $2.3 \times 10^{-4}$ $cm^{-1}$) amply demonstrates that there exist no dephasing processes on the picosecond time-scale in such crystals, in contrast to the case of $\alpha$-$N_2$ discussed in Sec. 6. This extremely small linewidth, still including an instrumental contribution, is in reasonable agreement with the assumption of scattering from ortho-$H_2$ impurities even in very pure crystals. This demonstrates that the coupling between the vibrons and the other excitations in the crystal is much smaller in para-$H_2$ than in $\alpha$-$N_2$. In any case, it would be of interest to measure the lineshape of the vibrons in para-$H_2$ at higher $c_{ortho}$ and higher temperature, which we are planning for the next future.

## Picosecond Relaxation in Pure and Isotopically Mixed $\alpha$-$N_2$

Recently, new time- and frequency-domain experiments have been performed on the dephasing relaxation of the vibrons in solid nitrogen by different research groups simultaneously. I will review the results, and make a detailed comparison between them.

**Time-domain Experiments.** The picosecond dephasing decay of the vibrons was measured using TR CARS by De Kinder *et al.* [57, 58] in $\alpha$-$N_2$ and $\beta$-$N_2$ crystals, and in $^{15}N_2$ doped crystals in the $\alpha$-phase. Further experiments on $Ar_x(N_2)_{1-x}$ mixed crystals [59] will be discussed in Sec. 7. The measured decay of the CARS signal of the vibrons in $\alpha$-$N_2$ at 2328 $cm^{-1}$ is illustrated in Fig. 13. Both Davidov components are simultaneously excited because the combined excitation linewidth ($\approx 8$ $cm^{-1}$) is larger than the $A_g$-$T_g$ splitting. The beating amplitude depends on the polarization of the laser beams with respect to the crystal axes, and on the exact tuning of the laser frequency difference $\omega_L - \omega_S$. Because of the much higher Raman cross-section of the $A_g$ compared to the $T_g$ transition, the former is usually dominating the TR CARS signal. Practically, it turned out to be impossible to get a signal with a large $T_g$ vibron contribution.

The decay curve shown in Fig. 14a (logarithmic scale) for the undoped crystal hardly exhibits any beating, because nearly only the $A_g$ vibron contributes. This measurement also shows that at short times the vibron decay is faster than exponential —the effective decay rate increases with time after excitation— which is the opposite of the observations made at longer times in $\alpha$-$N_2$ [47]. The pragmatic analysis of the decay of the coherent amplitude of each vibron excitation was performed using a product of an exponential and a gaussian, which in the traditional picture corresponds to the Fourier transform of the homogeneous and inhomogeneous spectral components, respectively. Taking into account the beating between the two modes, split by the frequency $\Omega_b$ one finds

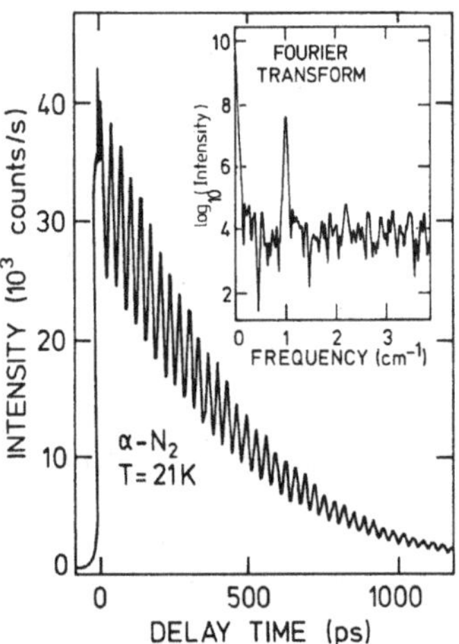


**Figure 13.** Dephasing decay of the vibrons in $\alpha$-$N_2$ at 21 K. The beating pattern between the $A_g$ and $T_g$ vibrons shows up in the Fourier transform (shown in the inset) as a narrow peak at the frequency of the factor group splitting (from Ref. [58]).

$$\begin{aligned} S(t) &= \mathcal{A}_A e^{-2t/T_{2A}-\Delta_A^2 t^2} + \mathcal{A}_T e^{-2t/T_{2T}-\Delta_T^2 t^2} \\ &+ \sqrt{\mathcal{A}_A \mathcal{A}_T} cos(\Omega_b t + \Phi_b) e^{-(1/T_{2A}+1/T_{2T})t-(\Delta_A^2+\Delta_T^2)t^2}, \end{aligned} \tag{18}$$

in which $T_{2A}$ and $T_{2T}$ are the total dephasing times, and $\Delta_A$ and $\Delta_T$ the inhomogeneous widths of the $A_g$ and $T_g$ transitions, respectively, and $\Phi_b$ is the phase of the beating signal. Different values of the decay parameters ($T_2/2$ and $\Delta$) for the two vibrons would show up most clearly, in decay curves such as Fig. 13, in a faster decay of the beating amplitude than of the overall signal. This could not be resolved in our experiments, and therefore the two vibron modes were assumed to have similar decay characteristics, which yields,

$$S(t) = (\mathcal{A}_A + \mathcal{A}_T + \sqrt{\mathcal{A}_A \mathcal{A}_T} cos\Omega_b t) e^{-2t/T_2-\Delta^2 t^2}, \tag{19}$$

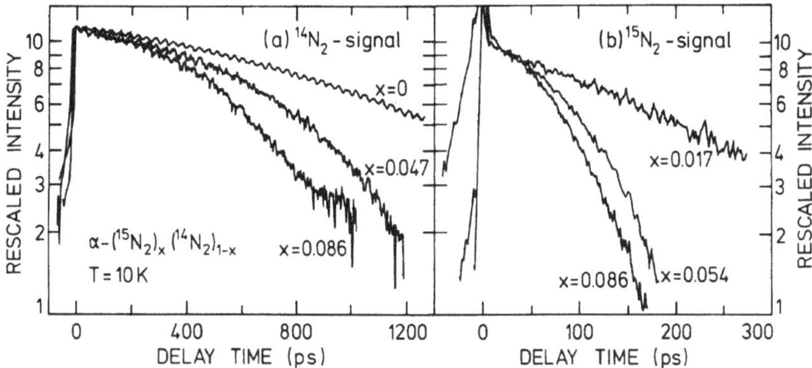


**Figure 14.** Normalized decay spectra measured at $T = 10$ K (a) for the $A_g$ $^{14}N_2$ vibrons and (b) for the $^{15}N_2$ stretching vibration, in isotopically mixed $\alpha$-$(^{15}N_2)_x(^{14}N_2)_{1-x}$ crystals. Note the reduced time scale in (b) compared to (a). (From Ref. [58].)

for the decay function. The $T_2/2$ and $\Delta$ parameters obtained from fitting the decay curves with (19) can be attributed to the $A_g$ mode because it is contributing most strongly.

The inhomogeneous width $\Delta \approx 0.007$ cm$^{-1}$ was found to depend on the specific sample, and on the interaction volume in the crystal. As in the case of the nanosecond decay measurements [47] this points to random strains in the crystals, probably resulting from cooling through the $\beta$-$\alpha$ phase transition. The dephasing time $T_2/2$ did not show this dependence on local conditions in the crystals, and is found to be a reliable parameter. It was found to decrease from about 800 ps at 7 K to 113 ps (see Fig. 15) just below the phase transition ($T = 35.6$ K), which corresponds to an homogeneous width of 0.0066 cm$^{-1}$ and 0.047 cm$^{-1}$, respectively. It is worth noting that the vibron relaxation becomes slower in the $\beta$-phase, $T_2/2 = 168$ ps (linewidth $\approx 0.032$ cm$^{-1}$) than in the $\alpha$-phase [57]. In the previous TR SRS experiments [47] these relaxation phenomena could not be investigated because of the low time resolution ($\approx$ 10 ns). Possibly, the larger "instantaneous" portion of the decay in the slowly grown crystal (Fig. 12b) —probably containing a weaker static disorder— is indicative of a larger contribution of the fast relaxation now studied in the picosecond time range.

The $A_g$-$T_g$ factor group splitting, $\Omega_b$, has been accurately determined from the beating pattern, either from the direct fitting procedure with (19), or from the position of the peak in the Fourier transform of the decay function (see inset of Fig. 13). The line splitting is shown as a function of temperature in Fig. 16. It is decreasing from 1.04 cm$^{-1}$ at low temperature to about 0.9 cm$^{-1}$ just before the phase transition, where it jumps to zero discontinuously. This behavior is very similar to that of the order parameter $\langle \cos \Theta \rangle$, which indicates that the renormalization over the increasing range of molecular orientations is at the origin of the decrease of the vibron coupling parameters [53].

**Frequency-domain Experiments.** Two Raman investigations in the frequency-domain [21, 82] yield results which are compatible with ours, and are complementary in the sense that separate lineshapes have been measured for the $A_g$ and $T_g$ vibrons.

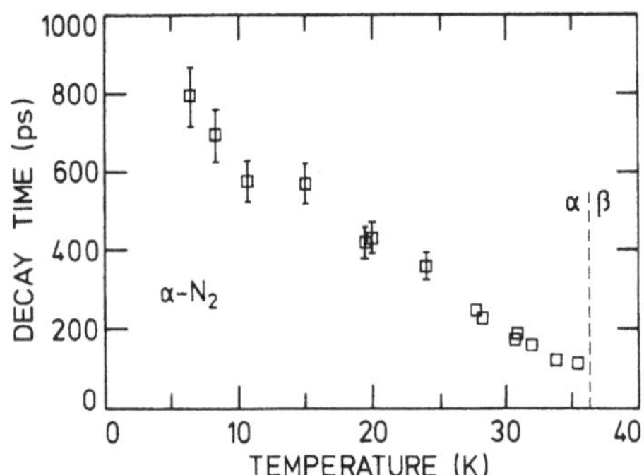


**Figure 15.** The homogeneous dephasing time, $T_2/2$, of the vibrons in $\alpha$-$N_2$ as a function of temperature. The $\alpha$-$\beta$ transition is indicated. (From Ref. [58].)

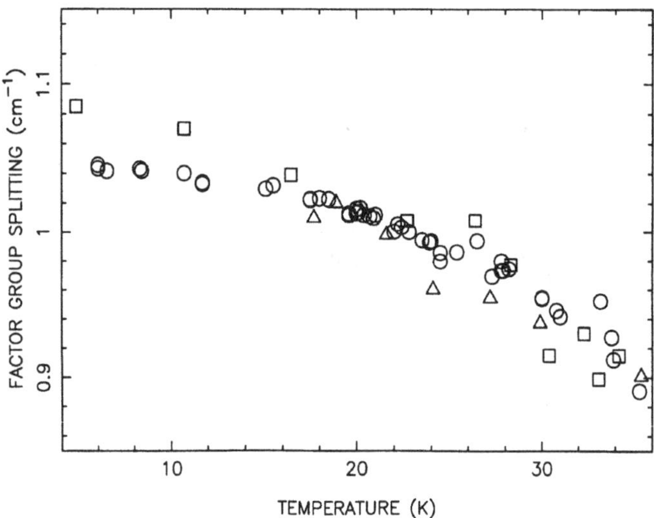


**Figure 16.** The $A_g$-$T_g$ factor group splitting of the vibrons in $\alpha$-$N_2$ as a function of temperature. Data are included from experiments using TR CARS (○), HR spontaneous Raman scattering (□) and HR stimulated Raman scattering (△) reported in Refs. [58], [21] and [82], respectively.

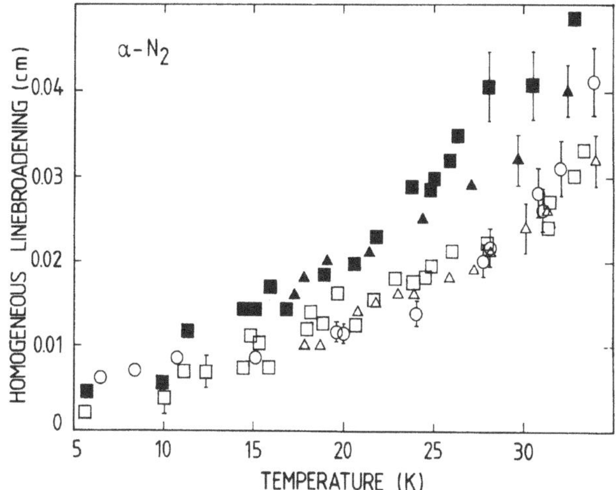


**Figure 17.** The measured linewidths of the $A_g$ and $T_g$ vibrons in $\alpha$-$N_2$, indicated by open and filled symbols, respectively, are shown as a function of temperature. Comparison is made between experiments using TR CARS (○), HR spontaneous Raman scattering (□) and HR stimulated Raman scattering (△) reported in Refs. [58], [21] and [82], respectively. The homogeneous widths were reported in the first case [58], but only total linewidths are available in the others [21, 82].

Very recently, Beck *et al.* [82] have measured the widths and the splitting of the vibron transitions by cw SRS at high resolution ($\approx 0.003$ cm$^{-1}$). The crystals were grown in a closed cell from the liquid in conditions close to the triple point, and then slowly cooled to the measuring temperature. These experiments were performed in order to obtain the precise solid state frequencies as a function of temperature, in relation to experiments with very large clusters of nitrogen molecules. The linewidths of and the splitting between the vibron transitions, measured between 17 K and the $\alpha$-$\beta$ phase transition, are shown in Figs. 16 and 17. However, only total linewidths have been reported without further information about the lineshape. In this paper no further interpretation of the data was presented.

Spontaneous Raman experiments were performed by Ouillon *et al.* [21] on $\alpha$-$N_2$ crystals, slowly grown by sublimation in the $\beta$-phase and then cooled to low temperature. The limiting resolution of their monochromator–Fabry-Perot tandem apparatus was $\approx 0.006$ cm$^{-1}$. Even narrower linewidths could be determined by careful fitting with a convolution of two Voigt profiles, one for the (independently known) instrumental transmission and one for the Raman spectrum. The Voigt profile $\mathcal{V}(\Delta_L, \Delta_G; \omega) = \mathcal{L}(\Delta_L; \omega) * \mathcal{G}(\Delta_G; \omega)$, is the convolution of a lorentzian, $\mathcal{L}$, and a gaussian, $\mathcal{G}$, with corresponding partial linewidths (FWHM), $\Delta_L$ and $\Delta_G$. Such a lineshape analysis is completely equivalent with the decay curve (18) adopted for fitting the TR CARS spectra, with $\Delta_L = 2/T_2$.

The data for both vibrons are in good agreement with those of Nibler *et al.* (see Fig. 17) in the T> 17 K range, although some discrepancies are found above 25 K for the weaker $T_g$ transition. Although full widths are reported, and not the homogeneous widths $\Delta_L$, the values for the $A_g$ vibron are in good agreement with those obtained from the exponential part of the decay in TR CARS [58], except maybe for the low temperatures ($< 10$ $K$) where the latter are somewhat larger. It should be noted that the linewidths determined by Ouillon *et al.* [21] at temperatures up to 15 K, are smaller than the announced resolution, which causes large error bars on these data, and probably large uncertainties concerning the details of the lineshape. Exactly in this region the signal is decaying slowly and small relative errors are found on the data in TR CARS, which in this case seem more reliable.

The Voigt line analysis of the HR spectra was performed, but only partial data about the lorentzian and gaussian fractions ($\Delta_L/\Delta$ and $\Delta_G/\Delta$, with $\Delta$ the Voigt FWHM) are discussed in this work: The gaussian fraction decreases between 15 K and 25 K from about 0.6 to 0.25, in favor of the lorentzian fraction, which is increasing from about 0.6 to 0.9. An interpretation of the gaussian contribution is given in terms of an homogeneous mechanism: The coupling with the librons would yield a stochastic dephasing mechanism which, at low temperatures, would yield a gaussian lineshape as is found in the slow modulation limit of motional narrowing. In order to support this proposal, a complete temperature dependence of the two components would have been desirable. The evolution of the gaussian versus lorentzian contribution is in qualitative agreement with the TR CARS measurements: At the lowest temperature they also have demonstrated comparable lorentzian and gaussian widths. However, in the TR CARS experiments the latter were observed to derive from genuine inhomogeneous effects. The gaussian linewidth was essentially temperature independent, as far as could be determined, but was found to vary with local crystalline conditions.

As seen from Fig. 16, the results for the factor group splitting determined with three experimental methods are in reasonably good agreement with each other. Again at low temperature discrepancies are observed, which can only result from unidentified systematic errors. There is no obvious reason why inaccuracies would occur in the measurements of the line splitting in HR Raman scattering. Also, in this range the decay is slow and the conditions are good for the TR CARS measurements, which is demonstrated by the very small scattering on these data. It should not be excluded that the local temperature in the interaction volume is raised in the TR CARS measurements due to the relatively high laser intensities. Possibly, this would also explain part of the discrepancies between the time- and frequency-domain determinations of the $A_g$ linewidth at the lowest temperatures.

**Dephasing decay of the $^{15}N_2$ Stretching Vibration** Experiments were undertaken in $^{15}N_2$ doped $\alpha$-$N_2$ crystals in order to investigate the influence of these impurities. In Fig. 14a the dephasing decay of the $A_g$ vibron is shown for crystals with different $^{15}N_2$ concentration up to 8.6 %. The decay becomes faster with increasing concentration, but a detailed analysis shows that only the inhomogeneous relaxation is becoming faster, and that no increase of the homogeneous decay rate $2/T_2$ could be observed. Therefore the faster dephasing cannot be attributed to scattering or trapping by these impurities. From this observation one can also expect a negligible effect of trapping at the $^{15}N^{14}N$ impurities, which are present at a concentration of 0.74 mol % in natural $N_2$, on the vibron relaxation in $\alpha$-$N_2$. In isotope enriched crystals the inhomogeneous width is not as much dependent on the crystal and on

**Table 5.** The inhomogeneous (gaussian) contribution to the linewidth of the vibrons and of the $^{15}N_2$ stretching vibration in $(^{15}N_2)_x(^{14}N_2)_{1-x}$ mixed crystals, as obtained from TR CARS measurements performed at 10 K.

| Concentration $^{15}N_2$ (mol %) | gaussian FWHM ($cm^{-1}$) $^{14}N_2$ | $^{15}N_2$ |
|---|---|---|
| 0.0 | 0.007[a] | |
| 1.7 | $0.018 \pm 0.004$ | $< 0.05$ |
| 4.7 | $0.029 \pm 0.005$ | — |
| 5.4 | $0.030 \pm 0.007$ | $0.16 \pm 0.04$ |
| 7.8 | $0.034 \pm 0.007$ | $0.19 \pm 0.06$ |
| 8.6 | $0.040 \pm 0.009$ | $0.20 \pm 0.05$ |

[a] Depending on specific crystal and interaction volume.

the interaction volume, and is increasing regularly with concentration (see Table 5). This is accounted for by the relative difference in molar volume between the two nitrogen isotopes, which amounts to $(0.74 \pm 0.20)$ % [53]. This small misfit of the $^{15}N_2$ impurity molecules in the $^{14}N_2$ host creates local strains which increase with concentration.

In contrast with the vibron spectra, which are subject to the exciton narrowing effect, those of the dilute impurities are expected to exhibit the width of the distribution of site energies, $D$ in equations (16) and (17). The TR CARS signal of the isotopic impurities, at 2249 $cm^{-1}$, could be detected at $^{15}N_2$ concentrations down to 1.7 mol %, which illustrates the sensitivity of our TR CARS experiments. This also provides an upper limit $D < 0.05$ $cm^{-1}$ for the undoped $\alpha$-$N_2$ crystals (Table 5). This is smaller than the values derived by Abram and Hochstrasser [47] from the nanosecond experiments ($D = 0.15$ $cm^{-1}$), but this could simply point to a better crystal quality. In heavily doped crystals, in which the isotopic substitution itself determines the strain distribution, their upper limit for the distribution width ($D < 0.25$ $cm^{-1}$ for 8 % [47]) is close to our result (Table 5).

These experiments further revealed a surprising result: Not only the inhomogeneous, but also the homogeneous relaxation of the localized vibrational states was found to be much faster than those of the $A_g$ vibrons. The exciton narrowing also affects the homogeneous linewidth! A dephasing time of the $^{15}N_2$ vibration of $T_2/2 = (160 \pm 10)$ ps was measured at 10 K, and no dependence on the $^{15}N_2$ concentration could be seen. Only a small decrease to $T_2/2 = (118 \pm 20)$ ps was found in measurements at 30 K. The low temperature value corresponds to an homogeneous width $\Delta_L = 0.033$ $cm^{-1}$.

The dephasing time of the vibrons tends towards that of the localized vibrations when the temperature approaches the $\alpha$-$\beta$ transition, above which the orientational order is disappearing. Anticipating on the interpretation of the vibron dephasing, the coherence of the vibrons seems to vanish close to the phase transition, and so is the exciton narrowing effect.

The inhomogeneous width is listed in Table 5 and one observes an increase which is parallel with that determined for the vibrons, with a factor $\approx 5$ between them. This inhomogeneous width can be considered to be the real spread $D$ of site energies, and, for the typical undoped $\alpha$-$N_2$ crystal, is bracketed by $0.007 < D < 0.05$ cm$^{-1}$. An increase of $D$ by at least a factor four was found at the highest concentration. It is surprising that in the nanosecond TR SRS measurements [47] the doping of the crystals with 10 mol % had no detectable influence on the shape of the decay curve.

**Static Disorder and Vibron-Libron Coupling.** On the basis of the new results of time- and frequency-domain investigations, it is worth reconsidering the model for the dephasing relaxation by scattering off weak static disorder, proposed by Abram *et al.* [47]. The picosecond dephasing processes are more complex than assumed in this model. The lineshapes with mixed lorentzian and gaussian character, are symmetrical, as far as could be determined. This is not unexpected because even the lowest measuring temperature, $T = 6$ K, was much larger than the vibron bandwidth. In such circumstances, however, lorentzian lines would be expected. Of course, the assumption of uncorrelated diagonal disorder is probably not realistic. The strong variations of the gaussian components with crystal and specific volume within it, for the undoped samples, points to large areas with different amounts and magnitudes of strain. Relatively slow variations of the vibrational frequency in space can be the source of genuine inhomogeneity. In the doped crystals the distance to an impurity can eventually cause the change in vibrational frequency of a host molecule, which would yield a strongly correlated disorder.

The difference in linewidth between the $A_g$ and $T_g$ vibrons becomes very small at low temperature, smaller than the experimental error on the frequency-domain data, and in fact so small that it cannot be reliably extracted from an analysis of the beating pattern in the time-resolved measurements (see Fig. 13). The small difference and its increase with temperature can as well result from a difference in efficiency of the dephasing of the vibrons by interaction with the librons, which was proposed by two research groups [21, 57, 58] as the main relaxation process. The vibron-libron interactions lead to intraband scattering processes which were described in detail by Ouillon *et al.* [21]. Also in this processes one expects higher probabilities for the $T_g$ compared to the $A_g$ vibron, because it can make transitions in a region with a much higher density of states. The temperature dependence of the linewidth was fitted on the basis of two fourth order diagrams, involving only librons with an effective frequency $\Omega = 32$ cm$^{-1}$, corresponding to the $E_g$ librons. Very recently, a sophisticated calculation [20] of the vibron and libron relaxation in $\alpha$-$N_2$ was performed using an accurate description of the anharmonic coupling between the modes, which was derived from a realistic intermolecular potential for the crystal. The perturbation calculation was performed to much higher order than the usual second or fourth order treatment. A reasonable agreement with experimental observations for the vibrons is thus obtained without fitting parameters.

The picture of stochastic perturbation of the vibrons by the librons was put forward by Ouillon *et al.* [21] in order to explain the gaussian component in the low temperature lineshapes, corresponding to the slow modulation case of motional narrowing. This type of process should be homogeneous, and is not expected to depend on the specific interaction volume, as was observed in the picosecond TR CARS experiments.

The measurements of the $^{15}N_2$ dephasing decay led us to the following, more pictorial description of the vibron dephasing [58]. The homogeneous linewidth of the vibrons at low temperature is strongly reduced, compared to that of the $^{15}N_2$ stretching vibration, by the exciton narrowing effect. This derives from the coherent nature of the quantum state over a large number of molecules. The interactions with the thermally excited librons destroy the coherence of the vibron state. For increasing temperatures towards the $\alpha$-$\beta$ phase transition, the relaxation rate of the remaining reduced vibron approaches that of the isolated molecule, that was independently investigated in the $^{15}N_2$ doped crystals.

An unsolved puzzle is presented by the observation of both short- and long-living contributions to the dephasing relaxation, demonstrated by the experiments on different time scales, which has to be confronted with the short lifetime of the vibrations of the localized vibrations of the isotopic impurities. The first observation points to the existence of a class of vibron states which at low temperature (at least up to 4.2 K) cannot evolve into the relatively fast decaying states, and one would be tempted to propose near-to-localized low lying states in the vibron band for this. It is hard to understand why such states would persist on a nanosecond time-scale, taking into account the $\approx$ 100 ps decay time of the $^{15}N_2$ stretching vibration.

# 7 VIBRATIONAL RELAXATION IN $Ar_x(N_2)_{1-x}$ MIXED CRYSTALS

## Experimental Results

A series of measurements of the vibrational relaxation was undertaken in $Ar_x(N_2)_{1-x}$ crystals in the $\alpha$-phase for concentrations up to 15 %, in order to investigate the influence of an atomic impurity —a "passive" entity for the exciton motion— on the vibron dynamics [59]. The decay of the TR CARS signal at the vibron frequency, $\Omega_v = 2328\ \mathrm{cm}^{-1}$, is shown in Fig. 18 for crystals with different Ar concentrations. The vibron dephasing is becoming faster with higher doping level and the decay is faster than exponential for each of the measurements. Therefore the spectra were analyzed using the same model decay curve (Eq. 18) as for those in pure and isotopically mixed crystals. In most of the measurements it was not possible to separately determine the decay parameters for the $A_g$ and $T_g$ mode, and the decay was described by (18), assuming $T_{2A} = T_{2T} = T_2$ and $\Delta_A = \Delta_T = \Delta$. Again, the derived parameters are more representative for the $A_g$ mode, which yields the dominant contribution to the signal. In Fig. 19 the inhomogeneous and homogeneous linewidths, and the $A_g$-$T_g$ factor group splitting of the vibrons are given as a function of temperature in crystals with Ar concentration ranging from zero to 15 %.

The factor group splitting between the two vibron transitions was determined in the same way as in the case of pure and isotopically mixed $\alpha$-$N_2$, using the Fourier transform method (as shown for pure $\alpha$-$N_2$ in Fig. 13). Especially at higher Ar concentrations, where a spread in beating frequencies was observed, this was more efficient than the direct fitting technique. At low temperature a near-to-linear increase of the line splitting with concentration was observed, resulting in a $\approx$ 17 % relative increase in a crystal containing 5 % Ar impurities. In the isotope enriched $\alpha$-$N_2$ samples the opposite effect was observed: a small decrease of the factor group splitting by $\approx$ 3 % in the 8.6 % doped crystals.

The homogeneous component is becoming faster with increasing concentration. The low-temperature dephasing time $T_2/2$ reduces from 800 ps in pure $N_2$ to 70 ps in the 7.5 % Ar crystal (see Fig. 19). The dephasing time was also found to decrease with increasing temperature, which in the pure crystals was ascribed to vibron-libron coupling. In one crystal, containing 2.75 % Ar, it was possible to obtain the separate dephasing times $T_2/2$ of the $A_g$ and $T_g$ vibrons (see Fig. 20), because exceptionally good signal-to-noise ratios were obtained as a result of favorable experimental circumstances (crystal quality and laser stability). At low temperature, a markedly shorter decay time was found for the $T_g$ vibron in this crystal, while that of the $A_g$ mode closely follows the results of the previous simplified analysis, as would be expected. The faster dephasing of the $T_g$ compared to $A_g$ mode is possibly related to the corresponding density of states, which is finite for the former and vanishing for the latter vibron in pure $\alpha$-$N_2$.

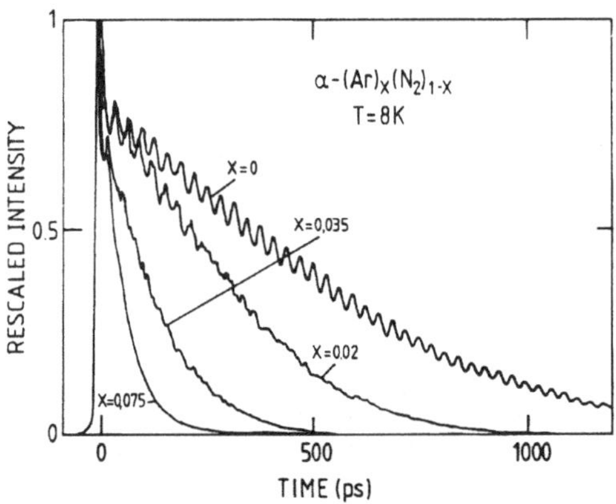


**Figure 18.** Normalized dephasing decay spectra of the vibrons in $\alpha$-$Ar_x(N_2)_{1-x}$ mixed crystals for different Ar concentrations ($T$ = 8 K) (from Ref. [59]).

The inhomogeneous linewidth at low temperature, $T = 8$ K, is increasing close to linearly from about 0.01 $cm^{-1}$ in pure $\alpha$-$N_2$ crystals, to 0.12 $cm^{-1}$ for the highest concentration, $\alpha$-$(Ar)_{15\,\%}(N_2)_{85\,\%}$. The broadening effect of Ar impurities is much larger than that of $^{15}N_2$, for the same concentration (see Table 5). In pure $\alpha$-$N_2$ and in crystals with Ar concentrations up to 2 %, the inhomogeneous width was observed to be independent of temperature. This was also the case for the $^{15}N_2$ isotope enriched crystals [58] at all measured concentrations. However, in the highly doped $\alpha$-$Ar_x(N_2)_{1-x}$ crystals the situation is drastically altered. The inhomogeneous width derived from the fitting analysis of the decay curves is increasing with temperature, and the temperature at which this sets in is decreasing with increasing Ar concentration (see Fig. 19c).

## Suppression of Vibron States by Coupling to the Librons

The argon atom fits quite well in the nitrogen crystal which preserves its structure in the concentration range of the experiments: However, the lattice parameter, $a$, of solid argon is smaller by nearly 7 % compared to $\alpha$-$N_2$. For $^{15}N_2$ the difference is much smaller, only a 0.24 % decrease of $a$. This rationalizes the much higher impact

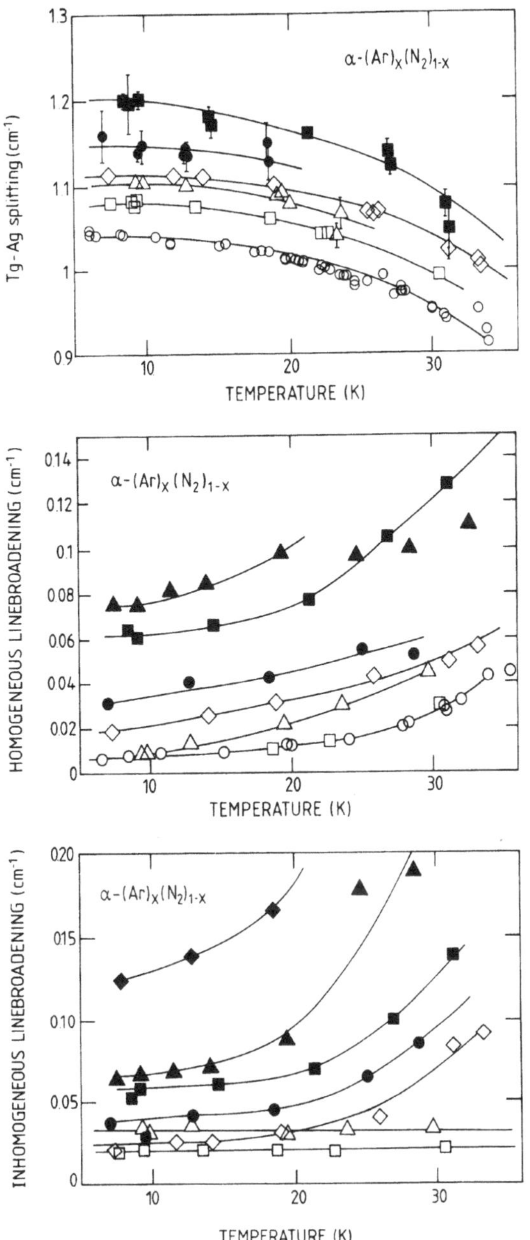


**Figure 19.** The beating frequency between the $A_g$ and $T_g$ vibrons, and their homogeneous and inhomogeneous linewidths as a function temperature and Ar concentration. The open symbols refer to samples of pure $\alpha$-$N_2$ for circles, and to Ar concentrations of 1 % for squares, 2 % for triangles and 2.75 % for diamonds. Similarly, closed symbols correspond to 3.5 % for circles, 5 % for squares, 7.5 % for triangles and 15 % for diamonds. The solid lines are only a guide to the eye. (From Ref. [59].)

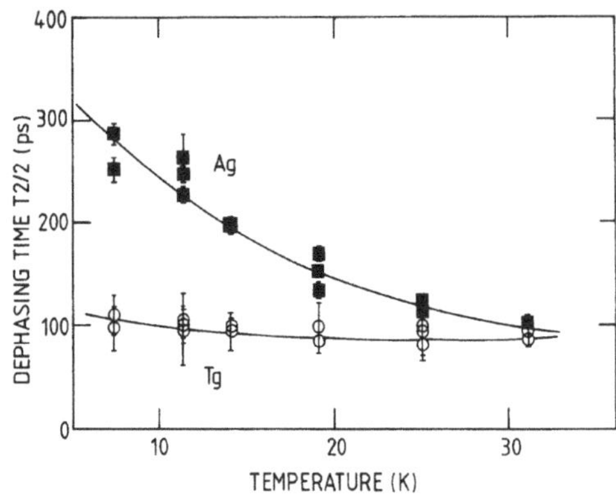


**Figure 20.** The dephasing times for the $A_g$ and $T_g$ vibron in $\alpha$-$(Ar)_{2.75\,\%}(N_2)_{97.25\,\%}$ could be separately determined using equation (18). (From Ref. [59].)

of Ar impurities on the different decay parameters. The size mismatch introduces strain effects in the crystal which consequent line broadening effects. For the vibron relaxation in $\alpha$-$N_2$ crystals, the effects of weak Ar doping or of $^{15}N_2$ enrichment are quite similar: The homogeneous decay time $T_2/2$ is not affected, and the inhomogeneous width, $\Delta$, is increasing with concentration but is independent of temperature. Above about 2 % Ar concentration, this drastically changes: The dephasing time decreases with concentration and the inhomogeneous broadening becomes temperature dependent (see Fig. 19). This points to the onset of very different mechanisms of dephasing relaxation.

An important property, which has not yet been underlined, is the spherical shape of the Ar atom, which is not participating in anisotropic interactions —electrostatic and repulsive— with the neighboring molecules. As a result, the Ar impurities, much more than isotopic impurities, introduce a disorder in the equilibrium orientations of the $N_2$ molecules in the mixed crystals, and this will also affect the $N_2$ librational motion. A direct indication for this was found in recent spontaneous Raman measurements of the low-frequency excitations in $\alpha$-$Ar_x(N_2)_{1-x}$ mixed crystals [60]. The narrow lines of the low-temperature librons, $E_g$ and $T_g$, are broadened with increasing Ar concentration and merge in a single broad and slightly asymmetric line. This is very similar to the evolution of this spectrum in pure crystals when the temperature increases and the amplitude of the vibrational motion is gradually increasing to large values towards the $\alpha$-$\beta$ phase transition. An analogous behavior was observed for the vibron-weighed density of states of the librons, which was measured through the vibron-libron combination Raman spectra. The large amplitude molecular librations can in turn affect the vibron dephasing, taking into account that the librons were found to be responsible for the homogeneous dephasing in $\alpha$-$N_2$.

Although the problem is a complicated one, we could draw the following picture of the relaxation effects in $\alpha$-$Ar_x(N_2)_{1-x}$ mixed crystals. At low temperature, the coherence of the vibrons is only weakly influenced by the librons, which still have a low thermal population, and the exciton narrowing effect is still operative to a certain extent. Increasing the Ar content increases the linewidth of the vibrons in two ways. At equal efficiency of the narrowing effect, the broadening of the site energy distribution is reflected in the apparent inhomogeneous vibron width (see Fig. 19c at $T = 8$ K). But also, the larger librational motion of $N_2$ molecules in the neighborhood of the smaller and spherical Ar impurities can affect the homogeneous relaxation of the vibrons through the vibron-libron coupling (see Fig. 19b at $T = 8$ K). The decrease of the $\alpha$-$\beta$ phase transition temperature with increasing Ar concentration could be related to the increasing librational amplitudes. Also with increasing temperature, the librational amplitude becomes larger and this enhances the efficiency of the vibron-libron scattering. The libron coherence is now disappearing at even lower temperatures than in the pure crystal, due to the static and dynamical disorder and the larger librational motions.

An as yet unexplained effect is the large increase of the factor group splitting for increasing Ar concentration (Fig. 19a). For decreasing concentration of the host molecules one would expect a reduction of the vibron bandwidth, roughly proportional to their concentration [52, Chap. 4]. Such results were recently also found for $N_2$:CO mixed crystals in a HR Raman investigation of the $N_2$ vibrons [105]. We first considered that the coupling [72, 73] between the molecular vibrations, which gives rise to the vibron states, could be enhanced by the slightly smaller lattice distance in the mixed crystals induced by the smaller size of Ar compared to $N_2$. The simple scaling of these parameters according to their dependence on the on the n.n. distance was completely insufficient to explain the observed increase of the line splitting. However, an important change of the interaction parameters could come from changes in the librational motion of the $N_2$ molecules. The intermolecular interactions must be averaged over the orientational motion. In the pure $\alpha$-$N_2$ crystal each of the molecules is librating around its order-defined $\langle 111 \rangle$ axis, with the same angular distribution. This situation is now strongly perturbed. Possibly, the changes in the distribution of equilibrium orientations, and in the librational motion around them, are at the origin of appreciably enhanced coupling between the stretching vibrations on neighboring molecules. It should be noted that this is not the case in $N_2$:CO crystals, in which the CO molecules help to preserve the orientational order, because they possess similar anisotropic properties as the $N_2$ molecules. Only two types of disorder occur: substitutional disorder as in $Ar_x(N_2)_{1-x}$, and head-tail disorder for the orientation of the heteronuclear impurity.

# 8 CONCLUDING REMARKS

The rotational and vibrational excitations in para-$H_2$ and $\alpha$-$N_2$ crystals are very suitable systems for the study of intraband dephasing processes, because of the exceptional properties of the quasi-decoupled excitations and because of the large variety of the phenomena involved. It would be very interesting to obtain comparable experimental information on other diatomic molecular crystals, e.g., CO and $O_2$, in which analogous relaxation processes are expected to occur.

In the case of the rotons in para-$H_2$ the Raman-active excitations are situated at the center of the band, and can elastically scatter into many modes in the Brillouin zone. The resulting single-exponential decay rates were measured in pure crystals and in crystals with varying impurity concentrations. The impurity scattering rates could even be accurately calculated, but the description of the relaxation in the pure crystal is still a matter of controversy.

In para-$H_2$ crystals the dephasing of the vibrons is shown to be extremely slow, as demonstrated by both nanosecond time-resolved and by frequency-domain experiments. No influence of thermal excitation of the lattice modes on the vibron dephasing has yet been observed, probably because the coupling with the vibrons is to weak, but further experiments are needed to confirm this observation. Again, scattering by ortho-$H_2$ impurities was demonstrated, but two factors complicate the description: The vibron density of states at the energy of the Raman-active vibron is zero and the impurity state is close in energy to the band, a situation intermediate between the separate band and the amalgamation limit. Additional information about the vibron relaxation in the picosecond time range would be of much interest.

The vibron relaxation in $\alpha$-$N_2$ has been investigate both in the nanosecond and in the picosecond time range, by time- and frequency-domain techniques. At low temperature, scattering by crystal imperfections is the dominant dephasing process, which leads to non-exponential decay and to non-lorentzian lineshapes for the Raman modes. A unified description for the short- and long-time behavior is however still missing. The effect of exciton narrowing was demonstrated by measurements on the $(^{15}N_2)_x(^{14}N_2)_{1-x}$ crystals by the much faster dephasing of the $^{15}N_2$ stretching vibration. At increasing temperatures the coupling with the librational modes increases the relaxation rate by intraband scattering, and this completely eliminates the exciton narrowing effect towards the $\alpha - \beta$ phase transition. Reasonable agreement is obtained in a very recent calculation of the relaxation as a function of temperature [20]. Extensive information has been gathered about the vibron relaxation in Ar-doped $\alpha$-$N_2$ crystals. Although this atom fits relatively well in the host crystal, it is found that it acts as a strong perturbation for the vibron dynamics, by the absence of an internal mode at the impurity site, by the smaller volume of the atom compared to the molecules, and by the absence of anisotropy in its interactions with the neighbors. A strong suppression of the formation of vibron states occurs even at relatively low concentrations.

## Acknowledgments

At his point I want to thank those who have contributed to our research on the various subjects reported here: J. De Kinder for the work on the vibrons in $N_2$, X.-Y. Chen, M. Leblans and C. Sierens for the investigations on rotons in para-$H_2$, A. Bouwen for the dedicated growing of perfect single crystals, and D. Schoemaker for many discussions and constant support. Furthermore, interesting discussions with M. Vanhimbeeck, H. De Raedt and A. Lagendijk, and excellent technical support by P. Casteels are gratefully acknowledged.

The author thanks the National Fund for Scientific Research (NFWO) for personal financial support. This work was made possible by further financial support from the Inter-University Institute for Nuclear Sciences (IIKW), and from the Geconcerteerde Acties (Ministerie van Wetenschapsbeleid, Belgium).

## References

1. P. C. M. Planken, L. D. Noordam, J.T.M. Kennis, and A. Lagendijk, Femtosecond time-resolved study of phonon–polaritons in $LiNb_3$, *Phys. Rev. B* 45:7106 (1992).

2. H. J. Bakker, P. C. M. Planken, L. Kuipers, and A. Lagendijk, Ultrafast infrared saturation spectroscopy of chloroform, bromoform, and iodoform, *J. Chem. Phys.* 94:1730 (1991).
3. E. J. Heilweil, M. P. Cassassa, R. R. Cavanagh, and J. C. Stephenson, Picosecond vibrational energy transfer studies of surface adsorbates, *Ann. Rev. Phys. Chem.* 40:143 (1989).
4. J. D. Beckerle, R. R. Cavanagh, M. P. Cassassa, E. J. Heilweil, and J. C. Stephenson, Subpicosecond transient infrared spectroscopy of adsorbates — vibrational dynamics of Co/Pt(111), *J. Chem. Phys.* 95:5403 (1991).
5. H. Graener, T. Q. Ye, and A. Laubereau, Transient-hole burning in the infrared spectrum of a polymer with intense picosecond pulses, *Phys. Rev. B* 41:2597 (1990).
6. A. Laubereau and W. Kaiser, Vibrational dynamics of liquids and solids investigated by picosecond light pulses, *Rev. Mod. Phys.* 50:607 (1978).
7. S. Califano, Phonon lifetimes in molecular crystals, *in:* "Applied Physics Spectroscopy", page 269, W. Demtröder and M. Inguscio, eds., Plenum Press, New York - London (1990).
8. P. Ranson, R. Ouillon, J.-P. Lemaistre, and G. M. Gale, Incoherent and coherent high resolution Raman spectroscopy: Complementary tools to study bound states in molecular crystals, *in:* "Recent Trends in Raman Spectroscopy", page 127, S. B. Bauergie and S. S. Jha, eds., World Scientific, Singapore (1989).
9. M. De Mazière, C. Sierens, and D. Schoemaker, Analysis of dephasing signal in picosecond stimulated-Raman-gain experiments, *J. Opt. Soc. Amer. B* 6:2376 (1989).
10. M. van Exter and A. Lagendijk, What is measured in a time-resolved stimulated Raman experiment? *Opt. Commun.* 56:191 (1985).
11. A. J. Sievers and W. E. Moerner, Persistent infrared hole-burning for impurity vibrational modes in solids, *in:* "Persistent Spectral Hole-Burning: Science and Applications", vol. 44 of *Topics in Current Physics*, W. E. Moerner, ed., Springer-Verlag, Berlin (1988).
12. M. D. Levenson and S. S. Kano, "Introduction to Nonlinear Laser Spectroscopy", Academic Press, New York–London (1988).
13. S. Califano, V. Schettino, and N. Neto, "Lattice Dynamics in Molecular Crystals", vol. 26 of *Lecture Notes in Chemistry*, Springer Verlag, Berlin (1981).
14. S. Velsko and R. M. Hochstrasser, Studies of vibrational relaxation in low-temperature molecular crystals using coherent Raman spectroscopy, *J. Phys. Chem.* 89:2240 (1985).
15. D. D. Dlott, Optical phonon dynamics in molecular crystals, *Ann. Rev. Phys. Chem.* 37:157 (1986).
16. S. Califano and V. Schettino, Vibrational relaxation in molecular crystals, *Intern. Rev. Phys. Chem.* 7:19 (1988).
17. M. Vanhimbeeck, H. De Raedt, A. Lagendijk, and D. Schoemaker, Calculation of rotational $T_2$ relaxation in solid parahydrogen and orthodeuterium, *Phys. Rev. B* 33:4264 (1986).
18. J. i. Igarashi, Calculation of Raman linewidth due to roton in solid parahydrogen with orthohydrogen impurities, *J. Phys. Soc. Jpn.* 56:4586 (1987).
19. X. Y. Chen, E. Goovaerts, and D. Schoemaker, Scattering-model calculation of the impurity-induced dephasing relaxation rates of the Raman-active J=2 rotons in parahydrogen, *Phys. Rev. B* 38:1450 (1988).

20. P. Procacci, G. F. Signorini, and R. G. Della Valle, Efficient calculation of high order self-energy corrections to the phonon linewidths: application to $\alpha$-nitrogen, to be published.

21. R. Ouillon, C. Turc, J.-P. Lemaistre, and P. Ranson, High resolution Raman spectroscopy in the $\alpha$ and $\beta$ crystalline phases of $N_2$, *J. Chem. Phys.* 93:3005 (1990).

22. G. F. Signorini, P. F. Fracassi, R. Righini, and R. G. Della Valle, Energy decay mechanisms and anharmonic lattice dynamics: The case of solid nitrogen, *Chem. Phys.* 100:315 (1985).

23. F. Ho, W.-S. Tsay, J. Trout, S. Velsko, and R. M. Hochstrasser, Picosecond time-resolved CARS in isotopically mixed crystals of benzene, *Chem. Phys. Lett.* 141:97 (1983).

24. S. Velsko, J. Trout, and R. M. Hochstrasser, Quantum beating of vibrational factor group components in molecular solids, *J. Chem. Phys.* 79:2114 (1983).

25. J. Trout, S. Velsko, R. Bozio, P. L. Decola, and R. M. Hochstrasser, Nonlinear Raman study of line shapes and relaxation of vibrational states of isotopically pure and mixed crystals of benzene, *J. Chem. Phys.* 81:4746 (1985).

26. B. H. Hesp and D. A. Wiersma, Vibrational relaxation in neat crystals of naphthalene by picosecond CARS, *Chem. Phys. Lett.* 75:423 (1980).

27. K. Duppen, B. M. M. Hesp, and D. A. Wiersma, A picosecond CARS study of librons in naphthalene, *Chem. Phys. Lett.* 79:399 (1981).

28. D. D. Dlott, C. L. Schosser, and E. L. Chronister, Temperature-dependent vibrational dephasing in molecular crystals: A picosecond CARS study, *Chem. Phys. Lett.* 90:386 (1982).

29. R. Ouillon, P. Ranson, and S. Califano, Temperature dependence of the bandwidths and frequencies of some anthracene phonons. High-resolution Raman measurements, *Chem. Phys.* 91:119 (1984).

30. C. L. Schosser and D. D. Dlott, A picosecond CARS study of vibron dynamics in molecular crystals: Temperature dependence of homogeneous and inhomogeneous linewidths, *J. Chem. Phys.* 80:1394 (1984).

31. T. J. Kosic, R. E. Cline Jr., and D. D. Dlott, Picosecond coherent Raman investigation of the relaxation of low frequency vibrational modes in amino acids and peptides, *J. Chem. Phys.* 81:4932 (1984).

32. J. Kalus, Temperature dependence of phonon-frequencies and -linewidths for weakly anharmonic molecular crystals, *J. de Chimie Physique* 82:137 (1985).

33. R. Righini, L. Angeloni, E. Castellucci, and P. Foggi, Temperature dependent dephasing of internal phonons in ionic crystals: Picosecond CARS measurements in $K_2SO_4$ and $KClO_4$ crystals, *J. Molecular Struct.* 175:123 (1988).

34. R. Righini, L. Angeloni, E. Castellucci, P. Foggi, and S. Califano, Relaxation of vibrational excitons in molecular-ionic crystals measured by picosecond time-resolved CARS, *Croatica Chemica Acta* 61:621 (1988).

35. L. Angeloni and R. Righini, Anomalous temperature dependence of the vibrational exciton lifetime in $NaNO_3$ crystal, *Chem. Phys. Lett.* 154:115 (1989).

36. R. Righini, L. Angeloni, P. Foggi, E. Castellucci, and S. Califano, Temperature-dependent (10–240 K) relaxation of vibrational excitons in $KClO_4$ crystal, *Croatica Chemica Acta* 61:621 (1988).

37. R. Bini, P. Foggi, P. R. Salvi, R. Simone, and V. Schettino, Vibrational relaxation of Davidov components of the 940 $cm^{-1}$ mode in $KClO_4$, *J. Molecular Struct.* 219:43 (1990).
38. G. M. Gale and P. Schanne, Time-resolved relaxation of one- and two-vibron states in the rubidium perchlorate crystal, *Chem. Phys.* 151:127 (1991).
39. R. Ouillon, P. Ranson, and S. Califano, High resolution Raman spectra of the Fermi diad in crystalline $CO_2$ at 6 K, *J. Chem. Phys.* 83:2162 (1985).
40. G. M. Gale, P. Guyot-Sionnest, W. Q. Zheng, and C. Flytzanis, Time-resolved nonlinear spectroscopy of a Fermi doublet: The $\{\nu_1, 2\nu_2\}$ Fermi resonance in $CO_2$ solid, *Phys. Rev. Lett.* 54:823 (1985).
41. R. Ouillon and P. Ranson, Time resolved study of the Fermi component $\omega(A_g)$ in solid $CO_2$ at $T \approx 5$ K, *J. Chem. Phys.* 85:1295 (1986).
42. G. M. Gale, P. Schanne, and P. Ranson, Vibrational dephasing of biphonon satellites in solid $CO_2$, *Chem. Phys.* 131:455 (1989).
43. P. Ranson, R. Ouillon, and S. Califano, Relaxation processes in molecular crystals. Temperature dependence of Raman active lattice phonons of crystalline $CO_2$, *J. Chem. Phys.* 89:3592 (1988).
44. L. Angeloni, R. Righini, P. R. Salvi, and V. Schettino, Relaxation dynamics of Fermi doublets in $CS_2$ crystals, *Chem. Phys. Lett.* 154:432 (1989).
45. F. Vallée, G. M. Gale, and C. Flytzanis, Time- and space-resolved study of a dressed polariton: The polariton Fermi resonance in ammonium chloride, *Phys. Rev. Lett.* 61:2102 (1988).
46. M. Baggen and A. Lagendijk, Time-resolved study of vibrational relaxation in solid $CO_2$ at high pressures, *Chem. Phys. Lett.* 177:361 (1991).
47. I. I. Abram, R. M. Hochstrasser, J. E. Kohl, M. G. Semack, and D. White, Coherence loss for vibrational and librational excitations in solid hydrogen, *J. Chem. Phys.* 71:153 (1979).
48. I. I. Abram, R. M. Hochstrasser, J. E. Kohl, and M. G. Semack, Decay of vibrational coherence in ortho-para mixtures of solid hydrogen, *Chem. Phys. Lett.* 71:405 (1980).
49. S. R. J. Breuck and R. M. Osgood, Vibrational energy relaxation in liquid $N_2$–CO mixtures, *Chem. Phys. Lett.* 39:568 (1976).
50. H.W. Löwen, K. D. Bier, and H. J. Jodl, Vibron-phonon excitations in the molecular crystals $N_2$, $O_2$, and CO by Fourier transform infrared and Raman studies, *J. Chem. Phys.* 93:8565 (1990).
51. C.-Y. Kuo, R. J. Kerl, N. D. Patel, and C. K. N. Patel, Nonradiative relaxation of excited vibrational states of solid hydrogen, *Phys. Rev. Lett.* 53:2575 (1984).
52. J. Van Kranendonk, "Solid hydrogen", Plenum Press, New York, London (1983).
53. T. A. Scott, Solid and liquid nitrogen, *Phys. Reports* 27:89 (1976).
54. E. Goovaerts, X. Y. Chen, A. Bouwen, and D. Schoemaker, Relaxation times of $\boldsymbol{k} = 0$ rotons in pure parahydrogen crystals and roton scattering by orthohydrogen impurities, *Phys. Rev. Lett.* 57:479 (1986).
55. C. Sierens, A. Bouwen, E. Goovaerts, M. De Mazière, and D. Schoemaker, Roton relaxation in parahydrogen crystals measured by time-resolved stimulated Raman gain, *Phys. Rev. A* 37:4769 (1988).

56. M. Leblans, A. Bouwen, C. Sierens, W. Joosen, E. Goovaerts, and D. Schoemaker, Dephasing relaxation of J=2 rotons in parahydrogen crystals doped with hydrogen-deuterium impurities, *Phys. Rev. B* 40:6674 (1989).

57. J. De Kinder, E. Goovaerts, A. Bouwen, and D. Schoemaker, Dephasing times of the stretching vibration in liquid $N_2$ and of the vibrons in the $\alpha$ and $\beta$ crystalline phases, *J. Lumin.* 45:423 (1990).

58. J. De Kinder, E. Goovaerts, A. Bouwen, and D. Schoemaker, Dephasing times of the vibrons in $\alpha$-$N_2$ and in $\alpha$-$(^{15}N_2)_x(^{14}N_2)_{1-x}$, *Phys. Rev. B* 42:5953 (1990).

59. J. De Kinder, A. Bouwen, E. Goovaerts, and D. Schoemaker, Suppression of vibron state formation in $Ar_x(N_2)_{1-x}$ mixed crystals, *J. Chem. Phys.* 95:2269 (1991).

60. J. De Kinder, E. Goovaerts, A. Bouwen, and D. Schoemaker, Raman study of the librational states in $\alpha$-$Ar_x(N_2)_{1-x}$ mixed crystals, *J. Lumin.* 53:72 (1992).

61. T. Momose, D.P. Weliky, and T. Oka, The stimulated Raman gain spectrum of the $Q_{1\leftarrow 0}$ transition of solid parahydrogen, *J. Mol. Spectr.* 153:760 (1992).

62. R. S. Knox, "Theory of Excitons", vol. 5 of *Solid State Physics*, Academic Press, New York–London (1963).

63. G. W. Robinson, Electronic and vibrational excitons in molecular crystals, *Ann. Rev. Phys. Chem.* 21:429 (1970).

64. C. B. Harris and D. A. Zwemer, Coherent energy transfer in solids, *Ann. Rev. Phys. Chem.* 29:473 (1978).

65. J. Klaftner and J. Jortner, Effects of structural disorder on the optical properties of molecular crystals, *J. Chem. Phys.* 68:1513 (1978).

66. I. I. Abram and R. M. Hochstrasser, Theory of the time evolution of exciton coherence in weakly disordered crystals, *J. Chem. Phys.* 72:3617 (1980).

67. I. I. Abram and R. M. Hochstrasser, Optical coherence properties of collective molecular systems: A comparison of the statistical and quantum approaches, *J. Chem. Phys.* 75:337 (1981).

68. V. M. Kenkre and P. Reineker, "Exciton Dynamics in Molecular Crystals and Aggregates", vol. 94 of *Springer Tracts in Modern Physics*, Springer-Verlag, Berlin-Heidelberg-New York (1982).

69. I. F. Silvera, The solid molecular hydrogens in the condensed phase: Fundamentals and static properties, *Rev. Mod. Phys.* 52:393 (1980).

70. K. Kobashi and T. Kihara, Molecular librations and the $\alpha$-$\gamma$ phase transition in solid nitrogen based on the Kihara potential, *J. Chem. Phys.* 72:378 (1980).

71. T. N. Antsygina, V. A. Slusarev, Y. A. Freiman, and A. I. Erenburg, Anharmonic effects in librational motion of $N_2$-type crystal, *J. Low Temp. Phys.* 56:331 (1984).

72. G. Zumofen and K. Dressler, *A priori* calculation of the vibron band and of the vibron-phonon and two-phonon combination bands in the Raman spectrum of solid $\alpha$-$N_2$, *J. Chem. Phys.* 64:5198 (1976).

73. G. Zumofen, K. Dressler, M. M. Thiéry, and V. Chandrasekharan, On the *a priori* calculation of the vibron band in the Raman spectrum of solid $\alpha$-$N_2$, *J. Chem. Phys.* 67:5983 (1978).

74. A. P. J. Jansen, W. J. Briels, and A. van der Avoird, *Ab initio* description of large molecule motions in solid $N_2$. I. Librons in the ordered $\alpha$ and $\gamma$ phases, *J. Chem. Phys.* 81:3648 (1984).

75. A. van der Avoird, W. J. Briels, and A. P. J. Jansen, *Ab initio* description of large molecule motions in solid $N_2$. II. Librons in the $\beta$-phase and the $\alpha - \beta$ phase transition, *J. Chem. Phys.* 81:3658 (1984).
76. W. B. J. M. Janssen, and A. van der Avoird, Dynamics and phase transitions in solid ortho/para hydrogen and deuterium from an *ab initio* potential, *Phys. Rev. B* 42:838 (1990).
77. R. W. Hellwarth, Third-order optical susceptibilities of liquids and solids, *Progr. Quant. Electr.* 5:1 (1977).
78. M. van Exter, "Ultra-fast Spectroscopy of Vibrational Excitations and Surface Plasmons", Ph.D. thesis, Universiteit van Amsterdam, The Netherlands (1988).
79. M. De Mazière and D. Schoemaker, Phase modulation technique for eliminating phase noise in picosecond $T_2$ measurements based on stimulated Raman gain, *J. Appl. Phys.* 58:1439 (1985).
80. M. van Exter and A. Lagendijk, Converting an AM radio into a high-frequency lock-in amplifier in a stimulated Raman experiment, *Rev. Sci. Instrum.* 57:390 (1986).
81. M. van Exter, A. Lagendijk, and E. Spaans, Interference phenomena in time-resolved stimulated Raman measurements, *Opt. Commun.* 59:411 (1986).
82. R. D. Beck, M. F. Hineman, and J. W. Nibler, Stimulated Raman probing of supercooling and phase transitions in large $N_2$ clusters formed in free expansions, *J. Chem. Phys.* 92:7068 (1990).
83. G. Herzberg, "Molecular Spectra and Molecular Structure: I. Spectra of Diatomic Molecules", Van Nostrand Reinhold, second edition (1950).
84. I. F. Silvera, "Simple Molecular Systems at Very High Density", Plenum Press, New York and London (1989).
85. W. Press, B. Janik, and H. Grimm, Frozen-in orientational disorder in mixtures of solid nitrogen and argon, *Z. Phys. B - Cond. Matter* 49:9 (1982).
86. H. Klee, H. O. Carmesin, and K. Knorr, Quadrupolar glass state of $Ar_{1-x}(N_2)_x$, *Phys. Rev. Lett.* 61:1855 (1988).
87. H. Klee and K. Knorr, Orientational-ordering transition fcc-$Pa_3$ of $Ar_{1-x}(N_2)_x$, *Phys. Rev. B* 43:8658 (1991).
88. J. De Kinder, E. Goovaerts, A. Bouwen, and D. Schoemaker, Evidence for the orientationally disordered cubic phase of $Ar_{0.15}(N_2)_{0.85}$ from librational and vibrational Raman scattering, *Phys. Rev. B* 44:10369 (1991).
89. A. Lagendijk, R. J. Wijngaarden, and I. F. Silvera, A model for the determination of the crystal field potential in $H_2$ and $D_2$ from high pressure experiments, *Phys. Lett.* 99A:97 (1983).
90. J. Van Kranendonk and G. Karl, Theory of the rotational and vibrational excitations in solid parahydrogen, and the frequency analysis of the infrared and Raman spectra, *Rev. Mod. Phys.* 40:531 (1968).
91. V. Soots, E. J. Allin, and H. L. Welsh, Variation of the Raman spectrum of solid hydrogen with ortho-para ratio, *Can. J. Phys.* 43:1985 (1965).
92. J. P. McTague, I. F. Silvera, and W. N . Hardy, Rotational Raman scattering in single crystal parahydrogen, orthodeuterium and HD, *in: Proc. Intern. Conf. on Light Scattering in Solids*, page 456, M. Balkanski, ed., Flammarion, Paris (1971).
93. P. J. Berkhout and I. F. Silvera, Mixing of rotational states, breakdown of the independent polarizability approximation and renormalized interactions in solid hydrogens under pressure, *Communications on Physics* 2:109 (1977).

94. L. Lassche, I. F. Silvera, and A. Lagendijk, Observation of the J=4 roton band in solid deuterium under pressure, *Phys. Rev. Lett.* 65:2677 (1990).

95. A. Lagendijk and I. F. Silvera, Roton softening in the solid hydrogens, *Phys. Lett.* 84A:28 (1981).

96. J. i. Igarashi, An anharmonic theory of the rotational excitation in solid ortho-deuterium under high pressure, *J. Phys. Soc. Jpn.* 53:2629 (1984).

97. N. H. Rich, M. J. Clouter, H. Kiefte, and S. F. Ahmad, Low frequency Raman spectra of the liquid and plastic crystal phases of $O_2$, $N_2$, CO, and of Ar crystals with these molecules as guests, *Can. J. Phys.* 60:1358 (1982).

98. G. Pangilinan, G. Guelachvili, R. Sooryakumar, K. N. Rao, and R. H. Tipping, Raman study of the $\alpha \rightarrow \beta$ structural phase transition of solid $N_2$, *Phys. Rev. B* 39:2522 (1989).

99. M.-C. Chan and T. Oka, Observation of the $U_{1\leftarrow 0}(1)$ transition of solid deuterium, *J. Chem. Phys.* 93:979 (1990).

100. J. De Kinder, "Dephasing of the Vibrons in Solid Nitrogen", Ph.D. thesis, Universiteit Antwerpen (U.I.A.), Belgium (1991).

101. R. Beck and J. W. Nibler, High resolution Raman loss spectra of solid $\alpha$-nitrogen and of matrix-isolated molecules, *Chem. Phys. Lett.* 159:79 (1989).

102. U. Buontempo, S. Cunsolo, P. Dore, and L. Nencini, Far infrared absorption spectra in liquid and solid $H_2$, *Can. J. Phys.* 60:1422 (1982).

103. S. Velsko and R. M. Hochstrasser, Theory of vibrational coherence and population decay in isotopically mixed crystals, *J. Chem. Phys.* 82:2180 (1985).

104. P. J. Berkhout, "Properties of Solid Molecular Hydrogens as a Function of Density and Ortho-Para Concentration as Studied by Raman Scattering", Ph.D. thesis, Universiteit van Amsterdam, The Netherlands (1978).

105. C. Turc, B. Perrin, R. Ouillon, and P. Ranson, Effect of orientational and substitutional disorder on the $N_2$ vibrational exciton in $N_2$/CO mixed crystals, *J. Chem. Phys.* 161:39 (1992).

# QUANTUM TRANSIENT TRANSPORT

Carlo Jacoboni and Fausto Rossi

Dipartimento di Fisica
Università di Modena
Via Campi 213/A
41100 Modena, Italy

## 1. INTRODUCTION

Charge transport in semiconductors is a subject of great technological importance, owing to its relevance for microelectronic devices. Also from a more basic point of view, this phenomenon has received large interest, since it allows the analysis of many fundamental aspects of solid-state physics, such as electron states, both localized and extended, surface and interface structures, electron-phonon interactions, impact ionization, etc. These two aspects of the discipline are not independent. In fact, the same technological improvements induced by, and devoted to, microelectronic interests, have allowed the fabrication of sophisticated physical systems with extremely well controlled geometries and compositions, where basic-physics research can be performed with great accuracy. Let us just mention, as example, the discovery of the Quantum Hall Effect[1] or the astonishing achievements of quantum interference devices[2].

From a microscopic point of view, charge transport is not a simple phenomenon to investigate, because of the large complexity of the processes involved. When a mobility experiment is performed, the value finally obtained in the measurement depends upon a number of physical properties of the material under investigation, such as its band structure, impurity content, and electron-phonon couplings. Such quantities, in particular the last ones, are in general not well known, and charge transport experiments do not give enough information, as a rule, to allow their unambiguous determination. Furthermore, the theory suffers of particular approximations (such as analytical expressions for the electron band structure or the phonon-dispersion curves, or the coupling constants that should not be constants), and it seems illusory to hope to overcome them within a reasonable amount of time.

The basic equation that describes electron transport in semiclassical terms is the Boltzmann equation (BE). Such an equation does not help too much in making the picture simpler, since it has a complicated integro-differential form that does not offer analytical solutions, except for very few simple cases, not applicable to real systems.

The situation has greatly improved with the introduction of powerful numerical techniques. In particular, the Monte Carlo (MC) method yields an exact numeri-

*Ultrashort Processes in Condensed Matter*, Edited by
W.E. Bron, Plenum Press, New York, 1993

cal solution of the BE[3–4] with essentially arbitrary band structures and scattering mechanisms. MC is a statistical numerical means used, in general, for solving mathematical problems; as such, it was born long before[5] its application to transport in semiconductors[6] and has been applied to a large number of scientific fields. In the case of charge transport, however, the statistical numerical approach to the solution of the BE may consist of a direct simulation of the dynamics of charge carriers inside the crystal, so that, while the solution of the equation is being built up, any physical information required can be easily extracted. After its introduction at the Kyoto Semiconductor Conference in 1966[6], MC has been widely and successfully applied to all sorts of physical situations[3]. It has also been extended to include the analysis of space and time-dependent problems[3–4], to include carrier-carrier interaction[7] and the exclusion principle[8], and to obtain many transport quantities such as diffusivity, autocorrelation functions, etc.[9], besides, of course, the electron distribution function and mean velocity and energy.

Today the MC method is considered the standard method to which other techniques are to be compared in order to estimate their validity. Furthermore, this method allows the simulation of particular physical situations unattainable in experiments, or even the investigation of nonexisting materials in order to emphasize special features of interest. In a sense, these ways of using the MC method make it similar to an experimental technique: the simulated experiment can in fact be compared with analytical theories.

The technological advancements have created experimental devices that, with respect to the situation of a few decades ago, are much better defined and "clean". As a consequence, the dialogue between experimental engineers and theoretical physicists has reached a level unpredictable until a few years ago, and the MC simulation is a reliable tool for device design[4]. Its industrial application is limited only by the large amount of computer time still necessary for precise results.

During the last few years, the technological achievements have gone much farther: systems are being realized where, as we shall see, the assumptions necessary for the applicability of semiclassical theory of transport fail, and a new challenge is facing theoretical physicists: the solution of transport problems in quantum theory[10]. Once more, many difficulties are present both from the mathematical and the physical point of view since many-body quantum equations are quite unmanageable, and the physics involves basic problems, such as open systems and measurements processes, at the frontier of our knowledge of quantum mechanics. For the very same reasons, however, the challenge is extremely exciting and enjoyable.

The need for a quantum treatment of transport in new devices is essentially due to the small space and time scales involved in their functioning. As it regards the short time scales, however, since several years ultrafast spectroscopy has reached the experimental limit[11] (below 1 ps) where a quantum analysis of the process is required, on the basis of a very simple application of the uncertainty principle.

In the following pages we shall try to give an overview of transient transport, focussing on those aspects of particular relevance from a quantum-theory point of view. However, in trying to give a self consistent presentation of the subject, we shall follow the historical development sketched above. First the semiclassical theory of transport will be reviewed (Sec. 2); then the fundamentals of quantum transport will be presented in its general lines (Sec. 3) and a more detailed analysis of transient phenomena will be discussed (Sec. 4). The problem of the destruction of quantum coherence of photoexcited carriers (the "dephasing" process) will finally be approached with a quantum kinetic formulation (Sec. 5), before a brief concluding section (Sec. 6).

## 2. SEMICLASSICAL ANALYSIS OF TRANSIENT TRANSPORT

### 2.1 Fundamentals of charge transport. The Boltzmann equation

In semiclassical transport theory an electron is assumed to be described by a wave-packet such that a well defined wavevector $\mathbf{k}$ and a well defined position $\mathbf{r}$ can be associated to the electron. Its momentum $\hbar\mathbf{k}$ changes with time according to

$$\hbar\frac{d\mathbf{k}}{dt} = e\Big(\mathbf{E} + \mathbf{v}\times\mathbf{B}\Big) \tag{2.1}$$

$$\mathbf{v}(\mathbf{k}) = \frac{1}{\hbar}\nabla_{\mathbf{k}}\epsilon(\mathbf{k}) \tag{2.2}$$

where $e$, $\mathbf{v}(\mathbf{k})$, and $\epsilon(\mathbf{k})$ are the electron charge (with its sign), velocity, and energy, respectively; $\mathbf{E}$ and $\mathbf{B}$ are applied electric and magnetic fields. In the following only electric fields will be considered.

The free flights of the electrons are supposed to be interrupted by instantaneous collisions with impurities and phonons. The scattering cross sections for these electronic interactions are derived from the quantum theory of scattering.

The fundamental quantity in classical transport theory is the carrier distribution function $f(\mathbf{r},\mathbf{k},t)$, defined as proportional to the density of electrons in the six-dimensional space $(\mathbf{r},\mathbf{k})$. The normalization constant can be chosen in such a way that

$$\frac{2}{(2\pi)^3}\int d\mathbf{k}\int d\mathbf{r} f(\mathbf{r},\mathbf{k},t) = N\ , \tag{2.3}$$

where $N$ is the total number of electrons in the crystal. The factor in front of the integral accounts for the density of electron states (including spin). Normalized in this way, $f$ indicates the occupation number of the electron states, so that $(1-f)$ is the probability to find the state $\mathbf{k}$ unoccupied. Since spin interactions are usually not involved in transport phenomena in presence of electric field alone (and in problems of interest for our purpose), the spin variables are not explicitly considered.

The equation that describes transport phenomena in the semiclassical approach is the BE[12]:

$$\frac{\partial f}{\partial t} + \mathbf{v}\cdot\nabla_{\mathbf{r}} f + \dot{\mathbf{k}}\cdot\nabla_{\mathbf{k}} f = \frac{\partial f}{\partial t}\Big)_{coll}$$

$$= \frac{V}{(2\pi)^3}\int d\mathbf{k}'\Big\{f(\mathbf{r},\mathbf{k}',t)P(\mathbf{k}',\mathbf{k})(1-f(\mathbf{r},\mathbf{k},t)) - f(\mathbf{r},\mathbf{k},t)P(\mathbf{k},\mathbf{k}')(1-f(\mathbf{r},\mathbf{k}',t))\Big\}\ , \tag{2.4}$$

where $V$ is the volume of the crystal and $P(\mathbf{k},\mathbf{k}')$ is the probability per unit time that an electron in $\mathbf{k}$ makes a transition to an empty state $\mathbf{k}'$. The spin multiplicity in the density of states has been omitted since it is assumed that no spin-flip transitions occur. If, for the sake of simplicity we assume that the electron gas is homogeneous and non degenerate, the BE may be simplified to

$$\frac{\partial f}{\partial t} + \dot{\mathbf{k}}\cdot\nabla_{\mathbf{k}} f = \frac{V}{(2\pi)^3}\Big\{\int d\mathbf{k}' f(\mathbf{k}',t)P(\mathbf{k}',\mathbf{k}) - \int d\mathbf{k}' f(\mathbf{k},t)P(\mathbf{k},\mathbf{k}')\Big\}\ , \tag{2.5}$$

Once the distribution function is known, all transport quantities of interest can be evaluated.

As it can be seen from the above, the basic knowledge required for writing down the BE and therefore for the analysis of any transport phenomenon is concerned with the band structure $\epsilon(\mathbf{k})$ of the material and the scattering mechanisms that electrons undergo within the crystal.

## 2.2 Nonlinear transport. Hot carriers

Non-linear transport deals traditionally with problems which arise when an electric field is applied, so intense that the current density deviates from the linear-response regime. Nonlinearity and its relation to carrier heating can be easily understood in a straightforward way. If, for simplicity, a simple model semiconductor with a spherical and parabolic band is considered, then the presence of the electric field **E** inside the crystal modifies the paths of the electrons between two successive scattering events by adding a velocity component in the direction of the field:

$$\mathbf{v}^{(i)}(t) = \mathbf{v}_{\circ}{}^{(i)} + \frac{e\mathbf{E}}{m} t^{(i)} \tag{2.6}$$

where $\mathbf{v}^{(i)}$ is the instantaneous velocity of the i-th electron, $t^{(i)}$ is the time elapsed after the last scattering event of the i-th electron, and $\mathbf{v}_{\circ}{}^{(i)} = \mathbf{v}^{(i)}(t^{(i)} = 0)$ is the velocity of the i-th electron immediately after its last scattering event (m and e are the electron effective mass and charge, respectively).

The drift velocity of the electron gas is obtained by averaging Eq. (2.6) at a given time t over all the electrons, corresponding to a different value of $t^{(i)}$ for each electron. If there is no accumulation of the effect of the electric field on the initial velocities $\mathbf{v}_{\circ}{}^{(i)}$ of the successive free-electron paths, all these initial velocities reproduce the electron distribution function in the absence of field, so that the average value of the first term on the r.h.s. of Eq. (2.6) is zero. Under the same weak-field hypothesis, the mean value of the electron energy differs from its zero-field value only by the amount related to the energy gained during each single flight. As long as this energy is negligible with respect to the electron energy, the distribution of the $t^{(i)}$'s at any given time t is independent of the field since the electron scattering probabilities are functions of the electron energies. The drift velocity, obtained by averaging Eq. (2.6) over all electrons, is therefore linear in the electric field, according to the Ohm's law.

If, on the contrary, the field is sufficiently high, the energy gained by the electrons during each single flight is not negligible; the electrons cannot fully dissipate this energy at each scattering event, and the effect of the field accumulates on the initial velocities $\mathbf{v}_{\circ}{}^{(i)}$; the distribution of the $t^{(i)}$'s also depends upon the field, and the linearity of the Ohm's law is lost. The mean energy of the carriers then increases, together with the dissipation capacity of the scattering mechanisms, and a new stationary state is reached in which the average electron energy is higher than at equilibrium. This is the origin of the expression "hot-electron effects" applied to non-linear transport in semiconductors. Fig. 1 shows a hydraulic analogue of the concept of hot electrons.

The concept of hot electrons is often associated with an electron temperature $T_e$, higher than the lattice temperature T. It is now known that this idea does not correctly describe reality since the electron distribution function does not always allow an unambiguous definition of an electron temperature (i.e. it is not Maxwellian). It must be recognised however that the concept of an electron temperature has greatly helped the understanding of non-linear transport and it still gives a useful terminology for a heuristic investigation of these problems.

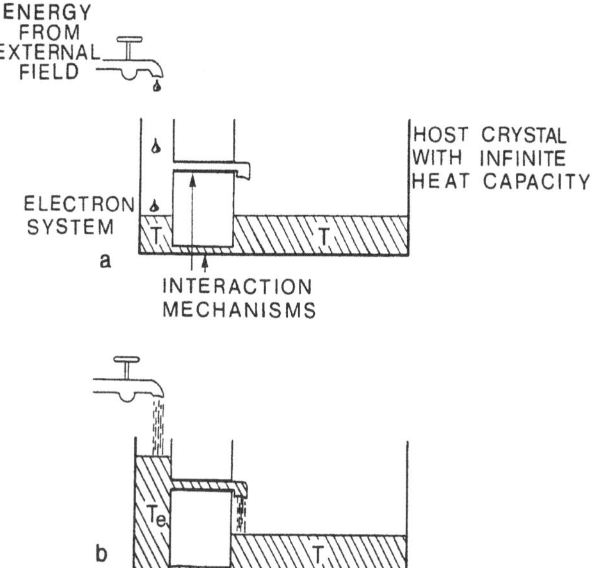


Fig. 1. Hydraulic analogue of the concept of hot electrons. Energy is transferred from the external electric field to the electron system and to the host crystal. In the low-field case (a) some interaction mechanisms are capable to mantain the temperature of the electron system equal to that of the thermal bath. At high fields (b) the energy supplied to the electrons is higher, and a new stationary state is attained by an increased efficiency of the scattering mechanisms already active at low fields and, sometimes, by the onset of new mechanisms.

### 2.3 The Monte Carlo method

The MC numerical approach for the solution of the Boltzmann equation in hot-electron conditions dates from the Kyoto Semiconductor Conference of 1966[6].

The simplest structure of a MC code, for a homogeneous, stationary system, consists of the determination of one possible path of an electron subject to the accelerative force of an external field and to given scattering mechanisms. By selecting random numbers evenly distributed between 0 and 1, we make these mechanisms to act stochastically in accordance with the probabilities determined by the total and differential cross sections, known from the theory of the corresponding processes. The electron path is recorded for a time sufficiently long for it to be representative of the behavior of all the carriers which contribute to the transport process.

The electron wavevector changes continuously between two successive scattering events. Thus, if $P(\mathbf{k}(t))dt$ is the probability that an electron in state $\mathbf{k}$ makes a collision during a time $dt$, then the probability that the electron which made a collision at time $t = 0$ will make its next collision in the interval $dt$ around the time t is given by

$$e^{\{-\int_0^t P(\mathbf{k}(t'))dt'\}} P(\mathbf{k}(t))dt \tag{2.7}$$

It is very impractical to generate flights with this distribution starting from evenly distributed random numbers. A very simple method has been devised to overcome this difficulty[13]. If $\Gamma = \frac{1}{\tau_o}$ is the maximum value of $P(\mathbf{k})$ in the region of $\mathbf{k}$ space of interest, a new fictitious "self-scattering" is introduced such that the total scattering probability, including the self-scattering is constant and equal to $\Gamma$. If the carrier undergoes such a self-scattering, its state $\mathbf{k}'$ after the collision is taken to be equal to the state $\mathbf{k}$ before the collision, so in practice the electron path continues unperturbed. Now, with a constant $P(\mathbf{k})$, Eq. (2.7) reduces to

$$\frac{1}{\tau_o} e^{\{-\frac{t}{\tau_o}\}} \tag{2.8}$$

and random numbers r can be used very simply to generate random flights

$$t_r = -\tau_o ln(r) \tag{2.9}$$

which have the distribution of Eq. (2.8). At the time $t_r$ the electron has then performed a collision. Since we know the action of the external field during the free-flight we also know the state of the electron at the instant of the collision; in particular, we know its energy $\epsilon$, and therefore all the scattering probabilities for the different scattering mechanisms considered. With a new random number r we can choose among these possible mechanisms the one which puts an end to the electron flight.

Finally, once we know which scattering determined the end of the flight, the electronic state immediately after the collision can be chosen in accordance with the differential cross-section of that particular process. This can be done in many different ways that will be not described here. We refer the reader to the specialistic literature on the subject[3].

The entire process is then repeated until, as we said above, the path of the electron is sufficiently long.

From each single flight we extract and record all the information necessary to the determination of the transport properties of interest , such as electron mean energy and its mean velocity. Furthermore, given an arbitrary mesh in $\mathbf{k}$ space, we may set up a histogram which contains the time spent by the electron inside each cell of the mesh. For sufficiently fine subdivisions of the mesh and sufficiently long simulation times, the histogram will represent the electron distribution function[3].

The principles of the MC technique are illustrated in Fig. 2.

In order to determine when the simulation time is sufficiently long for the purpose in view, we can split the simulation into a number of successive sub-histories of equal time durations t and make a determination of the quantity of interest on each of them. We may obtain the mean value of all these partial "measurements" as most probable "true" value, and their standard deviation as an estimate of its statistical uncertainty. Furthermore, it is possible to carry out, on the partial results, those tests which verify the statistical nature of the fluctuations.

When the purpose of the analysis is the investigation of a steady- state, homogeneous phenomenon, it is sufficient in general to simulate the motion of a single particle; from ergodicity we may assume that a sufficiently long path of this sample electron will give information on the behavior of the entire electron gas. When, on the contrary, the transport process under investigation is not homogeneous or is not stationary, or when inter-carrier Coulomb interaction is considered, then it is necessary to simulate a large number of electrons and follow them in their dynamical histories in order to obtain the desired information on the process of interest.

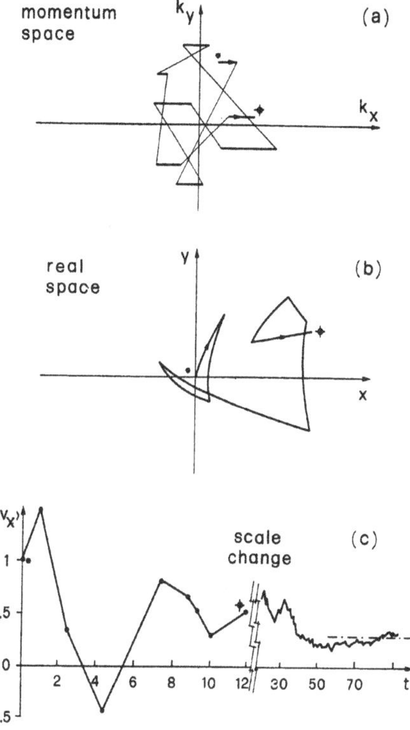


Fig. 2. The principles of the Monte Carlo method; for simplicity a two-dimensional model is considered here. Part (a) of the figure shows the simulation of the sampling electron in momentum space subject to an accelerating force due to an electric field oriented along the positive $x$ direction. The heavy segments are due to the effect of the field during the free flights, while light lines represent discontinuous variations of $k$ during the scattering processes. Part (b) shows the path of the particle in real space. It is composed of eight fragments of parabolas corresponding to the eight free flights in part (a) of the figure. Part (c) shows the average velocity of the particle obtained as a function of the simulation time. The left section of the figure ($t < 12$) is obtained by the simulation illustrated in the parts (a) and (b). The horizontal dot-dashed line represents the "exact" drift velocity obtained with a very long simulation time. Special symbols indicate corresponding points in the three parts of the figure (* is the starting point). All units are arbitrary.

It should be noted that, once a numerical solution of a given problem is obtained, its subsequent physical interpretation is still very important in gaining an understanding of the phenomenon under investigation. The MC method shows itself to be a very useful tool towards this end, since it permits simulation of particular physical situations unattainable in experiments, in order to emphasize special features of the physical system.

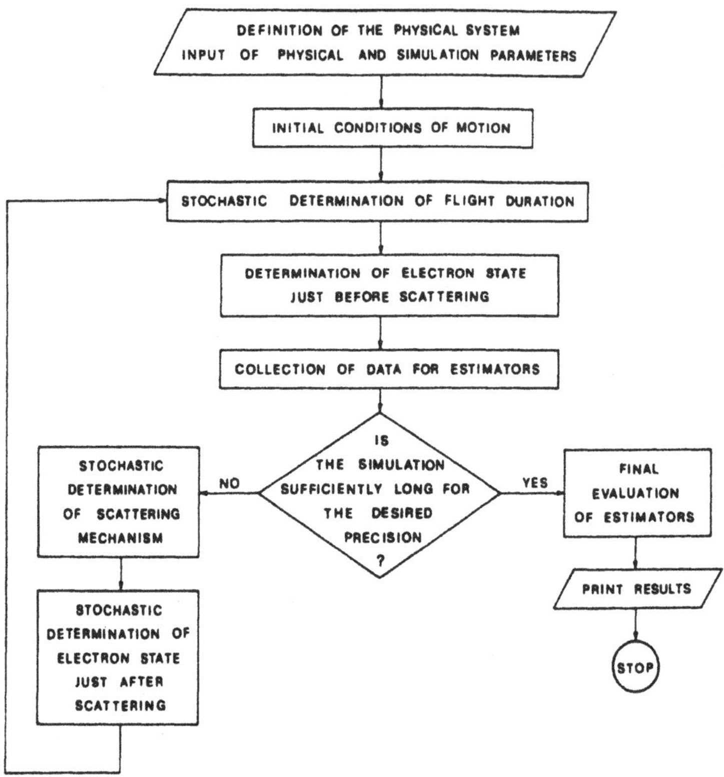


Fig. 3. Flowchart of a typical Monte Carlo program

## 2.4 Transient transport

For our present purpose, the transient dynamic behavior of a physical system under constant external forces can be defined as its evolution from given initial conditions to the steady state.

In the case of the response to an external field, the duration of the transient response is not known apriori and, in general, it will be at least of the order of the longest of the characteristic times of the carrier system. This time may be called the "transient-transport time" and it depends upon the values of the applied field and temperature. In the case of high-field transport in semiconductors it may roughly correspond to the energy relaxation time or to the time necessary for the repopulation of the different valleys in the band. Furthermore, the transient response depends strongly upon the initial conditions of the carriers.

Transient transport may occur in different physical systems. The first rather "academic" situation we may think of, is that of an infinite homogeneous system with a sudden change of the applied field, for example, from zero to a high value: the electron distribution requires some time to reach its steady-state nonequilibrium value. The second system of interest is a much more realistic one and can be present in very small devices: in a steady-state situation, particles may enter a small region with a very high applied field and reach the end of this region so quickly that the distribution function does not reach the form equivalent to a steady-state homogeneous system.

As a third example, more relevant to the present study, is that of an ensemble of carriers created with an initial distribution far from the steady-state distribution corresponding to the external forces. More precisely, photoelectrons may be generated with average energy much higher than at equilibrium.

For space or time dependent problems the analytical solution of the BE is even more difficult than for homogeneous and stationary problems, while for MC programs little work needs to be added to treat such cases. For these problems we cannot rely on ergodicity, and an ensemble of particles must be explicitly simulated, in what is now known as the ensemble Monte Carlo (EMC) technique. Provided the number of simulated particles is sufficiently large, the average value of a quantity of interest, obtained in this sample ensemble as a function of time, will be representative for the average on the entire gas. The corresponding standard deviation is again a measure of the statistical uncertainty of the results.

The transient dynamic response which will be obtained by means of the simulation, will of course depend upon the initial conditions of the carriers, and these must be assumed according to the physical situation that has to be explored.

As a typical example of transient-transport quantities, we show in Fig. 4 the mean velocity $\langle v \rangle$, the valley populations, and the mean energy $\langle \epsilon \rangle$, of electrons in GaAs as functions of time, after the application of a static field. The model used in the Monte Carlo simulation is given in Ref. 14. The initial distribution has been assumed to be an equilibrium Maxwellian. A general feature of high-field transient transport that can be recognized in Fig. 4 is an overshoot of $\langle v \rangle$ and $\langle \epsilon \rangle$, before they reach the steady-state values. The overshoot of $\langle v \rangle$ is usually associated with a larger momentum relaxation time of "cold" electrons[14] with respect to its steady-state value. The overshoot of $\langle \epsilon \rangle$ is a direct consequence of the overshoot of $\langle v \rangle$.

In the particular case of GaAs, however, the transient transport is strongly influenced, at high fields, by the electron transfer to the upper valleys, which further reduces the average velocity and kinetic energy. In the upper valleys, in fact, $\langle v \rangle$ and $\langle \epsilon \rangle$ have values approximately constant in time, and much lower than the corresponding values in the central valley, as can be seen in Fig. 4.

In order to show the importance of initial conditions, Fig. 5 reports EMC calculations for the case of holes in Si at $T = 300K$ and $E = 50kV/cm$[15]. The transient behavior is here investigated at different initial mean energies. An overshoot in the drift velocity can be seen that is larger for colder initial conditions and disappears for the hottest case. When the carrier energy is higher in the initial conditions than in the steady state (as is the case for photoexcited carriers) the mobility of the charges is first lowered by a higher scattering and reaches its stationary value more slowly.

As regards diffusion, a transient diffusion coefficient $D(t)$ can be defined as[16]

$$D = \frac{1}{2}\frac{d}{dt}\langle (z - \langle z \rangle)^2 \rangle = \langle \delta z(t) \delta v(t) \rangle = \int_0^t C_t(\tau)\, d\tau \tag{2.10}$$

where $C_t(\tau)$ is the transient velocity autocorrelation function

$$C_t(\tau) = \langle \delta v(t) \delta v(t+\tau) \rangle, \qquad 0 \leq \tau \leq t. \tag{2.11}$$

$$\delta z(t) = z(t) - \langle z(t) \rangle, \qquad \delta v(t) = v(t) - \langle v(t) \rangle$$

For simplicity, in Eq. (2.10) we have considered only the diffusivity along a given direction $z$, and $v$ represents the velocity along that direction. If we let $t$ go to infinity, more familiar steady-state definitions are recovered.

Equation (10) gives a clear physical insight into the diffusion phenomenon: the spreading of a carrier ensemble comes from the particle space-velocity correlations which arise during the time evolution of the system. Particles with velocities higher than average, having accumulated this positive fluctuation in the near past, are more

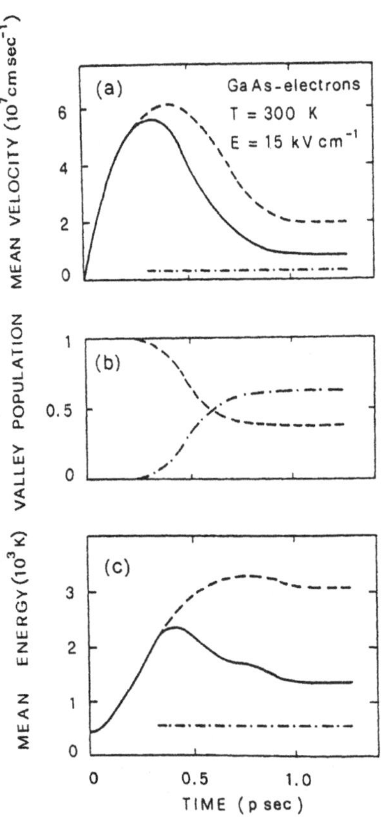


Fig. 4. Monte Carlo results for the mean velocity (a), the valley populations (b), and the mean energy (c), for an ensemble of electrons in GaAs at $300K$ as functions of the time elapsed after the application of a field $E = 15kV/cm$. Dashed and dot-dashed curves refer to quantities averaged over the central and upper valleys, respectively. Continuous curves refer to quantities averaged over all the particles.

probably in the front of the carrier space distribution. Similarly, the slower particles are in the rear of the distribution. Thus, particles in the front will tend to run away, and particles in the rear will tend to be left behind, causing a farther spreading of the distribution.

From the above discussion it results that in transient transport the diffusivity is different from its steady-state value for two reasons: at short times i)the electron distribution in the velocity space has not yet reached its steady-state value, and ii) the position-velocity correlations are not yet fully developed [9–16].

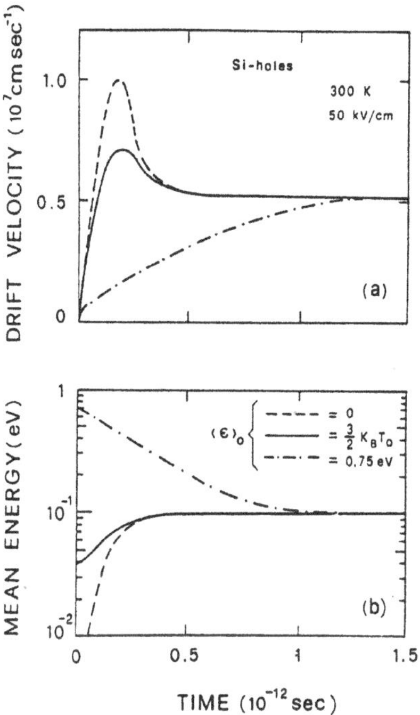


Fig. 5. Drift velocity (a), and mean energy (b) as functions of time for the case of holes in Si. Different curves refer to the reported different values of the initial mean energy with velocities randomly distributed .

## 3. FUNDAMENTALS OF QUANTUM TRANSPORT

### 3.1 The need of quantum transport

In the previous section we have reviewed the basic concepts of semiclassical electron transport in semiconductors. In that theory several assumptions were more or less explicitly made that require some considerations. As we shall see, they are not always justified in present-day research, so that a more rigorous quantum transport theory is needed.

In order to decide if semiclassical transport theory is good enough, it is necessary to decide how good is good enough. In particular, an acceptable electron energy uncertainty $\delta\epsilon$ is to be established. Its definition is somewhat arbitrary and it depends upon the particular phenomenon under investigation: It may be a quantity much smaller than the average electron energy, or much smaller than a threshold energy for a particular electronic transition, or much smaller than the precision of a particular experimental apparatus (e.g. in optical measurements).

In order to proceed with the analysis, a certain number of physical quantities should be introduced. Their definitions are, for the moment, not rigorous; only their intuitive meaning is necessary for the present discussion. They are the mean time $\tau$ between two successive electron collisions, the duration of the collisions $\tau_c$, the electron mean free path $\ell$, and an electron temperature $T_e$, defined here in such a way that $KT_e$ is of the order of the mean electron energy, where $K$ is the Boltzmann constant. Furthermore, in ultrafast spectroscopy two other experimental times are relevant: the observation time, that is the time elapsed from the preparation of the initial condition to the time of the measurement, and the time resolution of the experimental apparatus.

Let us now consider the assumptions made in semiclassical theory of electron transport in semiconductors.

a) During the free flight between two successive collisions electrons are considered as classical particles with well defined positions and momenta. For this approximation to be acceptable, we must be able to conceive wave-packets with a momentum uncertainty $\Delta p$ much less than their average momentum $p$ and, at the same time, a position uncertainty $\Delta x$ much less than the mean free path $\ell$:

$$\Delta p \ll p, \quad \Delta x \ll \ell \,. \tag{3.1}$$

From the uncertainty relations we then have, if $\ell \sim v\tau = p\tau/m$,

$$\hbar \leq \Delta p \Delta x \ll p\ell \sim \frac{p^2}{m}\tau \sim KT_e\tau, \tag{3.2}$$

or

$$\tau \gg \frac{\hbar}{KT_e} \sim 10^{-14}\ s \tag{3.3}$$

at ordinary electron temperatures.

b) Apart from isolated scattering events, electrons are described by single-particle states. To neglect many-body effects, such as collisional broadening (CB), is justified in the low-coupling limit. The order of magnitude of the CB $\Delta\epsilon$ can be estimated from the uncertainty relation as

$$\Delta\epsilon \approx \hbar/\tau. \tag{3.4}$$

$\Delta\epsilon$ should be smaller than the energy resolution $\delta\epsilon$ described above. Thus, if we require that $\delta\epsilon$ is much smaller than the electron mean energy, the same condition as in Eq.(3.3) is recovered. If, instead, we want $\delta\epsilon$ to be smaller than an energy resolution, for example, of 5 meV, then it should be

$$\tau \gg 10^{-13}s \ . \tag{3.5}$$

c) In the traditional treatment of the Boltzmann equation, collisions are, in general, assumed instantaneous in time and point-like in space. Since the interactions between the particles and the scattering agents have finite durations, this assumption is not correct in general, not even in classical theory. In the case of weak coupling, when scattering events are sufficiently rare, the duration of a collision may be negligible with respect to the free-flight time between successive collisions, and the assumption may be reasonable. In order to estimate the requirement for such a condition to be fulfilled, the collision duration has to be estimated. This estimate is somewhat arbitrary because the concept of collision duration itself is ill-defined. From first-order perturbation theory, we know that the final state of a transition is defined within an energy range of the order of $\hbar/t$, if $t$ is the time elapsed after the initial condition of the transition. A collision duration can then be defined, in general, with reference to the desired precision $\delta\epsilon$ of the electron energy:

$$\tau_c = \frac{\hbar}{\delta\epsilon} \tag{3.6}$$

Thus, the requirement that the collision duration is smaller than the time between collisions coincides with the requirement that the collisional broadening is smaller than the desired energy resolution, a result which is quite intuitive from a physical point of view.

d) During the collision time an external field $E$ may act on the initial and final electron wavevectors, changing them linearly with time. However, since the energy is not a linear function of k, the energy difference between the two states varies with time, and the interference effects that generate the transition are modified by the field. Such a process, called "intracollisional field effect" (ICFE), may modify the transition rates. In order to estimate this effect, we may compare the energy $\Delta\epsilon_E$ imparted to the electron by the field during the collision with the average electron energy:

$$\frac{\Delta\epsilon_E}{KT_E} \sim v\tau_c \frac{(eE)}{KT} \sim eE\sqrt{\frac{2}{mKT_E}}\frac{\hbar}{\delta\epsilon} \tag{3.7}$$

which is of the order of one for $E \sim 4 \cdot 10^4$ V/cm, if $\delta\epsilon = 5meV$ and $KT_E \sim 100meV$.

To the above discussion we should add that modern technology provides physical systems and devices of the order of $0.1\mu m$[17], where local fields as high as $10^6 V/cm$ can be present. Furthermore, ultrafast spectroscopy has reached a time resolution of the order of $10^{-14}s$. Several considerations then follow:

i) Fields are attainable for which the ICFE has to be taken into account.

ii) At such fields electrons reach energies of the order of or larger than $1eV$. At these energies the average time between collisions may be of the order of $10^{-14}$ s, so that most of the classical approximations fail. iii) The time resolution in ultrafast spectroscopy is shorter than the characteristic time-scale at which energy broadening is relevant.

As a consequence it seems clear that a quantum approach to the problem is required if we want to obtain results reliable for device applications and if we want to take advantage of the opportunity that technology offers us to study new physical phenomena of great interest.

A final important comment should be made about the space scale of the so called "mesoscopic" systems. The confining (or quasi-confining) dimensions of these structures (particularly for heterostructures), are of the order of the de Broglie wavelength of the electrons, so that a quantization (or resonance behavior) of at least one degree of freedom occurs. For the purposes of the present Institute, however, we shall limit ourselves to the consideration of quantum transport in homogeneous systems, where the quantum features are brought about by the short time scales involved or the interference phenomena between external fields and collision processes.

### 3.2 The general problem

The general problem under investigation is that of a quantum system described by some time-independent hamiltonian $\mathbf{H}^{(0)}$ and an external term $\mathbf{H}^{(ext)}$.

The system is often divided into two parts. One "small" part is the actual system of interest, for example the conduction electrons inside the crystal. The other "large" part is a system in thermal contact with the small system and it provides the necessary dissipation to bring or maintain the system in stationary conditions (for example: the phonon gas and the heat bath). "Small" and "large" refer to the fact that under the effect of the external perturbation the "small" system can get out of equilibrium, while usually the "large" part is considered to have such a large heat capacity (i.e. so many degrees of freedom) that it is not driven out of equilibrium by the perturbation. Hot-phonon effects are not considered here; if such a phenomenon is to be considered, however, phonons may be included in the "small" sub-system, and the "large" sub-system is the heat bath in contact with the boundary of the crystal.

The response of the small subsystem to the external perturbation is the goal of our investigation. If the perturbation is weak enough, the linear-response theory provides very general and elegant results, but it has been well known for about forty years that in semiconductor structures we often go over the limits of the linear-response regime and we have to look for solutions of the non-linear transport (or hot-electron) problem.

As we shall discuss in the following, the major theoretical difficulty in such a many-body problem is the reduction of the whole complexity of the system to a simplified version of it, which is analysed in terms of a reduced number of variables of interest. The dynamical properties of the reduced system are usually described in terms of a so called "kinetic equation", which can be defined as an equation for a function of only a subset of the variables of the system. The problem of obtaining a closed equation for the reduced system is one of the major problems of quantum transport theory.

In the reduction process part of the information is to be given up and when, in doing so, we substitute some detailed deterministic information with some average or probabilistic quantity, in general we introduce irreversibility in the description of the process. This fundamental step in transport theory is performed, for example, when we assume a random distribution of impurities and substitute their exact positions with the probability for the electron to encounter an impurity in its path, or when we substitute the exact phonon field (described by the dynamics of the "large" subsystem) with the probability of having the phonon system in given states according to the thermal distribution.

In this Section we will try to present in a simple way, the definitions of the main theoretical tools used in quantum-transport theories: density matrix, Green's functions, Wigner function, and path integral. We will also summarize their fundamental properties, the equations they satisfy, and how they are related to measurable quantities. Furthermore we will try to point out the connections among these major techniques.

### 3.3 The density-matrix formalism

As starting point of our discussion, we briefly recall some basic ideas of quantum statistical mechanics.

Given a physical system, described by the state vector $\mid \phi\rangle$, the expectation value of a measurement of a given quantity $A$ is given by:

$$\langle A\rangle = \langle \phi \mid \mathbf{A} \mid \phi\rangle, \tag{3.8}$$

where $\mathbf{A}$ is the operator related to the physical quantity of interest.

On the other hand, if the state of the system is not completely specified, we have to use the concepts of statistical physics to deal with our incomplete knowledge, and the expectation value, given in Eq. (3.8), is replaced by its ensemble average:

$$\overline{\langle A\rangle} = \overline{\langle \phi \mid \mathbf{A} \mid \phi\rangle}, \tag{3.9}$$

where the overbar indicates an average to be performed over a suitable statistical ensemble that accounts for our partial knowledge of the system. If $\{\mid \alpha\rangle\}$ is a complete set of basis vectors, the above equation can also be written as

$$\overline{\langle A\rangle} = \sum_{\alpha,\alpha'} \overline{\langle \phi \mid \alpha\rangle\langle \alpha \mid \mathbf{A} \mid \alpha'\rangle\langle \alpha' \mid \phi\rangle} = Tr(\overline{\mid \phi\rangle\langle \phi \mid}\mathbf{A}). \tag{3.10}$$

From the above result it is easy to recognize that the basic mathematical instrument in quantum statistical physics, is the density matrix operator

$$\rho = \overline{\mid \phi\rangle\langle \phi \mid}. \tag{3.11}$$

By means of this operator, the direct link in Eq. (3.10) between experiments and quantum statistical mechanics can be easily expressed as

$$\overline{\langle A\rangle} = Tr(\rho\mathbf{A}). \tag{3.12}$$

In a given representation $\{|\alpha\rangle\}$ the matrix elements of $\rho$ are called "the density matrix" of the system[18]. It is easy to show that the diagonal elements $\rho_{\alpha\alpha}$ of the density matrix give the probabilities $P_\alpha$ of finding the system in the various states $|\alpha\rangle$. In fact, if $P^{(i)} = \frac{1}{N}$ is the probability of selecting at random the i-th system, and $P(\alpha|i) = |\langle\alpha|\phi^{(i)}\rangle|^2$ the probability of finding the i-th system in the state $\alpha$, we have:

$$P_\alpha = \sum_i P^{(i)}P(\alpha|i) = \sum_i \frac{1}{N}\langle\alpha|\phi^{(i)}\rangle\langle\phi^{(i)}|\alpha\rangle = \langle\alpha\overline{|\phi\rangle\langle\phi|}\alpha\rangle = \rho_{\alpha\alpha}. \tag{3.13}$$

The time evolution of the average quantities in Eq. (3.12) can be obtained starting from the time evolution of the state vector in the Schrödinger picture

$$\mid \phi(t)\rangle = \mathbf{U}(t,t_i) \mid \phi(t_i)\rangle, \tag{3.14}$$

where $\mathbf{U}$ is the evolution operator of the whole ("large" plus "small") system, that verifies the following equation and initial condition:

$$i\hbar\frac{d}{dt}\mathbf{U}(t,t_i) = \mathbf{H}(t)\mathbf{U}(t,t_i), \qquad \mathbf{U}(t_i,t_i) = 1. \tag{3.15}$$

where $\mathbf{H}(t)$ is the Hamiltonian of the system.

Let us now consider the equation of motion of the density matrix operator in the different pictures of quantum mechanics. The time evolution of $\rho$ in the Schrödinger picture is obtained immediately by its definition (Eq. (3.11)) and the time evolution of the state vector given in Eq. (3.14). The result is the following:

$$\rho_S(t) = \mathbf{U}(t,t_i)\rho_S(t_i)\mathbf{U}^\dagger(t,t_i). \tag{3.16}$$

By differentiating with respect to time, we obtain the Liouville–von Neumann equation for the density matrix in the Schrödinger picture:

$$i\hbar\frac{\partial}{\partial t}\rho_S(t) = [\mathbf{H}(t),\rho_S(t)]. \tag{3.17}$$

In the Heisenberg picture the density matrix does not depend on time, since the state vectors are constant.

If the total Hamiltonian $(\mathbf{H}^{(0)} + \mathbf{H}^{(ext)})$ is split into two parts $(\mathbf{H}_0 + \mathbf{H}')$, where $\mathbf{H}'$ is considered as a perturbation, the interaction picture can be used, and the time evolution of the state vector is given by:

$$\mid \phi_I(t)\rangle = \mathbf{S}(t,t_i) \mid \phi_I(t_i)\rangle \ , \tag{3.18}$$

where the evolution operator $\mathbf{S}$ for the state vectors in the interaction picture verifies the following differential equation:

$$i\hbar\frac{\partial}{\partial t}\mathbf{S}(t,t_i) = \mathbf{H}'(t)\mathbf{S}(t,t_i). \tag{3.19}$$

Starting from Eq. (3.18) and following the same theoretical development used to derive Eq. (3.17), we obtain the Liouville–von Neumann equation in the interaction picture:

$$i\hbar\frac{\partial}{\partial t}\rho_I(t) = [\mathbf{H}'(t),\rho_I(t)]. \tag{3.20}$$

The above equation of motion seems to be more convenient because only the perturbation Hamiltonian appears explicitly in the commutator.

The time evolution of an average quantity $\overline{\langle A\rangle}$, which is always given by Eq. (3.12), is due to the time evolution of $\rho$ in the Schrödinger picture and by the time evolution of $\mathbf{A}$ in the Heisenberg picture. In the interaction picture, $\mathbf{A}$ carries the time evolution due to the known, unperturbed Hamiltonian, and $\rho$ the time evolution due to the perturbation Hamiltonian.

In order to discuss the reduction problem for a given sub-system of interest, let us consider again the example of $N$ electrons interacting with phonons in a crystal. The state vectors, and therefore the density matrix, will be functions of the electron coordinates $\mathbf{x} \equiv (\mathbf{r}_1, \mathbf{r}_2, \ldots, \mathbf{r}_N)$ and the phonon variables $\xi$:

$$\rho = \rho(\mathbf{x}, \xi, \mathbf{x}', \xi'). \tag{3.21}$$

If an observable $A^{(el)}$ acts only on the electron variables, then its matrix elements in the full space can be written as $A^{(el)}(\mathbf{x}',\xi',\mathbf{x},\xi) = A^{(el)}(\mathbf{x}',\mathbf{x})\ \delta(\xi-\xi')$ and its mean value is given by

$$\overline{\langle A^{(el)}\rangle} = Tr(\rho\mathbf{A}^{(el)}) = \int d\mathbf{x} \int d\mathbf{x}' \int d\xi \rho(\mathbf{x},\xi,\mathbf{x}',\xi) A^{(el)}(\mathbf{x}',\mathbf{x}) = Tr(\rho^{(el)}\mathbf{A}^{(el)})\ , \tag{3.22}$$

where

$$\rho^{(el)}(\mathbf{x},\mathbf{x}') \equiv \int d\xi \rho(\mathbf{x},\xi,\mathbf{x}',\xi) = Tr_\xi(\rho) \tag{3.23}$$

is the reduced electronic density matrix.

In practice, the observable $\mathbf{A}^{(el)}$ is often the average over the particles of the system of a single–particle quantity $\mathbf{A}^{(sp)}$. This is the case, for example, of the drift velocity or the mean energy of the electron gas. In such a case a single–particle density matrix $\rho^{(sp)}$ can be defined as the integral of the reduced electronic density matrix in Eq. (3.23) over all the coordinates but one. The average value is then given by:

$$\overline{\langle A^{(el)}\rangle} = \int d\mathbf{r} \int d\mathbf{r}' \rho^{(sp)}(\mathbf{r},\mathbf{r}') A^{(sp)}(\mathbf{r}',\mathbf{r}) = Tr(\rho^{(sp)}\mathbf{A}^{(sp)}). \tag{3.24}$$

The above two reduction processes are usually introduced in a more general and elegant way in the second-quantization formalism. If the observable of interest is a single-electron quantity, the corresponding single particle operator in second quantization is given by

$$\mathbf{A}^{(sp)} = \int d\mathbf{r} \int d\mathbf{r}' \mathbf{\Psi}^\dagger(\mathbf{r}') A^{(sp)}(\mathbf{r}',\mathbf{r}) \mathbf{\Psi}(\mathbf{r}), \tag{3.25}$$

where $\mathbf{\Psi}^\dagger(\mathbf{r})$ and $\mathbf{\Psi}(\mathbf{r})$ are the second quantized creation and annihilation operators, respectively. Therefore, the corresponding average value in Eq. (3.12) can also be written as:

$$\overline{\langle A^{(sp)}\rangle} = \int d\mathbf{r} \int d\mathbf{r}' Tr\{\rho\mathbf{\Psi}^\dagger(\mathbf{r}')\mathbf{\Psi}(\mathbf{r})\} A^{(sp)}(\mathbf{r}',\mathbf{r}). \tag{3.26}$$

From the comparison of this expression with Eq. (3.24), we obtain the form of the single-particle density matrix in second quantization:

$$\rho^{(sp)}(\mathbf{r},\mathbf{r}') \equiv Tr\{\rho\mathbf{\Psi}^\dagger(\mathbf{r}')\mathbf{\Psi}(\mathbf{r})\} = \overline{\langle \mathbf{\Psi}^\dagger(\mathbf{r}')\mathbf{\Psi}(\mathbf{r})\rangle}, \tag{3.27}$$

This second-quantized approach is more general since it allows to investigate physical systems with a variable number of particles.

The single-particle density matrix is often the only quantity we need in order to describe our transport problem. However if we try to write down an equation of motion for $\rho^{(sp)}$ starting from the Liouville–von Neumann equation, we obtain from Eqs. (3.17) and (3.27):

$$i\hbar\frac{\partial}{\partial t}\rho^{(sp)}(\mathbf{r},\mathbf{r}') = Tr\{[\mathbf{H},\rho]\mathbf{\Psi}^\dagger(\mathbf{r}')\mathbf{\Psi}(\mathbf{r})\}. \tag{3.28}$$

Since the Hamiltonian in general contains terms corresponding to electron–electron and electron–phonon interaction, the trace operation does not commute with $\mathbf{H}$, and we do not obtain a closed equation for $\rho^{(sp)}$ . In particular, using the explicit form of $\mathbf{H}$ in terms of second quantization, it can be shown that the equation of motion for the single-particle density matrix contains a two–particle density matrix, the equation of motion for the two-particle density matrix contains a three–particle one, and so on.

In order to obtain a closed equation of motion, a "truncation" in this infinite hierarchy of equations is required, and this is generally obtained by substituting the average value of a product of field–operator pairs with the product of their average values (mean field approximation)[19].

In Sects. 4 and 5 we shall see some applications of the density matrix formalism.

### 3.4 Green functions and Dyson equation

The general concept of Green function is that of a quantity that indicates how a "cause", or "source", in $q_i$ at the initial time $t_i$ determines an "effect" in $q$ at time $t$. In the case of a physical field the "causes" are given, in general, by field sources and boundary conditions. As it regards the Schrödinger equation, of direct interest to us, the wave function of the system $\phi(q,t)$ at time t is determined by the wave function itself at the initial time $t_i$, so that the corresponding Green function is defined by the relation

$$\phi(q,t) = i\hbar \int dq_i G^{(r)}(q,t;q_i,t_i)\phi(q_i,t_i) \ , \tag{3.29}$$

where $q$ stands for the set of variables of the system. The label $r$ indicates that we are dealing with the "retarded" Green function, which describes the "forward" propagation of the wave function, and it is assumed to have non-vanishing values only for $t > t_i$. The constant $i\hbar$ is introduced in Eq. (3.29) in order to have the $\delta$-function in the following Eq. (3.31).

The above equation, however, apart from the numerical factor $i\hbar$, is just Eq. (3.14) written in the $q$ representation. Therefore we may define a corresponding "retarded" Green operator as

$$\mathbf{G}^{(r)}(t,t_i) = \frac{\theta(t-t_i)}{i\hbar}\mathbf{U}(t,t_i), \tag{3.30}$$

where $\theta$ is the usual step function. Starting from the dynamical equation for the evolution operator (Eq. (3.15)), we obtain that the operator $\mathbf{G}^{(r)}$ is a solution of the "Green equation" corresponding to the differential equation in Eq. (3.15):

$$\left\{ i\hbar\frac{\partial}{\partial t} - \mathbf{H} \right\} \mathbf{G}^{(r)}(t,t_i) = \delta(t-t_i). \tag{3.31}$$

A similar "advanced" Green operator can be defined as

$$\mathbf{G}^{(a)}(t,t_i) = -\frac{\theta(t_i-t)}{i\hbar}\mathbf{U}(t,t_i), \tag{3.32}$$

that verifies the same Green equation (Eq. (3.31)) and, with an equation similar to Eq. (3.29), yields the state of the system at a time $t$ previous to the initial time $t_i$.

The retarded and advanced Green operators carry information on the dynamics, but not on the state, of the system as it results from their definitions in Eqs. (3.30) and (3.32).

As we shall see shortly, other Green functions can be defined that carry information also on the state of the system.

If we consider a single particle, in terms of second quantization we may write the $r$ matrix elements of the evolution operator as

$$U(\mathbf{r},t;\mathbf{r}',t') \equiv \langle \mathbf{r}|U(t,t')|\mathbf{r}'\rangle = \langle 0|\Psi(\mathbf{r})U(t,t')\Psi^\dagger(\mathbf{r}')|0\rangle \tag{3.33}$$

where $|0\rangle$ is the vacuum state. In the Heisenberg picture the last scalar product is $\langle 0|\Psi(\mathbf{r},t)\Psi^\dagger(\mathbf{r}',t')|0\rangle$, where $\Psi(\mathbf{r},t)$ is the field operator in the same picture, so that the retarded Green function for a system containing one particle can be written as

$$G^{(r)}(\mathbf{r},t;\mathbf{r}',t') = \frac{1}{i\hbar}\langle 0|\Psi(\mathbf{r},t)\Psi^\dagger(\mathbf{r}',t')|0\rangle. \tag{3.34}$$

A physical interpretation of the above expression can be given as follows: after a particle has been created in $\mathbf{r}'$ at time $t'$, the probability amplitude is evaluated of finding the particle in $\mathbf{r}$ at time $t$.

This interpretation suggests an alternative expression for the retarded Green function. We may start with the state containing already a particle, and then we may annihilate the particle in $\mathbf{r}'$ at $t'$ and evaluate the probability amplitude that the lack of the particle is "felt" in $\mathbf{r}$ at time $t$. The scalar product in Eq. (38) would be, for this case

$$\langle \Phi_{\mathcal{H}}|\Psi^\dagger(\mathbf{r},t)\Psi(\mathbf{r}',t')|\Phi_{\mathcal{H}}\rangle, \tag{3.35}$$

where $|\Phi_{\mathcal{H}}\rangle$ is the state vector of the system in the Heisenberg picture, or, following the most popular convention, its complex-conjugate

$$\langle \Phi_{\mathcal{H}}|\Psi^\dagger(\mathbf{r}',t')\Psi(\mathbf{r},t)|\Phi_{\mathcal{H}}\rangle, \tag{3.36}$$

This expression, however, is not proportional to the evolution operator as one could expect from the considerations above. The product $\Psi^\dagger\Psi$ in Eq. (3.36) must be substituted with the commutator (for bosons) or anticommutator (for fermions) of the two field operators. (In the case of Eq.(3.34), that is correct, the missing half of the commutator would give a vanishing contribution.)

In order to understand in physical terms why such a substitution is necessary, let us recall that the expectation value of the product $\Psi^\dagger(\mathbf{r},t)\Psi(\mathbf{r},t)$ at equal positions and times gives the intensity of the field (i.e. the average number of particles in $\mathbf{r}$) for the state under consideration. The same product at different arguments gives, in the same way, the correlation between the two amplitudes at different positions and times. If we look for the dynamical correlation, that is the propagator, without the information on the field intensity, we must subtract the product in reverse order. We may compare this result with the more familiar relations for harmonic-oscillator creation, annihilation, and number operators:

$$a^\dagger a = N, \qquad aa^\dagger = N+1, \qquad \textit{so that} \qquad [a,a^\dagger] = 1\ . \tag{3.37}$$

From the above considerations we obtain the Green functions for a single-particle system:

$$G^{(r)}(\mathbf{r},t;\mathbf{r}',t') = \frac{\theta(t-t_i)}{i\hbar}\langle \Phi_{\mathcal{H}} \mid [\Psi(\mathbf{r},t),\Psi^\dagger(\mathbf{r}',t')]_\pm \mid \Phi_{\mathcal{H}}\rangle, \tag{3.38}$$

$$G^{(a)}(\mathbf{r},t;\mathbf{r}',t') = -\frac{\theta(t_i-t)}{i\hbar}\langle \Phi_{\mathcal{H}} \mid [\Psi(\mathbf{r},t),\Psi^\dagger(\mathbf{r}',t')]_\pm \mid \Phi_{\mathcal{H}}\rangle, \tag{3.39}$$

In the case of a many body system, the Green operators defined above (Eqs. (3.30) and (3.32)) depend upon all the coordinates of the system and, therefore, there is no possibility to use them for practical purposes. As for the density matrix operator, however, when single-particle properties of the system are investigated, the mathematical instrument to be used is given by the single–particle Green functions defined by the same Eqs. (3.38) and (3.39), where now the vector $|\Phi_{\mathcal{H}}\rangle$ describes the many-particle state[19]. In this case, however, we deal with single-particle Green functions defined in a many-particle system, and therefore they contain a reduction of many degrees of freedom and include the effect, on each particle, of the interaction with all the other ones in the system.

As already seen above for the single-particle system, the average values of the single products of field operators in the commutators of Eqs. (3.38) and (3.39) carry information about the state of the many-particle system. They are also defined as Green functions, or, more properly, as the correlation functions $G^>$ and $G^<$:

$$G^>(\mathbf{r},t;\mathbf{r}',t') = \frac{1}{i\hbar}\langle\phi_{\mathcal{H}} \mid \Psi(\mathbf{r},t)\Psi^\dagger(\mathbf{r}',t') \mid \phi_{\mathcal{H}}\rangle, \tag{3.40}$$

$$G^<(\mathbf{r},t;\mathbf{r}',t') = \frac{1}{i\hbar}\langle\phi_{\mathcal{H}} \mid \Psi^\dagger(\mathbf{r}',t')\Psi(\mathbf{r},t) \mid \phi_{\mathcal{H}}\rangle. \tag{3.41}$$

If the system under consideration is only partially known, the tools of statistical physics must be used, and the Green functions have to be taken as ensemble averages of the quantities defined above. It is then clear that for this case the definition of $G^<$ at equal times coincides (a part from a numerical factor) with the single-particle density matrix defined in Eq. (3.27), and therefore it can be used for the evaluation of any single-particle quantity of interest. Thus the Green function $G^<$ contains all the single–particle properties of the many–body system.

The single-particle Green functions defined in Eqs. (3.38-3.41) still depend on the interaction of each particle with all the other particles in the system, and a simple closed equation of motion as Eq. (3.31) cannot be written in this case, since the propagation of a single particle depends on the entire many-body system. In particular, as we have seen for the density matrix, it is possible to define a set of hierarchical equations of motion[20], where the equation for the single-particle Green function contains also a two-particle Green function; the equation for the two-particle Green function contains a three-particle Green function, and so on.

The analysis of the single-particle Green functions can be performed, however, by means of a general technique, based on perturbation theory. The starting point of this technique is a perturbative expansion of the various single-particle Green functions written in the interaction picture in powers of the perturbation Hamiltonian. The various terms contain the unperturbed Green functions (i.e. the single-particle Green functions related to the unperturbed Hamiltonian) and are commonly expressed by means of Feynman diagrams. From the analysis of such diagrams it is possible to derive a set of equations (i.e. a matrix equation), called Dyson equation, in terms of a so called self–energy matrix $\Sigma$[10,21] that describes the effect of the perturbation on the particles of interest. They have the form

$$G = G_\circ + G_\circ \Sigma G. \tag{3.42}$$

where, $G_\circ$ is the Green-function matrix related to the unperturbed Hamiltonian.

The self–energy is, in general, hard to find and very often it is derived from a simplified model of interaction.

The Dyson equation has been the starting point for many approaches to quantum transport problems. As an example, we may mention the so called "quantum Boltzmann equation"[22].

### 3.5 The Wigner function

The Wigner function was introduced as an extension of the concept of distribution function to the quantum case, and it constitutes the more direct link between the quantum density matrix and the "classical" description of the evolution of the system in phase space through a distribution $f(q,p,t)$, where $q$ and $p$ represent a general set of conjugate variables. In quantum terms, we cannot define a probability function $\mathcal{P}(q,p)$ such that $\mathcal{P}(q,p)dqdp$ is equal to the probability of finding the system in $dq$ around $q$ and $dp$ around $p$, since this probability is ill-defined in quantum mechanics owing to the incompatibility of the two necessary measurements.

The Wigner function tries to approach this concept and is defined as the Weyl transform of the single-particle density matrix[23]:

$$f_W(\mathbf{r},\mathbf{k},t) = \frac{1}{(2\pi)^3}\int d\mathbf{r}' e^{i\mathbf{k}\cdot\mathbf{r}'}\rho(\mathbf{r}-\frac{\mathbf{r}'}{2},\mathbf{r}+\frac{\mathbf{r}'}{2},t) \tag{3.43}$$

Several interesting properties of $f_W$ so defined suggest to think of this function as a quantum extension of the concept of distribution function[24]. In particular

$$\int f_W(\mathbf{r},\mathbf{k},t)d\mathbf{k} = \langle \mathbf{r} \mid \rho^{(sp)}(t) \mid \mathbf{r}\rangle \equiv \rho^{(sp)}(\mathbf{r},\mathbf{r},t) \tag{3.44}$$

(position probability density);

$$\int f_W(\mathbf{r},\mathbf{k},t)d\mathbf{r} = \langle \mathbf{k} \mid \rho^{(sp)}(t) \mid \mathbf{k}\rangle \equiv \rho^{(sp)}(\mathbf{k},\mathbf{k},t) \tag{3.45}$$

(momentum probability density).

For an observable $A^{(sp)}(\mathbf{r},\mathbf{k},t)$ it can be shown that[23]:

$$\overline{\langle A\rangle}(t) = Tr(\mathbf{A}\rho) = \frac{1}{(2\pi)^3}\int d\mathbf{r}\int d\mathbf{k} A_W(\mathbf{r},\mathbf{k},t) f_W(\mathbf{r},\mathbf{k},t), \tag{3.46}$$

where $A_W(\mathbf{r},\mathbf{k},t)$ is a function obtained from the operator $\mathbf{A}^{(sp)}$ by applying the Weyl's rule in Eq. (3.43).

It is not possible however to attribute a direct probabilistic interpretation to $f_W$ since $f_W$ is, in general, not positive definite.

If we consider, for the sake of simplicity, a single-particle system, the Wigner function takes the form

$$f_W(\mathbf{r},\mathbf{k},t) = \frac{1}{(2\pi)^3}\int d\mathbf{r}' e^{i\mathbf{k}\cdot\mathbf{r}'}\phi^*(\mathbf{r}+\frac{\mathbf{r}'}{2},t)\phi(\mathbf{r}-\frac{\mathbf{r}'}{2},t). \tag{3.47}$$

where $\phi^*(\mathbf{r},t)$ is the particle wavefunction.

A physical insight into the meaning of the Wigner function can be obtained by observing that its value in a region $(\mathbf{r},\mathbf{k})$ gives information on the k-Fourier components of the autocorrelation of the wave function around $\mathbf{r}$. Thus, $f_W$ has large values in regions of phase-space where the presence of the particles can be detected within the uncertainty principle. Some more information related to the quantum mechanical phases are added, that can result in negative values for $f_W$.

Concluding this brief introduction to the Wigner function we note that, if the interaction of the particle with the rest of the system can be described by a perturbation potential $V(\mathbf{r})$, an equation for the Wigner function can be written[25]:

$$\frac{\partial f_W(\mathbf{r},\mathbf{k},t)}{\partial t} + \frac{\hbar \mathbf{k}}{m} \frac{\partial f_W(\mathbf{r},\mathbf{k},t)}{\partial \mathbf{r}}$$

$$= \frac{1}{i\hbar} \frac{1}{(2\pi\hbar)^3} \int d\mathbf{k}' \int d\mathbf{r}' e^{i\mathbf{k}'\cdot\mathbf{r}'} [V(\mathbf{r} - \frac{1}{2}\mathbf{r}') - V(\mathbf{r} + \frac{1}{2}\mathbf{r}')] f_W(\mathbf{r}, \mathbf{k} - \mathbf{k}', t) \; . \quad (3.48)$$

By performing a series expansion of the $V$ terms in the rhs of Eq.(3.48), an alternative form of the Wigner equation can be obtained, which reduces to the classical Liouville equation for $\hbar \to 0$.

In order to extend the above theory to the case where scattering is present, an "ad hoc" collision term $(\frac{\partial f_W}{\partial t})_{coll}$ has been added to the rhs of Eq. (3.48)[24], but it may not necessarily express the same phenomenology as the corresponding term in the Boltzmann equation.

Some interesting results have been obtained by considering moments of Eq. (3.48) in the relaxation-time approximation. Equations are obtained which reduce to the corresponding moment equations of the Boltzmann theory in the semiclassical limit, and contain quantum corrections which can be included in the classical picture[24].

### 3.6 The path-integral approach

The Feynman path-integral theory[26] is an alternative approach to quantum mechanics. It starts from the idea that all possible paths of a system from an initial state to a final one are to be considered as simultaneously realized, and their amplitudes add up, rather than their probabilities as it would be in classical concepts, to give the probability amplitude of finding the system in the final state. The fundamental equation in this approach is an explicit expression for the evolution operator as an integral over all possible paths of the exponential of the classical action. If $q$ represents a set of classical Lagrangian variables of the system, and $q(\tau)$ one given trajectory, or path, from the initial values $q_i = q(t_i)$ to the final values $q_f = q(t)$, the evolution operator in Eq. (3.14) can be written as

$$U(q_f, q_i, t, t_i) = \int_{q_i, t_i}^{q_f, t} \mathcal{D}q(\tau) \; e^{\frac{i}{\hbar} S[q(\tau)]}. \quad (3.49)$$

Here $\int \mathcal{D}q(\tau)$ indicates the integral over all paths that connect the initial state $\{q_i, t_i\}$ to the final state $\{q_f, t\}$ and $S[q(\tau)]$ is the classical action evaluated over each given trajectory in the integral:

$$S[q(\tau)] = \int_{t_i}^{t} L(q(\tau), \dot{q}(\tau), \tau) d\tau, \quad (3.50)$$

where $L(q, \dot{q}, \tau)$ is the classical Lagrangian of the system.

If a path-integral approach is followed and the integral in Eq. (3.49) evaluated, the resulting evolution operator can then be used to obtain the evolution of a state wavefunction or of the density matrix. However, even though in this method an explicit form for the evolution operator is directly given , its practical evaluation presents similar difficulties as the solution of the Schrödinger equation , or the Liouville-von Neumann equation.

By applying the expression for the evolution operator given in Eq. (3.49) to the density matrix, the following expression for $\rho$ is obtained:

$$\rho(q,q',t) = \int dq_i \int dq_i' \int_{q_i,t_i}^{q,t} \mathcal{D}q(\tau) \int_{q_i',t_i}^{q',t} \mathcal{D}q'(\tau) e^{\frac{i}{\hbar}[S(q(\tau))-S(q'(\tau))]} \rho(q_i,q_i',t_i). \quad (3.51)$$

Eq. (3.51) may be elaborated in a useful way by factorizing the effect of "external agents" with respect to the system of interest. In our case if $x$ are the variables of the "small" system of interest (for example an electron), and $X$ the variables of the interacting "large" system (for example the phonon bath), the exponential in Eq. (3.51) can then be factorized as follows:

$$\int ...e^{\frac{i}{\hbar}[S(x)-S(x')]} e^{\frac{i}{\hbar}[S(X)-S(X')+S(x,X)-S(x',X')]} \mathcal{D}x(\tau)\mathcal{D}x'(\tau)\mathcal{D}X(\tau)\mathcal{D}X'(\tau), \quad (3.52)$$

where $S(x)$ and $S(X)$ are the actions for the small and the interacting systems, respectively, and $S(x,X)$ the action of interaction between the two systems. The integrals over the paths of the interacting system involve only the second exponential, and an "influence functional" can be defined[27],

$$\mathcal{F}(x(\tau),x'(\tau)) = \int_{X_i,t_i}^{X,t} \int_{X_i',t_i}^{X',t} \mathcal{D}X(\tau)\mathcal{D}X'(\tau) e^{\frac{i}{\hbar}[S(X)-S(X')+S(x,X)-S(x',X')]}, \quad (3.53)$$

such that the evolved density matrix is written as

$$\rho(q,q') = \int dq_i \int dq_i' \int \mathcal{D}x(\tau) \int \mathcal{D}x'(\tau)\mathcal{F}(x(\tau),x'(\tau)) e^{\frac{i}{\hbar}[S(x)-S(x')]} . \quad (3.54)$$

Here for any given path of the system of interest, $\mathcal{F}$ carries the information of the influence on that path of the integral of all paths of the interacting system. The theoretical step performed by introducing the influence functional is not trivial since it includes all the influence of the interacting system on the behavior of the system of interest.

However, the explicit evaluation of the influence functional and the subsequent evaluation of the path integral in Eq. (3.54) are in general prohibitively difficult, and approximations must be made as in the more standard formulations of the problem.

## 4. QUANTUM ANALYSIS OF TRANSIENT TRANSPORT

As an application of the density matrix formalism discussed in Sect. 3.3, in the following sections we shall briefly summarize the work performed by the Modena group[28-30] aimed at the solution of the Liouville–von Neumann equation for the electronic density matrix in semiconductors. As we shall see, in principle, the method allows to evaluate the electronic density matrix as a function of time without any assumptions on the intensity of the electron-phonon interaction and on the strength of the applied field. The quantum equation is solved through a random generation of all possible quantum interactions at the various perturbative orders, in the same way as the usual classical Monte Carlo (CMC) generates classical scattering events.

### 4.1 Physical system and iterative expansion

A non-interacting electron gas in a semiconductor crystal, coupled to the phonon gas is considered. The system is assumed to be homogeneous, and its Hamiltonian is given by

$$\mathbf{H} = \mathbf{H}_e + \mathbf{H}_E + \mathbf{H}_p + \mathbf{H}_{ep}, \tag{4.1}$$

where $\mathbf{H}_e$ is the term corresponding to an electron in a perfect crystal, $\mathbf{H}_E = e\mathbf{E}\cdot\mathbf{r}$ describes the coupling to the electric field, and $\mathbf{H}_p$ is the Hamiltonian of the free-phonon system. The electron-phonon interaction Hamiltonian $\mathbf{H}_{ep}$ has the general form

$$\mathbf{H}_{ep} = \sum_{\mathbf{q}} i\hbar F(\mathbf{q})\{\mathbf{a}_{\mathbf{q}}e^{i\mathbf{q}\cdot\mathbf{r}} - \mathbf{a}_{\mathbf{q}}^{\dagger}e^{-i\mathbf{q}\cdot\mathbf{r}}\} = \mathbf{H}_{ab} + \mathbf{H}_{em}, \tag{4.2}$$

where $\mathbf{H}_{ab}$ and $\mathbf{H}_{em}$ refer to phonon absorption and emission, respectively, and $F(\mathbf{q})$ is a function of the phonon momentum $\mathbf{q}$ whose explicit form depends on the particular scattering mechanism considered.

The set of time-dependent basis vectors $\mid \mathbf{k}_\circ, \{n_{\mathbf{q}}\}, t\rangle$ represented by[31]

$$\frac{1}{\sqrt{V}}e^{i[\mathbf{k}(t)\cdot\mathbf{r}]}e^{-i\int_0^t d\tau\omega[\mathbf{k}(\tau)]} \mid \{n_{\mathbf{q}}\}, t\rangle, \tag{4.3}$$

is introduced, where $\mathbf{k}(t) = \mathbf{k}_\circ - \frac{e\mathbf{E}}{\hbar}t$, and $\omega(\mathbf{k}(t))$ represents the electronic band structure. They are direct products of electronic accelerated plane waves, normalized to 1 over the crystal volume V, and the phonon states $\mid \{n_{\mathbf{q}}\}, t\rangle$.

It can be shown[28] that the density matrix $\rho$ of the system in the representation of the set in Eq. (4.3), has a time evolution determined only by the perturbation Hamiltonian:

$$i\hbar\frac{\partial}{\partial t}\rho(x, x'; t) = [\mathbf{H}_{ep}(t), \rho(t)](x, x'), \tag{4.4}$$

where we have introduced the symbolic compact notation $x = (\mathbf{k}_\circ, \{n_{\mathbf{q}}\})$. We want to stress the strong similarity between Eq. (4.4) and the Liouville-von Neumann equation (3.20) in the interaction picture. The reason for such similarity is that the time-dependent basis vectors (4.3) are just the solutions of the time-dependent Schrödinger equation for the unperturbed Hamiltonian $\mathbf{H}_0 \equiv \mathbf{H}_e + \mathbf{H}_E + \mathbf{H}_p$ and, therefore, they describe the unperturbed dynamics. As a consequence, in such representation the matrix elements of $\rho$ evolves in time only because of the perturbation Hamiltonian $\mathbf{H}_{ep}$, as for the case of the interaction picture.

If we now formally perform a time integration of Eq. (4.4) from an initial time 0 to a final time $t$, we obtain:

$$\rho(x,x';t) = \rho(x,x';0) + \int_0^t dt' [\mathcal{H}_{ep}(t'),\rho(t')](x,x'), \tag{4.5}$$

where $\mathcal{H}_{ep} = \frac{1}{i\hbar}\mathbf{H}_{ep}$.

Starting from the above integral version of the Liouville-von Neumann equation, a perturbative expansion for $\rho$ is easily obtained by iterative substitutions:

$$\begin{aligned}\rho(x,x';t) &= \rho(x,x';0) + \int_0^t dt_1 [\mathcal{H}_{ep}(t_1),\rho(0)](x,x') \\ &+ \int_0^t dt_1 \int_0^{t_1} dt_2 [\mathcal{H}_{ep}(t_1),[\mathcal{H}_{ep}(t_2),\rho(0)]](x,x') + \dots \\ &= \rho^{(o)}(x,x';t) + \Delta\rho^{(1)}(x,x';t) + \Delta\rho^{(2)}(x,x';t) + \dots\ .\end{aligned} \tag{4.6}$$

For the evaluation of expectation values of electron quantities which are diagonal in the electronic part of the states in Eq. (4.3) we need and, therefore, we focus our attention on the diagonal elements $\rho(x,t) \equiv \rho(x,x;t)$ of the density matrix. Furthermore, we assume a diagonal initial condition for $\rho$ decoupled in electron and phonon coordinates. Under such conditions, for the diagonal terms of the density matrix $\rho(x,x;t)$ we can write down the following expansion:

$$\begin{aligned}\rho(x,t) &= \rho(x,0) + \int_0^t dt_1 [\mathcal{H}_{ep}(t_1),\rho(0)](x,x) \\ &+ \int_0^t dt_1 \int_0^{t_1} dt_2 [\mathcal{H}_{ep}(t_1),[\mathcal{H}_{ep}(t_2),\rho(0)]](x,x) + \dots \\ &= \rho^{(o)}(x,t) + \Delta\rho^{(1)}(x,t) + \Delta\rho^{(2)}(x,t) + \dots\ .\end{aligned} \tag{4.7}$$

Several considerations follow from the analysis of the above perturbative expansion. Taking into account the explicit form of the perturbation Hamiltonian given in Eq. (4.2), we easily recognize that only even-order contributions in the above expansion result to be different from zero. This is due to the fact that, for each perturbation Hamiltonian, we deal with a creation (annihilation) phonon operator and, therefore, looking to diagonal elements of $\rho$, only products of an even number of phonon creation and annihilation operators can result to be different from zero. Furthermore, in order to have a non-vanishing term in the above expansion, for any creation (annihilation) operator related to a given phonon mode $\mathbf{q}$, a corresponding annihilation (creation) operator related to the same mode $\mathbf{q}$ must be present ("phonon pairing"). As a consequence, each of these pairs can be regarded as an electron-phonon interaction process. We deal with two different types of processes: real ones for which the same phonon with mode $\mathbf{q}$ is absorbed (emitted) by each density-matrix index, and virtual ones which correspond to the absorption (emission) of a phonon mode $\mathbf{q}$ by one index and a subsequent emission (absorption) of the same mode $\mathbf{q}$ by the same density-matrix index. Therefore, each term in the expansion (4.7) can be regarded as a sequence of real and virtual processes.

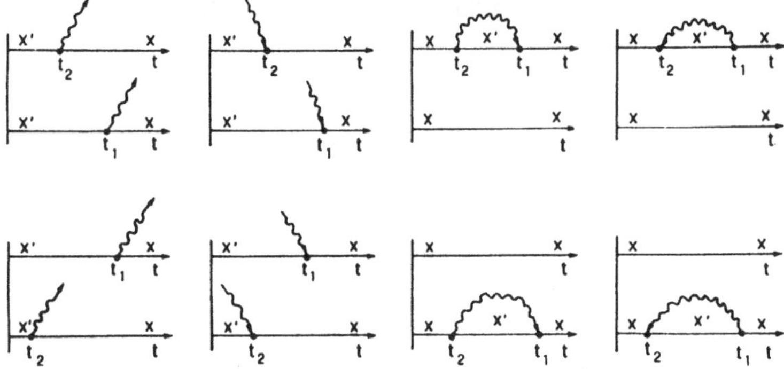


Fig. 6. Diagrams representing the second-order contribution to the density matrix. The horizontal axes represent the time for the two arguments of the density matrix, and arrows indicate phonon absorption and emission processes.

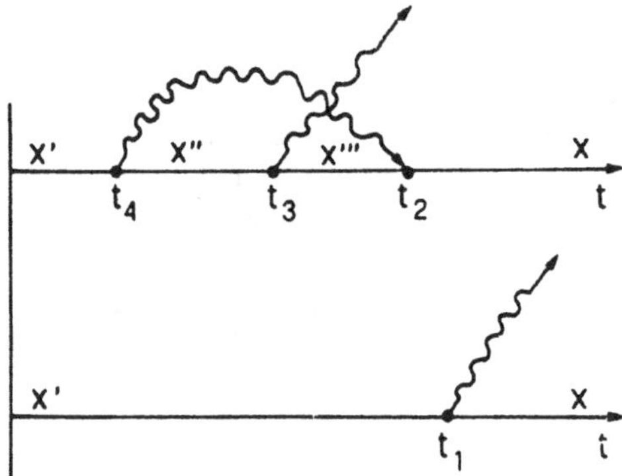


Fig. 7. Diagram representing the fourth-order contribution to the density matrix of the type shown in Eq. (18).

There is a simple and useful diagrammatic representation[28] of such terms. Such diagrams contain "real" emissions and absorptions, corresponding to the scattering "in" in the collision term of the semiclassical Boltzmann equation, and "virtual" processes, corresponding to scattering "out". Examples of this type of diagrams are shown in Figs. 6 and 7.

As discussed in Sect. 3.1, if we are only interested in electron properties, the mathematical instrument to be used is the reduced electron density matrix. It has been introduced in Eq. (3.23) as the trace of the total density matrix $\rho$ over the phonon coordinates. Such trace operation can be applied to Eq. (4.7) and an equivalent expansion in powers of the electron-phonon coupling for the diagonal terms of the electronic density matrix is obtained. The effect of this trace operation is a replacement of the phonon occupation numbers with the corresponding Bose occupation numbers.

In this way, we obtain a perturbative expansion for the electronic density matrix which can be regarded as a sequence of quantum processes in the same way as in the semiclassical simulation we derive the electron distribution function in terms of sequences of scattering events. This constitutes the starting point of the numerical QMC algorithm devised for the solution of the Liouville–von Neumann equation, which is based on random generations of all possible sequences of processes associated with the different perturbative corrections, in the same way as classical scattering events are generated in a semiclassical Monte Carlo.

Results obtained with the QMC procedure for different materials and physical conditions have been obtained at very short times after the initial conditions, when quantum features are expected to be more relevant. This choice allows to include only few terms of the perturbation expansion given in Eq. (4.7), which in turn limits the computer time to affordable values.

In the remaining part of this Section, an overview of some physical problems will be presented with special emphasis to the investigation of typical quantum effects, such as intracollisional field effect and collisional broadening, by comparison with the semiclassical case.

## 4.2 The Quantum Monte Carlo procedure

The numerical QMC algorithm devised for the solution of Eq.(4.4) is essentially based on a Monte Carlo evaluation of the sum in Eq.(4.7) by means of random generations, with arbitrary probabilities, of all possible processes associated with the different perturbative corrections. Such solution is based on the general Monte Carlo technique for the evaluation of an infinite sum of multiple integrals discussed in Ref. 32.

For example, a suitable algorithm could be the following:

i. The order $2n$ of the perturbative correction is chosen;
ii. $n$ processes (**q**-modes and absorption/emission) are selected;
iii. for each process, two times are generated, that correspond to the times at which the integrand functions are sampled;
iv. starting from the value $\mathbf{k}(t)$ at which $\rho$ is to be evaluated, both indices of the density matrix are translated backwards in time, if we are in presence of an electric field, to the time of the latest vertex. At this point the matrix element $\mathcal{H}_{ep}$ of the interaction is evaluated, and the current value of $\mathbf{k}$ is changed accordingly. This last step of the procedure is repeated until the time of the initial condition is reached.

Due to momentum conservation of the $\mathcal{H}_{ep}$ matrix elements, these selections determine the argument $\mathbf{k}_{in}$ of $\rho$ at t=0.

The quantity:

$$\frac{\mathcal{H}_{ep}(t_1)...\mathcal{H}_{ep}(t_n)}{\mathcal{P}}\rho(x_{in}, t = 0), \tag{4.8}$$

is then evaluated, where $\mathcal{P}$ is the probability of all the selections which have been made (given by the product of the probabilities of the single choices). An average of

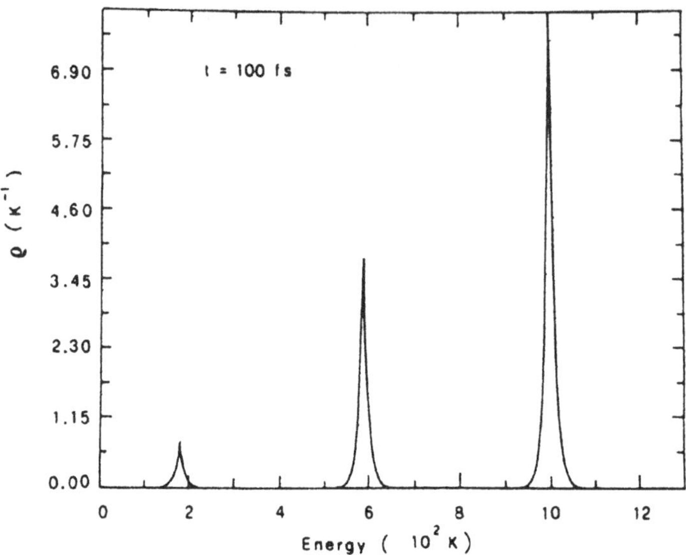


Fig. 8. Classical electron distribution as a function of energy for a simplified GaAs model at $t = 100fs$ after exitation. The highest peak at $1000K$ is the initial distribution at $t = 0$.

the estimator in Eq.(4.8) is finally obtained through many iterations of the procedure, and it gives an estimate of $\rho(\mathbf{k})$ at time t.

The order of the simulation can be chosen at will: it can be backwards in time, as in the example above, or forwards, as used in most practical cases.

Also the self-scattering technique can be extended to the quantum case, where it coincides with a zero-order approximation to the electron-phonon self energy[33].

In the next section we shall present some typical results obtained with the QMC technique for short-transient phenomena.

### 4.3 Quantum simulation of transient-transport phenomena

Different problems have been investigated with the QMC procedure at very short times after the initial conditions, when quantum features are expected to be more relevant.

A. Energy relaxation of photoexcited carriers

The QMC method has been applied to the case of photoexcited electrons in bulk GaAs[28]. The semiconductor model was simplified to a single spherical and parabolic band, and the interaction Hamiltonian included only polar coupling to optical phonons. Electrons were generated at $t = 0$ according to a distribution proportional to $exp\{-\alpha|\epsilon - \epsilon_\circ|\}$, where $\epsilon$ is the electron energy, $\alpha$ and $\epsilon_\circ$ are appropriate constants.

In order to obtain a better understanding of the physics involved, let us first consider the results of semiclassical theory.

Fig. 8 shows the results obtained with the traditional EMC at $t = 100fs$ after excitation. The highest peak at 1000 K represents what is left of the initial distribution

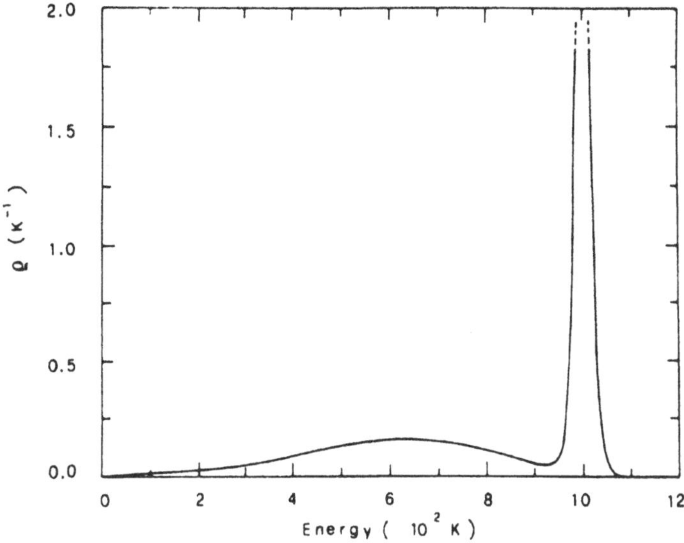


Fig. 9. Quantum electron distribution as a function of energy for a simplified GaAs model at $t = 100fs$ after exitation.

at $t = 0$. Two secondary peaks are clearly seen (phonon replicas), corresponding to electrons having emitted one or two optical phonons.

Fig. 9 shows the corresponding result obtained with the QMC (note the scale change). The initial distribution is diminished of a quantity very similar to that of the classical case. However electrons are spread, at $t = 100fs$, over a very wide range of energies, since energy needs not be conserved at these short times because of the uncertainty relation. The secondary peaks are not yet well formed and electrons can be found with energies not allowed in a classical description. If we go towards longer times the secondary peaks appear also in the quantum result when the conditions for the validity of the energy conservation are approached.

It should be noted that for realistic electron densities, carrier-carrier interaction would produce a spreading of the phonon replicas also in a semiclassical theory.

B. Quantum analysis of drift-velocity overshoot

Another significant application of the QMC procedure is the study of the drift-velocity overshoot in GaAs and Si[29,30]. For the case of silicon[30], a simplified semiconductor model was again used. Numerical results were obtained for different values of applied electric field and temperature.

Figure 10 shows a comparison between the quantum and the semiclassical drift-velocity overshoot at a low temperature ($T = 10K$). Here, for increasing values of the external field, we can see a corresponding enhancement of the quantum effect on the drift-velocity overshoot. This behavior is mainly due to the intracollisional field effect (ICFE)[34]. Such an effect depends inversely upon the scalar product $\mathbf{E} \cdot \mathbf{q}$ and therefore it favors transitions with momentum transfer normal to the electric field, thus decreasing the drift-velocity relaxation. The effect is still present at room temperature.

A similar behaviour was not found in the previous analysis for GaAs[29] because

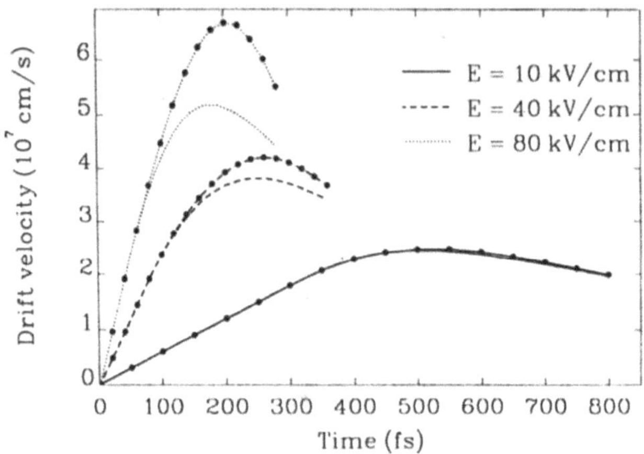


Fig. 10. Quantum drift velocity (curve marked with circles) compared with the semiclassical one for different values of the applied electric field at low temperature ($T = 10K$).

the explicit form of the electron-phonon interaction for a polar semiconductor ($F(\mathbf{q}) \propto q^{-1}$) favors by itself low momentum transfers even in the semiclassical case.

Finally, even though the scattering mechanism considered in the model (optical phonons) has a classical threshold energy for emission, the quantum corrections due to below-threshold transitions were not found to be relevant, both in Si and in GaAs. The reason is that the threshold energy is reached by the carriers at very short times when the effect of the perturbation on the electron dynamics is still very small.

The theoretical approach and the numerical procedure discussed above for the analysis of electron-phonon interaction has been also applied to the study of both electron-impurity interaction[35] and carrier-carrier interaction (especially for the analysis of impact-ionization processes[36]).

### 4.4 Wigner-function solutions

A. Finite-difference solution for steady-state conditions

The Wigner Equation (WE) in Eq. (3.48) can be written as

$$\frac{\partial}{\partial t} fw(q,p,t) + \frac{p}{m}\frac{\partial}{\partial q} fw(q,p,t) = \int_{-\infty}^{+\infty} dp'\, V_w(q,p-p') fw(q,p',t), \tag{4.9}$$

where

$$V_w(q,p) = -\frac{i}{\hbar}\int_{-\infty}^{+\infty} \frac{d\eta}{2\pi\hbar} e^{-ip\eta/\hbar}\left[V\left(q+\frac{\eta}{2}\right) - V\left(q-\frac{\eta}{2}\right)\right]. \tag{4.10}$$

A finite-difference approach for the steady-state solution of this equation has been developped by Frensley[37], and applied to the study of electronic devices. The approach is very efficient from the point of view of the required CPU time. On the other hand, its applicability is strongly limited by the large amount of computer mem-

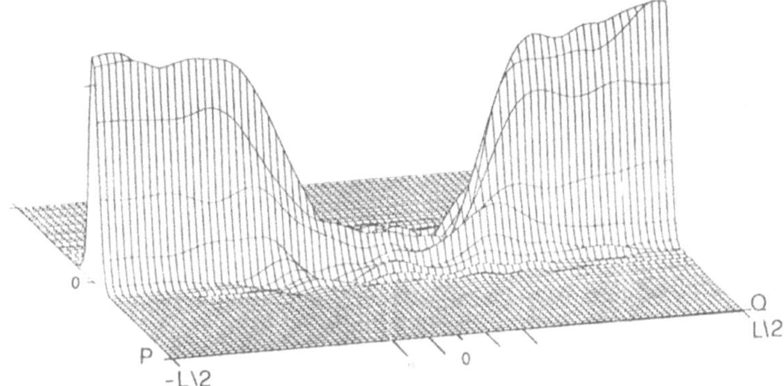


Fig. 11. Three-dimensional plot of the time-dependent Wigner function obtained with finite-difference algorithm at a time $t = 100fs$. At this time the WF is already very similar to the steady-state shape, which is reached at times $t > 400fs$.

ory required. As a consequence it is nowaday limited to the study of one-dimensional systems.

B. Finite-difference time-dependent solution

The finite-difference approach mentioned above can be applied to Eq. (4.9) also for the study of time-dependent phenomena. The resulting algorithm is characterized by longer CPU times but now the memory occupation is enormously reduced because the matrix elements are computed every time they are required in the calculation.

Fig. 11 shows the results obtained in a practical case. The device simulated consists of a quantum well bounded by barrier layers, thin enough to permit tunneling. Outside the barrier layers are thick layers of lower effective potential, which are doped so as to produce a free-electron density of 2 $10^{16}$ $cm^{-3}$. The size of the simulated region is $4.52 \times 10^{-6} cm$. The temperature was taken to be 300 K and the applied voltage 0.13V. The simulation has been performed up to the time $t = 100$ $fs$ after an initial condition when an equilibrium distribution at $T = 300K$ was assumed throughout the device. For a longer time $t > 400$ $fs$ the result obtained with the solution for steady-state condition is recovered.

C. Iterative expansion and Monte Carlo solution

If we introduce in the WE the following set of free-path variables

$$\{\tilde{q}, \tilde{p}, \tilde{t}\} \equiv \{q - \frac{p}{m}(t - t_\circ), p, t\}, \tag{4.11}$$

where $m$ is the particle mass and $\tilde{q}(q,p,t)$ and $\tilde{p}(q,p,t)$ represents the position and momentum of a classical particle that, without any applied field, will be in $q$ with momentum $p$ after a time $t$, and a transformed Wigner function $\tilde{f}$ such that

$$\tilde{f}(\tilde{q}, \tilde{p}, \tilde{t}) = fw(q,p,t) \ , \tag{4.12}$$

the Wigner transport equation (Eq.(4.9)) can be simply written as:

$$\frac{\partial}{\partial \tilde{t}}\tilde{f}(\tilde{q},\tilde{p},\tilde{t}) = \int_{-\infty}^{+\infty} dp' \, V_w(\tilde{q} + \frac{\tilde{p}}{m}(\tilde{t} - t_\circ), \tilde{p} - p') \, \tilde{f}(\tilde{q}, p', \tilde{t}) \, . \tag{4.13}$$

We are typically interested in solving Eq.(4.13) in a given region of our $(q, p, t)$ space according to some given boundary conditions. In particular, for our case the domain is defined by

$$-\frac{L}{2} < q < \frac{L}{2}, \qquad -\infty < p < \infty, \qquad t_\circ < t < \infty \, . \tag{4.14}$$

For this case, the boundary conditions are given by the Wigner function inside the region of interest at time $t = t_\circ$ and the Wigner function in one of the points $q = -\frac{L}{2}(p > 0)$ and $q = \frac{L}{2}(p < 0)$ for each $t \geq t_\circ$.

With the change of variables in Eq.(4.11) the domain is accordingly modified. In order to understand the physical meaning of the new boundary we may proceed as follows: For any given point $q, p, t$ inside the region of interest, we can consider the corresponding ballistic trajectory:

$$q(t') = q + \frac{p}{m}(t' - t), \qquad p(t') = p \, . \tag{4.15}$$

Going backward in time along this ballistic trajectory, at a given time $t^b$ we will encounter the boundary of our domain. There are two different situations: at time $t^b > t_\circ$ the trajectory exits the $q$-domain; at time $t^b = t_\circ$ the trajectory is still within the $q$-domain. For each of these two different cases we have to use the appropriate boundary condition for the Wigner function $f_W$.

Equation (4.13) can then be formally integrated from time $t^b$ (at which we know the boundary condition) to time $\tilde{t}$; Coming back to the original set of variables $\{q, p, t\}$ the resulting equation is

$$f_W(q,p,t) = f_W(q - \frac{p}{m}(t - t^b), p, t^b) +$$

$$+ \int_{t^b}^{t} dt' \int_{-\infty}^{+\infty} dp' \, V_w(q - \frac{p}{m}(t - t'), p - p') \, f_W(q - \frac{p}{m}(t - t'), p', t') \, . \tag{4.16}$$

We can now derive a series expansion of $f_W$ in powers of $V_w$ by means of iterative substitutions of Eq.(4.16) into itself, in the same way as was done for the Liouville equation for the density matrix. The result is

$$f_W(q,p,t) = f_W(q - \frac{p}{m}(t - t^b), p, t^b) +$$

$$+ \int_{t^b}^{t} dt_1 \int_{-\infty}^{+\infty} dp_1 \, V_w(q - \frac{p}{m}(t - t_1), p - p_1) \, f_W(q - \frac{p}{m}(t - t_1) - \frac{p_1}{m}(t_1 - t_1^b), p_1, t_1^b) +$$

$$+ \int_{t^b}^{t} dt_1 \int_{-\infty}^{+\infty} dp_1 \int_{t_1^b}^{t_1} dt_2 \int_{-\infty}^{+\infty} dp_2 \, V_w(q - \frac{p}{m}(t - t_1), p - p_1)$$

$$V_w(q - \frac{p}{m}(t - t_1) - \frac{p_1}{m}(t_1 - t_2), p_1 - p_2) f_W(q - \frac{p}{m}(t - t_1) - \frac{p_1}{m}(t_1 - t_2) - \frac{p_2}{m}(t_2 - t_2^b), p_2, t_2^b)$$

$$+ \ldots =$$

$$= f_W^{(0)}(q,p,t) + \Delta f_W^{(1)}(q,p,t) + \Delta f_W^{(2)}(q,p,t) + \dots \quad (4.17)$$

The above series expansion provides the solution of our Wigner transport equation for each point $q,p,t$ of the domain in terms of the given boundary conditions for $f$.

A Monte Carlo algorithm for the solution of Eq. (4.9) can then bedevised that consists of a simulation of a set of "sampling particles" from the initial time $t_o$ to the final time $t$. During each carrier "history", a corresponding estimator for the WF is built up in the same way as it is done for the semiclassical distribution function in the MC solution of the Boltzmann equation and for the density matrix in the MC solution of the Liouville-von Neumann equation. The physical system is the same as in Fig.11. We have performed simulations only for short physical times, starting from a uniform

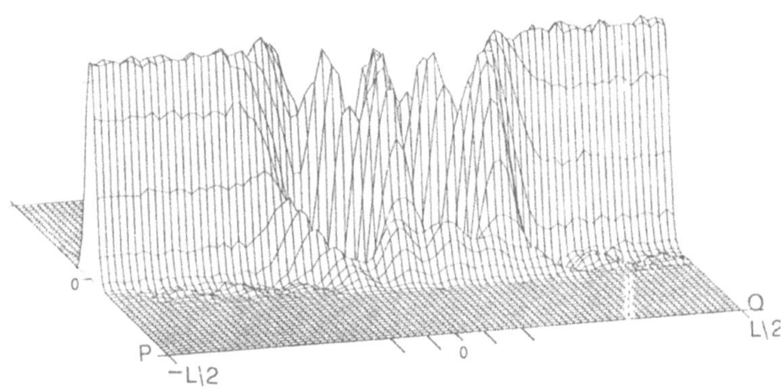


Fig. 12. Wigner distribution function at $t = 5fs$, obtained with a Monte Carlo procedure developed within the research group. The bias voltage is $0.130V$.

thermal equilibrium distribution of carriers inside the device. For simulative times of order of 10 $fs$ it is sufficient to evaluate the iterative expansion up to second order in the transfer function $V_w$.

Figure 12 shows the Wigner distribution function evaluated with Monte Carlo procedure at $t = 5\ fs$.

Finally, in order to show the effect of the potential barriers only on the shape of Wigner function, a simulation was performed with zero bias voltage (thermal equilibrium). The result at $t = 5\ fs$, is shown in Fig. 13.

A generalization of the above theory that incorporates phonon scattering is in progress.

# 5. QUANTUM KINETIC APPROACH FOR THE ANALYSIS OF COHERENT PHENOMENA IN PHOTOEXCITED SEMICONDUCTORS

## 5.1 Theoretical approach

The study of optically excited semiconductors on ultrashort time-scales has attracted large interest during the last years due to the improvements in the techniques to generate laser pulses in the femtosecond region and to perform measurements on these time-scales. The phenomena occurring in the subpicosecond regime can be di-

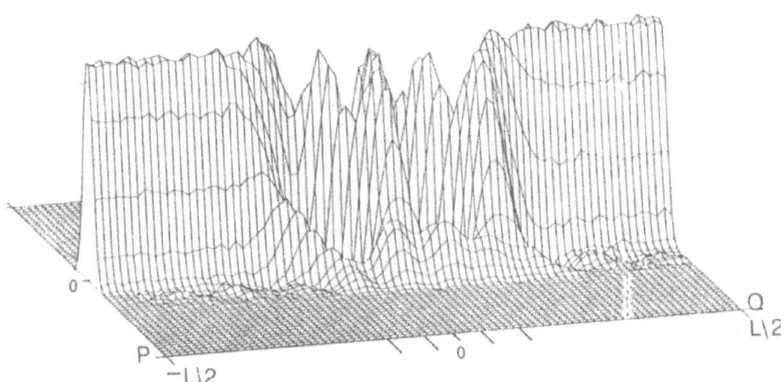


Fig. 13. Wigner distribution function at zero bias voltage (thermal equilibrium) evaluated by Monte Carlo algorithm. As it can be seen, after only $5fs$ the shape of the Wigner function in the region within the energy barriers is significantly changed with respect to the homogeneous distribution taken as initial condition.

vided into two classes: incoherent processes which lead to a stochastic dynamics, and coherent processes, where the phase-relation between different types of quasiparticles as well as with the coherent light field are important. Typical experiments referring to the first class are concerning the investigation of energy relaxation of photoexcited carriers[38–41] while typical for the second class are the experiments investigating photon echoes or quantum beats[42–44]. This division is reflected also in different methods usually taken for the theoretical analysis. The former ones have been widely studied using mainly the technique of Monte Carlo simulations[40,45–47] which turned out to be best suited to include on a microscopic level a large variety of scattering processes. On the contrary, the latter ones require a quantum kinetic description[48–52] which cannot simply be treated by means of a traditional Monte Carlo simulation. The reason is that for these phenomena the distribution functions, which are the basic quantities for semiclassical transport theory and which are calculated within a Monte Carlo simulation, do not provide sufficient information to describe the physical situation. In addition, information concerning the interband electron state mixing is needed. This is created by the coherence of the external light field interacting with the semiconductor, as well as by the Coulomb interaction between electrons and holes. This latter effect is of particular importance in energy regions close to the band gap.

In particular, on short time-scales, the two classes of phenomena cannot be separated. The light of an ultrashort laser pulse introduces coherence in the carrier system. If the physical phenomena under investigation occur on a time-scale shorter than the dephasing time, i.e. the time after which coherence effects are damped out, they should be taken into account in a theoretical analysis even if the measured quantities, e.g. energy relaxation rates, are variables which are purely determined by distribution functions.

In this section, we shall review a method introduced by Kuhn and Rossi[53,54] which retains the advantages of the Monte Carlo method in treating scattering processes and at the same time to generalize it in order to take into account coherent effects. This requires in addition to a simulation of the distribution functions of the various carrier types, as is done in the semiclassical case, a calculation of the polarizations introduced by the coherent light field. The use of the polarizations as dynamic variables then enables to take into account also intra- and interband effects of the carrier-carrier interaction which lead in Hartree-Fock approximation to an internal field and a renormalization of the energies. Furthermore, the generation process is treated in a self-consistent way, where the broadening due to the finite pulse duration does not have to be introduced as a phenomenological parameter, but comes out automatically with its full time-dependence.

Let us consider a bulk semiconductor with two parabolic bands, the conduction band and the heavy hole band. The system is described by a Hamiltonian which, as usual, can be decomposed into a part $H_0$ which is treated exactly, and a part $H_1$ describing interactions, which is treated within some approximation scheme.

### *A. Unperturbed Hamiltonian*

For the analysis of coherent phenomena the light-matter interaction has to be introduced in the unperturbed Hamiltonian $H_0$ which, in an electron-hole picture, for the system under investigation is then given by

$$H_0^c = \sum_{\mathbf{k}} \epsilon_{\mathbf{k}}^e c_{\mathbf{k}}^\dagger c_{\mathbf{k}} + \sum_{\mathbf{k}} \epsilon_{\mathbf{k}}^h d_{\mathbf{k}}^\dagger d_{\mathbf{k}} + A_0(t) \sum_{\mathbf{k}} \left[ M_{\mathbf{k}} e^{-i\omega_L t} c_{\mathbf{k}}^\dagger d_{-\mathbf{k}}^\dagger + M_{\mathbf{k}}^* e^{i\omega_L t} d_{-\mathbf{k}} c_{\mathbf{k}} \right] . \quad (5.1)$$

Here, $c_{\mathbf{k}}^\dagger$ and $c_{\mathbf{k}}$ denote the creation and annihilation operators of an electron in state $\mathbf{k}$, respectively; $d_{\mathbf{k}}^{\dagger}$ and $d_{\mathbf{k}}$ are the corresponding operators for holes; $\epsilon_{\mathbf{k}}^{e} = E_G + \hbar^2 k^2/2m_e$ and $\epsilon_{\mathbf{k}}^{h} = \hbar^2 k^2/2m_h$ the energies of electron and hole states, $E_G$ the energy gap, $m_e$ and$m_h$ the electron and hole effective masses. The external light field is treated on a semiclassical level. $M_{\mathbf{k}}$ is the dipole matrix element and $A_0(t)$ the pulse shape of the vector potential of the light field centered at the frequency $\omega_L$. In the simulations we will review, a Gaussian pulse $A_0(t) = A_L exp[-(t-t_0)^2/\tau_L^2]$ has been chosen.

To describe the kinetics of the system, we first need the distribution functions (intraband density matrices) of electrons and holes, defined as

$$f_{\mathbf{k}}^{e} = \langle c_{\mathbf{k}}^{\dagger} c_{\mathbf{k}} \rangle \qquad \text{and} \qquad f_{\mathbf{k}}^{h} = \langle d_{\mathbf{k}}^{\dagger} d_{\mathbf{k}} \rangle \ . \tag{5.2}$$

In order to describe the coherence phenomena we need additionally the two contributions to the interband polarization (interband density matrix), given by

$$p_{\mathbf{k}} = \langle d_{-\mathbf{k}} c_{\mathbf{k}} \rangle \qquad \text{and} \qquad p_{\mathbf{k}}^{*} = \langle c_{\mathbf{k}}^{\dagger} d_{-\mathbf{k}}^{\dagger} \rangle \ . \tag{5.3}$$

Using the Heisenberg equations of motion for the electron and hole operators, we obtain a closed set of equations of motion for the distribution functions and the polarization

$$\frac{d}{dt} f_{\mathbf{k}}^{e} = \frac{d}{dt} f_{-\mathbf{k}}^{h} = g_{\mathbf{k}}^{0}(t) \ , \tag{5.4}$$

$$\frac{d}{dt} p_{\mathbf{k}} = \frac{1}{i\hbar}\left[(\epsilon_{\mathbf{k}}^{e} + \epsilon_{-\mathbf{k}}^{h}) p_{\mathbf{k}} + M_{\mathbf{k}} A_0(t) e^{-i\omega_L t}(1 - f_{\mathbf{k}}^{e} - f_{-\mathbf{k}}^{h})\right] \ , \tag{5.5}$$

and the complex conjugate equation for $p_{\mathbf{k}}^{*}$. The generation rate $g_{\mathbf{k}}^{0}$ is given by

$$g_{\mathbf{k}}^{0} = \frac{1}{i\hbar}\left[M_{\mathbf{k}} A_0(t) e^{-i\omega_L t} p_{\mathbf{k}}^{*} - M_{\mathbf{k}}^{*} A_0^{*}(t) e^{i\omega_L t} p_{\mathbf{k}}\right] \ . \tag{5.6}$$

The main difference between the generation rate (5.6) and the standard result obtained from Fermi's Golden Rule is the fact that in this approach the generation process is retarded, i.e. by eliminating from the above system of equations the polarizations as kinetic variables, the generation rate at time $t$ does not only depend on the light field and the distribution functions at time $t$ but also on earlier times. In addition, due to the energy-time uncertainty relation, we obtain in a self-consistent way the broadening of the generation rate which, in a semiclassical treatment, has to be introduced in a phenomenological way. Finally, it turns out that the generation rate is not strictly positive; thus, the simulation we will review has to include also the process of recombination stimulated by the external field.

The above Eqs. (5.4) and (5.5), which describe a fully coherent dynamics, are modified by the various interaction mechanisms leading to additional contributions in the equations of motion:

$$\frac{d}{dt} f_{\mathbf{k}}^{e} = g_{\mathbf{k}}^{0}(t) + \sum_i \left.\frac{d}{dt} f_{\mathbf{k}}^{e}\right|^{i} \ , \qquad \frac{d}{dt} f_{\mathbf{k}}^{h} = g_{-\mathbf{k}}^{0}(t) + \sum_i \left.\frac{d}{dt} f_{\mathbf{k}}^{h}\right|^{i} \ , \tag{5.7}$$

$$\frac{d}{dt}p_{\mathbf{k}} = \frac{1}{i\hbar}\left\{(\epsilon^e_{\mathbf{k}} + \epsilon^h_{-\mathbf{k}})p_{\mathbf{k}} + M_{\mathbf{k}}A_0(t)e^{-i\omega_L t}(1 - f^e_{\mathbf{k}} - f^h_{-\mathbf{k}})\right\} + \sum_i \frac{d}{dt}p_{\mathbf{k}}\bigg|^i , \tag{5.8}$$

where the summation is over the various types of interactions.

*B Carrier-carrier interaction*

The carriers which are created by the laser field interact via the Coulomb potential $V_{\mathbf{q}}$. In a two-band model the total interaction ($H_1^{cc}$) can be separated into three parts: electron-electron, hole-hole, and electron-hole interaction. The Hamiltonian is given by

$$H_1^{cc} = \sum_{\mathbf{k},\mathbf{k}',\mathbf{q}} V_{\mathbf{q}}\left[\frac{1}{2}c^\dagger_{\mathbf{k}+\mathbf{q}}c^\dagger_{\mathbf{k}'-\mathbf{q}}c_{\mathbf{k}'}c_{\mathbf{k}} + \frac{1}{2}d^\dagger_{\mathbf{k}+\mathbf{q}}d^\dagger_{\mathbf{k}'-\mathbf{q}}d_{\mathbf{k}'}d_{\mathbf{k}} - c^\dagger_{\mathbf{k}+\mathbf{q}}d^\dagger_{\mathbf{k}'-\mathbf{q}}d_{\mathbf{k}'}c_{\mathbf{k}}\right] . \tag{5.9}$$

Using the Heisenberg equations of motion, the resulting carrier-carrier contributions in Eqs. (5.7) and (5.8) involve expectation values of products of four creation and annihilation operators. The lowest order approximation is obtained by factorizing such expectation values into distribution functions and polarizations. This constitutes the Hartree-Fock approximation and the results are:

$$\frac{d}{dt}f^e_{\mathbf{k}}\bigg|^{cc} = \frac{d}{dt}f^h_{-\mathbf{k}}\bigg|^{cc} = \frac{1}{i\hbar}\left[\Delta_{\mathbf{k}}p^*_{\mathbf{k}} - \Delta^*_{\mathbf{k}}p_{\mathbf{k}}\right] , \tag{5.10}$$

$$\frac{d}{dt}p_{\mathbf{k}}\bigg|^{cc} = \frac{1}{i\hbar}\left[\Sigma_{\mathbf{k}}p_{\mathbf{k}} + \Delta_{\mathbf{k}}(1 - f^e_{\mathbf{k}} - f^h_{-\mathbf{k}})\right] , \tag{5.11}$$

where

$$\Sigma_{\mathbf{k}} = -\sum_{\mathbf{k}'} V_{\mathbf{k}-\mathbf{k}'}(f^e_{\mathbf{k}'} + f^h_{-\mathbf{k}'}) \qquad \text{and} \qquad \Delta_{\mathbf{k}} = -\sum_{\mathbf{k}'} V_{\mathbf{k}-\mathbf{k}'}p_{\mathbf{k}'} . \tag{5.12}$$

Comparing with Eqs (5.4,5.5), we see that $\Sigma_{\mathbf{k}}$ and $\Delta_{\mathbf{k}}$ play the roles of a self energy and of an internal field, respectively. The carrier-carrier interaction in Hartree-Fock approximation leads therefore to a renormalization of the carrier energies and of the external field. The self energy is a real quantity, thus there is no incoherent scattering. The effect of the internal field on the dynamics of the distribution functions is obtained by adding this internal field to the external one in the generation rate (5.6). We will refer to this modified rate as full generation rate $g_{\mathbf{k}}$.

*C Carrier-phonon interaction*

Introducing the creation (annihilation) operators $b^\dagger_{\mathbf{q}}$ ($b_{\mathbf{q}}$) of a phonon with wave vector $\mathbf{q}$, the Hamiltonian describing electron-phonon and hole-phonon interaction is given by

$$H_1^{cp} = \sum_{\mathbf{q},\mathbf{k}} \left[ \gamma_\mathbf{q}^e c_{\mathbf{k}+\mathbf{q}}^\dagger b_\mathbf{q} c_\mathbf{k} + \gamma_\mathbf{q}^{e*} c_\mathbf{k}^\dagger b_\mathbf{q}^\dagger c_{\mathbf{k}+\mathbf{q}} + \gamma_\mathbf{q}^h d_{\mathbf{k}+\mathbf{q}}^\dagger b_\mathbf{q} d_\mathbf{k} + \gamma_\mathbf{q}^{h*} d_\mathbf{k}^\dagger b_\mathbf{q}^\dagger d_{\mathbf{k}+\mathbf{q}} \right] , \quad (5.13)$$

where $\gamma_\mathbf{q}^{e,h}$ are the interaction matrix elements. Using the Heisenberg equations of motion, we obtain

$$\frac{d}{dt} f_\mathbf{k}^e \bigg|^{cp} = \frac{1}{i\hbar} \sum_\mathbf{q} \left\{ \gamma_\mathbf{q}^e \left[ \langle c_\mathbf{k}^\dagger b_\mathbf{q} c_{\mathbf{k}-\mathbf{q}} \rangle - \langle c_{\mathbf{k}+\mathbf{q}}^\dagger b_\mathbf{q} c_\mathbf{k} \rangle \right] + \gamma_\mathbf{q}^{e*} \left[ \langle c_\mathbf{k}^\dagger b_\mathbf{q}^\dagger c_{\mathbf{k}+\mathbf{q}} \rangle - \langle c_{\mathbf{k}-\mathbf{q}}^\dagger b_\mathbf{q}^\dagger c_\mathbf{k} \rangle \right] \right\} , \quad (5.14)$$

$$\frac{d}{dt} p_\mathbf{k} \bigg|^{cp} = \frac{1}{i\hbar} \sum_\mathbf{q} \Big\{ \gamma_\mathbf{q}^e \langle d_{-\mathbf{k}} b_\mathbf{q} c_{\mathbf{k}-\mathbf{q}} \rangle + \gamma_\mathbf{q}^{e*} \langle d_{-\mathbf{k}} b_\mathbf{q}^\dagger c_{\mathbf{k}+\mathbf{q}} \rangle$$

$$+ \gamma_\mathbf{q}^h \langle b_\mathbf{q} d_{-\mathbf{k}-\mathbf{q}}^\dagger c_\mathbf{k} \rangle + \gamma_\mathbf{q}^{h*} \langle b_\mathbf{q}^\dagger d_{-\mathbf{k}+\mathbf{q}} c_\mathbf{k} \rangle \Big\} . \quad (5.15)$$

and the corresponding equation for the distribution function of holes. Again, as for the case of carrier-carrier interaction, the first-order contributions are obtained by factorizing the expectation values on the right hand side of Eqs. (5.14,5.15). These expectation values, however, always contain a single phonon creation or annihilation operator. If we neglect the possible existence of coherent phonon states, only operators corresponding to a well defined phonon occupation number have non-vanishing expectation values. Thus, the first-order terms vanish. In order to obtain the second-order contribution in terms of the carrier distribution functions and the polarizations, we need to perform a series of consecutive steps and to introduce at different levels suitable approximations. For the sake of brevity, we can not repeat here the details of the derivations, which can be found in Ref. 19.

The final result concerning the carrier-phonon contributions to the equations of motion for the distribution functions then turn out to be the standard Boltzmann scattering terms

$$\frac{d}{dt} f_\mathbf{k}^{e,h} \bigg|^{cp} = \frac{2\pi}{\hbar} \sum_{\mathbf{q},\pm} |\gamma_\mathbf{q}^{e,h}|^2 \delta(\epsilon_\mathbf{k}^{e,h} - \epsilon_{\mathbf{k}+\mathbf{q}}^{e,h} \pm \hbar\omega_\mathbf{q})$$

$$\left[ (N_\mathbf{q} + \tfrac{1}{2} \pm \tfrac{1}{2}) f_{\mathbf{k}+\mathbf{q}}^{e,h} (1 - f_\mathbf{k}^{e,h}) - (N_\mathbf{q} + \tfrac{1}{2} \mp \tfrac{1}{2}) f_\mathbf{k}^{e,h} (1 - f_{\mathbf{k}+\mathbf{q}}^{e,h}) \right] , \quad (5.16)$$

where $\omega_\mathbf{q}$ is the phonon dispersion relation. The contribution to the polarization can be expressed in terms of a complex carrier-phonon self energy $\Sigma_\mathbf{k}^{cp} = \Sigma_\mathbf{k}^e + \Sigma_\mathbf{k}^h$ according to

$$\frac{d}{dt} p_\mathbf{k} \bigg|^{cp} = \frac{1}{i\hbar} \Sigma_\mathbf{k}^{cp} p_\mathbf{k} , \quad (5.17)$$

The real and imaginary parts (with $\Sigma_\mathbf{k}^{e,h} = \hbar(\Omega_\mathbf{k}^{e,h} - i\Gamma_\mathbf{k}^{e,h})$ and $\mathcal{P}$ denoting the principal value) are given by

$$\Omega_\mathbf{k}^{e,h} = \frac{1}{\hbar} \sum_{\mathbf{q},\pm} |\gamma_\mathbf{q}^{e,h}|^2 \frac{\mathcal{P}}{\epsilon_\mathbf{k}^{e,h} - \epsilon_{\mathbf{k}+\mathbf{q}}^{e,h} \pm \hbar\omega_\mathbf{q}} \left[ (N_\mathbf{q} + \tfrac{1}{2} \pm \tfrac{1}{2}) f_{\mathbf{k}+\mathbf{q}}^{e,h} + (N_\mathbf{q} + \tfrac{1}{2} \mp \tfrac{1}{2})(1 - f_{\mathbf{k}+\mathbf{q}}^{e,h}) \right] , \quad (5.18)$$

$$\Gamma_\mathbf{k}^{e,h} = \frac{\pi}{\hbar} \sum_{\mathbf{q},\pm} |\gamma_\mathbf{q}^{e,h}|^2 \delta(\epsilon_\mathbf{k}^{e,h} - \epsilon_{\mathbf{k}+\mathbf{q}}^{e,h} \pm \hbar\omega_\mathbf{q}) \left[ (N_\mathbf{q} + \tfrac{1}{2} \pm \tfrac{1}{2}) f_{\mathbf{k}+\mathbf{q}}^{e,h} + (N_\mathbf{q} + \tfrac{1}{2} \mp \tfrac{1}{2})(1 - f_{\mathbf{k}+\mathbf{q}}^{e,h}) \right] . \quad (5.19)$$

*D Carrier-photon interaction*

The interaction with the coherent laser mode has already been taken into account in $H_0$. In addition, there is the interaction with the background photon field which gives rise to the spontaneous emission and thus to the luminescence spectrum. Since the band gap is tipically of the order of $1\,eV$, stimulated processes due to thermal photons are negligible at all relevant temperatures. The Hamiltonian for carrier-photon interaction can be rewritten in terms of a quantized foton field and results to be similar to that for carrier-phonon interaction, however it connects electron and hole states. Within the dipole approximation, it is given by

$$H_1^{c\gamma} = \sum_{\mathbf{q},\mathbf{k},\nu} \left[ \mu_{\mathbf{k}} c_{\mathbf{k}}^{\dagger} a_{\mathbf{q},\nu} d_{-\mathbf{k}}^{\dagger} + \mu_{\mathbf{k}}^{*} d_{-\mathbf{k}} a_{\mathbf{q},\nu}^{\dagger} c_{\mathbf{k}} \right] , \tag{5.20}$$

where $a_{\mathbf{q},\nu}^{\dagger}(a_{\mathbf{q},\nu})$ denote the creation (annihilation) operators of a photon with wave vector $\mathbf{q}$ and polarization index $\nu$, and $\mu_{\mathbf{k}}$ the coupling matrix element which, except for a normalization constant, coincides with $M_{\mathbf{k}}$ introduced in Eq. (5.1).

Spontaneous recombination processes typically occur on much longer times than other scattering processes. Thus, for the ultrafast dynamics of the carrier system they are of minor importance. However, in luminescence experiments the photon spectrum is the directly measured quantity. The recombination processes act as a detector of the state of the system and create the link between the theoretical calculation and the experimental output. Consequently, the influence of carrier-photon interaction on the carrier dynamics is neglected and this mechanism is only used to calculate the spectrum of the emitted photons. The rate of emitted photons $R_{\mathbf{q},\nu}$ with wave vector $\mathbf{q}$ and polarization $\nu$ is given by

$$R_{\mathbf{q},\nu} = \frac{d}{dt} \langle a_{\mathbf{q},\nu}^{\dagger} a_{\mathbf{q},\nu} \rangle = \frac{1}{i\hbar} \sum_{\mathbf{k}} \left[ -\mu_{\mathbf{k}} \langle c_{\mathbf{k}}^{\dagger} d_{-\mathbf{k}}^{\dagger} a_{\mathbf{q},\nu} \rangle + \mu_{\mathbf{k}}^{*} \langle a_{\mathbf{q},\nu}^{\dagger} d_{-\mathbf{k}} c_{\mathbf{k}} \rangle \right] . \tag{5.21}$$

We now have to distinguish two cases: In the presence of an interband polarization there is a coherent emission of photons. This is obtained by factorizing the products on the r.h.s. of Eq. (5.21). If the polarization is created by the external field, the coherently emitted photons have the same direction as the incident field. This is the basis for the analysis of four-wave mixing and photon-echo experiments. When observing the luminescence in a direction different from that of the generating beam, this contribution vanishes. The polarization created by the internal field leads to a photon emission in all directions.

If there is no coherent interband polarization, this first-order contribution to the luminescence spectrum vanishes. Then, the right-hand side can be treated in analogy with carrier-phonon interaction, and the incoherent emission spectrum $I_{em}^0(\omega_{\mathbf{q}})$ in second-order and Markov approximation is given by the Golden Rule formula

$$I_{em}^0(\omega_{\mathbf{q}}) = R_{\mathbf{q},\nu}^{inc} = \frac{2\pi}{\hbar} \sum_{\mathbf{k}} |\mu_{\mathbf{k}}|^2 \delta(\epsilon_{\mathbf{k}}^e + \epsilon_{-\mathbf{k}}^h - \hbar\omega_{\mathbf{q}}) f_{\mathbf{k}}^e f_{-\mathbf{k}}^h . \tag{5.22}$$

where $\omega_q = cq$, $c$ being the light velocity. This is the emission spectrum of a free electron-hole gas. It does not contain renormalization and correlation effects which are of particular importance in regions close to the band gap. They can be included by taking into account in the equations of motion for the terms on the r.h.s. of Eq. (5.21) the other types of interaction. In adiabatic and Markov approximation the spectrum is given by

$$I_{em}(\omega_{\mathbf{q}}) = R^{inc}_{\mathbf{q},\nu} = \frac{1}{i\hbar}\sum_{\mathbf{k},\mathbf{k}'}\left[\mu_{\mathbf{k}}\mu^*_{\mathbf{k}'}\left(\alpha^{-1}(\omega_{\mathbf{q}})\right)_{\mathbf{k}\mathbf{k}'} - \mu^*_{\mathbf{k}}\mu_{\mathbf{k}'}\left(\alpha^{*-1}(\omega_{\mathbf{q}})\right)_{\mathbf{k}\mathbf{k}'}\right] f^e_{\mathbf{k}'} f^h_{-\mathbf{k}'} \quad (5.23)$$

where

$$\alpha_{\mathbf{k}\mathbf{k}'}(\omega) = (\epsilon^e_{\mathbf{k}} + \epsilon^h_{-\mathbf{k}} - \hbar\omega + \hbar\Omega_{\mathbf{k}} - i\hbar\Gamma_{\mathbf{k}})\delta_{\mathbf{k}\mathbf{k}'} - (1 - f^e_{\mathbf{k}} - f^h_{-\mathbf{k}})V_{\mathbf{k}-\mathbf{k}'} \ . \quad (5.24)$$

The derivation of the absorption spectrum relevant for excite and probe experiments proceeds along the same lines and the result is obtained by replacing in Eq. (5.23) the product of the distribution functions $f^e_{\mathbf{k}'} f^h_{-\mathbf{k}'}$ by $(1 - f^e_{\mathbf{k}'} - f^h_{-\mathbf{k}'})$. Thus, the calculation of the spectra requires for each frequency the inversion of the matrix $\alpha(\omega)$ which can be performed numerically[48,52] in a suitable discretized k-space.

## 5.2 Numerical procedure

In order to see the limitations of a "conventional" Monte Carlo simulation and to discuss the problems arising when trying to generalize it for the study of coherent phenomena, let us briefly discuss how the traditional semiclassical transport theory is recovered from the present approach.

The semiclassical limit in this case is obtained by eliminating the polarizations as independent variables. Since the internal field is directly related to the polarization, it has to be neglected. Excitonic effects cannot be described in this limit. The polarization is eliminated by solving Eq. (5.5) within the adiabatic and Markov approximation as has been described for the case of carrier-phonon interaction. Thus, the result does not include the broadening due to the energy-time uncertainty relation. It can be included by keeping the full time-dependence of the light field and performing the Markov approximation only for the distribution functions and, if taken into account, the self energies. With the Markovian generation rate, the Boltzmann equations (5.7) represent a set of rate equations. They are the basis for the traditional Ensemble Monte Carlo (EMC) technique[3,4] which simply provides a Monte Carlo solution of the above system of rate equations. Since all quantities which describe coherent properties are eliminated, this approach is obviously limited to the study of incoherent phenomena. Within the Monte Carlo method, an ensemble of $N$ carriers is simulated (the total number $N$ not necessarily being constant) and each carrier $i$ having its wave vector $\mathbf{k}_i$ (in space dependent problems also its position vector $\mathbf{r}_i$). The solution of the equations can then be simply mapped onto a sequence of random "free flights", interrupted by random "scattering events".

Now let us look at the problems which we encounter when trying to treat coherent phenomena. First, of course, we have to solve in addition to the Boltzmann equations also the equation of motion for the polarization. Second, we have to deal with recombination processes due to the negative parts of the full generation rate. This is quite complicated within a "conventional" EMC. For the recombination process we need simultaneously an electron with wave-vector $\mathbf{k}$ and a hole with wave-vector $-\mathbf{k}$. Thus, in general for each recombination process we have to check the states of all carriers if there is a pair with the desired wave-vector (within a certain range in k-space), and this pair can recombine. The origin of this problem is related to the fact that we have specified much more information than necessary. Implicitly, we have treated the carriers as distinguishable particles. However, the only relevant information would be to say, e.g., there is one carrier with $\mathbf{k}_1$ and not to say carrier 1 is in the state $\mathbf{k}_1$, etc., as is done in the traditional simulation.

In order to discuss the approach proposed by Kuhn and Rossi, let us focus our attention on the system of quantum kinetic equations (5.7,5.8), which is a system of nonlinear equations. The nonlinearities are present even in absence of carrier-carrier interaction due to the Pauli exclusion principle. The basic idea of this approach is to perform a Monte Carlo simulation for the evaluation of the various distribution functions and a direct integration for the evaluation of the polarizations.

Due to the reasons discussed above, the usual representation of the carrier system is not appropriate for the problem under investigation. Therefore, within such approach we describe the state of the ensemble of carriers in the simulation by means of a "number representation" (which corresponds more directly to a "second-quantization" picture). It is obtained by introducing a phase-space discretization. For each cell characterized by a given wave-vector $\mathbf{k}$ and a given band index $\nu$, we specify the number $n_{\mathbf{k}}^{\nu}$ of carriers in the cell.

Given the set of "occupation numbers" $\{n_{\mathbf{k}}^{\nu}\}$ at time $t$, the distribution function $f_{\mathbf{k}}^{\nu}$ is obtained according to

$$f_{\mathbf{k}}^{\nu} = \frac{n_{\mathbf{k}}^{\nu}}{N_{\mathbf{k}}^{\nu}} \tag{5.25}$$

for each cell of the phase-space, where $N_{\mathbf{k}}^{\nu}$ is the total number of available states in the cell. Thus, we can easily check the Pauli exclusion term $(1 - f_{\mathbf{k}}^{\nu})$ in the scattering rates.

The basic lines of the simulation now proceed as follows: The total time is divided into time-steps $\Delta t$. The simulation starts at the initial time $t = 0$ before the laser has been switched on. The system is chosen to be in its fundamental state, that is the vacuum of electron-hole pairs. This choice corresponds to initial distribution functions and polarizations equal to zero all over the phase-space and, therefore, the total number of simulated carriers is initially equal to zero. The simulation then results in a loop over the sequence of time steps from the initial time up to the desired final time. At the beginning of each time-step, $t = t_i$, the system is completely specified by the distribution functions $f_{\mathbf{k}}^{\nu}$ and polarizations $p_{\mathbf{k}}$. From these quantities the self energy $\Sigma_{\mathbf{k}}$, the field renormalization $\Delta_{\mathbf{k}}$, and the generation rate $g_{\mathbf{k}}$ are evaluated. Equation (5.8) for the polarization is then directly integrated from $t_i$ to $t_i + \Delta t$. In this integration, the above mentioned variables are approximated by their values at the beginning of the time-step, the exponential functions occurring in the formal solution, however, are integrated analytically.

The transport equations describing the distribution functions are solved by means of the standard Monte Carlo technique, that is, a stochastic estimate of the distribution function $f_{\mathbf{k}}^{\nu}$ at the time $t_i + \Delta t$ is performed using, as polarization and distribution functions for the non-linear terms, the corresponding values at the initial time $t_i$. First, new electron-hole pairs are stochastically generated in the regions where $g_{\mathbf{k}} > 0$, and electron-hole pairs are stochastically removed where $g_{\mathbf{k}} < 0$. Then, for each carrier an initial wave-vector is determined at random with a constant probability distribution inside the cell. This is equivalent to assume a constant distribution function in the cell and introduces no additional approximation than already done when discretizing. A sequence of random free flights and random scattering events is generated in the standard way, until the carrier reaches the end of the time-step.

### 5.3 Applications

In this section we review and discuss the results of some "simulated experiments" performed with the generalized Monte Carlo procedure described above[53,54]. Different physical conditions have been investigated: The following Sect. A is concerned with applications characterized by a central laser energy $\hbar\omega_L$ far from the band gap, while Sect. B describes some "simulated experiments" characterized by a laser energy close to or within the band gap. For all these applications, a simplified GaAs semiconductor model has been employed: The band-structure is given by spherical and parabolic bands for electrons and holes; carrier-carrier interaction has been introduced up to the first order in perturbation theory within the screened Hartree-Fock approximation; and, finally, carrier-phonon interaction is introduced within the Markov approximation for polar optical and acoustic deformation potential coupling.

*A. Excitation far from gap*

Let us first examine some numerical results concerning simulations characterized by a laser energy far from the band gap. This is the typical situation for energy-relaxation experiments.

In Fig. 14 the self-consistent generation rates obtained from the Monte Carlo simulation are shown as function of the wave-vector $k$ for different times during the laser pulse. The plotted rate is obtained by integrating the generation rate over the angular variables. Figure 14 (a) shows the generation rates for the full generation model while in (b) the corresponding rates for the Markovian (classical) model are plotted. The latter ones are the rates used in a conventional Monte Carlo. Due to the Markovian limit, they do not contain regions with negative values. On the contrary, the rates in (a) (which reflect the effects of the retardation in the generation/recombination process) exhibit a strong time-dependence also in the shape. In particular, at short times (less or of the order of the pulse width $\tau_L$), the shape of the generation rate is found to be much broader than estimated from the uncertainty principle using the pulse width as uncertainty of time. The reason is that we have to use the "observation time" (i.e. the time since the start of the pulse) to estimate the line width. For longer times we note a narrowing of the generation rates, this narrowing, however, is accompanied by the build-up of negative regions off-resonance which can be interpreted as a stimulated emission process. Thus, the distribution of the generated carriers does not only become narrower with increasing time due to a generation mainly in resonance but also due to a recombination of those carriers which have been generated performing energy non-conserving transitions at short times. These results show that a self-consistent treatment of the generation process can be important if either the evolution is analyzed already during the pulse or if some scattering mechanism is so strong that it can remove those carriers generated with the "wrong" energy before they can recombine. They also show that for this self-consistent model it is essential to treat recombination processes within the simulation.

In Fig. 15 the mean kinetic energies of electrons and holes (a) and the polarizations (b) are shown as functions of time for the different models. The energy of the holes is practically independent of the model while the electron energy, in particular during the pulse, is increased in the retarded model. This can be understood from the energy uncertainty in the generation rate. The broadening at short times is symmetric in energy with respect to the laser frequency. The density of states, however, increases with increasing energy, thus it is more probable to generate carriers above resonance than below. Of course, also the stimulated emission then preferably removes these carriers (see Fig. 14(a), dotted line). In Fig. 15(b) the solid and the dash-dotted curves refer to the absolute value of the total polarization, $P^{coh}$, defined as

$$P^{coh} = |\sum_{\mathbf{k}} p_{\mathbf{k}}| \,. \tag{5.26}$$

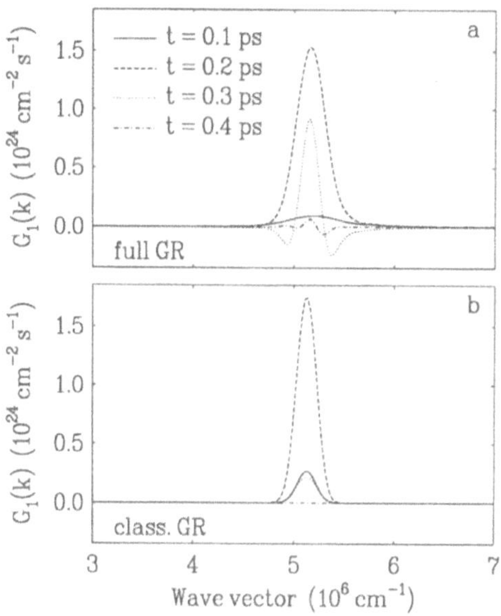


Fig. 14. Generation rate as a function of the wave-vector $k$ for different times, (a) taking into account the retardation and Hartree-Fock effects, and (b) in the classical limit without Hartree-Fock.

It decays due to the inhomogeneous broadening in $k$-space since each contribution $p_{\mathbf{k}}$ in the sum rotates with a different frequency. The dotted and dashed curves refer to the incoherently summed polarization $P^{incoh}$, given by

$$P^{incoh} = \sum_{\mathbf{k}} |p_{\mathbf{k}}| \,. \tag{5.27}$$

It is a measure of the degree of coherence still present in the system and it decays due to incoherent scattering processes. The relaxation times are of the order of $0.5\,ps$.

In fact, this is the typical time related to the total scattering rate of the incoherent processes, which, at low temperatures, is mainly due to optical-phonon emission. On the contrary, the coherent polarizations (solid and dash-dotted curves) decay with typical times that depend on the carrier distribution and, therefore, on the properties of the laser pulse. They reflect the coherent nature of both the carrier-carrier and carrier-light interaction and their time scale is typically much shorter than that related to incoherent phenomena.

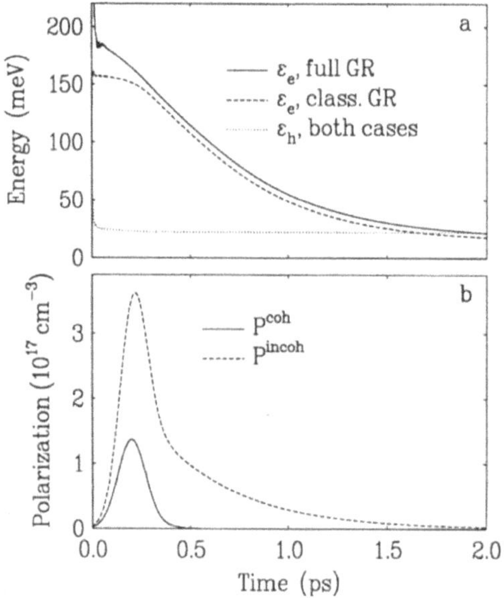


Fig. 15. (a) Kinetic energies of electrons and holes and (b) polarizations as functions of time for the retarded generation model with and without Hartree-Fock terms and for the Markovian (classical) generation model.

All the quantities discussed above help in understanding the physics of the carrier dynamics in a photoexcited semiconductor. However, they are not directly accessible by experiments. Therefore, in Fig. 16 the luminescence spectrum at different times is plotted as a function of the photon energy. Since we are far from gap, the effect of electron-hole interaction for the spectrum is not very large and, therefore, the off-diagonal terms in Eq. (5.24) have been neglected. The real and imaginary part of the self energy, however, have been taken into account. The calculated spectra refer to the luminescence measured not in the direction of the incident laser beam, thus only the incoherent part is shown.

We have shown the very early stage of the process, when the laser is still active, since at these times the differences between the models are most pronounced. Furthermore, all spectra are normalized with respect to their maximum in order to emphasize the line-shape. In a semiclassical treatment, the generation rate has a constant line-width. Thus, also the line-width of the luminescence spectrum is independent of time (solid lines in Fig. 16). From the full generation model, on the other hand, we observe a strong time-dependence of the luminescence line-shape. During the pulse, the spectrum is remarkably broader, and it approaches the semiclassical result towards the end of the pulse. We do not see any effects of relaxation in the shape of the spectra because the holes practically do not relax. Thus, electrons which have emitted an optical phonon cannot contribute to the spectrum, which is sensitive only to the product of the distribution functions. Luminescence experiments are performed already in this very early stage, when the laser is still active[40], therefore coherent effects in the generation process should not be neglected in a theoretical analysis.

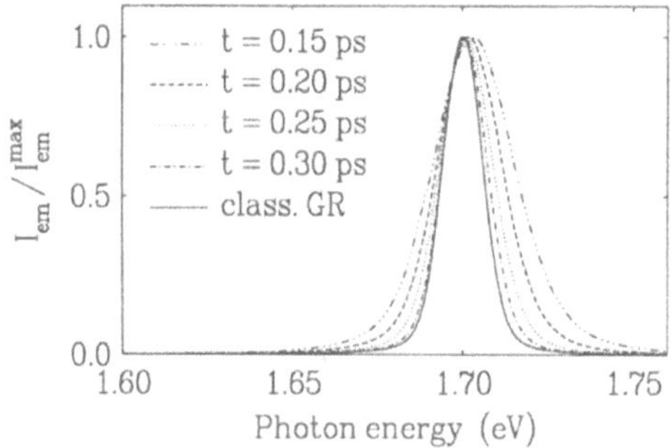


Fig. 16. Luminescence spectra at different times during the presence of the laser pulse (the center of the pulse is at $0.2\,ps$). The spectra are normalized to their maxima in order to emphasize the changes in their shape.

### *B. Excitation close to gap*

Let us now come to the analysis of some numerical results concerning laser excitations close to and within the band gap. This is the region where the effects due to Hartree-Fock terms are most pronounced and thus, besides giving insight in the physics, it has been used as a check of the method. A series of simulations with different values of the excess-energy has been performed, keeping constant all other simulation parameters.

Figure 17 shows the carrier density generated by a laser pulse with fixed amplitude, specified in terms of its Rabi frequency $\Omega_R = M_{\mathbf{k}} A_L / \hbar$, and pulse width $\tau_L = 1\,ps$ as a function of the excess energy. Both curves are obtained within the full generation model, for the solid curve the Hartree-Fock terms have been included while the dashed curve refers to a simulation without Hartree-Fock terms. As expected, without taking into account carrier-carrier interaction, the density is practically zero

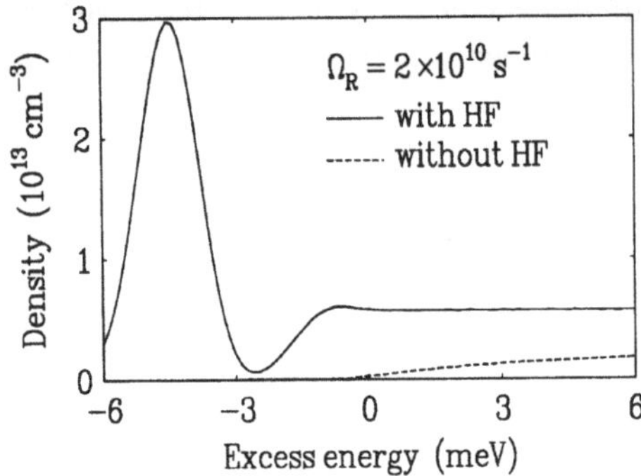


Fig. 17. Carrier density created by a laser pulse with fixed amplitude and a pulse width of 1 *ps* as a function of the excess energy of the laser. Except for the excess energy, all other parameters are unchanged.

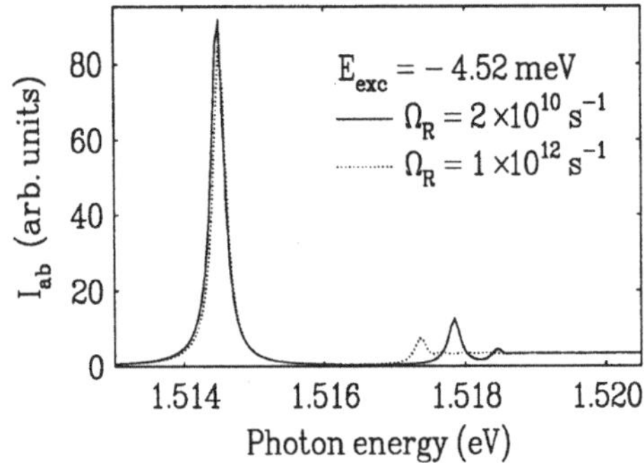


Fig. 18. Absorption spectra obtained from simulations with two different laser intensities performed in resonance with the exciton ground state.

for excitation below the gap. It starts slightly below the gap due to the finite pulse duration. Including Hartree-Fock terms, the strong n=1 exciton line and a subsequent minimum between the $n = 1$ and the $n = 2$ exciton lines is recovered. The higher excited levels are not resolved due to the energy uncertainty associated with the short pulse duration.

Finally, let us return to the analysis of the spectrum. Figure 18 shows the absorption spectrum after the end of the laser pulse for two different laser intensities.

In the case of the weak excitation (solid line) we practically recover the equilibrium absorption spectrum and the lowest three exciton lines are resolved. For the higher excitation the position of the exciton ground state luminescence is unchanged, while the excited state is shifted towards smaller energies. This confirms the treatment of electron-hole interaction within this approach.

**References**


1. K.Von Klitzing, G.Dorda, and M. Pepper, Phys. Rev. Lett. **45**, 494 (1980).
2. See, for example, C.W.J. Beenakker and H. van Houten, Solid State Phys, Vol.44, 1, 1991.
3. C.Jacoboni and L.Reggiani, Rev. Mod. Phys., **55**, 645 (1983).
4. C.Jacoboni and P.Lugli, *The Monte Carlo Method for Semiconductor Device Simulation*, Springer-Verlag, Wien, New York, 1989. Referenza generale sul MC da JoL(85).
5. See, for example, N.P.Buslenko, D.I.Golenko, Y.A.Shreider, I.M.Sobol', and and V.G. Sragovich, *The Monte Carlo Method*, Pergamon, London, 1966. Referenza MC 1966.
6. T.Kurosawa, Proc. 8th Int. Conf. Phys. Semic. (Kyoto), J. Phys. Soc. Japan, Suppl. 21, 424 (1966).
7. R. Brunetti and C. Jacoboni, A. Matulionis and V. Dienys, Phisica 134B, 369 (1985).
8. S. Bosi and C. Jacoboni, J. Phys. C(, 315 (1976).
9. R. Brunetti and C. jacoboni: "Transient and Stationary Properties of Hot-Carrier Diffusivity in Semiconductors", in "Semiconductors Probed by Ultrafast Laser Spectroscopy", vol.1 edited by R.J.J. Alfano, Academic Press (1984).
10. Proc. NATO A.S.I. on "Granular Nanoelectronics" edited by D.K. Ferry, J.R. Barker and C. Jacoboni, NATO Series B: Physics Vol. 251 (1991).
11. See, for example, Sect. IV in "Hot Carriers in Semiconductor Nanostructures - Physics and Applications" edited by Jagdeep Shah, Academic Press (1992).
12. F. Reif: "Fundamentals of Statistical and Thermal Physics", Mc Graw Hill (1965).
13. H.D. Ress, Phys. Lett. A26, 416 (1968).
14. H.D. Ress, J. Phys. Chem. Solids 30, 643 (1969).
15. L. Reggiani, J.C. Vassiere, J.P. Nougier, and D. Gasquet, Proc. 3rd Int. Conf. on Hot Carriers in Semiconductors, J. Physique Coll., Tome 42 C7, 357 (1981).
16. P. J. Price, in "Fluctuation Phenomena in Solids", ed. by R. E. Burgess, Academic Press (1965).
17. "Physics of Quantum Electron Devices" (F. Capasso, ed.), Springer Series in Electronics and Photonics 28, Springer-Verlag, Heidelberg, 1990.
18. D. Ter Haar, Reports on progress in Physics, 24, 304 (1961).
19. L.P. Kadanoff and G. Baym, " Quantum Statistical Mechanics", Benjamin/Cummings, Reading, Mass., 1962 .
20. G. Rickayzen, "Green's functions and condensed matter", Academic Press, New York, 1980.
21. G.D. Mahan, "Many-particle Physics", Plenum, New York, 1981.
22. G.D. Mahan, "Quantum transport in semiconductors" (D.K. Ferry, J.R. Barker, and C. Jacoboni, eds.), Plenum, New York, 1990 (in press).
23. E. Wigner, Phys. Rev. 40, 749 (1932).
24. G.J. Iafrate, in "The Physics of Submicron Semiconductor Devices" (H.L. Grubin, D.K. Ferry, and C. Jacoboni, eds.), NATO-ASI Series B, vol. 180, p. 521, 1988.
25. M. Toda, R. Kubo, and N. Saito, "Statistical Physics I", Springer-Verlag, Heidelberg, 1983.

26. R.P. Feynman, and A.R. Hibbs, "Quantum Mechanics and Path Integrals", McGraw-Hill, New York, 1965.
27. R.P. Feynman and F.L. Vernon, Annals of Phys. , 24, 118 (1963).
28. R. Brunetti, C. Jacoboni, and F. Rossi, Phys.Rev. B 39, 10781 (1989).
29. F. Rossi, and C. Jacoboni, Sol.St. Electr. 32, 1411 (1989).
30. F. Rossi and C. Jacoboni, Semicond. Sci. Technol. **7**, B383-B385 (1992).
31. W.V. Houston, Phys. Rev. 57, 184 (1940).
32. F. Rossi, P. Poli, and C. Jacoboni, Semicond. Sci. Technol., to be published.
33. F. Rossi and C. Jacoboni, Europhys. Lett. **18**, 169 (1992).
34. J.R. Barker, Sol. St. Electr. 21, 267 (1978).
35. P. Menziani, F. Rossi, and C. Jacoboni, Sol. State Electr. 32, 1807 ( 1989).
36. W. Quade, F. Rossi, and C. Jacoboni, Semicond. Sci. Technol. **7**, B502-B505 (1992).
37. W. R. Frensley, Rev. Mod. Phys. **62**, 3 (1990).
38. J. Shah, "Photoexcited hot carriers: from cw to 6 fs in 20 years," *Solid State Electron.* **32**, pp. 1051-1056, 1989.
39. K. Leo, W. W. Rühle, H. J. Queisser, and K. Ploog, "Reduced dimensionality of hot-carrier relaxation in GaAs quantum wells," *Phys. Rev. B* **37**, pp. 7121-7124, 1988.
40. T. Elsaesser, J. Shah, L. Rota, and P. Lugli, "Initial thermalization of photoexcited carriers in GaAs studied by femtosecond luminescence spectroscopy," *Phys. Rev. Lett.* **66**, pp. 1757-1760, 1991.
41. J. A. Kash, "Carrier-carrier scattering in GaAs: Quantitative measurements from hot (e,$A^0$) luminescence," *Phys. Rev. B* **40**, pp. 3455-3458, 1989.
42. P. C. Becker, H. L. Fragnito, C. H. Brito Cruz, R. L. Fork, J. E. Cunningham, J. E. Henry, and C. V. Shank, "Femtosecond photon echoes from band-to-band transitions in GaAs," *Phys. Rev. Lett.* **61**, pp. 1647-1649, 1988.
43. K. Leo, M. Wegener, J. Shah, D. S. Chemla, E. O. Göbel, T. C. Damen, S. Schmitt-Rink, W. Schäfer, "Effects of coherent polarization interactions on time-resolved degenerate four-wave mixing," *Phys. Rev. Lett.* **65**, pp. 1340-1343, 1990.
44. E. O. Göbel, K. Leo, T. C. Damen, J. Shah, S. Schmitt-Rink, W. Schäfer, J. F. Müller, and K. Köhler, "Quantum beats of excitons in quantum wells," *Phys. Rev. Lett.* **64**, pp. 1801-1804, 1990.
45. M. A. Osman and D. K. Ferry, "Monte Carlo investigation of the electron-hole-interaction effects on the ultrafast relaxation of hot photoexcited carriers in GaAs," *Phys. Rev. B* **36**, pp. 6018-6032, 1987.
46. P. Lugli, P. Bordone, L. Reggiani, M. Rieger, P. Kocevar, and S. M. Goodnick, "Monte Carlo studies of nonequilibrium phonon effects in polar semiconductors and quantum wells. I. Laser photoexcitation," *Phys. Rev. B* **39**, pp. 7852-7865, 1989.
47. P. Lugli, P. Bordone, S. Gualdi, L. Rota, and S. M. Goodnick, "Femtosecond phenomena in III-V semiconductors," *NASECODE VI, Proc. 6th Int. Conf. on the Numerical Analysis of Semiconductor Devices and Integrated Circuits*, ed. by J. J. H. Miller, pp. 238-243, Boole Press, Dublin, 1989.
48. H. Haug in *Optical Nonlinearities and Instabilities in Semiconductors*, ed. by H. Haug, pp. 53, Academic Press, San Diego, 1988.
49. S. Schmitt-Rink, D. S. Chemla, and H. Haug, "Nonequilibrium theory of the optical Stark effect and spectral hole burning in semiconductors," *Phys. Rev. B* **37**, pp. 941-955, 1988.

50. M. Lindberg and S. W. Koch, "Effective Bloch equations for semiconductors," *Phys. Rev. B* **38**, pp. 3342-3350, 1988.
51. R. Zimmermann, "Transverse relaxation and polarization specifics in the dynamical Stark effect," *phys. stat. sol. (b)* **159**, pp. 317-326, 1990.
52. A. V. Kuznetsov, "Coherent and non-Markovian effects in ultrafast relaxation of photoexcited hot carriers: A model study," *Phys. Rev. B* **44**, pp. 13381-13392, 1991.
53. T. Kuhn and F. Rossi: "Analysis of coherent and incoherent phenomena in photoexcited semiconductors: A Monte Carlo approach", *Phys. Rev. Lett.* **69**, 977-980 (1992).
54. T. Kuhn and F. Rossi: "Monte Carlo simulation of ultrafast processes in photoexcited semiconductors: Coherent and incoherent dynamics", *Phys. Rev.* **B46**, 7496 (1992).

# STUDY OF IRREVERSIBLE PROCESSES IN CONDENSED MATTER BY NONLINEAR TIME AND SPACE RESOLVED TECHNIQUES

Christos Flytzanis

Laboratoire d'Optique Quantique du C.N.R.S.
Ecole Polytechnique
91128 Palaiseau cedex, France

## INTRODUCTION

The propagation, storage and redistribution of excitations in condensed matter underlie many fundamental processes there. For processes related to the dynamics and relaxation of excitations in the molecular or atomic level the optical techniques that exploit the different features of the lasers, in particular their high selectivity and resolution, in frequency or time domain, have provided a unique wealth of information and constitute the most powerful tools that we have presently at hand to selectively investigate these processes. This information is essential for both fundamental and technological reasons since the relaxation processes constitute the central problem for understanding[1] the irreversible processes in general and also set in particular the ultimate limits for the implementation and exploitation of several optical processes in devices. Many of these techniques have reached a very high degree of technical sophistication and others are still under assessment.

Below we will content ourselves to classify the most important ones that are currently in use and focus our attention on some of the most important aspects ; these techniques address in particular those related to the investigation of propagating and hybrid excitations. We shall mainly concentrate on the evolution of the coherence of these excitations as this is an essential feature in nonlinear interactions in matter. Before doing this we pause to define and clarify certains elementary aspects of the relaxation processes in molecular or atomic excitations in condensed matter.

## ELEMENTARY RELAXATION PROCESSES. MARKOFFIAN REGIME

Our central problem is the relaxation processes and temporal evolution of an assembly of two level systems embedded in a condensed matrix which will also play the role of thermal reservoir. We formulate the problem and derive the equations for the standard case, the so called Bloch equations[2,3], and proceed to cursively discuss the complications introduced by more realistic cases.

### General Case

We designate by a and b the low and high energy levels respectively and by $h\omega_o$ their energy spacing when the system is isolated ; inside the matrix in principle $h\omega_o$ depends on the local environment as well where the systems are embedded. These systems are uniformly distributed with number density N and are incessantly subject to random perturbations $V_r(t)$ through their coupling to the fluctuating degrees of freedom of the condensed environment. We will assume these perturbations to be stationnary markoffian processes[1,4,5,6] with

$$\langle V_r(t)\rangle = 0 \tag{1}$$

$$\langle V_r(t)\, V_r(t')\rangle = D^2\, e^{-|t-t'|/\tau_c} \tag{2}$$

where D is a measure of their strength and $\tau_c$ is a correlation time which for most purposes will be very short, much shorter than any relevant time of the problem so that the correlation function (2) practically can be replaced by a delta function of strength $\sigma = D^2\tau_c$ , $\langle\ \rangle$ designates ensemble averages. The 2-level systems are brought off equilibrium by an external coherent perburbation $V_c(t)$ not correlated with $V_r(t)$ ; in addition these systems may interact mutually through $V_t$ that provides excitation transfer from system to system. The perturbation $V_c(t)$ can be steady state, for instance sinusoidal, or pulsed and the relaxation processes accordingly can be studied in frequency or time domain respectively. In the latter case one has a direct measure of relaxation times while in the former one measures spectral widths[7] and line forms from which one can also extract relaxation times. The two approaches in principle are related through a Fourier transform ; in practice each one has its own advantages and range of usefulness. Here we limit ourselves in techniques related to real time domain which give direct information about the relaxation times.

The evolution of the system and all its observables can be evaluated with the density matrix operator $\rho_T$ which also includes the reservoir degrees of freedom. We may assume that $\rho_T$ can be factorized as

$$\rho_T = \rho \otimes \rho_r$$

where $\rho$ refers to the assembly of the two-level systems and $\rho_r$ to the reservoir and concentrate our attention only on $\rho$.The temporal evolution of the system is described by the density matrix operator equation

$$\frac{d\rho}{dt} = \frac{1}{ih}\left[h_o + V_r + V_t + V_c, \rho\right] \tag{3}$$

where $h_o$ is the unperturbed hamiltonian which may depend on the environment. The relaxation processes are reflected in the temporal evolution of different physical quantities

$$A = \mathrm{Tr}\rho\, A \tag{4}$$

where A is the quantum mechanical operator representing this quantity. The complexity of equation (3) can be greatly reduced[8,9] by appealing to the statistical features of the random potential $V_r$ and its strenght relative to other perturbations.

## Bloch Equations

The standard case[2,3] corresponds to the systems being localized in identical environments and not mutually interacting : $V_t = 0$ and $\omega_o$ identical for all systems (in space and time). In this case the infinite order matrix $\rho$ is diagonalized in 2 x 2 identical matrices and equation (3) reduces to

$$\frac{d\rho}{dt} = \frac{1}{ih}\left[h_o + V_c, \rho\right] + \left.\frac{d\rho}{dt}\right|_R \tag{5}$$

with

$$\left.\frac{d\rho_{ab}}{dt}\right|_R = -\frac{\rho_{ab}}{T_2} \tag{6}$$

$$\left.\frac{d\rho_{aa}}{dt}\right|_R = -\frac{\rho_{aa} - \rho_{aa}^{(o)}}{T_1} \tag{7}$$

with $\rho_{aa} + \rho_{bb} = \rho_{aa}^{(o)} + \rho_{bb}^{(o)} = 1$, $\rho_{ii}^{(o)}$ being the equilibrium values of the diagonal elements of $\rho$. For all purposes we assume that

$$V_c(t) = \mu\, F(t) \tag{8}$$

where $\mu$ is a material operator a function of a "generalized coordinate" q of the system that couples to the externally applied "force" F(t) ; we also simplify by assuming that only the nondiagonal element of $\mu$ is nonvanishing and we define the instantaneous generalized Rabi frequency $\Omega_R(t) = \mu_{ab}F(t)/\hbar$, its average value $\bar{\Omega}_R$ and the pulsed force area $\theta = 2\pi\int \Omega_R(t)dt$. Clearly the generalization to several coordinates can be done along the same lines.

In (6) and (7) $T_2$ and $T_1$ are the transverse and longitudinal relaxation times [3] respectively with $T_2 << T_1$.They are expressed[3,8,9] in terms of certain correlation functions and are related to two quite distinct processes, the coherence and energy relaxation processes respectively. The former, which is essentially classical, is the decay of the induced polarisation (or induced transition "dipole") $\mathcal{P} = \mu_{ab}\,(\rho_{ab} + \rho_{ba})$ and the latter which is essentially quantum mechanical corresponds to the decay of the population difference $\Delta\rho = \rho_{aa} - \rho_{bb}$ or equivalently the decay of the energy $\varepsilon = h\omega_0\,\Delta\rho$ stored into the system through the excitation process. The inequality $T_2 << T_1$ implies that first the matrix $\rho$ becomes diagonal in a time scale $T_2$ and then its diagonal elements revert to their equilibrium values in a time scale $T_1$. Intuitively one expects $T_2 < T_1$ because all random fluctuations can contribute to dephase the dipole but among them only those that can also accomodate the energy quantum $h\omega_o$, as required by energy conservation, are involved in the energy relaxation as well.

Equations (5) (6) and (7) can also be rewritten [10] in vector form in the space of 2x2 Pauli mattrices. One can also show that (5) (6) and (7) can be replaced by the following system of two coupled equations

$$\frac{d^2\mathcal{P}}{dt^2} + \frac{2}{T_2}\frac{d\mathcal{P}}{dt} + (\omega_0^2 + \frac{1}{T_2^2})\mathcal{P} = \mu_{ab}\,F\,\Delta\rho \tag{9}$$

$$\frac{d\Delta\rho}{dt} + \frac{\Delta\rho}{T_1} = \frac{1}{h\omega_o}\,F\,.\,\frac{d\mathcal{P}}{dt} \tag{10}$$

a damped harmonic oscillator and a diffusion equation respectively. Since $T_2 << T_1$ the damping of the harmonic oscillator is essentially determined by $T_2$ ; furthermore $T_2\omega_o >> 1$ and the $T_2^{-2}$ shift in $\omega_o^2$ can be neglected. It is important to notice in equation (9) that the $T_2$ is related to the classical damping of an oscillating dipole while the right hand side member is a quantummechanical force.

The validity conditions for equations (5) (6) and (7), or equivalently (9) and (10) are[3,8,9,11]

- the bath is unaffected by the relaxation processes (infinite energy and phase sink reservoir)

- the correlation time $\tau_c$ is much shorter than any relevant time or

$$\Gamma\tau_c,\ \tau_c/T_i,\ \Delta\omega\tau_c,\ \bar{\Omega}_R\tau_c/h << 1$$

where $\Gamma$ is the occurence rate of the individual random events ("collisions"), $\Delta\omega = \omega - \omega_o$, $\omega$ being any characteristic frequency of the external force F.
These conditions are essentially the same as those used in the "impact approximation" and in particular insure that the random perturbations are markoffian.

**Extended Cases**

Despite the simplifying assumptions previously used to derive the Bloch equations in the standard case these seem to satisfactorily describe the relaxation in a very wide class of real quantum systems and over a wide spectral range : with slight modifications it can be extended to cover a far wider class of relaxation processes encountered in realistic complex systems. We shall attempt below to classify the most important cases.

- independent systems (homogeneous case) : this is essentially the case where the standard model strictly applies. We recall that the two-level systems are assumed localized in identical environments and do not mutually interact. By identical environment we understand that all systems are modified by the same amount irrespective of their position in the condensed matrix and furthermore $\tau_c$ D << 1 (fast modulation[6]) ; this case is also termed homogeneous.

- independent systems (inhomogenous case) : if the two-level systems are not situated in identical environments the $\omega_0$ will be shifted by amounts $\delta\omega_0$ that depend on their position inside the condensed matrix ; one has a distribution of $h_o$ in (5) or equivalently a distribution of $\omega_0$ in (9) over a range $\Delta\omega^* = 2/T^*$ and this is the inhomogeneous case. A similar situation also occurs when $\tau_c$ D >> 1 (the slow modulation limit[6]). Equations (5) or (9) can still be used after one has performed an averaging process over the $h_o$ in (5), or the $\omega_0$ in (9), and this essentially amounts in introducing an additional dephasing time T* so that the total dephasing time is $1/T'_2 = 1/T_2 + 1/T^*$ ; if $T^* < T_2$ one has the inhomogenous case while for $T^* > T_2$ one recovers the homogenous one.

- interacting systems (random case) : if the two-level systems are mutually interacting $V_t$ is nonvanishing and the infinite order matrix $\rho$ in (3) cannot be diagonalized in 2 x 2 matrices each satisfying (5). The 2-level systems are coupled and in particular the excitation on one an be transfered to another along different pathways involving several intermediate real or virtual transfer steps. The problem can become tractable only under some very drastic simplifications by introducing a cut off on the extent of these pathways and suppressing any memory effects : for instance the excitation may not return in the initial point. One then obtains a population relaxation time $T_3$, in addition to $T_1$ and $T_2$, related to the so called cross relaxation[3,12] ; the apparent population relaxation time is $1/T'_1 = 1/T_1 + 1/T_3$

- interacting systems (periodic case) : if the mutually interacting two-level systems form a periodic array the situation changes drastically. With Floquet's theorem one shows that the eigenstates of such a periodic assembly are distributed[13] in bands or branches labelled by $\sigma$ and within each such band or branch the states are labelled with a continuous index, the wave vector $\underline{k}$, that can take all values within the first Brillouin zone ; the latter has an extension of the order of $K = \pi/a$ where a is the smallest spatial period. One has propagating excitations with energies $h\omega_\sigma(k)$, which are analytic functions of $\underline{k}$ characterized by a density of states J $\approx |\partial\omega/\partial\underline{k}|$ and a group velocity

$$\underline{v}_g = \nabla_k\, \omega_\sigma\, (k)$$

which also measures the flatness of the dispersion relation, the relation between $\omega_\sigma$ and k. If the group velocity is very small over the whole of the B.Z., flat dispersion, one essentially recovers a localized state picture similar to the homogeneous case. If the group velocity is large, propagation effects may interfer with intrinsic relaxation processes and special techniques must be developped to disentangle them. This is in particular the case of the decay of the polariton[13], the polarization excitation mode close to a dipole allowed resonance of a phonon or exciton.

## TIME RESOLVED SPECTROSCOPY

The central goal of time resolved spectroscopy is the identification of the different relaxation regimes and determination of the corresponding relaxation times.

### Main Scheme

Despite the apparent complexity and diversity of the techniques that are being used for directly studying the relaxation processes in real time domain there are some underlying principles and patterns common to all of them[14-18]. Indeed all these techniques employ the following scenario.

- excitation stage : the system is perturbed by an "instantaneous" external force $F_e(t)$ that resonantly couples to a "'coordinate" q of the system and drives it off equilibrium at time "t" = 0 ; as a consequence the coordinate acquires a phase and an amplitude.

- free precession : the system may be left to "precess" freely for a time interval $t_d$ after the excitation stage till it settles in the relaxation regime we wish to study.

- interrogation stage : the system is probed at time $t_d$ with an external weak force $F_p(t)$ that couples to a coordinate of the system q' that acquired phase and amplitude following the excitation but this coordinate may not necessarily be the same as the one directly driven off equilibrium in the excitation stage ; by measuring the signal as a function of the delay $t_d$ one obtains the relaxation time appropriate to the chosen relaxation regime or channel.

The exciting and probing forces, $F_e(t)$ and $F_p(t)$ respectively, are powers of the electric field fixed by the degree of the nonlinear interaction involved in the excitation and probing stages respectively. Because the off equilibrium amplitude of the coordinate q' necessarily depends on the external force $F_e$ applied at the excitation stage and in addition the signal also depends on $F_p$, time resolved spectroscopy is by essence nonlinear. Furthermore because the excitation and probing stages must be of short duration one must use short light pulse techniques. We anticipate that the time resolution is not fixed by the pulse duration but by the decay of the appropriate correlation[17] function of $F_e$ and $F_p$ ; this is actually determined by the decay time of $F_e$ and rise time of $F_p$.

The excitation and the interrogation stages may or may not coincide in space and accordingly we distinguish between local and non local time-resolved techniques respectively; the later are used whenever relaxation and propagation effects must be separated as in the case for fast propagation excitations like the polaritons.

### Principles

We wish to give a more quantitative[17] but still rough description of the different time resolved techniques. For this we remind that the optical properties of an assembly of identical microscopic systems (molecules) are described with the dielectric constant tensor

$$\varepsilon = 1 + 4\pi N \alpha \tag{11}$$

where $\alpha$ is the molecular polarizability tensor which can be expressed as a sum of harmonic oscillator contributions of type (9) if $\mu$ is the dipole operator, $\mu = ex$, and $F(t)= E(t)$ is the applied electric field ; we disregard local field corrections. As a general rule the polarizability $\alpha$ is an analytic function of certain coordinates q, for instance the vibrational or rotational coordinates, and of the electric field E present in the medium. If these acquire amplitude and phase through an external agency (excitation stage) we may formally set

$$\alpha(q,E) \equiv \alpha + \delta\alpha = \alpha_0 + \alpha_q^{(1)} q + \beta_E E + \frac{1}{2}\alpha_q^{(2)} q^2 + \ldots \tag{12}$$

$\alpha_0$ is the electronic polarizability of the rigidly fixed molecule, $\alpha_q^{(1)} = \partial\alpha / \partial q$ and $\alpha_q^{(2)} = \partial^2\alpha / \partial q^2$ are the Raman first and second order tensors, $\beta_E = \partial\alpha/\partial E$ is the second order electric polarizability.

The probing field $E_p(t)$, applied at time $t_d$ after the excitation was switched off, sets up a polarisation

$$P = N\alpha_o E_p + N\delta\alpha E_p = P_o + \delta P \equiv P_o + P_{NL} \tag{13}$$

which in addition to the linear term $P_o = N\alpha E_p$ also contains oscillating terms in $P_{NL} = N\delta\alpha E_p \equiv \delta\varepsilon E_p$ which will radiate and generate fields $E_s$ with different characteristics than $E_p$ according to equation

$$\Delta E - \frac{1}{c^2}\frac{\partial^2}{\partial t^2}(\varepsilon_0 E) = \frac{4\pi}{c^2}\frac{\partial^2 P_{NL}}{\partial t^2} \tag{14}$$

One selects a particular component in $P_{NL}$

$$P_s(t) = \mathcal{P}_s(t)\, e^{ik_s r - i\omega_s t} \tag{15}$$

and solves equation (14) as a function of $t_d$ by setting

$$E_s = A_s(t)\, e^{ik_s r - i\omega_s t} \tag{16}$$

The decay of the intensity $I_s \approx |E_s|^2$ as a function of $t_d$ gives a direct measurement of the relevant relaxation times. The amplitude $A_s$ is appreciable in the direction of phase matching

$$\underline{k}_e \approx \underline{k}_s \tag{17}$$

and this is the essence of the coherent time resolved spectroscopy for the determination of $T_2$ or $T^*$ ; note that the same information can be obtained by conventional high resolution spectroscopy in frequency domain. On the other hand the determination of $T_1$, the much longer energy relaxation time, is made by measuring the scattered $A_s$ field at large angle with respect to (17), say 90° degrees and is the essence of the incoherent time resolved spectroscopy ; note that this information cannot be obtained by high resolution spectroscopy in frequency domain since there one measures the total width

$$\Delta\omega = \frac{1}{T_1} + \frac{2}{T_2} \approx \frac{2}{T_2} \tag{18}$$

since $T_1 >> T_2$.

For the coherent regime, which is essentially classical, unstead of using the quantum mechanical approach to derive equation (9) we may proceed along a classical approach by noticing that in the presence of an electric field one stores energy

$$W(Q) = -\frac{1}{2}\alpha(Q)\,E^2 \qquad (19)$$

per molecule which can be interpreted as a potential energy where Q is the classical quantity related to the coordinate q that affects the polarizability $\alpha$ ; its conjugated force being $F_Q = -\partial W/\partial Q$ the classical equation of motion of Q is

$$\ddot{Q} + \Gamma\dot{Q} + \Omega_0^2 Q = -\,\partial W/\partial Q \qquad (20)$$

where $\Omega_0$ is the frequency for small amplitude motion of coordinate Q and $\Gamma$ is its damping constant ; by proper identification of the different quantities and minor approximations equation (20) is identical to (9) and also gives a more intuitive description of the process. The above intuitive approach can actually be used for all coordinates Q that satisfy the criteria of the adiabatic approximation (Born-Oppenheimer approximation) and its generalizations.

## Nonlinear Time-Resolved Techniques

Following the scenario outlined in the previous section by appropriate choice of the excitation and interrogation mechanisms a multitude[17,18] of time resolved nonlinear techniques can be devised that allow one to address a most wide range of relaxation processes. Some of these techniques are more commonly used than others and have reached a high level of sophistification and flexibility and below we shall give a succint discussion of the latter. The time-resolved nonlinear optical techniques have several advantage which are continuously improved with the new ultrashort light pulse technology[19] ; some of them are

- extremely high selectivity achieved by exploiting appropriate nonlinear mechanisms, resonances and other means,

- high signal to noise discrimination,

- very clear separation of coherent and incoherent regimes by exploiting the phase matching conditions,

- high selectivity in time and space that in particular allows to study relaxation processes in any chosen space point inside the bulk or on the surface of the medium.

We classify the time resolved nonlinear optical techniques into local and non local.

**Local Time-Resolved Techniques.** These concern the study of relaxation processes in localized excitations ; the interrogation stage coincides in space with the excitation stage. The most developped and used[17] time resolved nonlinear technique for the study of vibrational relaxation proceses is the time resolved Coherent Antistokes Raman Scattering also designated CARS. Here two short intense light pulses of frequencies $\omega_1$ and $\omega_2$ interact simultaneously inside the medium and coherently drive a Raman active transition of frequency $\omega_R$ in the region of their spatial overlap ; the frequencies $\omega_1$ and $\omega_2$ are such that $\omega_1 - \omega_2 \approx \omega_R$. The evolution of the coherence and population difference between the two states involved in the transition is probed with a delayed weak short pulse of frequency

$\omega_3$ by measuring the antistokes intensity at frequency $\omega_p = \omega_3 + \omega_R$. In regard to the scheme outlined previously and the notation used there we have

$$V_e \approx -\frac{1}{2}\alpha_q^{(1)} q\, E_1(t)\, E_2(t) \tag{21}$$

at the excitation stage or with the notations previously introduced in equation (8) $\mu = -\frac{1}{2}\alpha_q^{(1)} q$ and $F(t) = E_1(t)\, E_2(t)$ and

$$V_p = -\frac{1}{2}\alpha_q^{(1)} q\, E_3(t)\, E_p(t) \tag{22}$$

at the interrogation stage. In the coherent regime the detection is made in the phase matching condition (17) and one measures $T_2$ while in the incoherent regime the detection is made at right angle to the phase matching condition and one measures $T_1$. At presently this constitutes the most powerful technique for measuring $T_1$. There are several parameters that can be varied to improve the selectivity and other performances of this technique. Note in particular that the excitation can be made at any point inside the bulk of the medium if the latter is transparent to the frequencies involved. In addition, in molecules with several coupled modes, one may wish to study[17] intermode energy and coherence transfer. Such information can be obtained by detecting the antiStokes from another mode than the one that has been coherently driven in the excitation stage.

There are other extensions of the CARS technique. One of them is the time resolved Coherent AntiStokes Higher Order Raman Scattering[20,21], also designated by CAHORS, to study vibrational overtones in liquids and bound two-phonon (biphonon) states in molecular crystals or more complex vibrational states like the Fermi resonances. All these compound states are manifestations of strong anharmonic interactions and the study of the loss of their coherence and desintegration channels allow a very selective study of the different anharmonic terms of the lattice potential. These studies are summarized below in a latter section.

Clearly the CARS technique can only be used for Raman active modes. For infrared active (dipole allowed) modes one may replace [22] the coherent excitation stage by one photon (infrared) excitation resonant with the material transition followed by a delayed excitation to a fluorescent state. The measure of the fluroescence as a function of the delay between the two excitations gives information about the relaxation of the infrared active mode. Because of the lack of tunable infrared sources with short pulses this technique or similar ones[23] has not been used to the same extent as the CARS. A limitation of this technique is that the excitation and probe states are quite often restricted into the optical penetration depth of the sample near the surface which can be different from the bulk. We also mention the time resolved luminescence technique which has been successfully[24] used to study fast photocarrier relaxation processes in semiconductors but can also be used elsewhere. Here the fast luminescence signal subsequent to a first ultrashort pulse is sampled at a given frequency by up converting it with a second delayed ultrashort pulse inside a crystal without inversion center ; the intensity of the up converted signal as a function of the delay between the two pulses gives important information about the decay.

The CARS technique can also be used[17,25] for the study of relaxation proccesses in inhomogeneous broadened Raman active transitions. For electric dipole transitions, however, techniques like photon echos or time resolved hole burning provide a more direct insight into this question. The photon echo technique[26] has been extensively developped to study coherent relaxation processes in the visible involving electronic transitions ; in the infrared where vibrational transitions occur the results have been scarce. There are several variants of the original photon echo technique like the stimulated echo[27] and accumulated echo[27] techniques. Some attempts[28] to observe Raman echoes have been made but they seem inconclusive The time resolved hole burning technique is straight forward. One measures the transmission characteristics of a weak short pulse tuned in the spectral range of a inhomogenous broadened transition which has been previously excited by a very intense short pulse. The decay of the hole burned and its width give informations about population and coherence relaxation times.

A very powerful time resolved technique to study relaxation processes is the transient optical gratings technique[28-30] which can also be designated as real time holography. In its simplest configuration two pulsed beams of same frequency $\omega$ close to a transition of the medium interact in the medium via a refractive index or photoinduced absorption change ; a third pulsed beam at the same or a different frequency $\omega'$ delayed with respect to the first two diffracts off this pattern ; and the evolution of the diffracted intensity as a function of the time gives information about the amplitude and phase relaxation of the grating and by the same token about energy migration and loss of coherence. There are several variants and extensions of the original transient grating technique described above. One such a variant is the transient degenerate four wave mixing[31,32] or the transient optical phase conjugation. Two intense counterpropagating pulsed beams $E_1$ and $E_2$ of same frequency $\omega$ interact inside a medium with a third pulsed beam $E_3$ of same frequency $\omega$ and set up a third order poalrization $\underline{P}_c^{(3)}$ which generates a fourth beam of frequency $\omega$ counterpropagating and phase conjugated to $E_3$. One can show that this phase conjugated beam results from diffraction off two spatial transient gratings and a temporal grating. Indeed one can show[31] that

$$\underline{P}_c^{(3)} = a(\underline{E}_1.\underline{E}_3^*)\,\underline{E}_2 + b(\underline{E}_2.\underline{E}_3^*)\underline{E}_1 + c(\underline{E}_1.\underline{E}_2)\,\underline{E}_3^*$$

and the three terms in the right hand correspond to the three gratings : by appropriate delays between the pulsed beams and judicious choice of beam polarizations one can study different relaxation and diffusion processes of the material excitation at frequency $\omega$. One may also use the so-called polychromatic optical phase conjugation[33], where $E_3$ has a freqency $\omega' \neq \omega$ in which case $E_c$ has frequency $\omega_c = 2\omega - \omega'$ ; furthermore $E_1$ and $E_2$ are not counterpropagating but their propagation directions make a angle $\theta$ that is bisected by that of $E_3$ and $E_c$ which are counterpropagating.

Another variant of transient optical grating technique is the impulsive stimulated Raman technique[34-37] with ultrashort pulses of duration $\tau_p$ in the few femtosecond range. Such pulses have a frequency width $\Delta\omega_p \approx 1/\tau_p$ which can extend up to few hundreds $cm^{-1}$ and under certain conditions one can, with a single ultrashort pulse, coherently drive vibrational or librational or other low lying states of frequencies that fall within $\Delta\omega_p$ ; diffraction of a single ultrashort pulse off the interference pattern set by the first pulse gives information about the relaxation processes of these low lying states. There have been several investigations[34-39] with this technique or similar ones. One has observed in particular Raman quantum beats and studied relaxation processes of low lying vibrational modes.

All the previous techniques employ coherent pulses. There has been recently proposed and demonstrated[40] the use of incoherent light to study relaxation processes and in particular extract information concerning the relaxation time $T_1$ and $T_2$. Here one exploits the fact that a temporally incoherent light of wide spectral width possesses a vary short correlation time $\tau_c$ much smaller than the light pulse duration $\tau_p$. In an autocorrelation measurement such a light appears like a single pulse of duration $\tau_c$. Accordingly this technique consists in spatially spliting a pulsed light beam of frequency $\omega$ into two beams and temporally delaying the one relative to the other by $\tau$ before mixing them in a resonant medium. The energy of the output beams in the phase matching direction is measured as a function of $\tau$ to obtain a kind of correlation profile associated with both the incident light and the resonant material response. The time resolution of this technique is set up by the light correlation time $\tau_c$ and not by the duration of the light pulse itself. This technique has been applied in some simple cases but its exploitation in a larger scale and in more complicated cases is not that straightforward.

**Nonlocal Time-Resolved Techniques.** The previous techniques are essentially designed to study relaxation processes of localized excitations in a molecular entity embedded in a condensed matrix. Here the energy and coherence relaxation occur within the immediate environment of the molecule. For delocalized excitations in more or less periodic molecular systems the problem can be fondamentally different and also becomes more complex because the relaxation process within the immediate molecular environment can take place concurrently with the escape of the excitation from the excited molecule and its resonant transfer to other identical molecules. At presently there are two techniques that allow to address this problem and desintangle the relaxation from the propagation : the electro-optic Cerenkov effect[41-43] and the nonlocal (space resolved) time resolved CARS[41,45]. In addition some of the previous techniques, for instance the transient optical grating technique[29-31], can be used to study relaxation and diffusion processes of delocalized excitations.

The electro-optic Cerenkov technique with femtosecond pulses in the visible exploits[41] the optical rectification or inverse electro-optic effect in a non centrosymmetric crystal to generate an extremely fast far infrared electromagnetic transient, few femtoseconds in duration, and excite propagating infrared active excitations in the same crystal ; these are, for instance, the polaritons. This produces a Cerenkov cone of pulsed radiation since the relevant far infrared polarisation is created with a group velocity $v'_g$ appropriate to the visible spectrum (dielectric constant $\varepsilon_\infty$) while its far infrared radiation propagates with a group velocity $v''_g$ approriate to the far infared (dielectric constant $\varepsilon_o > \varepsilon_\infty$) and one in general has $v''_g < v'_g$. The existence of this Cerenkov cone is subsequently exploited to study the propagation of the polariton in real space. This is achieved with a second femtosecond probe pulse parallel to the previous which measures the small birefringence that is due to the electrooptic effect induced by the electric field of the far infrared transient as it moves in synchronism with the Cerenkov wave front. This technique has been already demonstrated [43] for polaritons in ferroelectric crystals (Li Ta $O_3$) ; its selectivity is, however, limited and the separation of temporal and spatial features is not straight forward. Furthermore this method is restricted to low-frequency polaritons, with an upper frequency limit given by the infrared femtosecond pulse bandwidth.

The non local[44,45] time resolved CARS technique does not suffer from these limitations. This technique has been recently proposed[44] and demonstrated for the study of polariton pulses in crystal without inversion center. In this technique a picosecond-duration polariton wave packet of frequency $\omega_\pi$ is created at an initial instant and space point in the crystal by coherent Raman scattering using time-coincident picosecond pulses of frequencies $\omega_L$ and $\omega_s$, such that $\omega_L - \omega_s = \omega_\pi$. This polariton wave packet subsequently propagates freely inside the crystal in a direction fixed by overall wave-vector conservation, and its temporal and spatial evolution is followed by coherent anti-Stokes Raman scattering of a picosecond probe pulse ($\omega_p$) displaced in time (by $t_d$) and space (by $X_d$) with respect to the excitation. The measured dependence of the spatial displacement $X_d$, where the signal is maximum, on time delay $t_d$ is directly related to the energy propagation characteristics of the polariton wave packet. This technique has been to used[44] to study dephasing of polariton packets both in perfect and partially disordered crystals, dressed[45] polaritons (Fermi resonance) and can also be extended to study surface polaritons. Below we review these studies.

Nonlocal time resolved techniques have only recently been demonstrated but it is expected to undergo rapid developments in the near future.

## SPATIOTEMPORAL EVOLUTION OF POLARITON PULSES IN CRYSTALS: TRACKING OF POLARITONS

In this section we summarize the principles and main results of the non local time-resolved nonlinear techniques for the study of polariton pulses in crystals. Polaritons[46] play a very important role in crystals optics[47-54] and they determine the propagation, storage and redistribution of electromagnetic energy close to resonances. On must have precise first hand knowledge concerning their characteristics and in particular their coherence.

## Polariton Dispersion and Damping

We briefly recall the salients features of polariton modes[46,49,50]. We concentrate our attention on their dispersion relation and relaxation.

A polariton is a mixed electric dipole allowed material excitation and electromagnetic mode in a crystal. If we admit that for an isotropic nonmagnetic dielectric crystal the dielectric constant is :

$$\frac{c^2k^2}{\omega^2} = \varepsilon(\omega) = \varepsilon_o + \frac{\Omega_p^2}{\omega_T^2 - \omega^2} \tag{23}$$

where $\Omega_p^2 = 4\pi Ne^{*2}/m$, N,e* and m being the effective number density, charge and mass respectively of the material excitations one gets :

$$\omega_\pi^2(k) = \frac{1}{2}\left(\frac{c^2k^2}{\varepsilon_o} + \omega_T^2 + \Omega_p^2\right) \pm \frac{1}{2}\left[\left(\frac{c^2k^2}{\varepsilon_o} - \omega_T^2 - \Omega_p^2\right)^2 + \frac{4c^2k^2}{\varepsilon_o}\Omega_p^2\right]^{1/2} \tag{24}$$

for the eigenfrequencies of the polariton mode propagation ; thus for each value of k there are two frequencies that fall in the upper (+) and lower (-) polariton dispersion curves which are schematically depicted in Fig.1, where we also included the dispersionless nonpropagating longitudinal mode of frequency $\omega_L$ such that $\varepsilon(\omega_L) = 0$ or $\omega_L^2 = \omega_T^2 + \Omega_p^2 = \omega_s^2\,\varepsilon_s / \varepsilon_o$, the Lyddane-Sachs-Teller relation ; for $\underline{k} = 0$ the upper polariton frequency coincides with this frequency while for the lower branch goes to zero if no other electric dipole resonances lie below $\omega_T$. The form of the total energy $U_T$ stored in the polariton is also quite instructive as it is partly electromagnetic ($U_E$) and partly mechanical ($U_M$) :

$$U_T = U_E + U_M \approx \frac{1}{2}Nm(\omega^2 + \omega_T^2)\,u^2 + (\varepsilon_o + \varepsilon)\frac{E^2}{8\pi} \tag{25}$$

which propagates with group velocity :

$$v_g = v_p\left[1 + \frac{1}{2}\frac{\omega}{\varepsilon(\omega)}\frac{\partial\varepsilon(\omega)}{\partial\omega}\right]^{-1} \tag{26}$$

where $v_p = c/\sqrt{\varepsilon(\omega)}$ is the phase velocity ; in Fig. 2 we depict how the $U_E$ and $U_M$ vary with $\omega$. The above picture suffices for the description of the phonon polaritons because m is large and the phonon dispersion is flat[47]. For excitons[49-51] with their much smaller values of m (four orders of magnitude smaller), the situation is more complex because of the intrinsic

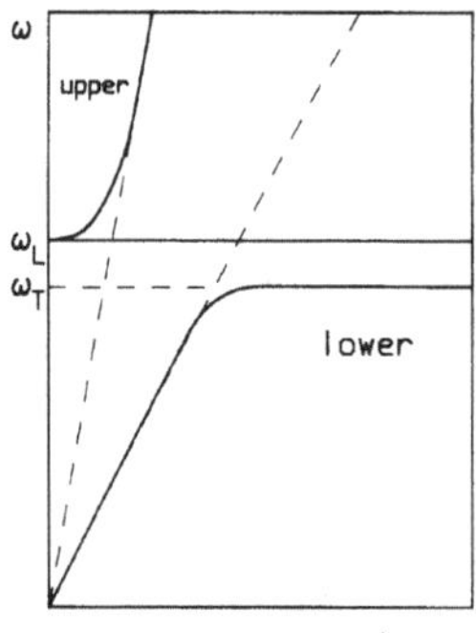


Fig.1. Dispersion curves for the upper and lower branches of a phonon polariton

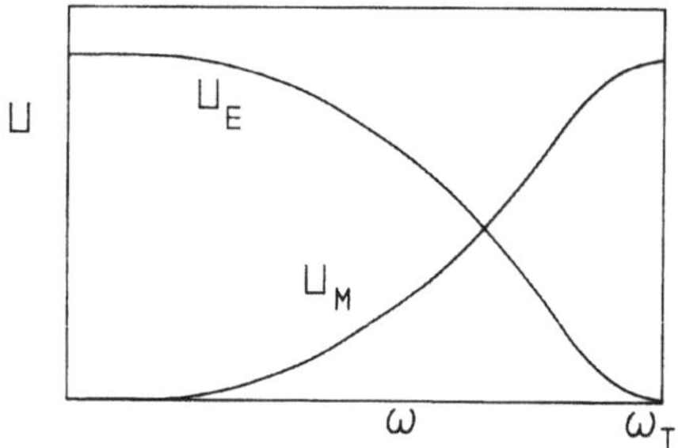


Fig.2. Variation of the electrical ($U_E$) and mechanical ($U_M$) components of the polariton as a function of frequency

exciton motion which introduces a nonlocality, also termed spatial dispersion, and is present in both transverse and longitudinal modes and leads in particular to the existence of two polaritons with widely different wavevectors but the same frequency. Polariton effects may also appear with more complex material excitations (bi-excitons, bi-phonons, plasmons, magnons) in interaction with electric dipole excitations to form dressed polaritons[53] and also in connection with resonances in the magnetic permeability $\mu(\omega)$ that also enters (23). Finally we wish to point out that there also exist surface polaritons[54] and their dispersion differs from that in three dimensions so that special provisions must be made to phase-match the two.

The previous outline overlooks the polariton damping brought in by their stochastic interaction with other degrees of freedom. Because of the composite character of the polariton this interaction proceeds through both its electromagnetic and mechanical components and the loss of coherence and energy cannot be reduced within the framework that prevails for localized two-level systems. Since the two parts are coupled coherently any damping mechanism will indiscriminately affect both parts and there is no clear-cut manner to locate the damping in either of them ; however under certain plausible assumptions one may differentiate their origin.

In an ordered crystal the damping of polaritons is mainly due to the finite lifetime of their mechanical component[55,56] which is coupled to other modes through electrical and mechanical anharmonicities[57,58] ; it is also related to the absorption losses in the reststrahlen region. If one sets

$$\varepsilon(\omega) = \varepsilon_0 + \frac{\Omega_p^2}{\omega_T^2 - \omega^2 + i\omega\gamma} \tag{27}$$

instead of (23), the polariton damping is given [55,52] by :

$$\Gamma_A = S_M \frac{\omega}{\omega_T} \gamma \tag{28}$$

where $S_M$ is the material component strength in the polariton. Actually in a more rigorous description[57-59] one obtains :

$$\varepsilon(\omega) = \varepsilon_0 + \frac{\Omega_p^2}{\omega_T^2 - \omega^2 - 2\omega_T \Pi_R(\omega) - 2i\omega \Pi_I(\omega)} \tag{29}$$

instead of (27) where $\Pi_R$ and $\Pi_I$ are the real and imaginary parts of phonon self energy implying non-Lorentzian damping that is strongly frequency dependent but only mildly temperature dependent ; the latter essentially enters through the Bose-Einstein statistics in the density of states.

In imperfect crystals along with the previous damping, which can be termed temporal disorder, an additional damping is introduced through the spatial disorder of the dielectric constant which mostly affects the electromagnetic component of the polariton. Its effect is similar to the one produced on the light scattered by a medium with a spatially random dielectric constant[60,61] ; light suffers attenuation in the forward propagation mode and a loss of coherence as a consequence to its scattering into other modes. This damping depends on the amplitude of the fluctuations of the random dielectric constant. Neglecting multiple scattering (see below) to a first approximation we may assume that the two damping

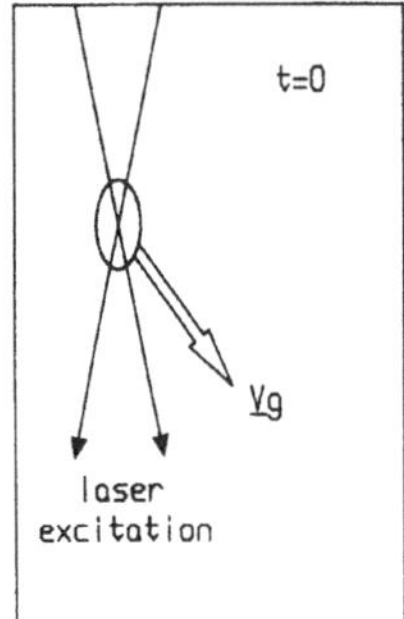

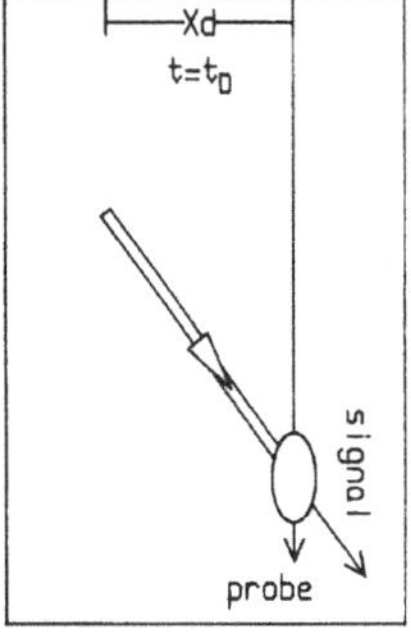


Fig.3. Excitation, propagation and detection of an ultra-short polariton pulse in the bulk crystal

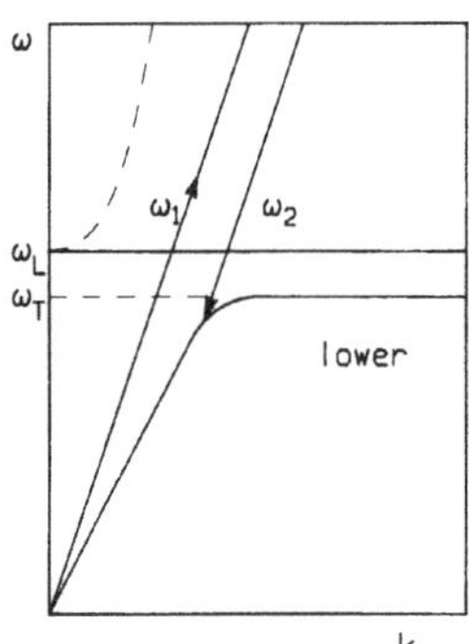

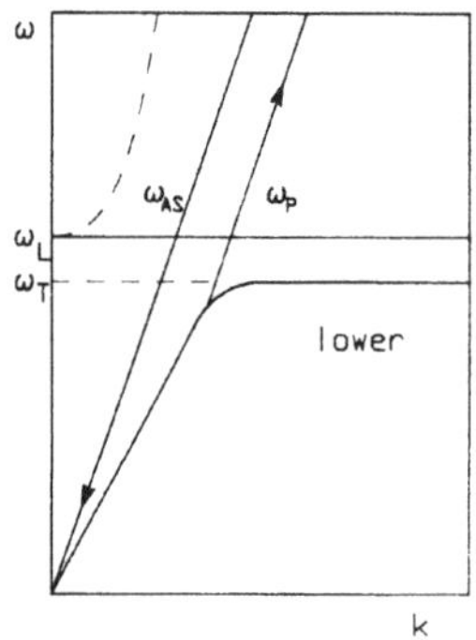


Fig.4. Coherent Raman excitation and anti-Stokes probing of the lower polariton branch

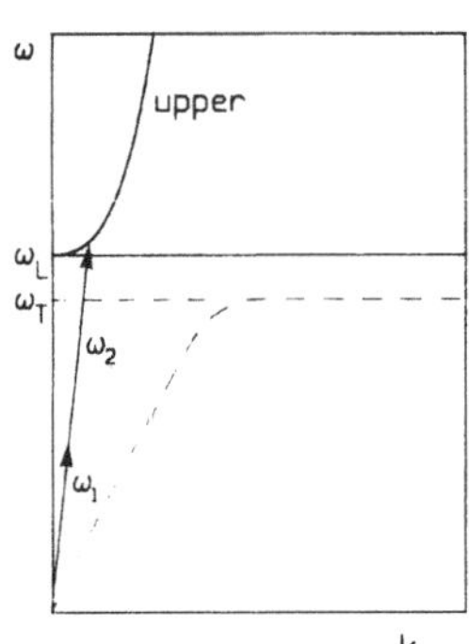

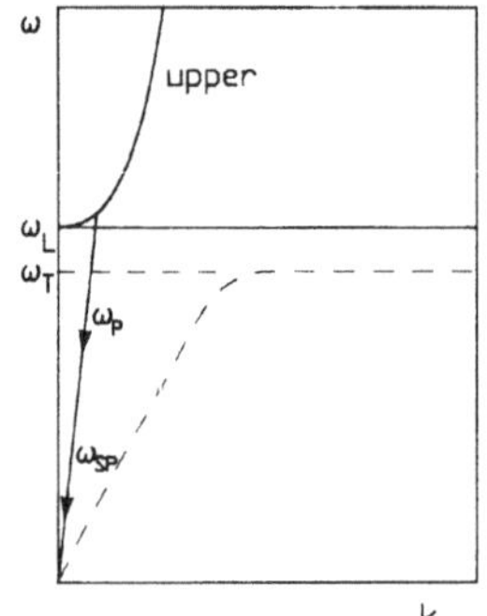


Fig.5. Two-photon excitation and phase-matched parametric probing of the upper polariton branch

processes, due to temporal and spatial disorders respectively, are statistically uncorrelated and hence we may write for the total damping rate of the polariton :

$$\Gamma_{\Pi} = \Gamma_A + \Gamma_D \tag{30}$$

where $\Gamma_D$ depends on the strength of the spatial fluctuations of the dielectric constant and is only mildly frequency dependent as expected from density of states considerations.

It is quite evident that the damping mechanisms play a crucial role in the polariton behavior. Linear optical techniques provide only meager information which is indirectly extracted by fitting the observations with (27) or (29) and the same is true for most incoherent spectroscopic techniques which only provide global information on the energy content and loss in the crystal subsequent to a few scattering events. The most serious drawback of these linear techniques and related incoherent ones is that one is restricted by the optical penetration length in the crystal and extraction of any information about damping and dispersion is complicated by the inherent polariton propagation and transmission at the crystal surfaces which may substantially alter the bulk polariton features. Clearly these difficulties can be only circumvented with nonlinear techniques that take into account the nonlocal character of the polariton and provide high selectivity ; the polariton frequency and wavevector along its dispersion curve can be obtained by nonlinear interaction of optical fields whose frequency and wavevector fall in the transparency region of the crystal.

## Tracking of Polariton Pulses

We present now the underlying principle and pattern of a nonlocal nonlinear optical technique[44,45,62-65] that allows one to determine the propagation and damping characteristics of polariton pulses by directly following their evolution in real space and time domain. It employs the following scenario (see also Fig.3).

Excitation stage. An external instantaneous force $F_e$ (t), nonlinear in the electric field amplitudes, resonantly couples to the polariton coordinate and drives it off equilibrium at "instant" t = 0 and at crystal "position" $\underline{X}$ = 0 ; as a consequence the polariton coordinate acquires a phase and an amplitude and its wave vector is fixed by the phase matching condition.

Free precession and propagation. The polariton pulse propagates freely in the direction fixed by the phase matching condition for a time interval $t_D$ after the excitation stage over a distance $\underline{R}_D$ such that :

$$\underline{R}_D = \underline{v}_g t_D \tag{31}$$

where $\underline{v}_g$ is its group velocity.

Interrogation stage. At "instant" $t_D$ the polariton pulse is probed at "position" $\underline{X}_D$ with an external weak force $F_p$ (t) of frequency $\omega_p$ and the signal generated at the combination frequency $\omega_\pi + \omega_p$ is measured in the matched configuration as a function of $t_D$.

The exciting and probing forces, $\underline{F}_e$ and $\underline{F}_p$ respectively, are products of electric fields whose order is fixed by the nonlinear interaction one exploits at each stage and to a large extent is fixed by symmetry considerations ; the spatial and temporal overlap of these fields determines the spatial and temporal resolution of the technique respectively. We also stress the fact that in the excitation stage the frequency and wavevector of the driving force $F_e$ is an algebraic combination of those of the fields that constitute this force and one can vary them at will across the polariton dispersion in any direction for both bulk and surface polaritons ; similarly in the interrogation stage the combination frequency can be chosen conveniently to provide best discrimination with the background.

The simplest case for applying this technique is in crystals that lack inversion symmetry in which case the polariton is both single and double photon allowed. Nonlinear coupling in the excitation stage can then be attained with two light fields $E_1$ and $E_2$ and is described by the phenomenological energy density[66] :

$$V = - d_E \underline{E}_1 \underline{E}_2 \underline{E}_\pi - d_M \underline{E}_1 \underline{E}_2 \underline{Q}_\pi \tag{32}$$

where $\underline{E}_\pi$ and $\underline{Q}_\pi$ represent the electric field and transverse mechanical amplitudes associated with the polariton and $d_E$ and $d_M$ are coupling parameters which vary slowly with the polariton frequency $\omega_\pi$ ; $d_E$ is simply the combination frequency susceptibility or second order susceptibility for sum or difference frequency generation, while $d_M$ is the Raman or two-photon transition amplitudes[67]. The polariton wavevector is obtained by the phase matching condition :

$$\underline{k}_\pi = \underline{k}_1 \pm \underline{k}_2 \tag{33}$$

and its frequency from energy conservation conditions :

$$\omega_\pi = \omega_1 \pm \omega_2 \tag{34}$$

where the choice of plus (+) or minus (-) sign is fixed by which of the two polariton branches is investigated. Indeed the polariton wavevector changes dramatically close to the resonance $\omega_T$ in contrast to $\underline{k}_1$ and $\underline{k}_2$ since the frequencies $\omega_1$ and $\omega_2$ are in the transparency region of the crystal and their vector sum and difference cannot indefinitely follow the polariton dispersion curve. These problems have been extensively exploited in frequency resolved nonlinear spectroscopy of polaritons[68,69]. Simple vector algebraic considerations indicate that for :

the lower branch, the excitation stage proceeds via coherent Raman excitation and the coupling term (16) is

$$V_- = - d_E E_1 E_2^* E_\pi^* - d_M E_1 E_2^* Q_\pi^* + \text{c.c.} \tag{35}$$

while the interrogation stage involves phase matched coherent anti-Stokes Raman scattering ; this technique allows the study of the lower branch of phonon polaritons in cubic crystals (Fig.4) : in anisotropic crystals this limitation is circumvented and actually the method can be used to study the upper polariton branch as well.

The upper branch
The excitation stage involves two photon absorption and the coupling term (16) is :

$$V_+ = - d_E \underline{E}_1 \underline{E}_2 \underline{E}_\pi - d_M E_1 E_2 Q_\pi + \text{c.c.} \tag{36}$$

while the interrogation stage is followed by phase matched parametric emission. With present-day short pulse laser sources this technique can be used to study the upper branch of exciton polaritons in cubic noncentrosymmetric crystals (see Fig.5) but with the advent of far infrared short pulse laser sources it can be used to study low frequency phonon-polaritons as well. The exciton-polariton dephasing has also been addressed in the time domain using time-resolved four wave-mixing[70] and transient gratings techniques[30]. These techniques however proceed via strong linear polariton excitation within the absorption layer, close to the surface layer, and, hence, only give access to the high excitation regime where the polariton dephasing is dominated by polariton-polariton interactions[70].

For crystals that possess inversion symmetry the polaritons are not accessible by double photon excitation but one can use the next higher order interaction scheme for instance coherent hyper-Raman excitation followed by coherent anti-Stokes hyper-Raman scattering or three photon absorption followed by the corresponding parametric emission process to study the low and upper polariton branches respectively.

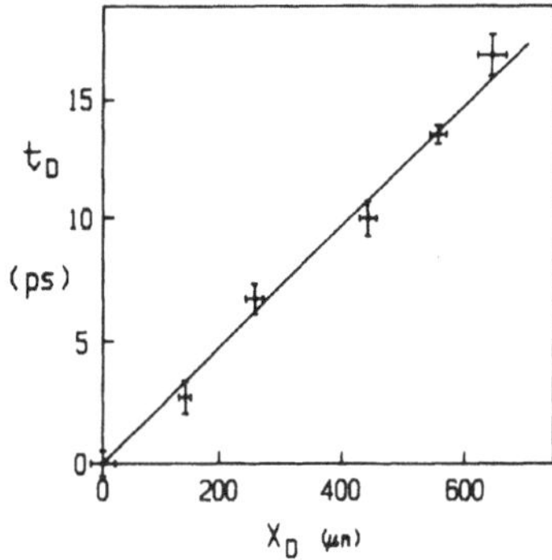


Fig.6. The probe delay at which signal is maximum as a function of lateral proble deplacement

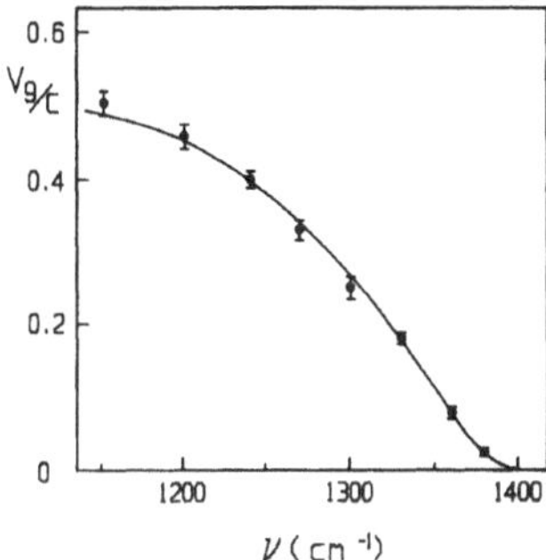


Fig.7. Measured polariton group velocity as a function of polariton frequency

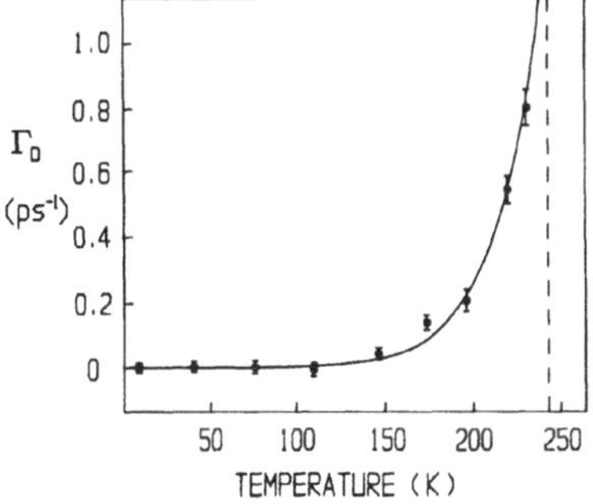


Fig.8. Behavior of the strongly temperature sensitive component $\Gamma_D$ of the polariton relaxation rate as a function of temperature. The dotted line indicates the order/ disorder transition temperature

## Bare Phonon-Polariton Propagation and Damping

**Polariton Propagation.** An example of the use of the technique to directly follow[44] the propagation of a polariton and determine its characteristics is that of the $NH_4Cl$ crystal, a cubic crystal obtained from the NaCl one by replacing $Na^+$ with $NH_4^+$ which is an intrinsically noncentrosymetric molecular entity. At low temperature the ammonium ions are all ordered the same way but at higher temperatures their orientation becomes disordered and the crystal undergoes an order/disorder phase transition. The technique was demonstrated and used to study the polariton related to the $\nu_4$ vibrational mode of the $NH_4^+$ radical with $\nu_{4T} = 1400$ cm$^{-1}$ and $\nu_{4L} = 1418$ cm$^{-1}$ for the transverse and longitudinal modes respectively : their dispersion curves and Raman spectra have been extensively studied.

The slope of the linear dependence of the optimum spatial displacement $X_D$ on probe delay $t_D$ precisely gives the polariton group velocity (Fig. 6). This was found to vary dramatically with polariton frequency (Fig.7), from $\approx c/2$ at low frequency down to $\approx c/50$ at high frequency, which reflects the change in polariton character from photon like to phonon-like as we sweep the dispersion relation. These directly determined experimental values were found to perfectly agree with the ones calculated from the dispersion law which has been extensively studied by Raman scattering. This is the first direct determination of a polariton group velocity.

**Polariton Damping.** If the spatial or temporal spreading (due to velocity dispersion) of the "tracked" polariton packet is small the intensity decrease of its peak coherent anti-Stokes signal with time delay and space displacement directly gives the loss of coherence time constant ($T_2$) of the polariton packet. Except for extreme cases, for instance when the polariton interacts with the crystal surface, the time decay of the polariton coherence was found to be exponential throughout its dispersion curve and for a wide range of temperatures up to and close to the order disorder transition temperature. The value of $T_2$ however exhibits a strong resonant behavior with frequency and a characteristically critical behavior with the temperature which we wish to discuss below.

The damping of the non propagating longitudinal component of the $\nu_4$-mode at 1418cm$^{-1}$ was also determined and was found to possess a similar behavior.

**Anharmonicity vs Disorder.** As previously stated in a perfect crystal the damping of the polariton is mainly due[55] to the finite lifetime of their mechanical component $Q_\pi$ which is anharmonically coupled to many phonon bands. In ammonium chloride ($NH_4Cl$) at low temperature this mechanism is the most important one and in particular explains the frequency variation of the polariton damping rate which can be traced to a many-phonon band degenerate with the polariton. This process gives only a weak temperature dependence of the relaxation rate $\Gamma_A$, via Bose-Einstein occupation numbers, as experimentally observed between 7 and 130 K.

In the presence of spatial disorder as previously stated an additional damping mechanism is introduced. In the case of $NH_4Cl$ such a spatial disorder can be gradually introduced for instance by increasing the temperature and approaching the order-disorder transition temperature $T_c \approx 243$ K. The effect of the disorder can be globally measured with the order parameter L which can be deduced from thermodynamic data. As can be seen in Fig. 8, the disorder induced part of the damping rate, $\Gamma_D$, can be described by the simplest law for a disorder related process : $\Gamma_D = \gamma_D(1\text{-}L^2)$. This gives an excellent fit to the experimental data and supports the assumption that dielectric disorder makes a strong contribution to the loss of coherence in polaritons.

## Coherence of Dressed Phonon-Polariton

**Polariton Fermi Resonance.** With this technique it is now possible to investigate some new aspects of the interaction of polaritons with other collective excitations in a crystal. A typical example is the polariton Fermi resonance[71] : this is produced whenever a polariton branch crosses a many-phonon band, most often a two-phonon band, of the same symmetry as the bare polariton. This situation is frequently encountered in crystals because both polaritons and two-phonon states span large regions in frequency and wave vector space. For a strong anharmonic coupling, the resulting new excitation can be described in terms of a dressed polariton with substantially different characteristics compared to the bare polariton.

The main consequences of polariton Fermi resonance are a partial localization, or slowing down, of the polariton which results from the opening of a new gap in the polariton dispersion curve and a strong, frequency dependent damping of the resulting dressed polariton. These manifestations can be theoretically analyzed by use of the Green's function method, and, in particular, it can be shown that the line shape of the dressed polariton is still lorentzian on the edges of the two-phonon band (where the density of states is relatively low) albeit strongly broadened by the opening of a new direct relaxation channel into the two-phonon continuum .

This lorentzian broadening, directly proportional to the two phonon density of states can be directly evidenced and measured[44] in the time and space resolved CARS experiments by the exponential behavior in time of the polariton-induced anti-Stokes signal.

**Polaritons in Uniaxial Crystals.** In cubic crystals, the wave-vector restrictions associated with Raman techniques limit the accessible polariton region to a part of its lower branch. This limitation can be circumvented in uniaxial crystals[72], where the birefringence allows the entire polariton dispersion curve to be observed. By exploiting this fact the time and space resolved CARS technique was extended[62,63] in uniaxial crystals, like $LiO_3$, to investigate the upper and lower ordinary $E_1$ polariton on both sides of the highest frequency reststrahlen band ($\omega_{TO} = 768$ cm$^{-1}$, $\omega_{LO} = 843$ cm$^{-1}$).

## Coherence of Exciton-Polaritons

The problem of the coherence of exciton-polaritons in polar semiconductors was also addressed with this technique[65,64] as it allows a direct determination of the dephasing time and is exempt of the drawbacks of previously used techniques : linear excitation within the absorption layer[73], time resolved luminescence or induced absorption[74,75], and four wave mixing in time domain[70] and frequency domain[76]. The demonstration was performed[65,77] with the measurement of the intrinsic dephasing time of transverse (polariton) and longitudinal components of the $Z_3$-exciton in cuprous chloride (CuCl).

The coherent excitation of the exciton-polariton of frequency $\omega_\pi$ and wavevector $k_{e\pi}(\omega_{e\pi})$ is realized in the bulk of the crystal by two photon absorption of two synchronized picosecond pulses with frequencies $\omega_1$ and $\omega_2$ and wave vector $\underline{k}_1$ and $\underline{k}_2$ such that $\omega_{e\pi} = \omega_1 + \omega_2$ and $\underline{k}_{e\pi} = \underline{k}_1 + \underline{k}_2$ (see Fig.5). These conditions introduce certain restrictions in the applicability of the technique since they can only be satisfied for the upper-branch. Close to the exciton resonance and neglecting spatial dispersion, the amplitude of the coherently driven exciton packet is proportional to :

$$d_\lambda = d_{E\lambda}\left(\omega_e^2 - \omega_\lambda^2\right) + d_{M\lambda} \qquad (37)$$

where $\omega_e$ is the bare exciton frequency at $\underline{k} = 0$ ($\omega_e \sim 3.202$eV in CuCl) and $\lambda=\pi$,L labels the exciton polariton ($\pi$) and longitudinal (L) exciton respectively. The coupling parameters $d_{M\lambda}$ and $d_{E\lambda}$ are related, respectively, to two-photon absorption and sum frequency generation close to $\omega_e$. The evolution of the exciton coherence, after the local excitation process has terminated, is followed by phase-matched parametric emission at $\omega_d = \omega_e - \omega_p$ stimulated by a third picosecond pulse delayed and spatially separated with respect to the excitation stage. We wish to stress here the fact that the excitation and probing can be done at will anywhere inside the crystal since all involved frequencies are in the transparency range of the medium : the spatial resolution is fixed by the overlap extension of the interacting beams.

The demonstration of this technique was performed[65,77] on two upper-branch polaritons in CuCl with energies $h\omega^{b}_{e\pi} \approx 3.208eV$ and $\omega^{f}_{e\pi} = 3.217eV$ corresponding respectively, to a backward, $\theta = 180°$ and a forward $\theta = 0°$ excitation geometry, and on the longitudinal exciton. As pointed out previously the technique allows the study of the spatiotemporal evolution of the polariton pulse by separating the excitation and probing stages in time and space. However, in the case of CuCl, relaxation was found to occur much faster than propagation and the polariton wave packet was probed only locally and similarly for the longitudinal exciton since it is not a propagating mode.

In Fig.9, a measurement is reproduced for the dephasing rate of the two investigated polaritons $\omega^{f}_{e\pi}$ and $\omega^{b}_{e\pi}$ at a crystal temperature of 7K. The low intensity ratio of the signals, $I_f/I_b \approx 10^{-4}$, is a consequence of the destructive interference for $\theta = 0°$ between the material and electric contributions in $d_\lambda$ in (19). In Fig.10, are reported the measured values of the dephasing rates $\Gamma$ for different temperatures in the range 7-60K and the calculated ones with an analytical expression of this rate based on the assumption that the dephasing is due to the three main exciton-phonon scattering processes namely the longitudinal optical phonon assisted scattering[78] through the Fröhlich interaction (LO), the longitudinal acoustic phonon (LA) scattering mediated by the deformation potential (DP) and the transverse acoustic phonon (TA) one mediated[79] by the pieroelectric effect (PE). The most probable processes for upper-branch polaritons are extraband down- and up- conversion into a lower-branch polariton, with, respectively, emission or absorption of a phonon[80]. These processes are strongly enhanced compared to intraband ones because of the higher density of final accessible states. At low temperatures scattering off acoustic phonons provide the dominant mechanism while for high temperatures the Fröhlich mechanism becomes dominant. Here, only the up-conversion process needs to be taken into account because of the very low density of accessible states for the down-conversion process. Their compound effect as depicted in Fig.10 satisfactorily reproduces the observed behavior of $\Gamma$.

The dephasing rate of the longitudinal exciton was also measured over the same temperature range ; the values are within the same range as those of the transverse (polariton) exciton and the overall temperature dependence is similar implying that the same exciton-phonon mechanisms are at work here too. The study allowed the determination of the different coupling constants which were also independently estimated[81].

## Polariton Transport and Localisation in a Disordered Dielectric

All types of polaritons are composite excitations partly material and partly electromagnetic and as such are expected to share features of both modes. Thus their damping and coherence is affected by both material anharmonicity and spatial disorder. The first has been extensively addressed in the literature both experimentally and theoretically but much less the second which however can be quite critical in certain circumstances as previoulsy shown[20]. Here we succintly analyse the problem of polariton transport[64] in a crystal with a real refractive index that varies randomly[60,61] in space in the light of the recent theories[82-87] that have been developped to take into account effects related to multiple scattering of waves in disordered media.

In the case of pure electromagnetic waves under certain conditions the phase correlations and the interference of multiply scattered waves off the randomly distributed spatial fluctuations of the refractive index can lead to a drastically new behavior in the overall wave propagation, the most prominent feature being the possible localization[82,83] of the waves : the effective light diffusion constant vanishes at a certain wavelength $\lambda^*$ which establishes a threshold separating localized from delocalized states of the electromagnetic field. The necessary conditions, however, are very stringent for transparent media[82,85] ; it turns out that the corresponding condition is easier to reach in the case of polaritons[64].

We introduce "dispersive" disorder in the crystal by letting the real background dielectric constant become space dependent $\varepsilon(\underline{r})$ and fluctuate randomly in space around an

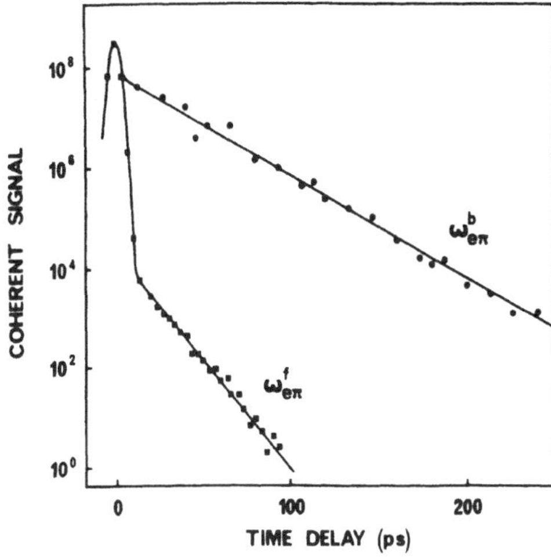


Fig.9. Dephasing rate for the two investigated polaritons in CuCl at 7K

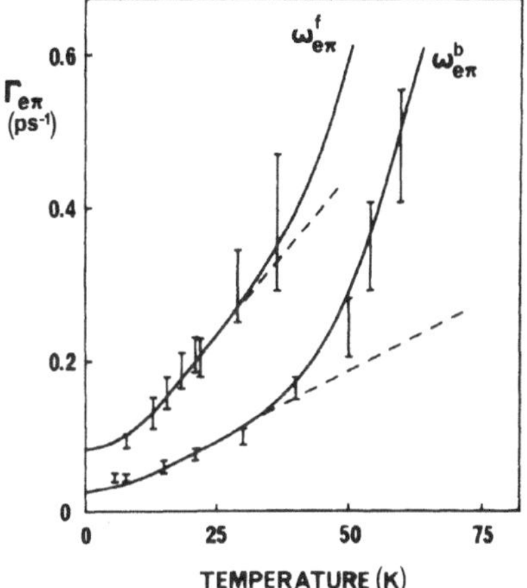


Fig.10. Temperature dependence of the $\omega^{f}_{e\pi}$ and $\omega^{b}_{e\pi}$ polariton dephasing rates $\Gamma_{e\pi}$ in CuCl

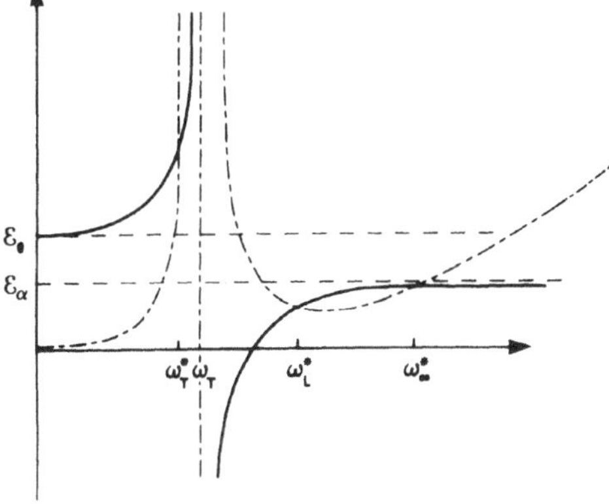


Fig.11. The left (full and right (dotted) members of eq.29 showing the existance of three mobility edges

average value $\varepsilon_\infty$ while the oscillator parameters are held constant[60] ; the fluctuating part $\delta\varepsilon(\underline{r}) = \varepsilon(\underline{r}) - \varepsilon_\infty$ with $<\delta\varepsilon(\underline{r})> = 0$ is assumed to be a gaussian random function with spatial correlation :

$$<\delta\varepsilon(\underline{r})\ \delta\varepsilon(\underline{r}')> = \mu^2\delta(\underline{r} - \underline{r}') \tag{38}$$

where $\mu$ is a constant and $< >$ everywhere denotes ensemble averages. For later use we also introduce the renormalized quantities $\tilde{\varepsilon}(\underline{r}) = \varepsilon(\underline{r}) / \varepsilon_\infty$, $\delta\tilde{\varepsilon}(\underline{r}) = \delta\varepsilon(\underline{r}) / \varepsilon_\infty$, $\tilde{\mu} = \mu/ \varepsilon_\infty$ and $\tilde{\varepsilon} = c/\sqrt{\varepsilon_\infty}$.

The polariton electric field $E_\pi$ and polarization $P_d$ then obey the coupled equations :

$$\left\{\nabla^2 + (\omega/\tilde{c})^2\left(1 + \delta\tilde{\varepsilon}(\underline{r})\right)\right\} E_\pi = c^{-2}\ddot{P}_c \tag{39}$$

$$\ddot{\underline{P}}_d + \omega_T^2 \underline{P}_d = \omega_p^2 E_\pi \tag{40}$$

where, anticipating the use of the hydrodynamic regimė in deriving the transport coefficients, we have neglected a term $\nabla.\{\delta\varepsilon(\underline{r})\ \underline{E}_\pi\}$ in (39). The polariton characteristics can be most conveniently introduced in terms of the Green function $G(\underline{q},\omega)$. Averaging over all disorder configurations one merely obtains an additional isotropic attenuation which can be lumped together with the one due to the material anharmonicity. In the presence of strong dispersive disorder, however, correlation and interference between multiple scattered waves drastically alter the situation. Such effects are suppressed in the short ranged average Green's function $<G(q,\omega)>$ so that the relevant quantity[86] is now the one squared over all disorder configurations :

$$P(\underline{k},k';\omega,\Omega) = < G_-(\underline{k},\omega + \Omega)\ G_+(\underline{k}',\omega\ ) > \tag{41}$$

where $G_-(k,\omega)$ and $G_+(k,\omega)$ respectively stand for the retarded and advanced one polariton Green's function ; the quantity $P(\underline{k},k';\omega,\Omega)$ in (41) is the averaged intensity propagator and can be expanded in terms of the averaged Green's functions and also averaged over pairs of scattering events which correlate the two Green's functions[86,87]. The expansion can be obtained by the Bethe-Salpeter equation along the same lines as for electrons or sound ; to lowest order in $\tilde{\mu}^2$ the expansion of the intensity propagator leads to a diffusion pole while higher order terms yield divergent integrals whose dominant contribution comes from the maximally crossed diagrams[88].

In the hydrodynamic regime, namely small $\underline{q} = \underline{k} - \underline{k}'$ and $\Omega$, these can be resumed and one obtains after rearragements of the terms the relation for the mobility edge $\omega^*$ for polariton transport

$$\left(\varepsilon_\infty + \frac{\omega_p^2}{\omega_T^2 - \omega^{*2}}\right)^3 = \kappa\omega^{*6}\left\{\varepsilon_\infty + \frac{\omega_T^2\omega_p^2}{\left(\omega_T^2 - \omega^{*2}\right)^2}\right\}^4 \tag{42}$$

where $\kappa = 12\ \sigma\mu^4/\pi^2c^6$.

In Fig.11 we give a graphic representation of the right and left members of this relation for a small value of the parameter $\kappa$ which is related to the disorder. We see that for small $\kappa$

there are three mobility edges $\omega_T^*$, $\omega_L^*$ and $\omega_\infty^*$ ; with $\omega_T < \omega_L < \omega_\infty$ , $\omega_T^* < \omega_T$ and $\omega_L^* > \omega_L$. The states are localized for $\omega > \omega_\infty$ and within $\{\omega_T^*, \omega_T\}$ and $\{\omega_L, \omega_L^*\}$ ; thus as a small disorder is introduced the polariton states become localized on either side of the reststrahlen region and in addition above a certain high frequency $\omega_\infty^*$ as in transparent media. As the disorder increases the localized regions increase and eventually $\omega_L^*$ merges with $\omega_\infty^*$ for a critical value $\kappa_c$. Beyond this value there is only one mobility edge $\omega_T^*$ and all polariton states with $\omega < \omega_T^*$ are delocalized and above it are localized.

In the region where the polariton is localized one can define a delocalization length $\xi(\omega)$ by

$$\xi_i(\omega) = \frac{2\pi^2 L}{(4\pi L q_0 - 1)} \tag{43}$$

where $L = 1/4\pi^3\rho(\omega)D_o$, $D_o = \ell v_g/3$, $\ell$ is the polariton coherence length, $\rho(\omega)$ its density of states and $q_0$ is a cutt off that eliminates high wave vectors that do not correspond to diffusion processes. Since $q_0 \sim 1/\ell$ one easily sees that $\xi(\omega) > \ell$ and close to the mobility edges $\xi$ diverges as $1/|\omega_i^* - \omega|$ namely with a critical exponent of unity.

In the above analysis we ignored the presence of the damping due to the phonon anharmonicity ; its effect is mainly felt close to the reststrahlen region and can be incorporated by introducing a friction term $\gamma\dot{P}_d$ in (40) which then destroys its time inversion symmetry and diffuses the sharp mobility edges (42). Clearly localization can occur if $\gamma\tau < 1$ ; otherwise the temporal phase coherence loss due to the anharmonicity washes[89] out the interference of the multiple scattered waves.There are no experiemental studies concerning the polariton localization.

## Polariton Optics

We have presented a nonlocal time-resolved technique that allows one to address and study all aspects of the spatio-temporal evolution of short polariton pulses. This technique opens up new possibilities for the investigation of fundamental problems associated with pulses of collective excitations in crystals regarding their propagation characteristics, dephasing and energy loss processes. In this respect the problem of the polariton pulses is of central interest since it is connected with electromagnetic signal propagation close to a resonance. Many nonlinear optical effects in crystals crucial for optoelectronic devices are greatly enhanced close to electric dipole allowed resonances and must be described in terms of coherent polariton pulse interactions. Of particular interest here are the parametric optical interactions[90,91] which are of both fundamental and technological interest : parametric optical amplifiers and oscillators parametric instabilities and chaos...

The space and time resolved CARS technique allows one to "track" a polariton pulse at any "point" inside the crystal and analyse its phase and energy content and assess the impact of temporal and spatial disorder and, in particular, the effect of boundaries. The latter is of particular importance for understanding polariton optics inside a crystal :

reflection and transmission by plane boundaries
polariton Fabry-Perot cavities
polariton total reflection and tunnelling
polariton nonlinear propagation.

One can observe reflected polaritons and polaritons transmitted from one crystal to another separated by air.

A problem of particular fundamental interest is the interaction of polaritons with random spatial disorder. This problem has only been addressed in the case of orientational disorder in $NH_4Cl$ where advantage was taken of the possibility to arbitrarily "tune" the crystal disorder by changing the crystal temperature, which allows a variable degree of disorder to be probed in the same sample. However, measurements in other disordered systems, such as isotopically disordered ones, are of particular interest for assessing the mutual coherence of the electromagnetic and mechanical parts of the polariton ; furthermore, since there is a close connection between polariton and photon scattering by disorder, one can expect to observe polariton localization.

As stated above, the intrinsic limitation of the Raman technique to non centrosymmetric crystals can be circumvented by the use of the hyper-Raman configuration for coherent excitation and probe in crystals like NaCl. Here the decrease in efficiency of the hyper-Raman processes can be counterbalanced by using the very high peak power delivered by picosecond lasers and the much higher damage threshold in shortening the light pulses in such ionic crystals. Furthermore, one could also address the problem of propagation and relaxation of vibrational polaritons in highly disordered media such as glasses or liquids .

## COHERENCE AND PARAMETRIC INSTABILITIES OF TWO PHONON-STATES. LARGE AMPLITUDE VIBRATIONAL MOTION

The vibrational overtones and multiphonon states in condensed media are essential for the understanding and the description of large amplitude vibrational motion[92] in condensed matter and the anharmonic forces[93,94] that come into play there. Through the study of their time evolution, much insight can also be gained about phonon breakdown[95] large wave vector phonon dynamics[96], lattice instabilities and phase transitions[97] or stereochemical reactions[98]. These problems are also relevant in high non equilibrium phonon systems[99], for instance high temperature pulses[100] and phonon hot spots[101] in dielectrics.

A central problem here is the coherence and population evolution of these multiphonon states. For single phonon states these processes are usually associated with the transverse ($T_2$) and longitudinal ($T_1$) relaxation times respectively but for the multiphonon states the situation is far more complex because of the compound character of these excitations and the additional relaxation channels that this introduces ; more importantly the anharmonic interactions, although much weaker than the harmonic ones, under certain conditions introduce profound changes in the multiphonon spectrum[102-106].

To fix our ideas we recall that in the harmonic approximation the excited energy of a lattice with two phonons in branches $\sigma'$ and $\sigma''$ with wave vectors $\underline{k}'$ and $\underline{k}''$ respectively is given by :

$$\hbar\Omega(\underline{k}) = \hbar\{\omega(\underline{k}') + \omega(\underline{k}'')\} \qquad 44)$$

with $\underline{k} = \underline{k}' + \underline{k}''$. The optically (infrared or Raman) accessible two-phonon states form a quasi continuum with $\underline{k} = 0$ or $\underline{k}' = -\underline{k}''$ and $\underline{k}'$ lies anywhere within the first Brillouin zone. The bandwith of this quasi continuum is equal to the sum or the widths W of the individual phonon branches ; W is a measure of the intermolecular coupling and phonon localization. When the anharmonicity is switched on the phonons interact with each other through the third and fourth order terms, $h^{(3)}$ and $h^{(4)}$ respectively, in the usual expansion of the lattice hamitonian[47,93,94]

$$h = h^{(0)} + h^{(3)} + h^{(4)} \qquad (45)$$

If the strength $\Gamma_B$ of the fourth order term $h^{(4)}$, which is essentially intramolecular, is much larger than the intermolecular coupling W a localized bound two-phonon state or bi-phonon splits off the free two-phonon continuum (see fig.12) with narrow linewidth and nonnegligible oscillator strength borrowed from the two-phonon quasi-continuum which is considerably reduced ; the residual quasi-continuum contains the free two-phonon states. In addition if such a bound (or quasi-bound two-phonon state) of energy $\Omega_B$ is nearly degenerate with a single phonon state of energy $h\omega_1$ a hybridization mediated through the third order term $h^{(3)}$ may occur resulting in a double peak[105] (Fermi doublet)[104] whose two components split off on either side of the free two-phonon residual continuum (see fig.12) ; the relative strength of the two components depends on the ration $\beta/(\Omega_2 - \omega_1)$ where $\hbar\beta$ is the strength of the third order term $h^{(3)}$.

Let us introduce[92,20,21] the expectation value of the optically accessible two-phonon coordinate :

$$<q_+q_-> = \mathrm{Tr}\rho q_+q_- \tag{46}$$

where $q_+$ and $q_-$ are the single phonon coordinates with wavevector $\underline{k}$ and $-\underline{k}$ respectively and $\rho$ is the density matrix operator in the optically accessible two-phonon state space which consists of the ground, the bound two-phonon and the quasi continuum of the free two phonon states, $|\psi_o>$, $|\psi_B>$ and $|\psi_F>$ respectively ; for simplicity we restricted ourselves to the phonons of a single optic branch. The dynamics of the bound two-phonon state which is singled out in the coherent excitation process in CAHORS, enter the calculation through the expectation value :

$$<Q>_B = \mathrm{Tr}\rho_B q_+q_- \tag{47}$$

where $\rho_B$ is the density matrix operator in the subspace spanned by $|\psi_o$and $|\psi_B>$. Note that because of different interactions with the bath $<Q>_B \neq <q_+> <q_->$ while for the expectation values in the free two-phonon subspace $<q_+q_->_F \equiv \mathrm{Tr}\rho_\Phi q_+q_- \approx <q_+> <q_->$. In the following we shall omit the brackets <> but keep in mind the real physical content of these coordinates. By explicitly introducing the bound two-phonon coordinate $Q_B$ and its frequency $\Omega_B$ we actually take into account the main effect of the fourth order term $h^{(4)}$ and are left with a residual term of the form :

$$h^{(4)}_{BF} = = \Gamma_{BF} Q_B q_+ q_- \tag{48}$$

the coherent interaction term between the bi-phonon and the free two-phonon states ; any other term in $h^{(4)}$ can be included in the damping of the two-phonon state. The term (48) plays a crucial role both for the transfer of coherence from the bound two-phonon state to the free two-phonon states and the parametric instabilities that may occur then.

The experimental studies[20,21] revealed that depending on the relative strength of the two anharmonic terms in (45) the loss of coherence of the bound two-phonon states mainly occurs through two channels :

- an <u>intrinsic</u>[20] one where the bound (or quasi bound two-phonon state) internally loses its coherence to the free two-phonon states from which it is formed through the term $h^{(4)}$ and the later are subsequently dissolved into single phonons : this mechanism which essentially leads to a temperature independent behavior occurs whenever the bound or quasi bound two-phonon state is near or on top of the quasi-continuum of the free two-phonon spectrum a common situation when only phonons of a single phonon branch are involved. The model has been described in ref. 92 where using Green's function techniques it was found that this mechanism in general leads to an exponential decay of the coherence with a lifetime $T'_B = \Gamma'^{-1}_B$

$$\Gamma'_B \approx \upsilon_o \tag{49}$$

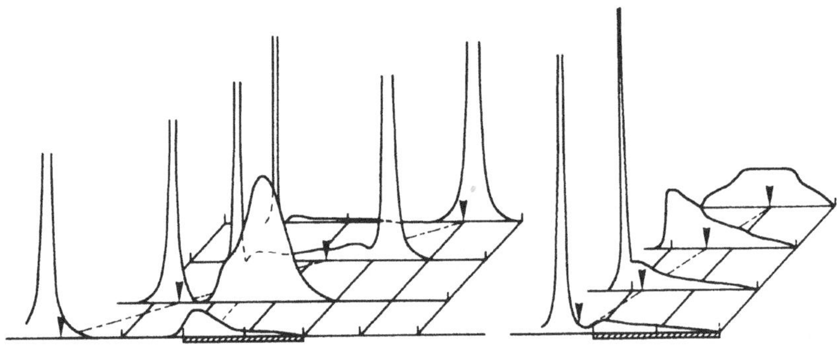

Fig.12. Left : dependence of two-phonon spectra on anharmonic strength $\Gamma_B$ (increasing from left rear to right front)
Right : production of two bound states as a one-phonon line approaches the free two-phonon band position is shaded

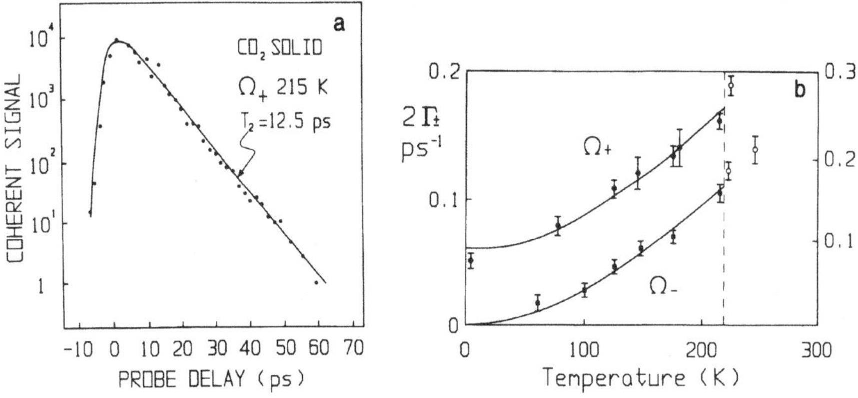


Fig.13. a) Coherent anti-Stokdes signal vs probe delay for the $\Omega_+$ line of $CO_2$ solid at 215k.
b) Variation of the relaxation rates $2\Gamma_+$ and $2\Gamma_-$ with temperature for the $\Omega_+$ line (circles) and the $\Omega_-$ line (squares) of the Fermi doublet in $CO_2$

where $\upsilon_o$ is the density of free two-phonon states calculated where the overlap with the bound two-phonon state is maximal.

- an extrinsic[21] one where the frequency of the bound (or quasibound) two-phonon state, which is essentially intramolecular, fluctuates in time as a result of the intermolecular degrees of freedom or other low lying states ; this mechanism which leads to a temperature dependent behavior (characteristically $\sim T^2$ is dominant whenever the bound (or quasibound) two-phonon state is far removed from the quasi continuum as a consequence of a mutual repulsion with a closely lying single phonon state through the $h^{(3)}$ term. It was shown in ref.21 that this mechanism too in general leads to an exponential decay (Fig.13) of the coherence with a damping constant

$$\Gamma_B'' \sim kT^2 \tag{50}$$

One generally expects that the intrinsic mechanism is dominant at low temperatures or in crystals with one or two simple molecules per unit cell while the extrinsic one prevails in crystals with many molecules per unit cell or large and easily deformable molecules. The two extreme situations are exemplified[20,21] with the $CS_2$ and $CO_2$ crystals respectively while $N_2O$ crystal constitutes[107] an intermediate case since the Fermi resonance is less pronounced than in $CO_2$ but more so than in $CS_2$.

The important point to notice form theses investigations is that the bound two-phonon states may maintain its coherence for relatively long times when the intrinsic mechanism is operative ; if the free two-phonon states into which this coherence is transfered can retain it long enough, under certain conditions, this long overlap in time may engender[92,108] parametric instabilities. Below we wish to address this important point in a more quantitative way.

The coherent excitation with two intense fields allows one to generate coherent bound two-phonon states with wave vector $k \approx 0$ and large amplitude. Their amplitude can be viewed as a field amplitude satisfying an inhomogeneous wave equation. Since the overlap $<\psi_F/\psi_B>$ of the bound and free two-phonon states can be substantial the bound two-phonon can break down with finite probability into two free phonons coherently propagating with wave vectors $\underline{k}'$ and $\underline{k}''$ such as $\underline{k}' + \underline{k}'' = 0$ and amplitudes q' and q'' that satisfy the one phonon propagation equation with a source term. The source term is related to $<\psi_F/\psi_B>$ whose square also measures the probability of transferring the coherence from the bound to the free two-phonon amplitudes with total wave vector conserved or $\underline{k}' + \underline{k}'' = \underline{k}$.

A similarity with the parametric amplifications and oscillation in three wave interactions in nonlinear optics[91] can be drawn as follows : the overlap $<\psi_F/\psi_B>$ corresponds to the second order susceptibility $\chi^{(2}$ and $Q_B$, $q_+$ and $q_-$ corresponds to the field amplitudes of the pump (p), signal (s) and idler (i) respectively[90,91]. The phonon propagation equation can be linearized within the envelop approximation similar to the one used in nonlinear optics. This striking analogy pointed in refs 92,108 can be the starting point of nonlinear phonon optics.

We present now a simple description of a class of parametric instabilities which are of relevance in many interesting situations when the intrinsic mechanism for coherence transfer is operative. We will assume that no Fermi resonance[104,105] occurs or $|\beta/(\Omega_2 - \omega_1)| << 1$ so that the third order term $h^{(3)}$ only introduces a mere renormalization of the one phonon spectrum. Furthermore since the main effect of $h^{(4)}$ is accounted for by the formation of the bound two-phonon state as stated previously the coherent interaction between the bound two-phonon amplitude Q and the single phonon amplitudes $q_+$ and $q_-$ is mediated through the residual term (48) where $q_+$ and $q_-$ are the amplitudes of the two phonons of opposite wave vectors that interact most strongly with the bound two-phonon state. Any other term in $h^{(4)}$ will be absorbed in the damping constants.

Using (48) and (30) the equations of motion of the coordinates are then given by

$$\ddot{Q}_B + \frac{1}{T_B}\dot{Q}_B + \Omega_B^2 Q + \Gamma_{BF} q_+ q_- = \frac{1}{4}\alpha'' E^2 \tag{51}$$

$$\ddot{q}_+ + \frac{1}{T_+}\dot{q}_+ + \omega_+^2 q_+ + \Gamma_{BF} Q_B q_- = 0 \tag{52}$$

$$\ddot{q}_- + \frac{1}{T_-}\dot{q}_- + \omega_-^2 q_- + \Gamma_{BF} Q_B q_+ = 0 \tag{53}$$

For a first approximation the term $\Gamma_{BF} q_+ q_-$ in (51) can be disregarded (it can be absorbed in the damping) and in $E^2$ only the cross term that oscillates at frequency $\omega_L - \omega_S \approx \Omega_B$ will be retained. For simplicity we may also safely assume that $\omega_+ = \omega_- = \omega_o$ and $T_+ = T_- = T$ for phonons of the same branch and opposite wave vectors. Introducing then $q_\Sigma = q_+ + q_-$ and $q_\Delta = q_+ - q_-$ one has

$$\ddot{Q}_B + \frac{1}{T_B}\dot{Q}_B + \Omega_B^2 Q_B = F\cos\Delta\omega t \tag{54}$$

$$\ddot{q}_\Sigma + \frac{1}{T}\dot{q}_\Sigma + \omega_o^2 q_\Sigma + \Gamma_{BF} Q_B q_\Sigma = 0 \tag{55}$$

and similarly for $q_\Delta$, where $F = \frac{1}{2}\alpha'' E_L E_S$ and $\Delta\omega = \omega_L - \omega_S \approx \Omega_B$. Solving for $Q_B$ in (54) one obtains $Q_B = \mathrm{Re}\{Q_o e^{i\Delta\omega t}\}$ where

$$Q_o = \frac{1}{4}\,\frac{(\alpha'')^2 E_L E_S}{\Omega_B^2 - \Delta\omega^2 + i\Delta\omega T_B^{-1}} \tag{56}$$

and replacing it in (55) one gets

$$\ddot{q}_\Sigma + \frac{1}{T}\dot{q}_\Sigma + \omega_o^2 (1 + \mu\cos\Omega_o t)\, q_\Sigma = 0 \tag{57}$$

with

$$\mu = \frac{\Gamma_{BF} Q_o}{\omega_o^2} << 1 \tag{58}$$

where it will be assumed that $\Delta\omega T_B > 1$ ; in general we may set

$$\Omega_o = 2\omega_o + \varepsilon \tag{59}$$

where $\varepsilon << \omega_o$ is the frequency mismatch between the bound and free two-phonon states. Equation (57) is that of the parametric oscillator which has been extensively studied in the literature[109,110]

$$Q_o > Q_{th} = \frac{\omega_o}{\Gamma_{BF}} \left(\varepsilon^2 + \frac{1}{T^2}\right)^{1/2} \tag{60}$$

which together with (56) introduce a threshold value for $E_LE_S$. The fields required to reach this condition are usually large but manageable in certain cases ; clearly for $\varepsilon >> \frac{1}{T}$ the threshold conditions simply becomes

$$Q_o > \frac{\omega_o}{\Gamma_{BF}} \varepsilon \tag{61}$$

which is easier to satisfy.

The occurence of parametric instabilities has some major implications :

- large densities of one-phonon states can be coherently driven even when they do not couple directly with the fields (Raman inactive modes)
- the parametric process favors creation of large wave vector phonons and their dynamics may be probed optically ; indeed since in general $\Omega_B$ overlaps or lies below the lower part of the free two-phonon quasi continuum the frequency mismatch $\varepsilon$ is smaller for pairs of phonons at the edge of the Brillouin zone than in the center
- new effects appear when the polariton character of the phonons is also taken into account
- when a structural phase transition occurs the point density of two-phonon states changes and according to (49) also $T_B$ will change which may affect the parametric instability
- one can have transition to chaotic behavior since the set of equations exhibit such a behavior
- all the above considerations can be extended to other types of elementary excitations in particular acoustic phonons, magnons or plasmons ; the case of magnons is of particular interest since bound two-magnons states care large oscillator strength.

## NON MARKOFFIAN REGIME. MEMORY EFFECTS

The previous discussion was essentially based on the relaxation time description where the thermal reservoir correlation time $\tau_c$ is assumed to be much shorter than any relevant time of the dynamical variable under investigation. In practice this amounts in replacing the correlation function by a delta function which results in a white noise and a markoffian process in the case of a single dynamical variable. The description is more complex if the relaxation proceeds through two or more coupled slow dynamical variables in which case under certain assumptions one may have still an effective relaxation time description for each variable but in general the corresponding effective relaxation time is frequency dependent as the noise that experiences each observable is filtered by the others. Actually this is the case with the coupled equations (9) and (10) for the polarization (coherence) and population difference $\mathcal{P}$ and $\Delta\rho$ respectively. It is also the case of the polariton mode ccompare (29))which is a mixture of material excitation and an electromagnetic mode and the case of the bound material modes (bi-phonons, bi-excitons, bi-magnons etc... or hybrid modes). All these problems can be treated by the coupled mode technique[112,67,21] which within the linear approximation results in the linear relaxation regime.

Actually the underlying fundamental reason[1] for such a behavior is the introduction of the temporal and spatial coarse-grain averaging procedure in the derivation of the relevant equations for the few observable variables from the microscopic ones that involve all variables. This results[1,113,114] in a projection of the complete process on the slow variable subspace and in a contraction of information concerning the fast variables. In this averaging procedure the time grains are much shorter than the time scales of evolution of the slow variables which also fix the observation time regime, but still larger than the correlation time $\tau_c$ so that one essentially samples an environment at instant equilibrium and all retardation effects are smeared out where by instant here we mean the time grains ; otherwise stated the dissipation can be related to fluctuations in thermal equilibrium. In the case of a brownian variable this results in a friction that is determined by the instantaneous velocity of the variable and its evolution is described by a Langevin equation :

$$\dot{\underline{Q}} + \underline{\underline{\gamma}}\,\underline{Q} = \underline{F}(t) + \underline{F}_c(t) \qquad (62)$$

where F(t) with <F> = 0 is the fluctuating force of the fast variables, $F_c$ is the external coherent force and $\gamma$ is the constant friction ; in (62) we implicitly assumed additive[114] stochastic processes which is the simplest to treat analytically. Multiplicative[114] stochastic processes can also occur but in general these can be treated analytically in some simple cases; for instance the case of the harmonic oscillator :

$$\dot{Q}(t) = i\Omega(t)\,Q(t)$$

where $\Omega(t)$ is a stochastic gaussian processes and ReQ and ImQ are the position and momentum of the oscillator can be treated[1] and used to describe the motional narrowing.

Thus the coarse graining procedure and the separation into slow (observable) and fast variables is conditionned by the positions relative to each other of the relaxation, the observation and the correlation times. In particular, if the relaxation time or the observation time is comparable to the correlation time the environment has not enough time to reach an equilibrium and retardation effects in the friction that experience the observables are felt. In this case equation (62) must be replaced[1] by an integrodifferential one, the generalized Langevin equation :

$$\dot{\underline{Q}} + \int_{-\infty}^{t} \underline{\underline{\gamma}}(t - t')\,\underline{Q}(t') = \underline{F}(t) + \underline{F}_c(t) \qquad (63)$$

where $\gamma$ (t) now includes retardation or memory effects ; these are also termed reaction effects. Similar provisions must be made in the case of coupled Langevin equations. To the extent that this equations are linear in the observables one can still use harmonic analysis to extract their time evolution but this is not in general exponential or even not effectively so. Numerous such cases are now encountered in physics[1,114].

In order to implement these aspects into the nonlinear response and study them with the time resolved nonlinear optical techniques one must go back to the Liouville equation[1,114]

$$ih\frac{d\rho(t)}{dt} = L(t)\,\rho(t) \qquad (64)$$

of the total density matrix operator for the relevant system of variables and those of the reservoir the two interacting strongly with each other. In (64) L (t) stands for the commutator [H (t),] and $H = H_o + H'$ with $H_o$ being the hamiltonian of the relevant system, a function of the reservoir variables and H' is the interaction of the relevant variables with an external intense coherent field ; provisions must also be made to include effective field corrections. For our purpose the relevant quantity[115,116] is :

$$\rho_I^{(3)}(t) = \left(\frac{1}{i\hbar}\right)^3 \int_{-\infty}^{t} dt_1 \int_{-\infty}^{t_1} dt_2 \int_{-\infty}^{t_2} dt_3\, L_I(t_1)\, L_I(t_2)\, L_I(t_3)\, \rho(-\infty) \qquad (65)$$

where $\rho_I$ and $L_I$ are defined in the interaction picture.

If we consider the case of two pulses of same frequencies $\omega$ electric field envelopes $E_1(t)$ and $E_2(t-t_s)$ where $t_s$ is their separation in time and $\underline{k}_1$, $\underline{k}_2$ are their respective wave vectors one finds[115] after some approximations

$$I^{(3)}(t) \approx \theta_1^2\,\theta_2^2\, e^{-2[2S(t-t_s) + 2S(t_s) - S(t)]} \qquad (66)$$

for the radiation emitted and detected in the direction $2\underline{k}_2 - \underline{k}_1$ where

$$S(t) = \int_{\infty}^{\infty} \frac{d\omega}{2\pi\omega^2} J(\omega)(1 - \cos\omega t) \tag{67}$$

and

$$J(\omega) = \int_{-\infty}^{\infty} dt\, e^{i\omega t} \langle V_r(t)\, V_r(0) \rangle \tag{68}$$

is the spectral density of the reservoir correlation function ; $\theta_i$ is the pulse area for i = 1,2.

When the observation time is much longer than $\tau_c$ one gets

$$S(t) \sim \frac{1}{2} J(0)\, t \tag{69}$$

and

$$I^{(3)}(t) \sim e^{-2t/T_2} \tag{70}$$

with $T_2 = 2/J(0)$ namely the exponential decay : this is the fast modulation case.

For short observation times one gets[115] :

$$S(t) \sim D^2 t^2 \tag{71}$$

with

$$D^2 \sim \int J(\omega)\, d\omega/2\pi = \langle V_r^2 \rangle \tag{72}$$

and

$$I^{(3)} \sim e^{-D^2(t-2t_s)^2} \tag{73}$$

which for $Dt_s > 1$ implies the formation of an echo at $2t_s$ namely a situation similar to that of an inhomogeneous broadened line : this is the slow modulation case. In between these two extreme cases the problem can only be treated[115] numerically ; the decay is not exponential and memory effects are clearly present with a feature that is precursor to the echo.

Actually nonexponential decay and memory effects can also result from interference of two otherwise completely random markoffian processes. We illustrate this with the defect diffusion limited dephasing[117,118]. Let us consider identical transition dipoles each supposed to dephase itself with a single dephasing time $T_2$ ; for simplicity we consider[117] a one dimensional array of such dipoles but the model can be generalized[118] to three dimensions. Let us now assume that along this axis a number of mobile point defects exist with number density $1/2\,\ell_o$, $\ell_o$ being a length ; their motion will be described by a diffusion equation with a diffusion constant D. We assume that a collision of a defect with a dipole leads to complete and instant dephasing of the latter so that the coherence relaxes according to

$$\pi(t) = e^{-t/T_2}\left(1 - \int_0^t P(t)\, dt\right) \tag{74}$$

where P(t) is the probability that a defect collides with a dipole for the first time at instant t. We will take into consideration only the nearest defect at t = 0 and we assume that this is at distances in the interval $\{\ell, \ell + d\ell\}$ with probability $\rho(\ell)d\ell = \exp(-\ell/\ell_o)d\ell/\ell_o$ then

$$P(t) = \int P(t,\ell)\,\rho(\ell)\,d\ell \tag{75}$$

where

$$P(t,\ell) = (\ell/4\pi D)^{1/2}\, t^{-3/2}\, e^{-\ell^2/4Dt} \tag{76}$$

as obtained from the solution of the diffusion equation. One has[117] :

$$P(t,) = (1/\tau_D)^{1/2}\left\{\left(\frac{1}{\pi t}\right)^{1/2} - \left(\frac{1}{\tau_D}\right)^{1/2} e^{t/\tau_D}\,\mathrm{erfc}\,(t/\tau_D)^{1/2}\right\} \tag{77}$$

where $\tau_D = \ell_o^2/D$ is a time characteristic of the diffusion. After inserting this expression in (74) differentiating $\pi$ and taking its Laplace transform, one gets :

$$\mathcal{L}\left(-\dot{\pi}(t)\right) = \frac{1}{1+i\omega T_2}\left[1 + \frac{i\omega T_2}{1+\lambda\,(1+i\omega T_2)^{1/2}}\right] \tag{78}$$

where $\lambda = \tau_D/T_2$. This expression strongly deviates from that of the Lorentzian form which one obtains for the defect free case. Although one cannot obtain analytically $\dot{\pi}(t)$ by inverse Laplace transform one easily sees from (78) that the coherence does not decay exponentially any longer but can be fitted quite closely with the stretched exponential decay function

$$\pi(t) \sim \exp(-t/T_2)^{\beta}$$

where $\beta<1$. In particular one finds that the dephasing of a dipole is far more likely immediately after one of its neighbors has been dephased than it is at an arbitrary time ; this is a cooperative memory effect brought up by the diffusion process which tends to render uniform the defect density distribution.

There are few studies of memory effects in relaxation processes. In fact a consistent procedure for characterizing different stages of memory from the markoffian memoryless one up to the completely deterministic case is lacking. With the progress in ultrashort laser pulse techniques these problems will come into the forefront, combined with spatial disorder and memory effects they will lead to a whole class of interesting effects.

## GENERAL REMARKS

Irreversible processes at the molecular level can be studied with nonlinear optical techniques. These techniques allow one to obtain very important informations concerning the spatial and temporal disorder that affects the coherence and energy content of the elementary excitations in condensed matter. In this respect nonlinear time and space resolved optical techniques are particularly powerful and selective. There are however many important irreversible processes that have not been studied yet with these techniques. Here we have in mind processes where memory effects play an important but also tunneling processes. Finally the study of processes that result from competition of spatial and temporal disorder will require much preliminary theoretical study before they can be tackled experimentally with these techniques.

## REFERENCES


1. See for instance R. Kubo, M. Toda and N. Hashitsume, Statistical Physics II, *Nonequilibrium Statistical Mechanics*, Springer Verlag, Berlin 1978
2. F. Bloch, *Phys.Rev.* **70**, 460 (1946)
3. See for instance, A. Abragam , *Principles of Nuclear Magnetism,* Oxford Univ.Press London, 1961, or C.P. Slichter, *Principles of Magnetic Resonance*, Springer Verlag Berlin, 1980.
4. N. Bloembergen, E.M. Purcell and R.V. Pound, *Phys.Rev.* **73**, 679 (1948)
5. P.W. Andersson and P.R. Weiss, *Rev.Mod.Phys.* **25**, 269 (1953)
6. R. Kubo in Fluctuations, *Relaxation and Resonance in Magnetic Systems*, D. Ter Haar, ed. (Plenum, N.Y. 1962) p.23
7. See for instance. M.D. Levenson, *Introduction to Nonlinear Laser Spectroscopy*, Academic Press, New-York, 1982.
8. R.V. Wagness and F. Bloch, *Phys.Rev.* **89**, 728 (1953)
9. A.G. Redfield, *Phys.Rev.* **98**, 1787 (1955)
10. R.P. Feynman, F.C. Vernon and R.W. Hellwarth, *J.Appl.Phys.* **28**, 49 (1957)
11. See for instance P.R. Berman, *J.Opt.Soc.* **B3**, 564 and 572 (1986)
12. G. Mourou, I.E.E.E. *J.Quant.Electr.* **11**, 1 (1975)
13. See for instance C. Kittel, *Introduction to Solid State Physics* John Wiley, New-York 1966 or W. Ashcroft and N.D. Mermin, *Solid State Physics*, Holt Saunders, Tokyo, 1961
14. F. de Martini and J. Ducuing, *Phys.Rev.Lett.* **17**, 117 (1966)
15. R. R. Alfano and S.L. Shapiro, *Phys.Rev.Lett.* **26**, 1247 ; ibid **29**, 1655 (1972)
16. A. Laubereau, D. van der Linde and W. Kaiser, *Phys.Rev.Lett.* **27**, 802 (1971); ibid **28**, 1162 (1972)
17. A. Laubereau and W. Kaiser, *Rev.Mod.Phys.* **50**, 607 (1978). This paper contains very thorough discussion of several aspects and applications of nonlinear time resolved techniques for the study of vibrational relaxation in liquids and crystals.
18. C. Flytzanis, in *Applied Laser Spectroscopy*, Eds. W. Demtröder and M. Inguscio, Nato ASI Series, Plenum Pres, N.Y. (1990)
19. For a very accessible and up to date review of optical pulse techniques see C.V. Shank, *in Ultrashort Light Pulses and Applications*, W. Kaiser Ed.,Springer Verlag, Berlin 1988
20. M.L. Geirnaert, G.M. Gale and C. Flytzanis, *Phys.Rev.Lett.* **52**, 815 (1984)
21. G.M. Gale, P. Guyot-Sionnest, W.Q. Zheng and C. Flytzanis, *Phys.Rev.Lett.* **54**, 823 (1985)
22. A. Laubereau, A. Seilmeier and W. Kaiser, *Chem.Phys.Lett.* **36**, 232 (1975)
23. D. Ricard and J. Ducuing, *J.Chem.Phys.* **62**, 3616 (1975)
24. See for instance J. Shah, T.C. Damen and B. Deveaud, *Appl.Phys.Lett.* **50**, 1307 (1987)
25. G.M. Gale, P. Guyot-Sionnest and W.Q. Zheng, *Opt. Comm.* **58**, 395 (1986)
26. A. Kurnit, I.D. Abella and S.R. Hartmann, *Phys.Rev.Lett.* **13**, 567 (1964)
27. W. Mossberg, A. Flusberg, R. Kachru and S.R. Hartmann, *Phys.Rev.Lett.* **42**, 1665 (1979)
28. H. Hesselink and D.A. Wiersma, *Phys.Rev.Lett.* **43**, 91 (1979)
29. D. Vanden Bont, L.J. Muller and M. Berg, *Phys. Rev. Lett.* **67**, 3700 (1991) ; S. R. Hartman, *I.E.E.E. J. Quant. Elec.* **4**, 802 (1968) ; K.P. Leung, T.W. Mossberg and S.R. Hartman, *Opt. Comm.* **43**, 145 (1982)
30. D.W. Phillion, D.J. Kuizenga and A.E. Siegman, *Appl.Phys.Lett.* **27**, 85 (1975).; J.R. Salcedo, A.E. Siegman, D.D. Dlott and M.D. Fayer, *Phys.Rev.Lett.* **41**, 131 (1978)
31. H.J. Eichler, *Opt.Acta.* **24**, 631 (1977)
32. M.D. Fayer, *Ann.Rev.Phys.Chem.* **33**, 63 (1982)
33. See for instance *Optical Phase Conjugation*, R. Fisher Ed. Acad.Press, New-York, 1985.
34. T. Yajima and Y. Taira, *J.Phys.Soc.Japan*, **47**, 1620 (1979)
35. G. Mannenberg, *J.Opt.Soc.Am.* **B3**, 853 (1986)

36. See for instance S. Ruhman, A.G. Joly, B. Kohler, L.R. Williams and K.A. Nelson, *Rev.Phys.Appl.(Paris)* **22**, 1717 (1987) ; Yan Y.X. Gamble, E.B. and K.A. Nelson, *J. Chem.Phys.* **83**, 5391 (1989)
37. J. Chesnoy and A. Mokhtari, *Phys.Rev.* **A38**, 3566 (1988) ; 37. A. Mokhtari and J. Chesnoy, *Europh. Lett.* **5**, 523 (1988)
38. M.J. Rosker, F.W. Wise and C.L. Tang, *Phys.Rev.Lett.* **57**, 321 (1986) and *J.Chem.Phys.* **86**, 2827 (1987)
39. M. Mitsunaga and C.L. Tang, *Phys.Rev.* **A35**, 1720 (1987)
40. N. Morita and T. Yajima, *Phys.Rev.* **A30**, 2525 (1984)
41. D.H. Auston, *Appl.Phys.Lett.* **43**, 713 (1983) ; D. Auston and K.P. Cheung, *J. Opt. soc. Am. B2*, 606 (1985)
42. D.H. Auston, K.P. Cheung, J.A. Valdmanis and D.A. Kleinman, *Phys.Rev.Lett.* **53**, 1555 (1984)
43. D. Auston and M.C. Nuss, *IEEE, J.Quant.Electr.* **24**, 184 (1988)
44. F. Vallée, G. Gale and C. Flytzanis, *Phys.Rev.Lett.* **61**, 2102 (1988)
45. G. Gale, F. Vallée, C. Flytzanis, *Phys.Rev.Lett.* **57**, 1867 (1986)
46. K. Huang, *Proc. Roy. Soc.* **A208**, 352 (1951)
47. M. Born and K. Huang, *Dynamical Theory of Crystal Lattices* Clarendon Press, Oxford (1954)
48. Th. Förster, *Ann. Phys.* (Leipzig) **2**, 55 (1948)
49. J.J. Hopfield, *Phys. Rev.* **112**, 1555 (1958) ; *J. Phys. Soc. Japan Suppl* .**21**, 77 (1966)
50. R.S. Knox *Theory of Excitons* in Solid State Physics Suppl. 3, Eds F. Seitz and D. Turnbull, Academic Press New York (1963)
51. E.I. Rashba and M.D. Sturge, *Excitons* North Holland, Amsterdam (1982)
52. D.L. Mills and E. Burstein, *Reps. Prog. Phys.* **37**, 817 (1974)
53. R. Loudon and J. Haynes, *Light Scattering in Solids* (1980)
54. V.M. Agranovich ,*Surface Polaritons* North Holland Amsterdam (1982)
55. R. Loudon, *J. Phys.* **A3**, 233 (1970)
56. Jr. A.S. Barker and R. Loudon, *Rev. Mod. Phys.* **44**, 18 (1972)
57. B. Szigeti, *Trans. Farad. Soc.* **4**, 155 (1949) ; *Proc. Roy. Soc.* **A252**, 217 and **A258**, 577 (1955)
58. C. Flytzanis, *Phys. Rev. Lett.* **29**, 772 (1972)
59. A. Maradudin and R.F. Wallis, *Phys. Rev.* **125**, 1277 (1962)
60. A. Ishimaru, *Wave Propagation and Scattering in Random Media* Academic Press, New York (1978)
61. V. Ginzburg, *Physique Théorique et Astrophysique*, Editions Mir, Moscou (1975)
62. F. Vallée, G.M. Gale and C. Flytzanis (1989)
63. F. Vallée and C. Flytzanis, *Phys. Rev.*, to appear in *Phys. Rev.*
64. C. Flytzanis, G.M. Gale and F. Vallée, *S. Akhmanov Memorial Volume*, Eds. H. Walther and N. Koroteev, M. Scully (to appear)
65. F. Vallée, F. Bogani and C. Flytzanis, *Phys. Rev. Lett.* **66**,1509 (1991)
66. C. H. Henry and C.G.B. Garett, *Phys. Rev.* **17**, 1058 (1968)
67. C. Flytzanis, in *Quantum Electronics, A Treatise, Vol I*, Eds C.L. Tang and H. Rabin Acad. Press (1975)
68. J.P. Coffinet and F. de Martini, *Phys. Rev. Lett.* **27**, 1506 (1971)
69. D.C. Haueisen and H. Mahr, *Phys. Rev. Lett.* **26**, 838 (1971) ; D. Frölich, I. Möhler and P. Wiesner, *Phys. Rev. Lett.* **31**, 369 (1971)
70. Y. Masumoto, S. Shionoya and T. Takagahara, *Phys. Rev. Lett.* **51**, 923 (1983)
71. V.M. Agranovich and I.I. Lalov, *Zh. Eksp. Teor. Fiz.* **61**, 656 (transl. 1972 Sov. Phys. JETP **34** 350) (1971)
72. A.A Anikliev A A, L.C. Reznik, B.S. Umarov and J.F. Scott, *J. Raman Spectr.* **15**, 60 (1984)
73. F. Askary and P.Y. Yu, *Phys. Rev.* **B31**, 6643 (1985)
74. Y. Oka Y K. Nakamura and H. Fujisaki, *Phys. Rev. Lett.* **57**, 2857 (1986)
75. Y. Masumoto and S. Shinoya, *J. Phys. Soc. Jpn* **51**,181 (1982)
76. M. Dagenais and W.F. Sharfin, *Phys. Rev. Lett.* **58**, 1776 (1987)
77. F. Vallée, F. Bogani and C. Flytzanis , Proceedings of VII th Interna. Symp. on *Ultrafast Processes in Spectroscopy* Ed. Laubereau A, Adam Hilger, Bristol (1992)

78. C. Weisbuch and R.G. Ulbrich in *Light Scattering in Solids III*, edited by M. Cardona and G. Guntherodt (Springer-Verlag, Berlin) p. 207 (1982)
79. J.D. Zook, *Phys. Rev.136*, A869 (1964)
80. V.V. Travnikov and V.V. Krivolapchuk *Zh. Eksp. Teor. Fiz.* **85**, 2087 (transl. *Sov. Phys. JETP***58**, 1210) (1983)
81. T. Takagahara, *Phys. Rev.* **B31**, 8171 (1985)
82. John Sajeev, *Phys. Rev. Lett.* **53**, 2169 (1984) ; *Comm. in Cond. Matter* **14**, 193 (1988)
83. P.W. Andersson, *Phil Mag.* **52**, 505 (1985)
84. Leung Tsang and A. Ishimaru, *J. Opt. Soc. Am.* **A1**, 836 (1984)
85. A. Genrack, *Phys. Rev. Lett.* **58**, 2043 (1987)
86. A.J. Kane and M. Stone, *Annals Phys.* **131**, 36 (1981)
87. T.R. Kirkpatrick and I.R. Dorfman, *in Fundamental Problems in Statistical Mechanics* VI, Ed. Cohen E G D Elsevier Science, p. 365 (1985)
88. D. Vollhardt and P. Wölfle, *Phys. Rev.* **22**, 4666 (1980)
89. A.H. Golubentsev, *Zh. Eksp. Teor. Fiz.* **86**, 47 (transl. Sov. Phys. JETP **59** 26) (1984)
90. S.A. Akhmanov and R.V. Khokhlov, *Zh. Eksp i Teor. Fiz* **43**, 352 (transl. Soviet Phys. JETP **16** 252) (1962)
91. Y.R. Shen, *Principles of Nonlinear Optics* (John Wiley, N.Y.) (1984)
92. C. Flytzanis, G.M. Gale and M.L. Geirnaert, in *Applications of Picosecond Spectroscopy to Chemistry*, edited by K.B. Eisenthal (Reidel, Higham, Mass.) p. 205 (1984)
93. see for instance *Lattice Dynamics and Intermolecular Forces* edited by S. Califano (Academic Press, N.Y.) (1975)
94. see for instance, J.H. Reisland, *The Physics of Phonons* (John Wiley, London) (1973)
95. R. Orbach and L.A. Vredevoe, *Phys. 1* , 91 (1968) ; R. Orbach, *Phys. Rev. Lett.* **16**, 15 (1966)
96. M.J. Colles and J.A. Giordnaine, *Phys. Rev. Lett.* **27**, 670 (1972)
97. A. Glass, *Phase Transitions,* Cambridge Univ. Press, Cambridge (1980)
98. A. Goldanski, preprint
99. I.B. Levinson, preprint
100. C.C. Ackerman and R.A. Guyer, *Am. Phys.* **50**, 128 (1968)
101.V.I. Kozub, *Sov; Phys. JETP* **67**, 1191 (1988) ; V.I. Kazhovtsev and I.B. Levinson, *Sov. Phys. JETP* **61**, 1318 (1985)
102. A. Ron and D.F. Hornig, *J. Chem. Phys.* **39**, 1129 (1963)
103. M.H. Cohen and J. Ruvalds, *Phys. Rev. Lett.***23**, 1378 (1969)
104. F. Fermi, *Z. Phys.* **71**, 250 (1931)
105. J. Ruvalds and A; Zawadowski, *Phys. Rev.* **B2**, 1172 (1970)
106. J. C. Kimball, C.Y. Tong and Y.R. Shen, *Phys. Rev.***B23**, 4946 (1981)
107. F. Vallée, G.M. Gale and C. Flytzanis, *Chem. Phys. Lett. 124*, 216 (1986)
108. G.M. Gale, F. Vallée and C. Flytzanis, in *Time Resolved Vibrational Spectroscopy*, Eds. A. Laubereau and M.S. Stockburger, Springer Verlag, p. 117 (1985)
109. see for instance, L. Landau and E. Lifshitz, *Mechanics* (Pergamon Press, London) p. 80 (1978)
110. K. Nishikawa, *J. Phys. Soc. Japan* **24**, 916 (1968)
111 J.M. Wersinger, J.M. Finn, E. Ott, *Phys. Rev. Lett. 44*, 453 (1980)
112. J. Baker and J. Hopfield, *Phys. Rev.* **135**, 1732 (1964)
113. R. Zwanzig, *J. Chem. Phys. 33*, 1338 (1960) ; H. Mori, *Progr. Theor. Phys.* **33**, 424 (1965)
114. see for instance, M.G. van Kampen, *Stochastic Processes in Physics and Chemistry,* North Holland, Amsterdam (1990)
115. M. Aihara, *Phys. Rev.***B25**, 53 (1982)
116. R. Loring and S. Mukamel, *J. Chem. Phys.* **83**, 4353 (1985)
117. S. Glarum, *J. Chem. Phys.* **33**, 639 (1960)
118. H. Scher, M.F. Shlesinger and J.T. Bendler, *Physics Today* January 91, p.26

CONTRIBUTORS

1. Prof. W. E. Bron
Department of Physics
University of California, Irvine
Irvine, CA 92717 USA
FAX: (714)725-2174
Phone: (714)856-4345

2. Prof. H. M. Van Driel
Department of Physics
University of Toronto
Toronto, Canada M5S 1A7
FAX: (416)978-3936
Phone: (416)978-4200

3. Prof. Chr. Flytzanis
Laboratoire D'Optique Quantique
Ecole Polytechnique
91128 Palaiseau Cédex
France
FAX: 33-169-33-30-17
Phone: 33-169-33-4124

4. Prof. E. Goovaerts
Department of Physics
University of Antwerp, UIA
B-2610 Antwerp
Belgium
FAX: 323-820-2245

5. Prof. Carlo Jacoboni
Dipartimento di Fisica
Univ. de Modena
Via Campi 213A
41100 Modena
Italy
FAX: 39-59-36-74-88
Phone: 39-59-586-046

6. Dr. Jagdeep Shah
AT&T Bell Laboratories
Holmdel NJ 07733 USA
FAX: (908)949-6010
Phone: (201)949-3691

7. Dr. J. Kuhl
MPI-FKF
Heisenbergstr 1
7000 Stuttgart 80 Germany
FAX: 49-711-68-74-371
Phone: 49-711-68-60-590

8. Prof. Dr. Ad Lagendijk
Van der Waals-Zeeman
Laboratorium
Valckenierstraat 65
1018 XE Amsterdam
The Netherlands

# HOT CARRIER RELAXATION IN GaAs/AlGaAs QUANTUM WELLS IN THE PRESENCE OF A QUASI-EQUILIBRIUM ELECTRON HOLE PLASMA

P. Brockmann[1,2], Jeff F. Young[1], P. Hawrylak[1], F. Chatenoud[1] and H.M. van Driel[2]

[1]Institute for Microstructural Sciences, National Research Council, Ottawa, Ontario, K1A 0R6, Canada

[2]Dept. of Physics and Ontario Laser and Lightwave Research Centre, University of Toronto, Toronto, Ontario, M5S 1A7, Canada

The relaxation of hot carriers in semiconductors is in general a complex many-body problem. One important aspect of this process is the interaction of hot carriers with coupled LO phonon/plasmon modes (CPPM). Very little is known about the detailed kinematics of this interaction especially in 2D semiconductor structures (e.g. quantum wells) in the presence of an ambient electron hole plasma.

Here we present results of time-resolved and time-integrated non-equilibrium Raman scattering experiments that address this issue. Hot carriers were produced by optical excitation in a $GaAs/Al_{.30}Ga_{.70}As$ multiple quantum well sample (well width 200 Å) using picosecond dye laser pulses which also act as a (time-delayed) Raman probe for CPPM with in-plane wavevector $q_{//} \sim 5*10^4 cm^{-1}$. Due to the known intersubband spacings and kinematic considerations, these modes can only be generated by hot electrons scattering from subband e3 to e2. Using a second picosecond pulsed dye laser tuned close to the e1-hh1 gap of the quantum well or a cw Ti:Saph laser for the time-integrated measurements a variable density, quasi equilibrium, electron-hole plasma (QEEHP) was also generated ~50 ps before the hot carrier injection pulse in order to study the effect of this plasma on the hot CPPM generation rate.

The time integrated spectra show a <u>reduction</u> of the non-equilibrium CPPM occupation number (determined from the anti-Stokes/Stokes ratio of the mode) by a factor of ~ 6 as the QEEHP density is increased to $\sim 3 * 10^{11}$ $cm^{-2}$. Part of this effect can be explained by a decrease of the CPPM lifetime from 4.5 ± 0.5 ps without QEEHP to 1.9 ± 0.5 ps at the highest density attained in the time-resolved experiments. The remaining decrease is attributed to direct hot carrier/cold plasma Coulomb scattering processes which compete with the LO phonon generation mechanism. A many body calculation of this carrier-carrier scattering rate, and of the renormalized carrier-phonon interaction, is combined with a simple kinetic model to reproduce the salient features of the observed reduction in the hot CPPM occupation number.

# MICROSCOPIC ANALYSIS OF THE ELECTROSTATIC PHONON POTENTIAL IN RECTANGULAR QUANTUM WIRES

Claudia Bungaro,[1] Paolo Lugli,[1] Fausto Rossi,[2] Lucio Rota,[2] and Elisa Molinari[2]

[1] Dipartimento di Ingegneria Elettronica, Università di Roma "Tor Vergata", via della Ricerca Scientifica 1, 00173 Roma, Italy

[2] Dipartimento di Fisica, Università di Modena, via Campi 213/A, 41100 Modena, Italy

The ultrafast carrier dynamics in polar semiconductors is mainly characterized by the scattering with polar-optical phonons and this interaction has been shown to be strongly dependent on the dimensionality of the system. For a detailed understanding of such process a study of both electron and phonon properties on a microscopic level is required. In contrast to the case of quantum wells (QW's), the dynamical properties of quasi-1D systems are not yet well understood.

We present a detailed study of optical phonons and the associated electrostatic potential in thin rectangular GaAs quantum wires embedded in AlAs based on a microscopic treatment of lattice dynamics. The results show that, as in the QW case,[1] it is possible to distinguish two different types of modes, the confined modes, for which the potential is confined within the wire, and modes with interface character, for which the potential has a maximum on the GaAs/AlAs interfaces. Fig. 1(a) shows the first confined mode and Fig. 1(b) a mode which has an interface character in one direction and a confined character in the other.

We have then compared the results of our microscopic model with the macroscopic dielectric continuum model (DCM) which in the case of the QW has been shown to be in good agreement with first principle calculations.[1] Also in this case the agreement for the confined modes is quite good while for the interface modes the simple DCM model has no analytic solution in a rectangular geometry.[2] Approximate analytic solutions, based on decoupling the two directions perpendicular to the wire, can only yield modes which have interface character both along $x$ and $y$, with the same symmetry in the two directions, while it cannot reproduce modes like the one in Fig. 1(b).

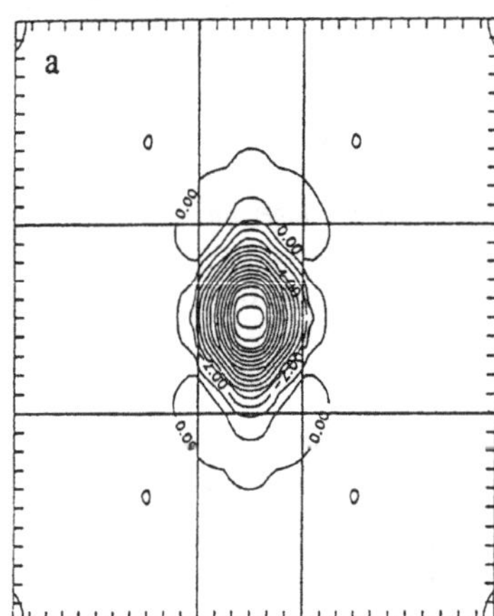


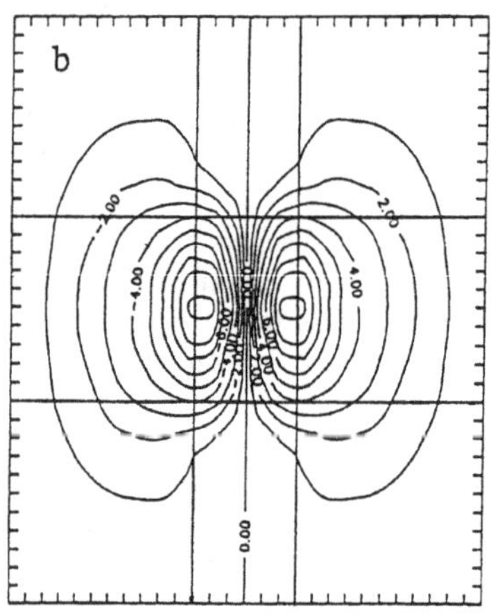


**Figure 1.** Contour plots of the phonon potential for a confined mode (a) and an interface one (b).

## REFERENCES

1. H. Rücker, E. Molinari, and P. Lugli, Phys. Rev. **B45**, 6747 (1992); ibid. **B45**, 3643 (1991).

2. K.W. Kim, M.A. Stroscio, A. Bhatt, R. Michevicius, and V.V. Mitin, J. Appl. Phys. **70**, 319 (1991).

# CARS MEASUREMENTS ON THE VIBRONS IN SOLID NITROGEN

J. De Kinder, A. Bouwen, E. Goovaerts, and D. Schoemaker

Physics Department
University of Antwerp (U.I.A.)
Universiteitsplein 1,
B-2610 Wilrijk (Antwerp), Belgium

Time-resolved coherent anti-Stokes Raman scattering (TR-CARS) was used to study the dephasing of the $\mathbf{k} \simeq \mathbf{0}$ vibrational states in solid $N_2$, and in mixed crystals of $\alpha$-$(^{15}N_2)_x(^{14}N_2)_{1-x}$ and $Ar_x(N_2)_{1-x}$. [1,2] In the $\alpha$ phase of these crystals, intramolecular excitations on different sites couple to form delocalized crystal states, which are called vibrons. The observed dephasing decay of these vibrons is faster than exponential and has a superimposed beating pattern, with a period determined by the factor group splitting between the $A_g$ and the $T_g$ mode. By fitting a trial function to the experimental data, three relevant parameters are obtained: the beat frequency, the exponential decay time, and the Gaussian contribution to the decay. The latter two are related to the homogeneous and inhomogeneous line broadening, respectively.

Our measurements [1] of the dephasing of the $^{14}N_2$ vibrons in $\alpha$-$N_2$ and in mixed crystals of $\alpha$-$(^{15}N_2)_x(^{14}N_2)_{1-x}$ and of the $^{15}N_2$ stretching vibrations in $\alpha$-$(^{15}N_2)_x(^{14}N_2)_{1-x}$ show that the vibron coupling process strongly reduces both the homogeneous and the inhomogeneous line broadening. The homogeneous line broadening for coupled and localized states become mutually equal towards the $\alpha$-$\beta$ phase transition. We relate the suppression of the vibron state formation with increasing temperature to the increasing thermal population of the libron states and to the resulting larger librational amplitude. The dephasing mechanism for the vibrons is found to be scattering off libron states. The factor group splitting between the $A_g$ and the $T_g$ vibron was determined with a high accuracy (0.007 $cm^{-1}$), and decreases towards the $\alpha$-$\beta$ phase transition, above which it vanishes.

By substituting Ar atoms for the $N_2$ molecules, the vibron state formation is even more strongly suppressed. [2] This can be seen from both the increase in homogeneous and inhomogeneous linewidth. For a substantial Ar concentration ($> 2$ %), the inhomogeneous linewidth becomes temperature dependent, clearly showing the equivalence between rising temperature and higher Ar content of the crystal. This is consistent with the results for the pure $\alpha$-$N_2$ crystals, since for an higher Ar concentration, one also expects an increase of the amplitude of the librational states. The $A_g$-$T_g$ factor group splitting increases markedly (17 %) for a small concentration (5 %) of Ar atoms. Low frequency spectra and vibron-phonon combination spectra, obtained by spontaneous Raman scattering [3], show also the same behavior for either rising temperature or higher Ar concentration.

## REFERENCES

1. J. De Kinder, E. Goovaerts, A. Bouwen, and D. Schoemaker, *Phys. Rev. B* **42**, 5953 (1990).
2. J. De Kinder, A. Bouwen, E. Goovaerts, and D. Schoemaker, *J. Chem. Phys.* **95**, 2269 (1991).
3. J. De Kinder, E. Goovaerts, A. Bouwen, and D. Schoemaker, *J. Lumin.* **53**, 72 (1992).

# TIME-RESOLVED SPECTROSCOPY OF EXCITONS AND BIEXCITONS IN GaAs QUANTUM WELLS

G.J. Denton[1], D.J. Lovering[1], R.T. Phillips[1] and G.W. Smith[2]

[1]Cavendish Laboratory
Madingley Road
Cambridge
CB3 0HE
[2]D.R.A. Malvern
Great Malvern
WR14 3PS
U.K.

The polarization dependence of four-wave-mixing in a single GaAs quantum well has been investigated using the self-diffraction geometry and different combinations of incident beam polarizations. The resulting signals were measured as functions of time delay between incident pulses and incident photon energy.

We find virtually no self-diffracted signals for opposite-circular incident beams; this is consistent with a model based upon the angular momentum of excitons in the quantum well.

The spectral lineshapes for same-circular and parallel-linear polarizations are similar. They display a single peak at low injection ( $6kWcm^{-2}$ peak intensity ) which develops a low-energy shoulder as the injection is increased up to $3MWcm^{-2}$ peak. When this measurement is repeated with perpendicular-linear polarizations, only a single peak is detected over the entire injection range and it is at approximately the same energy as the shoulder found with same-circular and parallel-linear polarizations. The energy splitting between the main peak and the shoulder is approximately equal to the biexciton binding energy of 1.1meV (Phillips 1992). It therefore appears possible that only biexciton signal is detected for perpendicular-linear polarizations, while same-circular and parallel-linear yield signals from both heavy-hole (hh) exciton and biexciton.

Crossed-linear polarizations produce a much smaller signal when the incident energy is tuned to the light-hole (lh) exciton than do same-circular and parallel-linear. Lh-hh beats are observed in time-resolved measurements when the laser is tuned to excite both lh and hh exciton transitions. In these measurements the signal intensity is an order of magnitude smaller for perpendicular-linear polarizations. Also, a 1/2 period phase shift is seen between the beats produced by perpendicular-linear polarizations and those produced by same-circular and parallel linear.

Exciton-biexciton beating is also measured in this sample for appropriate laser tuning (Lovering 1992). To confirm the biexcitonic contribution, continuous-wave, low-temperature spectroscopy was carried out and this showed a low-energy peak which varied super-linearly with respect to the hh exciton signal. A simple biexciton lineshape was developed which fitted these experimental results closely. Exciting the sample resonantly with the lh or hh exciton was found to enhance the biexciton population.

## REFERENCES

D.J. Lovering, R.T. Phillips, G.J. Denton and G.W. Smith, Phys. Rev. Lett. **68**, 1880 (1992)
R.T. Phillips, D.J. Lovering, G.J. Denton and G.W. Smith, Phys. Rev. **B45**, 4308 (1992)

# OPTICAL GAIN AND ULTRAFAST NONLINEAR RESPONSE IN GaAs/AlAs TYPE-II QUANTUM WELLS

Winston S. Fu,[a),b)] G. R. Olbright,[a)] J. F. Klem,[c)] J. S. Harris, Jr.[b)]

[a)]Photonics Research Incorporated, 4840 Pearl East Circle #200W, Boulder, CO 80301
[b)]Solid State Electronics Laboratory, Stanford University, Stanford, CA 94305
[c)]Sandia National Laboratory, Albuquerque, NM 87185

Understanding optical gain dynamics in semiconductor quantum wells (QWs) is motivated to a great extent by the current interest in semiconductor QW diode lasers. We have investigated the femtosecond optical gain and the associated nonlinearities for the unusual case of electrons which are distributed between the direct-GaAs and indirect-AlAs layers in GaAs/AlAs type-II multiple quantum-well (MQW) structures. The type-II QW is formed when quantum confinement of the Γ-electron state in the GaAs layers raises it above the X-electron state in the AlAs layer.[1] Consequently, the electron (*e*) and holes (*h*) are confined to different layers after optical excitation. We will refer to the energy splitting between the GaAs Γ-state and the AlAs X-valley as the "Γ-X splitting."

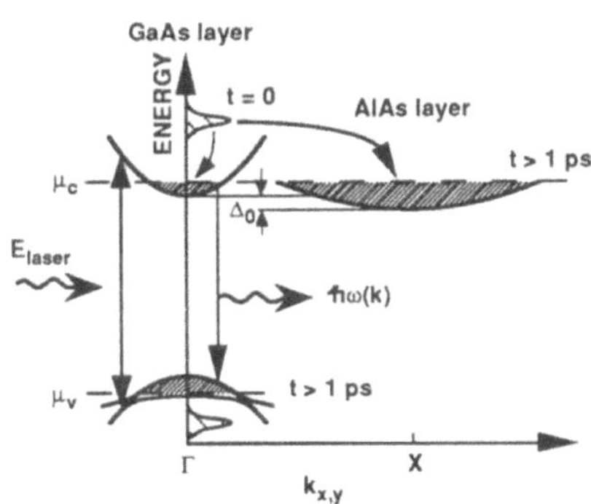


**Figure 1.** Schematic band diagram, in coordinate space, in the plane of a type-II structure. The Γ-X splitting, $\Delta_0$, is ≈80 meV for our 30-Å(GaAs)/80-Å(AlAs) type-II MQW sample.

One consequence of the spatial separation of the *e* and *h* plasmas in type-II QWs is the inhibition of optical gain (which results in a higher gain threshold).[2] The condition for gain in type-II heterostructures is difficult to satisfy because the AlAs X-valley lies energetically below the GaAs Γ-valley as shown in Fig. 1. However, we observed a significant increase in the gain lifetime (> 150 ps) in type-II MQWs, as compared to type-I MQWs (~ 50 ps). We attribute this increase in the gain lifetime to the presence of a "reservoir" of electrons in the AlAs X-valley (which has a very long lifetime; ~ 1 μs).

We also investigated the dependence of the optical (gain) nonlinearities on longitudinal static electric field. We were able to control the gain threshold (as well as bandwidth, amplitude and lifetime) by simply varying the applied reverse bias. Finally, at early times (≈1.5 ps following intense optical excitation) we observed an ultrafast (< 0.5 ps) nonlinear optical response in the gain/absorption spectra of GaAs/AlAs type-II MQWs which we have also observed in narrow type-I structures, and attributed to electron-hole scattering.[2]

## REFERENCES

1. see e.g. J. Ihm, *Appl. Phys. Lett.* **50**, 1068 (1987).
2. see the following and references therein: W. S. Fu, J. S. Harris, R. Binder, S. W. Koch, G. R. Olbright, *IEEE J. Quantum Electron.*, **28** 2404 (1992); and W. S. Fu, G. R. Olbright, J. F. Klem, and J. S. Harris, Jr., *Appl. Phys. Lett.*, **61** 1661 (1992).

**Massimo Gurioli, Juan Martinez-Pastor, and Marcello Colocci**

European Laboratory for Non Linear Spectroscopy
Department of Physics, University of Florence
Largo Enrico Fermi 2, 50125 Firenze, Italy

Semiconductor quantum wells (QW) are known to act as efficient traps for the carriers photogenerated by optical absorption. Less attention has been devoted to the inverse mechanism i.e. detrapping or escape of carriers out of the QWs.

We demonstrate, using both continuous wave and time resolved PL measurements, that carrier escape out of the QWs is the main intrinsic mechanism responsible for the thermal quenching of the photoluminescence (PL) in thin GaAs/AlGaAs quantum wells and double barrier quantum wells (DBQW). At high temperatures we find a strong quenching of the PL together with the decrease (down to few ps) of the PL decay time $T_L$ by several orders of magnitude, clearly denoting the relevance of efficient non radiative mechanisms. However due to the strong temperature dependence of the radiative time constant $T_R$ (that we extract from the joint analysis of the PL decay time $T_L$ and the integrated PL intensity $I_L$ ) the Arrhenius plot of $I_L$ cannot be used for finding the activation energy of the non radiative channels. Indeed it is the temperature dependence of $T_L$ that contains the information on the non radiative recombination time $T_{NR}$. From the Arrhenius plot of $T_L$ we find, as shown in Table 1, activation energies much smaller than the electron-hole pair confinement energies but in good agreement with the unipolar escape out of the wells of the less confined carrier species. In principle, the unipolar detrapping produces a space charge accumulation and therefore a band bending which will oppose to further carrier leakage into the barriers. However this effect is negligible in our case, due to the small excess of photogenerated carriers ($\approx 10^{10}$ $cm^{-2}$) at the excitation power used in our experiments.

Table 1 Parameters of the QWs investigated; all energies are measured in meV. The samples are DBQWs, i.e. GaAs/$Al_{0.3}Ga_{0.7}As$ QWs with 0, 1 or 2 of AlAs mono-layers (ML) embracing the well of width $L_W$ (0 ML samples are standard SQW). The activation energies $E_A$ are extracted from fitting the temperature dependence of $T_L$; the electron-hole pair confinement energies $\Delta_{e\text{-}h}$ are measured by the shifts of the PL peaks with respect to the PLE spectra of the AlGaAs barriers; the electron (hole) confinement energies $\Delta_e$ ($\Delta_h$) are calculated by fitting $\Delta_{e\text{-}h}$ with a effective mass model using two different band offset ratios : 1) $\Delta E_c$:$\Delta E_v$=70:30; 2) $\Delta E_c$:$\Delta E_v$=65:35.

| Sample | $L_W$ (Å) | $\Delta_{e\text{-}h}$ | $E_A$ | $\Delta_h{}^1$ | $\Delta_e{}^1$ | $\Delta_h{}^2$ | $\Delta_e{}^2$ |
|---|---|---|---|---|---|---|---|
| 2 ML | 20 | 45 | 20 | 31 | 14 | 43 | 3 |
| 1 ML | 20 | 105 | 50 | 49 | 56 | 60 | 46 |
| 0 ML-1 | 20 | 190 | 70 | 69 | 123 | 87 | 103 |
| 0 ML-2 | 20 | 200 | 65 | 71 | 128 | 90 | 107 |
| 2 ML | 40 | 240 | 100 | 90 | 150 | 110 | 130 |
| 1 ML | 40 | 270 | 90 | 97 | 172 | 116 | 173 |
| 0 ML-1 | 40 | 300 | 90 | 103 | 197 | 124 | 180 |
| 0 ML-2 | 40 | 300 | 90 | 103 | 197 | 124 | 180 |

The results obtained, on one hand, show that thermal escape is the main non radiative process, at least in the case of thin GaAs/AlGaAs wells; on the other hand, they demonstrate that the escape of free carriers is more efficient than that of excitons. Finally the comparison of the thermal activation energies with the confinement energies results in a better agreement if a band offset ratio $\Delta E_c$:$\Delta E_v$=70:30 is used.

# AUGER RECOMBINATION IN GaAs AT VERY HIGH CARRIER DENSITIES

Eli Glezer, Yakir Siegal, Juen-Kai Wang, Walter Mieher, and Eric Mazur

Division of Applied Sciences and Department of Physics
Harvard University
Cambridge, MA 02138

We have used 80 fs laser pump and probe pulses to investigate carrier relaxaion processes in GaAs at carrier densities approaching the disordering threshold. Very little is known about GaAs in the high carrier density regime, and our preliminary results suggest that the Auger recombination time in this regime may be as fast as 500 fs, significantly faster than predicted by static screening theory.[1]

In this experiment, the reflection of the probe pulse and the generated second harmonic signal served as probes of the dielectric constant. The advantage of the second harmonic generation (SHG) probe is that due to its shorter penetration depth it probes only the uniformly pumped outer layer. A change in the linear dielectric constant causes a change in the SHG efficiency due to a rotation of the transmitted wave with respect to the crystal axes, and also due to a change in the reflectivites at both the fundamental and second harmonic frequencies. The disadvantage of the SHG probe is that it is more difficult to interpret than the reflectivity measurement: the SHG signal is affected by changes in the linear dielectric constant at both the fundamental and second harmonic frequencies, as well as any changes in the second order susceptibility.

In order to arrive at a carrier relaxation rate, the simplest analysis presumes that the dominant effect of the optical excitation is the addition of a free electron contribution to the dielectric constant. This contribution changes the real part of the dielectric constant according to the Drude model, with the assumption $\omega\tau >> 1$. While this may be true at low carrier densities, we have found that our experimental results cannot be fully explained by this simple model. From our data, it seems likely that in order to accurately reproduce the experimental results, the theory will have to take into account the effect of the excited carrier density on the band structure of the crystal. In order to more fully understand the dynamics of the carrier population, we are currently perfoming an experiment to track the time evolution of both the real and imaginary parts of the dielectric constant, following intense optical excitation.

EG acknowledges a Ph.D. Fellowship from the Hertz Foundation. YS acknowledges a Ph.D. Fellowship from the National Science Foundation. This work was supported by the Joint Services Electronics Program and the Material Research Laboratory Program under contract with Harvard University.

## REFERENCES

1. Yoffa, E. J., Phys. Rev. B, **21**, 2415 (1980)

# GENERATION OF SHORT PULSE FAR INFRARED RADIATION FROM HIGH INTENSITY LASER PLASMAS

H. Hamster, A. Sullivan, S. Gordon, R. W. Falcone

Department of Physics
University of California at Berkeley
Berkeley, CA 94720

Femtosecond laser pulses with terawatt intensities are focused on gas and solid targets. A strong emission of coherent far infrared radiation is observed . We investigate the dependence of the pulselength and intensity on the plasma density. This effect promises to be a new source for high power far infrared generation.

The ponderomotive forces generated in the focus of a terawatt laser pulse are sufficient to create a very large density difference between ionic and electronic charges in the focal region. If an ultrashort pulse is used to excite the laser plasma, a powerful electromagnetic transient is expected.[1] We report the observation of this effect.

In our experiment we focus 120 fs laser pulses with intensities > $10^{18}$ W/cm$^2$ on solids and gases.[2] A liquid helium cooled bolometer is used in conjunction with a Fouriertransform spectrometer to characterize the emitted far infrared (FIR) radiation. We observe a strong resonant enhancement of the radiation if the plasma frequency is close to the inverse pulse length of the laser. At these plasma densities, the emission extends over several cycles of a plasma oscillation. For a non-resonant excitation of the plasma at high densities we observe the emission of an approximately half cycle electromagnetic wave form with a frequency centered around 1.5 THz. Figure 1 shows the autocorrelation of this pulse. The strongest FIR signal is observed from the interaction of lasers with solid density plasmas.

We have developed a linearized hydrodynamic model which accounts for the motion of a cold plasma driven by the strong ponderomotive forces present in the laser focus. We predict and observe the main emission to be in a direction ≈ 45 degrees from the direction of pulse propagation (Fig. 2). The emission of a strong FIR radiation indicates the existence of wake fields as strong $10^8$ V/cm. This effect promises to be a new source for high power, ultra short pulse far infrared generation.

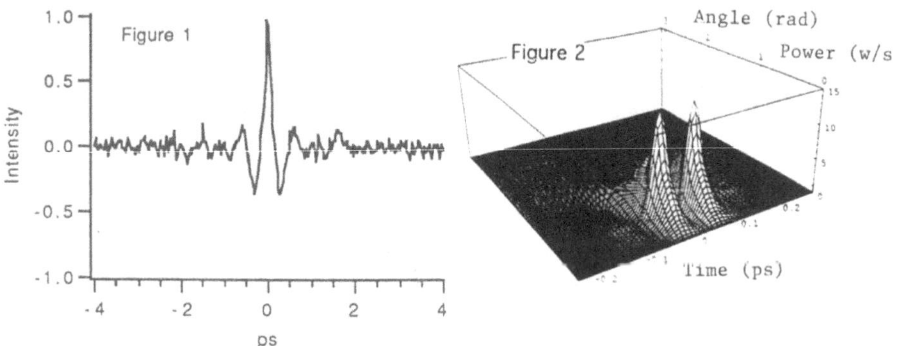


## REFERENCES

1. H. Hamster and R. Falcone, "Proposed source of sub-picosecond far infrared radiation" , *in*: "Ultrafast Phenomena VII", C.B. Harris et al., ed., Springer-Verlag, Berlin (1990)
2. A. Sullivan et al. , "Multiterawatt, 100-fs laser",*Opt. Lett.* **16,** 1406 (1991)

# ULTRAFAST PHOTOGRAPHY OF PHYSICAL PROCESSES

B. Hopp, Zs. Bor, B. Rácz, G. Szabó

Department of Optics and Quantum Eletronics
JATE University
H-6720 Szeged Dóm tér 9., Hungary

In order to get important information of the fast physical, chemical and biological phenomena, we have to observe and record theirs temporal changes, very quickly ($10^{-6}$-$10^{-9}$ second). Recording and photography of these processes need very comlicated, complex systems.

In our investigations we used the following stroboscopic arrangement for ultrafast photography: an ArF excimer laser was used as the excitation source. The sample to be investigated was locally illuminated by subnanosecond pulses from an $N_2$ laser pumped dye laser. The delay between the excimer and dye laser pulse could be varied in the range of 0 - 10 ms. The sample was observed and recorded by a microscope equipped with a color TV camera and videorecorder[1].

## APPLICATIONS

### I. Time resolved study of surface shock wave formation during excimer laser ablation

In this case the sample was a pig eye (in vitro) and we investigated the UV photoablation of it. We could observe shear waves on the surface of the eye (Fig. 1.a). The formation, the propagation and the gradual attenuation of the surface shear wave could clearly be seen on the videomonitor. Having these data we represented the propagation of this wave (Fig. 1.b) and calculated the wave velocity (Fig. 1.c).

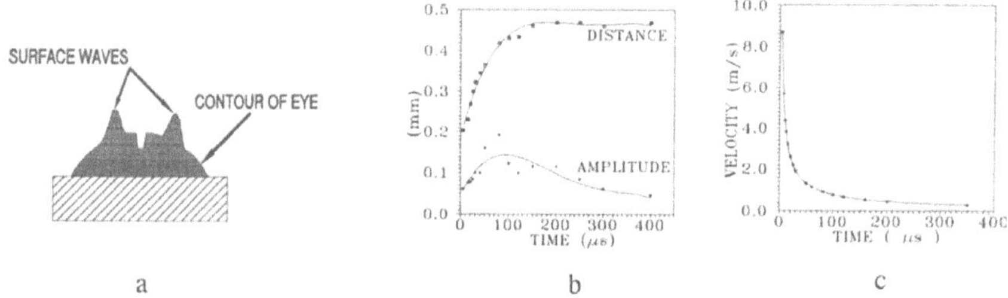


**Figure 1.** The propagation of the surface wave on pig eye

### II. Time resolved investigation of laser induced blow-off of thin metal films

Here the sample was a thin W film on glass substrate. We could observed the exploded metal fragments by the above mentioned arrangement and found that the metal detachment occures in instalments.

## REFERENCE

1. Longhurst RS. "The Foucault Test. Schlieren Systems", in: Geometrical and Physical Optics. 3rd ed. Longman London and New York. 1973; 369-370.

# MEMORY EFFECTS IN THE ULTRAFAST DYNAMICS OF PHOTOEXCITED SEMICONDUCTORS

Tilmann Kuhn,[1] Jürgen Schilp,[1] and Fausto Rossi[2]

[1]Inst. für Theoretische Physik, Universität, 7000 Stuttgart 80, Germany
[2]Dipartimento di Fisica, Università, 41100 Modena, Italy

In the usual semiclassical picture the dynamics of carriers is governed by instantaneous transition rates between different $k$-states. These rates are calculated by using Fermi's Golden Rule and thus imply energy-conserving transitions. On short time-scales, however, memory effects and the related energy-time uncertainty relation play a role.[1-3] We have investigated these non-Markovian effects for the case of carrier-light[3] and carrier-phonon interaction.

Figure 1(a) shows the generation rate as a function of wave-vector at different times for a 100 $fs$ Gaussian laser pulse. At short times the rate is much broader than estimated from the pulse duration. The subsequent narrowing is accompanied by a stimulated recombination off-resonance during the second half of the pulse. In Figure 1(b) the evolution of the distribution function of electrons, created by a 100 $fs$ pulse within the semiclassical model, is shown for the non-Markovian carrier-phonon coupling. At short times the replicas due to optical phonon emission are broadened, exhibiting a narrowing with increasing time. Thus, in the sub-picosecond regime energy-nonconserving transitions contribute to the carrier dynamics through both carrier-light and carrier-phonon interaction.

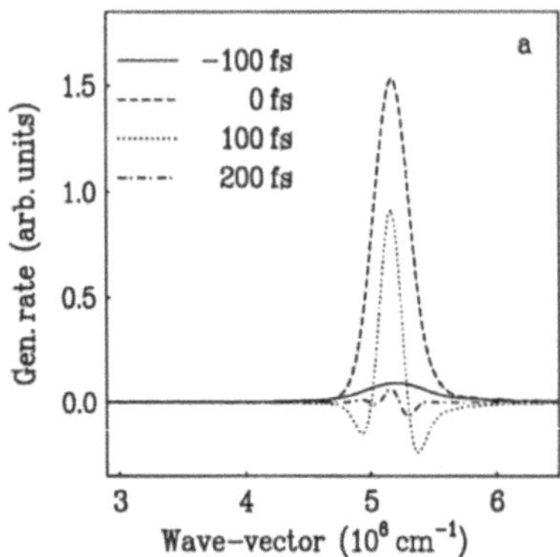


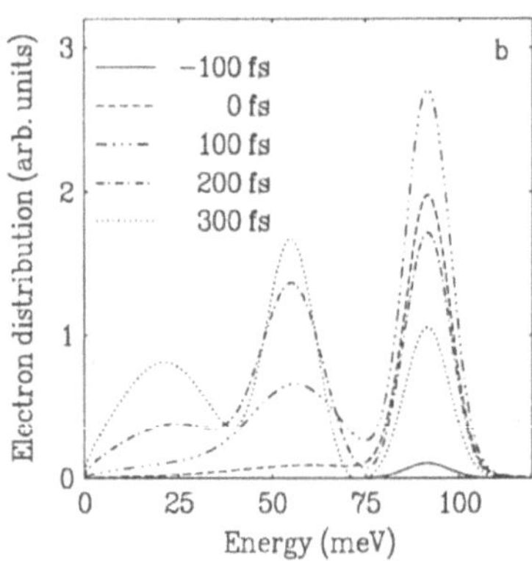


**Figure 1.** Non-Markovian broadening due to (a) carrier-light and (b) carrier-phonon interaction.

## REFERENCES

1. R. Zimmermann, Transverse relaxation and polarization specifics in the dynamical Stark effect, *phys. stat. sol. (b)* 159:317 (1990).
2. A. V. Kuznetsov, Coherent and non-Markovian effects in ultrafast relaxation of photoexcited hot carriers: A model study, *Phys. Rev. B* 44:13381 (1991).
3. T. Kuhn and F. Rossi, Analysis of coherent and incoherent phenomena in photoexcited semiconductors: A Monte Carlo approach, *Phys. Rev. Lett.* 69:977 (1992).

# RADIATIONLESS TRANSITION OF F CENTERS IN ALKALI HALIDES

M. Leblans, F. De Matteis, E. Gustin, A. Bouwen, and D. Schoemaker

Physics Department
University of Antwerp (U.I.A.)
Universiteitsplein 1,
B-2610 Wilrijk (Antwerp), Belgium

The absence of luminescence of the $F$ center in the lithium halides, NaI, and NaBr was demonstrated by Bartram and Stoneham[1] to be in agreement with the Dexter-Klick-Russell (DKR) criterion, proposing nonradiative deexcitation during lattice relaxation at the crossing-point of the potential energy curves of the ground and the excited state. Other authors[2] argued that the $F$ center reaches the relaxed excited state (RES) before the electronic transition occurs. The aim of this work is to discriminate between the two models on the basis of picosecond pump-probe measurements of the ground-state recovery after optical excitation[3] of the $F$ center in NaBr and NaI. Also, the influence of $H_s^-$ impurities on the $F$-center relaxation has been extensively checked. Time resolved transient absorption measurements at 10 K on ultrapure NaBr and NaI permitted us to establish a relaxation time of 6±1ns for NaBr, whereas the $F$ center in NaI relaxes on a timescale of 10 to 30 ns. Measurements on hydrogenated NaBr ($[H_s^-] \sim 10^{19}cm^{-3}$), after $H_s^-$ – $F$ center aggregation by means of $F$-light bleaching,show a faster decay with a time constant of about 200 ps. The resonant Raman spectrum yields the accepting modes relevant for the cross-over process. For NaBr and NaI it consists mainly of a narrow gapmode at $137cm^{-1}$ and $113cm^{-1}$, respectively. Their spectral width is related to the vibrational cooling rate. The FWHM of the gapmode is established to be 1±0.2 for NaI and 4.3±0.3 for NaBr with a resolution of $0.4cm^{-1}$ and $1cm^{-1}$, respectively.

The results on the decay of F centers in NaBr have been shown to provide several arguments in favor of nonradiative relaxation from the *relaxed excited state* instead of the cross-over transition.[3] Our present decay measurements in NaI and Raman spectra in both NaBr and NaI confirm the foregoing conclusions: (i) The measured relaxation time is reasonably close to the predicted 27ns. (ii) Within the cross-over model, the $F$-center relaxation is determined by the vibrational cooling rate. However, the inverse width of the gapmode ($\Delta\nu^{-1}$=7.7ps for NaBr; $\Delta\nu^{-1}$ = 33ps for NaI), appears to be much smaller than the time constant of the ground-state recovery measurements. The $F$-center relaxation in NaBr:$H_s^-$ can be explained by a time evolution of the form $\exp(-t/\tau - \sqrt{\gamma t})$. The square root dependence in the exponent is typical for energy transfer due to dipole-dipole interaction between randomly distributed centers.[5]This suggests electronic-vibrational energy transfer from the $F$ center in NaBr to the $H_s^-$ impurity. For the $F_H(H_s^-)$ center the additional promoting interaction strength due to dipole-dipole coupling is estimated to be comparable to the one for the pure $F$ center. Taking also into account that the high $H_s^-$ vibrational frequency ($\omega_H$=492$cm^{-1}$) effectively lowers the activation energy[2,4] for the radiationless transition from a thermalized excited state, a theoretical decay time $\tau_{d-d}$=2.7ns is predicted. Hence, the dipole-dipole energy transfer represents an additional decay channel able to compete with the pure F-center relaxation.

## REFERENCES

1. R.H. Bartram and A.M. Stoneham, *Solid State Commun.* **17** (1975) 1593.
2. G. Baldacchini, D.S. Pan, and F. Lüty, *Phys. Rev. B* **24** (1981) 2174.
3. F. De Matteis, M. Leblans, W. Joosen, and D. Schoemaker, *Phys. Rev. B* **45**, (1992) 10377.
4. R. Englman and J. Jortner, *Molecular Physics* **18** (1970) 145.
5. V.M. Agranovich and M.D. Galanin, *Electronic excitation energy transfer in condensed matter.* (North-Holland, Amsterdam,1982).

# ELECTRIC FIELD-INDUCED SHG SAMPLING

Guntis V.Liberts

Institute of Solid State Physics
University of Latvia
8 Kengaraga St.,Riga LV 1063
Latvia

For infrared and visible laser pulses, the pulse duration is usually obtained from measurements of the second-order (intensity) autocorrelation function using second harmonic generation (SHG) in nonlinear crystals. As for ultrafast electrical signals synchronized with the laser light pulses electro-optic sampling technique has been developed[1]. This technique has been based on the linear electro-optic (Pockels) effect, formally described with the second order nonlinear optical susceptibility. Higher order optical autocorrelation techniques based on the third-order nonlinear susceptibility and crossover of the static and light fields may be transferred to the analysis of ultrafast electrical signals.

We present here a new approach to nonlinear optical sampling measurements, based on the electric field-induced second harmonic (EFISH) generation. We propose centrosymmetric (CS) medium as a material for the sampling head while another details of measurement scheme (except the doubled frequency of the registrated light intensity) are close to conventional EO method. The application of an external electric field $E_l(0)$ together with two laser light fields $E_j(\omega)$ and $E_k(\omega)$ create in the sampling unit the nonlinear polarization source $P_i(2\omega)$:

$$P_i(2\omega) = \chi_{ijkl}(-2\omega, \omega, \omega, 0)\ E_j(\omega)\ E_k(\omega)\ E_l(0)$$

We have analyzed different configurations and regimes of EFISH sampling including collinear and non-collinear geometry, transmission and reflective modes and availability of external (sampling heads) and internal (built in) schemes. Particularly, using of dc bias field in the sampling unit and some advantages of the up-conversion of the sampling light due to EFISH generation has been discussed.

There are different classes of materials, suitable for the EFISH sampling. First, the temporal and spatial evolution of the electrical fields in gases and liquids near the resonance conditions should be determined. Secondly, solid state sampling tips from the CS perovskite type paraelectric crystals (for example $KTaO_3$, $SrTiO_3$ or $K(Ta,Nb)O_3$ solid solutions) may be efficient for the room temperature and cryogenic applications (compatible with HTSC transmission lines). Furthermore, near the ferroelectric phase transition temperature such an efficiency extremely increases. Finally, the third order nonlinear susceptibilities of the CS semiconductors Si and Ge offers new possibilities of internal and external probing of the ultrafast electrical transients in these practically important materials.

## REFERENCE

1. J.A.Valdmanis and G.Mourou, Subpicosecond electrooptic sampling: Principles and applications, *IEEE J.Quantum Electron.* 22:69 (1986).

# FEMTOSECOND INFRARED OPTICAL PARAMETRIC OSCILLATOR PUMPED SYNCHRONOUSLY BY A MODE-LOCKED DYE AND Ti:SAPPHIRE LASER

G. Mak, Q. Fu and H. M. van Driel

Department of Physics, University of Toronto and
Ontario Laser and Lightwave Research Centre
Toronto, Ontario, Canada M5S 1A7

There is great interest in generating ultrashort, tunable, infrared pulses for time-resolved spectroscopy and nonlinear optics experiments in physics, engineering and chemistry. An optical parametric oscillator (OPO) pumped synchronously by a continuously mode-locked femtosecond laser is a new scheme for generating truly repetitive, time-bandwidth limited, infrared pulses. Sources have recently been realized taking advantage of intracavity pumping within a colliding-pulse mode-locked dye laser [1], and extracavity pumping with a dye [2] or Ti:sapphire laser [3]. In particular, 62-fs pulses, at 76 MHz repetition rate, tunable from 1.20-1.34 $\mu$m and average power greater than 175 mW has been demonstrated by us [3].

The singly-resonant OPO uses a 1.5-mm thick KTP nonlinear crystal, cut for noncollinear type II phase-matching, which is mounted at the intracavity focus of a four mirror ring cavity. All mirrors are highly reflecting near $\lambda$=1.3 $\mu$m except for a 2% output coupler. In the first system [2], 150-fs pulses (76 MHz) at 645 nm, from a hybridly mode-locked dye laser, are focused into the OPO. The oscillator is pumped 2.8 times above the threshold of 110 mW. The OPO produced 220-fs pulses with an average power of 30 mW, tunable over the full mirror range from 1.20-1.34 $\mu$m. In principle, this OPO is tunable from 0.9-4.5 $\mu$m with additional mirror sets. The oscillator gain, threshold and pulse width are consistent with a model that takes into account pulse group velocity dispersion, Poynting vector walk-off and Gaussian beams.

Subsequently, we have taken advantage of a high peak power Kerr lens mode-locked Ti:sapphire laser to pump the OPO [3]. Even with an increased oscillation threshold of 180 mW, high average power of 185 mW is produced when the OPO is pumped by 110-fs pulses at 765 nm and 800-mW average power. There is large self-phase modulation of the pulses, because the peak intracavity intensity at the KTP crystal is 40 GW/cm$^2$, which allows them to be compressed with a prism pair to 62 fs. The idler pulses, with pulse widths $<$400 fs and average power of $\sim$100 mW, are tunable from 2.10-1.78 $\mu$m as the OPO signal is tuned. The OPO has been actively stabilized and long term operation with amplitude noise less than 1% has been observed.

## REFERENCES

[1] D. C. Edelstein, E. S. Wachman, and C. L. Tang, Appl. Phys. Lett. **54**, 1728 (1989); E. S. Wachman, W. S. Pelouch, and C. L. Tang, J. Appl. Phys. **70**, 1893 (1991).

[2] G. Mak, Q. Fu and H. M. van Driel, Appl. Phys. Lett. **60**, 542 (1992).

[3] Q. Fu, G. Mak and H. M. van Driel, Opt. Lett. **17**, 1006 (1992).

# SHORT-TIME ABSORPTION OF ELECTROMAGNETIC RADIATION IN ELECTRON GAS

S. Olszewski

Institute of Physical Chemistry of the
Polish Academy of Sciences,
44/52 Kasprzaka, 01-224 Warsaw, Poland

The energy absorbed in a short-time process by a free electron of energy $E_i$ is [Olszewski, S., 1992, Phil. Mag. B 66, 251]:

$$E_A^{(i)} = \left( \frac{e}{m_e c} \right)^2 \left( A_x^0 \right)^2 \left( m_e c^2 \right)^{-1} \frac{2m_e}{3} E_i; \tag{1}$$

simultaneously the frequency $\omega$ of the incident light should not be smaller than some critical frequency

$$\omega_c = \left[ \hbar^{-2} \left( \frac{e}{m_e c} \right)^2 \left( A_x^0 \right)^2 \frac{2}{3} m_e E_i \right]^{1/2}; \tag{2}$$

$A_x^o$ is the amplitude of the incident electromagnetic wave represented by the perturbation $(e/m_e c) i\hbar \vec{A} \nabla$ having the vector potential $\vec{A}$ for which $\nabla \cdot \vec{A} = 0$. We ask at which condition it can be obtained

$$E_A^{(i)} = W = \hbar \, \omega_{ph}, \tag{3}$$

where W is the energy and $\omega_{ph}$ the frequency in the normal absorption process. From (1)-(3) we find

$$\omega_c = \frac{1}{\hbar} \left( m_e c^2 \, W \right)^{1/2}, \tag{4}$$

so the frequencies $\omega$ of the incident light which satisfy the condition

$$\omega_{ph} > \omega \geq \omega_c \tag{5}$$

can be attained for $W > m_e c^2$. For $W < m_e c^2$ a relation opposite to (5), viz. $\omega \geq \omega_c > \omega_{ph}$, is obtained.

# MEASURING FRACTAL DIMENSION IN POLYMERIC SYSTEMS BY FLUORESCENCE METHOD

Önder Pekcan

Istanbul Technical University
Department of Physics
Maslak, 80626 Istanbul, Turkey

## ABSTRACT

The direct energy transfer method was used to study the internal morphology of polymer colloid particles produced by non-aqueous dispersion polymerization. These particles with diameters 0.2 to 1.0 μm contain ca.94 mol % of an amorphous glassy polymer (PVAc) and 6 % a rubbery polymer (poly(2-ethylhexyl methacrylate)[PEHMA] which forms both a surface covering and interconnected network within the particle. We employed a fractal analysis technique and the model of restricted geometries to interpret fluorescence intensity decays of excited donors to obtain the apparent dimension of the rubbery phase. The following equation for fluorescence intensity I(t) decay was used for the analysis[1]

$$I(t) = e^{-t/\tau_o - At^{d/6}}$$

where $\tau_o$ is the natural lifetime of donor,d is fractal dimension and A is constant proportional to number of acceptors. Anthracene, selectively doped into the rubbery phase was used as the energy acceptor. Napthalene, covalently bound to the PEHMA serve as the donor. Within the 38 Å maximum probing distance between donor and acceptor the apparent dimension d 2.3 was measured. When films are prepared by dissolving particles with chloroform and then dopind rubbery phase with anthracene, we find d = 1.3. This indicates that even in the film the PEHMA forms a continuos network, and second that the nascent morphology produced when the particle is formed is different from that in the film. In both cases if the PEHMA is swollen with hexadecane, d increases. These results can be interpreted by assuming that the PEHMA is present in the structure as thin cylindrical network.

## REFERENCES

1- J. Klafter and J.M. Drake, "Molecular Dynamics in Restricted Geometries" Wiley, New York (1989).

# DIRECT GENERATION OF SUBNANOSECOND PULSES BY A HIGH PRESSURE MINIATURE EXCIMER LASER

B. Rácz[1], A. Patócs[2], G. Szabó[1], Zs. Bor[1]
and F. Ignácz[2]

[1]Department of Optics and Quantum Electronics,
JATE University, H-6720 Szeged Dóm tér 9, Hungary

[2]Research Group on Laser Physics, H-6720 Szeged
Dóm tér 9, Hungary

## Introduction

A direct method for short excimer laser pulse generation seems to be possible by using small laser heads excited by very fast driving circuits. Sze and Seegmiller have observed[1] that the pulse duration in their miniature excimer laser was as short as 4 ns without any effort to achieve pulse shortening.
Direct generation of 1 ns excimer laser pulses has been described[2] using short electrode length discharge tubes and Blumlein type high voltage pulse generator. Using similar exciting circuit and extremely low cavity decay time, 1.2 ns long pulses were produced[3].

## Experimental & Results

The discharge tube of the laser had the following parameters: 95 mm electrode length, 4.5 mm separation, seven spark gaps for UV preionization. The resonator was formed by an aluminium coated total reflector and an uncoated quartz output coupler, and its length was set to 200 mm.
The driving circuit was a thyratron switched C-C transfer type high voltage pulse generator with $C_1=8.1$ nF primary, and $C_2=1.5$ nF secondary capacitors.
It is known[4] that when the current density is too large, the excimer molecules generated within the glow discharge will be quenched by the excess electrons. To avoid this, we used specially contoured electrodes[5], and the effective active volume was 2.0 mm(W) x 4.5 mm(H) x 90 mm(L) = 0.81 ccm.
The duration of the laser pulses strongly depends on the total gas pressure, the smallest values were measured at 6 bar in XeCl and KrF to be 670 ps and 870 ps respectively, while the output energies reached 1.2 mJ and 3.5 mJ. The corresponding peak powers were 1.8 and 4.0 MW, these are the highest values among the similar small active volume lasers. The maximum characteristic output powers (output power per unit active volume) were estimated to be 2.2 GW/l and 4.8 GW/l.

## References

1. R. C. Sze and E. Seegmiller: *IEEE J-QE* **QE-17** 81-91 (1982)
2. K. Yamada, K. Miyazaki, T. Hasama and T. Sato: *IEEE J-QE* **QE-24** 177-182 (1988)
3. A. Takahashi, M. Maeda and Y. Noda: *IEEE J-QE* **QE-20** 1196-1201 (1984)
4. R.S.Taylor, P.B. Corkum, S. Watanabe, K.E. Leopold, and A.J. Alcock: *IEEE J-QE* **QE-19** 416-425 (1983)
5. G. J. Ernst: *Opt. Commun.* **49** 275-277 (1984)

# INTERACTION OF A NONSTATIONARY ELECTRON-HOLE PLASMA WITH COHERENT OPTICAL PHONONS IN GAP

G.O. Smith*, T. Juhasz, Y.B. Levinson**, and W.E. Bron

*Department of Physics, University of California, Irvine, California, 92717*

The interaction of an optically induced nonstationary electron-hole plasma with coherently excited optical phonons in GaP has been investigated through time-resolved coherent anti-Stokes Raman spectroscopy (TR-CARS)[1]. Nonexponential optical-phonon dephasing is observed in the presence of the electron-hole plasma. A detailed presentation of the theoretical model describing the plasma-phonon interaction, the cooling of the plasma, and the time dependent damping of the plasma is presented. A comparison of the results from the theoretical model and from both the experimental data and earlier results indicate good agreement.

Investigations in the frequency domain of plasma-phonon interactions in GaP have previously been conducted for both a temperature induced (*n*-doping) one-component plasma [2] and an optically induced two-component plasma [3]. In the present case, an electron-hole plasma in formed through two-photon absorption (TPA) of incident laser radiation. Since this method of photoexcitation produces a plasma which is nonstationary during the observation period, it is necessary to use a time-domain technique to observe the dynamics of the interaction between the plasma and the phonons.

We observe an increase in the instantaneous dephasing rate of the LO phonons during the first 150 ps of the interaction after the excitation of the plasma. The dephasing rate increases with increasing plasma density. Furthermore, we observe that the increase in the dephasing rate becomes negligible after 600 ps. A comparison of the plasma density, as determined from a one-parameter fit to the experimental data, with the independently calculated plasma density, as determined assuming one- and two-photon absorption, show very good agreement. Furthermore, results on the plasma temperature as a function of time and the plasma damping as a function of temperature show good agreement with those reported earlier [2] with comparable samples and plasma densities.

## ACKNOWLEDGEMENT

Supported through NSF DMR-89-13289 and ARO DAAL 0389-K-0060

## REFERENCES

1. G.O. Smith, T. Juhasz, W.E. Bron, and Y.B. Levinson, *Phys. Rev. Lett.* 68:2366(1992).
2. W.E. Bron, S. Mehta, J. Kuhl, and M. Klingenstein, *Phys. Rev.* B 39:12642(1989).
3. J.E. Kardontchik and E. Cohen, *Phys. Rev. Lett.* 42:669(1979).

*Current address: Max-Planck-Institute für Festkörperforsch., Heisenbergstr. 1, 7000 Stuttgart 80, Germany.
**Current address: Dept. of Nuclear Phys., The Weizmann Institute, 76100 Rehovot, Isreal.

# INDEX

Zeitfracht Medien GmbH
Ferdinand-Jühlke-Straße 7
99095 Erfurt, Deutschland
produktsicherheit@kolibri360.de